Magnetic Recording

Volume I TECHNOLOGY

ABOUT THE EDITORS

Denis Mee is an IBM Fellow at the IBM General Products Division, San Jose, California. He is the author of *The Physics of Magnetic Recording* (North Holland Publishing Co., Amsterdam, 1984). Eric Daniel, consultant, was a Fellow of the Memorex Corporation, Santa Clara, California, until he retired in 1983. Each editor has had more than thirty years of experience in magnetic recording, including both computer data and analog applications, and has published twenty or more papers in scientific journals.

Magnetic Recording

Volume I   TECHNOLOGY

EDITORS

C. Denis Mee
IBM Corporation, San Jose, California

Eric D. Daniel
Woodside, California

McGraw-Hill Book Company

New York   St. Louis   San Francisco   Auckland   Bogotá
Hamburg   London   Madrid   Milan
Mexico   Montreal   New Delhi   Panama
Paris   São Paulo   Singapore
Sydney   Tokyo   Toronto

Library of Congress Cataloging-in-Publication Data

Magnetic recording.

  Includes bibliographies and index.
  Contents: v. 1.   Technology.
  1. Magnetic recorders and recording.   I. Mee, C. Denis.
II. Daniel, Eric D.
TL7881.6.M24   1987   621.38   86-10432
ISBN 0-07-041271-5

  234567890   DOC/DOC   89321098

ISBN 0-07-041271-5

The following figures are copyright © by the IEEE (author and copyright year are given in each caption; complete reference is in the end-
of-chapter list): Figs. 2.22, 2.25b, 2.25c, 2.26, 2.27, 2.28c, 2.30, 2.41,
2.42c, 2.44, 2.56, 2.57, 3.21, 3.22, 3.34a, 3.55, 3.58, 4.4, 4.5, 4.6, 4.7, 4.8,
4.9, 4.10, 4.12, 4.13, 4.14, 4.15, 4.16, 4.17, 4.19, 4.20, 4.21, 4.26, 4.27,
4.30, 4.31, 4.32, 4.34, 4.41, 4.42, 5.20, 5.21, 5.22, 5.23, 6.6, 6.10, 6.11,
6.24, 6.31, 7.28, 7.42.

Following are sources of figures that were not credited in captions:
Figs. 4.6 (after Lindholm, 1977); 5.20 to 5.23 (Mallinson, 1985); 7.31
and 7.32 (Bouchard et al., 1984); 7.34 (Miu, 1985); 7.37 (Bouchard et
al., 1985); 7.57 and 7.58 (Gross et al., 1980).

The editors for this book were Stephen G. Guty and David Fogarty,
the designer was Naomi Auerbach, and the production
supervisor was Thomas G. Kowalczyk. It was set in Century Schoolbook
by University Graphics, Inc.
Printed and bound by R. R. Donnelley & Sons Company.

Contents

Chapter 1. Introduction   1

Eric D. Daniel   *Woodside, California*
C. Denis Mee   *IBM Corporation, San Jose, California*

Chapter 2. The Recording and Reproducing Processes   22

Barry K. Middleton   *Manchester Polytechnic, Manchester, England*

Chapter 3. Recording Media   98

Eberhard Köster   *BASF, Aktiengesellschaft, Ludwigshafen, West Germany*
Thomas C. Arnoldussen   *IBM Corporation, San Jose, California*

Preface

Information storage applications continue to grow at a rapid rate in the 1980s due to the successful development of business and consumer products for processing data, video, and audio signals. Advances in storage products have occurred largely because of rapid progress in two key storage technologies—semiconductors and magnetic recording. The consequent reductions in cost of storage have all but wiped out competing storage technologies. While some of the alternative static-memory and beam-addressable storage technologies are very successful for specific applications, the versatility of magnetic recording in providing different storage media formats has resulted in its application in the form of tapes, rigid disks, flexible disks, cards, drums, and sheets. This has resulted in the spread of magnetic recording applications to data, video, and audio storage for both professional and consumer applications. Magnetic recording now dominates the video recording industry, and is moving rapidly into new areas such as 8-mm home movies and still-camera photography. All of the many products using magnetic recording utilize the same basic inductive-recording technology which has developed over the last 30 to 50 years. The technology has improved immensely as understanding has been gained of the physics of the recording and reproducing processes. Advances in recording materials and processes for fabricating components have also contributed in a major way to extending the performance of magnetic recording technology into different industries.

Despite the growth of the magnetic recording industry, the early interest of students in the relevant disciplines of solid-state physics, chemistry, mechanical engineering, and information theory has been hampered by the lack of university teaching emphasis on the multidiscipline technology of magnetic recording. In the past few years, new magnetic recording research centers have been established in several universities in the United States, and major recording research groups have emerged overseas, especially in Japan. The primary purpose of these books on magnetic recording is to provide a text for courses in university graduate schools and in industry. The first volume covers the basic technology in depth, and the companion covers the many and varied applications.

Magnetic Recording, Volume I: *Technology,* is concerned with establishing the underlying technologies that are common to all forms of magnetic recording. Separate chapters treat the processes by which recording and reproduction take place; the materials, design, and fabrication of media; the materials, design, and fabrication of heads; the limits on performance due to noise, interference, and distortion; the key magnetic and recording measurement techniques that have evolved; and, finally, the mechanical interface between the head and medium that is of critical importance in all but optically addressed media.

Magnetic Recording, Volume II: *Applications,* is concerned with the major applications of magnetic recording. These are broadly classified into data or analog recording categories, according to the nature of the signal to be recorded rather than the means of carrying out the recording. The data recording chapters cover the evolution of recording systems using rigid disks, flexible disks, and magnetic tape drives. The analog recording chapters cover video and audio recorders, including those in which the signals are stored in binary digital form. A separate chapter is devoted to coding and error control. The volume ends with a chapter on reversible optical recording, in which signals are recorded thermomagnetically and reproduced magnetooptically.

SI units (Système International d'Unités) are used throughout. Where other units are widely used, values expressed in these units are also listed in parentheses.

We are grateful to the authors and to a number of external reviewers in assisting us in our attempts to produce a uniform coverage of the subject matter. Special acknowledgement goes to T. C. Arnoldussen, G. Bate, A. E. Bell, H. N. Bertram, M. O. Felix, R. J. Gambino, J. M. Harker, A. M. Heaslett, M. K. Hill, R. E. Jones, J. G. McKnight, L. Rosier, R. Wood, and J. K. Wolf. We are also grateful to others in the recording industry who have kindly provided original materials to us. Finally, we wish to thank IBM for providing the environment and support necessary to make these volumes a reality, with special thanks to G. L. Cavagnaro, J. A. Harriss, and the members of the Technical Document Center for their patience and diligence in working with the manuscripts.

Contents of Volume II

List of Symbols

A	area; exchange energy constant
a	arctangent transition parameter; lattice constant
A_c	cross-sectional area of head core
A_g	cross-sectional area of head gap
a_x	transition parameter for longitudinal magnetization
a_y	transition parameter for perpendicular magnetization
B	magnetic induction (flux density); width (breadth) of slider; half-amplitude (0-peak) of zig-zag pattern
b	radius of particle; space between magnetoresistive head shields
b_0	critical particle radius for single-domain behavior
b_p	critical particle radius for superparamagnetic behavior
B_r	remanent induction (flux density)
B_s	saturation induction (flux density)
C	capacitance
c	damping constant
D	recording density (linear density); width of domain wall; flexural rigidity of a flexible medium
d	head-to-medium spacing
d_0	spacing corresponding to nominal "in-contact" conditions
D_{50}	linear density at which the output falls 50%
E	energy; Young's modulus
e	charge of electron; head output voltage
E_a	anisotropy energy
E_d	demagnetization energy
E_e	exchange energy
E_k	magnetocrystalline energy
e_n	noise output voltage
e_s	signal output voltage
e_x	output from longitudinal magnetization

e_y	output from perpendicular magnetization
F	force
f	frequency
f_0	Larmor frequency
f_s	signal frequency
G	gain; Green's function
g	gap length
g_{eff}	effective gap length
H	magnetic field; normalized spacing of a slider (h/h_{min})
h	reduced magnetic field; stripe height of magnetoresistive head; air bearing spacing
H_a	total anisotropy field
H_{appl}	applied field (e.g., of magnetometer)
H_b	bias field
H_c	coercivity
h_c	reduced coercivity (H_c/H_a)
H_d	demagnetizing field
H_g	deep gap field
H_k	magnetocrystalline anisotropy field
H_m	shape anisotropy field; maximum (applied) field
h_{min}	minimum air bearing spacing
$\mathbf{H}_{mrx}$, $\mathbf{H}_{mry}$	sensitivity-function fields for a magnetoresistive head
h_0	reduced nucleation field (H_0/H_a)
H_r	remanence coercivity
h_r	reduced remanence coercivity (H_c/H_a)
H_{rx}, H_{ry}	fields from a ring head (Karlqvist equations)
H_t	total field ($H_{\text{appl}} + H_d$)
H_1	positive field axis of the Preisach diagram
H_2	negative field axis of the Preisach diagram
$H_{0.25}$	reverse field required to switch 25% of particles after saturation $[M_r(H_{0.25})/M_r(\infty) = 0.5]$
$H_{0.75}$	reverse field required to switch 75% of particles after saturation $[M_r(H_{0.75})/M_r(\infty) = -0.5]$
$H_{0.5}$	dc field in ideal anhysteresis required to obtain one-half saturation $[M_{ar}(H_{0.5})/M_{ar}(\infty) = 0.5]$
I	intensity of electron beam
i	current
J	exchange integral
j	$\sqrt{-1}$

K	magnetic anisotropy constant
k	wave number $(2\pi/\lambda)$; Boltzmann constant
K_0, K_1, K_2	crystalline anisotropy constants
k_1, k_2, k_3	spring constants
K_u	uniaxial anisotropy constant
$K_\perp$	perpendicular anisotropy constant
$K(H_1, H_2)$	Preisach distribution function
L	inductance; length of dipole; length of head poles; length of slider
l	length of particle; length of head magnetic circuit
l_c	length of head core
L_d	spacing loss
L_g	gap loss
L_δ	thickness loss
L_1, L_2	dimensions of taper-flat slider
M	magnetization
m	reduced magnetization $[M(H)/M(\infty)]$; particle magnetic moment; dipole magnetic moment; mass; mass of electron
m_{ar}	reduced anhysteretic susceptibility, $M_{ar}(H)/M_{ar}(\infty)$
m_B	Bohr magneton
M_d	dc demagnetizing remanent magnetization
M_r	remanent magnetization (remanence)
m_r	reduced remanent magnetization $[M_r(H)/M_r(\infty)]$
M_s	saturation magnetization of a recording medium
M_{sb}	saturation magnetization of bulk material
M_0	peak value of sine-wave magnetization; value of magnetization in between transitions
$M(H)$	magnetization in a field H
$M(H_m)$	magnetization in a field of maximum value H_m
$M_r(H)$	remanent magnetization after applying a field H
$M_r(H_m)$	remanent magnetization after applying a field of maximum value H_m
$M_r(\infty)$	saturation remanent magnetization (retentivity)
N	demagnetization factor; number of particles per unit volume
n	number of turns
N_a, N_b, N_c	demagnetization factors along the axes of an ellipsoid
N_b	number of bits per byte
N_{eff}	effective demagnetization factor used for deskewing hysteresis loop

P	normalized pressure (p/p_a)
p	pole-tip length; volumetric packing density; pressure
p_a	ambient (atmospheric) pressure
p_{25}	pulse width at 25% amplitude
p_{50}	pulse width at 50% amplitude
q	mechanical load
R	resistance; reluctance
r	radius; radial coordinate
R_c	reluctance of core
R_g	reluctance of gap
r_h	radius of head
R_0	base resistance of magnetoresistive head
S	remanence squareness [$M_r(H)/M_s$]; element of surface; Sommerfeld number (ratio of viscous to pressure forces in a fluid)
s	distance between head and permeable layer
$S(H)$	remanence squareness in an applied field H
$S(H_m)$	remanence squareness in an applied field of maximum value H_m
T	temperature; thickness of pole in single-pole head; tension; torque
t	time; total thickness of medium including substrate; magnetoresistive sensor thickness
T_c	Curie temperature
U	magnetostatic potential; radial displacement velocity of air bearing surface
u	component of velocity
U_0	magnetostatic potential between pole tips
V	head-to-medium velocity
v	volume; particle volume; component of velocity
V_m	medium velocity (when different from head-to-medium velocity)
v_p	critical volume for superparamagnetic behavior
W	work; transverse displacement
w	track width; head width; component of velocity
w_{eff}	effective track width
w_t	tape width
X	normalized coordinate (x/B)
Y	normalized coordinate (y/L)
Z	atomic number; impedance
α, β, γ	direction cosines, angles

δ	thickness of a magnetic medium; thickness of a domain wall; eddy current penetration depth; stripe width in transmission line theory
δ_f	thickness of base film
Δh_r	switching field distribution
η	efficiency
Θ	angular coordinate;
Θ_c	Curie temperature
Λ	bearing number
λ	wavelength; magnetostriction coefficient; characteristic length in transmission line theory
λ_s	saturation magnetostriction coefficient
μ	magnetic permeability; fluid viscosity
μ_0	magnetic permeability of vacuum
μ_r	relative permeability
μ'	real permeability
μ''	imaginary permeability
ν	Poisson's ratio
ξ	nondimensional damping
ρ	density; resistivity; base resistivity of magnetoresistive head
ρ_m	magnetic pole volume density
σ	stress; squeeze number
σ_m	magnetic pole surface density
σ_r, σ_0	stresses along polar coordinates
σ_s	specific magnetic moment (saturation moment per unit mass)
$\sigma_{\rho\rho}, \sigma_{00}, \sigma_{\rho 0}$	stresses
τ	time constant; dibit peak shift
Φ	magnetic flux
ϕ	magnetic flux; angular coordinate
χ	magnetic susceptibility
χ_{ar}	anhysteretic susceptibility
χ_0	initial reversible susceptibility
ω	angular frequency $(2\pi f)$

Introduction

Eric D. Daniel

Woodside, California

C. Denis Mee

IBM Corporation, San Jose, California

This chapter provides a simple, nonmathematical review of magnetic recording principles and introduces some of the more frequently used terms. It also gives an indication of the variety of configurations and applications of magnetic recording, and outlines the direction and focus of present and future research and development efforts. The primary purpose is to provide a reader, relatively new to magnetic recording, with the background required to assimilate the more specialized chapters that follow. The introductory material may also assist readers skilled in some, but not all, of the various aspects of magnetic recording to put their specialties in perspective and expand their interests.

Since this introduction does not include a tutorial review of the principles of magnetism, a reader lacking a background in magnetism should consult one of the standard texts (Brown, 1978; Chikazumi and Charap, 1978; Wohlfarth, 1980–1982). For additional review and general reading in magnetic recording, reference can be made to a limited number of books or collections of key papers (Hoagland, 1963; Mee, 1964; Masumoto,

1977; Jorgensen, 1980; Camras, 1985; White, 1985) and to a recently published set of review articles (Arnoldussen, 1986; Bate, 1986; Bertram, 1986;
Jeffers, 1986; Wood, 1986).

1.1 Magnetic Recording Characteristics

1.1.1 Properties of a permanent magnet

A permanent magnet is composed of a ferromagnetic or ferrimagnetic
material that has the property of staying magnetized in a given direction
after the field that created this magnetization is removed. This property
is displayed by a hysteresis loop, such as that shown in Fig. 1.1a, in which

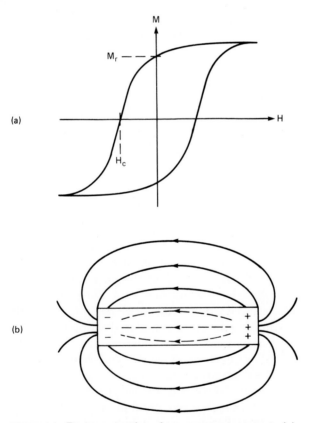

Figure 1.1 Basic properties of a permanent magnet. (a)
Hysteresis loop showing remanent magnetization M_r and
coercivity H_c; (b) bar magnet showing surface poles and the
fields these poles produce outside (full lines) and inside
(dashed lines) the bar.

the magnetization M is plotted against the applied field H (Chikazumi and Charap, 1978). The magnetization is not a unique function of the field but depends on the direction and magnitude of previous applied fields. There are two important parameters of the loop: first, the remanent magnetization M_r, the magnetization that remains after the field is removed; second, the coercivity H_c, the reverse field required to reduce the magnetization to zero. The remanent magnetization indicates the extent to which the magnet stays magnetized after the applied field is removed; the coercivity expresses the degree to which the magnet resists being demagnetized. The product $M_r H_c$ measures the strength of the magnet.

The external and internal fields of a bar magnet are sketched in Fig. 1.1b. If it is assumed that the magnetization of the bar is approximately uniform, the magnetic poles are surface poles on the left and right faces. The field is directed from the positive to the negative pole faces. That portion of the field that lies within the bar opposes the magnetization and constitutes a demagnetizing field H_d. As the strength of this field approaches that of the coercivity (it cannot exceed it), progressively larger losses in remanent magnetization occur.

The strength of the demagnetizing field is primarily dependent upon the geometry of the magnet, and can be expressed in terms of the demagnetization factor (Chikazumi, 1978; Wohlfarth, 1980)

$$N = \frac{-H_d}{M} \tag{1.1}$$

Except in certain geometries (ellipsoids), N varies throughout the body of the magnet, but simple arguments can be based on considering a mean value for a bar magnet. Two extremes are of interest. If the bar is made increasingly long in the x direction, as in Fig. 1.2b, the demagnetizing factor N_x in this direction decreases and eventually approaches zero, while the factors N_y and N_z in the y and z directions approach 0.5. If the bar is made increasingly short in the x direction and the dimensions in the y and z directions expanded, as in Fig. 1.2c, N_x approaches a maximum value of 1 ($H_d = -M_x$) in the central region, while N_y and N_z tend to zero.

1.1.2 Basic processes in magnetic recording

In principle, a magnetic recording medium consists of a permanent magnet configured so that a pattern of remanent magnetization can be formed along the length of a single track, or a number of parallel tracks, defined on its surface. Recording (or writing) takes places by causing relative motion between the medium and a recording transducer. A simple, one-

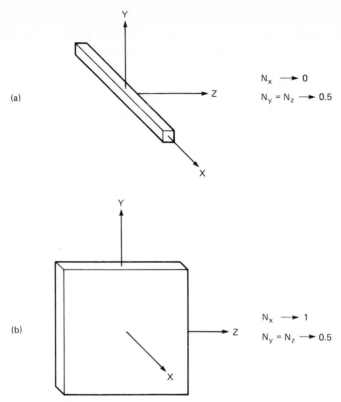

$$N_x \longrightarrow 0$$
$$N_y = N_z \longrightarrow 0.5$$

$$N_x \longrightarrow 1$$
$$N_y = N_z \longrightarrow 0.5$$

Figure 1.2 Demagnetizing factors of bar magnets of square cross section and different length-to-width ratios. (a) A long, thin bar; (b) a wide, thin sheet. The condition $N_x + N_y + N_z = 1$ always applies.

track example is given in Fig. 1.3a. The medium is in the form of a magnetic layer supported on a nonmagnetic substrate. The transducer, or recording head, is a ring-shaped electromagnet with a gap at the surface facing the medium. When the head is fed with a current representing the signal to be recorded, the fringing field from the gap magnetizes the medium as shown in Fig. 1.3b. For a constant medium velocity, the spatial variations in remanent magnetization along the length of the medium reflect the temporal variations in the head current, and constitute a recording of the signal. In the ordinary way the recording process is highly nonlinear because it relies on hysteresis.

The recorded magnetization creates a pattern of external and internal fields analogous, in the simplest case, to a series of contiguous bar magnets. When the recorded medium is passed over the same head, or a reproducing head of similar construction, at the same velocity, the flux emanating from the medium surface is intercepted by the head core, and a voltage is induced in the coil proportional to the rate of change of this flux. The voltage is not an exact replica of the recording signal, but it constitutes a reproduction of it in that information describing the recording signal can be obtained from this voltage by appropriate electrical processing.

In between recording and reproduction, the recorded signal can be stored indefinitely, provided the medium is not exposed to magnetic fields comparable in strength to those used in recording. At any time, however, a recording that is no longer required can be erased by means of a strong field applied by the same head as that used for recording, by a separate erase head, or by a bulk eraser that subjects the medium, in its entirety,

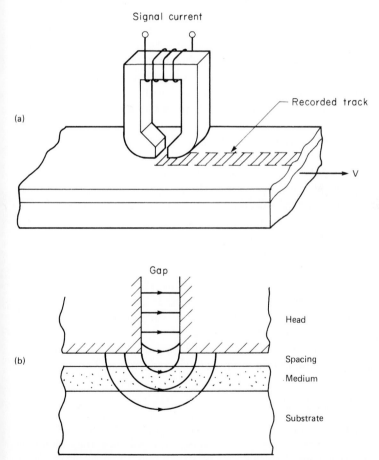

Figure 1.3 Illustration of the recording process using a single-track ring head. (*a*) Three-dimensional view; (*b*) cross section showing the magnetic field from the gap.

to a 50- or 60-Hz field. After erasure, the medium is ready for a new recording. Overwriting an old signal with a new one, without a separate erase step, is adequate in many circumstances. Overwriting has to be used when only a single head is available for writing and reading.

The recording and reproducing processes will be described more fully later in this Introduction and are analyzed in detail in Chap. 2.

1.1.3 Media configurations

With the exception of some early wire recorders, all magnetic recorders use media in the form of a magnetic layer (sometimes two layers) supported by a nonmagnetic substrate. The magnetic material of the medium may be a coating made of magnetic particles held in a plastic binder (Bate, 1986), or it may consist of a deposited film of metal or oxide (Arnoldussen, 1986). The traditional particulate material is a type of ferrimagnetic iron oxide, gamma ferric oxide, in the form of small, needle-shaped particles. Most deposited metal films use alloys made predominantly of cobalt, and have the advantage of having a much higher magnetization than particulate media, coupled with the facility of making the magnetic layer very thin. Coercivity is important in media for much the same reasons as it is in permanent magnets, and has tended to increase, particularly in recent years. Such increases have been necessitated by the development of more sophisticated recording devices with shorter minimum recorded wavelengths and consequently larger demagnetizing fields.

One of the advantages of magnetic recording is that it can be readily adapted to use many configurations of the recording media. Media using plastic film as a substrate are used mainly in the form of tapes and flexible disks, but also appear in a variety of card and sheet configurations coated over the whole surface, or in a magnetic stripe coated on a portion of the surface. Media using metal or ceramic substrates have been made in the form of drums and rigid disks. Tapes vary in width from less than 4 to over 50 mm, and in length from a cartridge roll of less than 1 m to a reel of over 3000 m. Flexible disks vary in diameter from about 75 to 200 mm. Rigid disks range from 75 to about 360 mm, and up to a dozen double-sided disks are rotated on a single spindle.

The properties, materials, design, and fabrication of particulate and deposited-film magnetic media are discussed in detail in Chap. 3.

1.1.4 Head configurations

There is great variety in the design and composition of heads and in the way they interface with media (Jeffers, 1986). The medium may be moved in nominal contact with fixed heads, or moved past a head, supported by an air bearing, which maintains a controlled spacing between the surfaces of the medium and the head. Heads may also be embedded in a rotating cylinder in contact with a moving medium, so that the heads scan across the medium at high velocity. A head may contain a single element and record or reproduce a single track, or it may contain many elements and record many tracks simultaneously. Reproduction may take place using an inductive head like the ring-shaped structure of Fig. 1.3. Inductive heads can also be made where a thin main pole is on the magnetic side of the medium, while an energized auxiliary pole is on the substrate side.

It is also possible to design heads that respond to magnetic flux rather than to its rate of change. Such flux-sensing heads use galvanomagnetic effects in which the magnetic field causes a change in electric field (the Hall effect) or in resistance (magnetoresistance), or make use of the principle of the flux gate. Finally, optical beams may be used to record on a heat-sensitive magnetic medium (thermomagnetic recording), and to reproduce such a recording using Kerr or Faraday rotation effects (magnetooptic reproduction) (Meiklejohn, 1986).

A variety of magnetic materials is used in making heads. Laminated, high-permeability metals are the traditional means of making the core, and are still used when the coercivity of the medium demands a high recording field from the head. Ferrite cores have the advantages of greater ease of fabrication and better wear characteristics. For rigid-disk and certain other applications, semiconductor techniques have been adapted to make heads by depositing multilayer films. Such film heads are particularly suited to applications where smallness of size is dictated or where many-element multitrack operation is required. Head assemblies vary in width from about 20 μm to 50 mm, while individual tracks range from some 10 μm up to 6 mm. Up to 100 head elements have been built in a single stack using film fabrication technology.

Details of the properties, materials, design, and fabrication of heads are given in Chap. 4. Magnetooptic recording is not treated in any detail in this book, but is the subject of Chap. 9 of Volume II.

1.1.5 Recording signals

Magnetic recording is used to store many different types of signals. Analog recording of sound was the first and is still a major application. Audio recording involves relatively low-frequency signals and low medium speeds, but is demanding in terms of linearity and signal-to-noise ratio. Digital recording of encoded computer data on disk and tape recorders has evolved as another major use. Rigid-disk drives use high signal frequencies coupled with high medium speeds, and emphasize small access times together with high data reliability. A third large application area is video recording, the recording of visual images in the form of video signals, for professional or consumer use. The high video frequencies are normally recorded using frequency-modulation (FM) encoding, and high head-to-medium velocities are obtained by the use of scanning-head drums. Instrumentation recording covers a wide range of applications and may use any of the recording techniques described above. The trend is to use digital encoding to record all types of signals because, in principle, it allows original signals to be reconstructed perfectly through the use of error detection and correction schemes. Digital audio is already used widely for professional purposes and is available for consumer use. Active development and standardization is under way for professional digital video recording.

As depicted in Fig. 1.4, there are essentially three types of signals used in magnetic recording. The linearity required in analog audio recording is obtained by using ac bias. This consists of a high-frequency current which

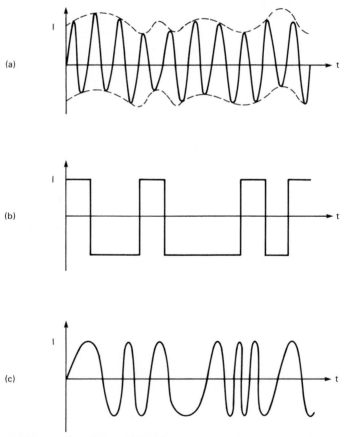

Figure 1.4   Three types of recording signal current. (*a*) An analog signal to which a high-frequency ac bias is added; (*b*) a binary digital signal; (*c*) a frequency-modulated signal.

is added to the signal current in the record head and, in effect, replaces normal hysteretic magnetization of the medium with a linear process known as *anhysteretic magnetization* (Westmijze, 1953*b*). In digital recording, a coded series of reversals in the direction of magnetization, or transitions, constitutes the recorded signal. Reproduction is effected by reconstructing the pattern by, for example, detecting "zero crossings" in the reproduced waveform. The situation in FM video recording is similar, except that the positions of the zero crossings are modulated by the video signal.

The way in which the signals are processed before and after recording differs greatly according to the application (Wood, 1986). However, the underlying magnetic recording process, in which the temporal variations of a signal or its encoded version are represented by spatial variations in magnetization along a recorded track, is common to all applications. This process, together with the inductive reproducing process, is expressed in elementary but general terms in the introductory review given here.

1.2 Magnetic Recording Principles

1.2.1 Recording field

There are two distinct types of inductive head: the ring head (such as that of Fig. 1.3), in which the recording field is the leakage field from the gap, and the single-pole (or probe) head, in which the recording field spreads out from the tip of a thin energized pole. The essential characteristics of these recording fields can be grasped by considering idealized models in which the length of the gap or the thickness of the single pole is infinitesimally small, and all other relevant dimensions, including the width of the head, are infinite. The last assumption reduces the problem to a two-dimensional one.

The field from the idealized ring head is shown diagrammatically in Fig. 1.5*a*. The lines of force are semicircles about the gap centerline. When an element of the recording medium passes over the head at a distance y, it experiences a field that starts from the perpendicular direction, rotates to a longitudinally directed maximum in the central plane of the gap, and then rotates toward the opposite perpendicular direction while falling to zero. The field vector follows the circular locus shown in Fig. 1.5*b*, where the diameter of the circle is inversely proportional to y.

The corresponding field from the idealized single-pole head is shown in Fig. 1.6*a*. The field contours spread radially outward from the pole tip, and the direction of the field experienced by an element of medium now rotates from longitudinal, through perpendicular, to longitudinal in the opposite sense. The locus of the field vector is again circular, but the center is now on the y axis, as shown in Fig. 1.6*b*.

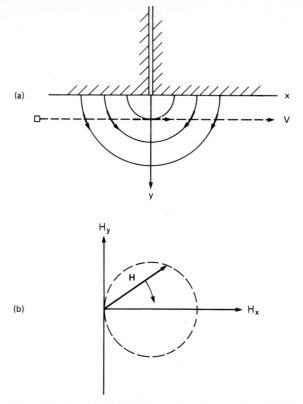

Figure 1.5   Recording field from a ring head with semi-infinite pole dimensions and an infinitesimally short gap. (*a*) Cross section of the pole tips showing the lines of force and the passage of an element of medium through the field at a spacing *y*; (*b*) the locus of the vector field *H* experienced by the element of medium.

An important conclusion is that either type of head gives rise to a rotating recording field that has both longitudinal and perpendicular components, and hence will tend to induce a recorded magnetization that has both longitudinal and perpendicular components (Tjaden and Leyten, 1964). Generally, the ring head promotes longitudinal, and the single-pole head perpendicular, components of field and magnetization. At small separations and high fields, the reverse situation can occur; the final magnetization is determined in the trailing region, where the element of medium exits the head field, rather than in the region of maximum field strength.

In practical forms of the ring head, the gap *g* is of finite length, and the poles are also finite in length. These dimensions complicate, but do not change, the basic nature of the head-field distribution except at very close spacings for which $y \ll g$ (Westmijze, 1953a; Karlqvist, 1954). In practical forms of the single-pole head, the thickness T of the pole is finite, which has an effect analogous to that of the gap in a ring head. Also, an efficient single-pole head must provide a low-reluctance return path for the flux, a path which, ideally, is situated on the other side of the recording medium opposite the main pole. This return path may consist of a large auxiliary pole, a high-permeability underlayer, or a combination of these. The field distribution is modified, particularly close to the auxiliary pole or underlayer, by an emphasis of the perpendicular component (Iwasaki and Nakamura, 1977).

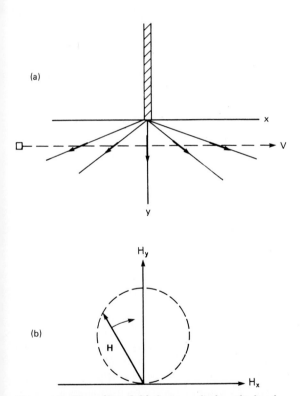

Figure 1.6 Recording field from a single-pole head with a pole of infinite depth and infinitesimal thickness. (a) Cross section of the pole tip showing the lines of force and the passage of an element of medium through the field at a spacing y; (b) locus of the vector field H experienced by the element of medium. (The pole is assumed to be uniformly wound with a coil, or energized inductively by a nearby auxiliary pole.)

1.2.2 Medium anisotropy

The direction of the recorded magnetization is also strongly influenced by the magnetic anisotropy of the medium. For example, a common type of medium consists of acicular (needle-shaped) particles with their long dimensions oriented parallel to the longitudinal direction during manufacture (Bate, 1986). Such a medium exhibits uniaxial anisotropy in that it has a much higher remanent magnetization in the longitudinal direction than in any other direction. It will therefore favor longitudinal recording, in which the magnetization is directed along the length of the track. Alloy films may also be made in such a way that the preferred directions of magnetization lie in the plane of the film, although the anisotropy in the plane is usually multiaxial rather than uniaxial (Arnoldussen, 1986)

Conversely, media can be made which favor magnetization normal to the plane. Thus, certain alloy films may be designed to have a crystal orientation and grain structure that produce uniaxial anisotropy perpendicular to the plane of the film (Iwasaki and Ouchi, 1978). Also, suitable particles can be oriented so that their easy axes lie perpendicular to the plane of a medium (Fujiwara, 1985). Either type of medium favors perpendicular recording, particularly if the perpendicular anisotropy field H_k is strong enough to counter the demagnetizing field associated with a medium in the form of a thin layer ($H_k > M_r$).

A third type of medium can be made, for example, using particles which individually have many (six or eight) different easy directions of magnetization (Jeschke, 1954; Lemke, 1979). If the shape anisotropy associated with a finite thickness is ignored, such a medium is intrinsically isotropic. The direction of its recorded magnetization is dominated by the geometry of the head-field distribution rather than by the magnetic properties of the medium.

1.2.3 Longitudinal recording

The combination of a ring head and a medium having longitudinal anisotropy tends to produce a recorded magnetization which is predominantly longitudinal. This combination has been the one used traditionally, and it still dominates all major analog and digital applications. Ideally, the pattern of magnetization created by a square-wave recording signal would be like that shown in Fig. 1.7a. If the square wave has a fundamental frequency f, and the medium moves with velocity V, the recorded pattern can be characterized by the wavelength

$$\lambda = \frac{V}{f} \tag{1.2}$$

by the bit length

$$b = \frac{\lambda}{2} \tag{1.3}$$

or by the linear density

$$D = \frac{2f}{V} \tag{1.4}$$

measured in flux reversals per millimeter (fr/mm).

Figure 1.7a represents a low-density recording. The bulk of the flux created by the magnetization reversals returns outside the medium, and the demagnetizing field is small. At higher densities, Fig. 1.7b, the demagnetizing field and the loss in magnetization associated with it become progressively larger. Eventually, the transitions lose their sharpness and the recorded pattern approaches a sine wave corresponding to the fundamen-

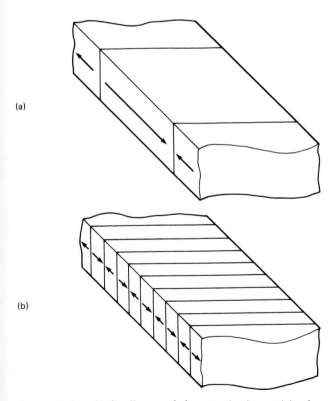

(a)

(b)

Figure 1.7 Longitudinally recorded magnetization at (a) a low and (b) a high density.

tal of the square wave that represents the recording signal. In the central plane of the medium, the demagnetization factor at high densities approaches the limiting value $N = 1$, and the demagnetizing field approaches the value $H_d = -M_0$, where M_0 is the demagnetized value of the magnetization between transitions, or the peak value if the magnetization is sinusoidal. At the surface of the medium, however, the demagnetization factor approaches the limiting value $N = 0.5$, and the demagnetizing field approaches the value $H_d = -0.5M_0$, one-half the field limit in the central plane (Wallace, 1951; Westmijze, 1953b). In Sec. 1.2.5 on reproduction it is shown that, as the recording density increases, the useful flux supplied to the reproducing head comes from a progressively shallower layer immediately beneath the surface. It can therefore be argued that the demagnetization conditions at the surface of a medium of appreciable thickness are of dominant importance compared with those in the center.

1.2.4 Perpendicular recording

The corresponding situation for a medium with perpendicular anisotropy is illustrated in Fig. 1.8. In the central plane of the medium the demagnetization conditions are the reverse of what they were for longitudinal recording. At the low density shown in Fig. 1.8a, the demagnetization factor in the perpendicular direction approaches the maximum value $N = 1$, corresponding to $H_d = -M_0$. At a high density, shown in Fig. 1.8b, the demagnetization factor and the demagnetizing field approach zero. From this central-plane analysis it would appear that perpendicular recording should be the ideal mode to use for high-density recording (Iwasaki and Nakamura, 1977). If, however, the demagnetization factor is evaluated at the surface rather than at the center, the limiting value is $N = 0.5$, corresponding to a demagnetizing field of $H_d = -0.5M_0$, exactly the same as for longitudinal recording (Wallace, 1951; Westmijze, 1953b). Therefore, at densities so high that conditions near the surface rather than in the central plane are of paramount importance, the effects of demagnetization in perpendicular recording should be little different from those in longitudinal recording (Westmijze, 1953b; Mallinson and Bertram, 1984).

Such arguments are on too limited a basis to warrant arriving at definitive conclusions concerning the relative merits of the longitudinal and perpendicular modes of recording. More complete analyses need to take into account a variety of complex macromagnetic and micromagnetic considerations. Important macromagnetic factors include the finite head-field gradients and demagnetization effects during the recording process; the keeping effect of heads (particularly large-pole ring heads) in modifying demagnetization fields; the enhancement of perpendicular recording

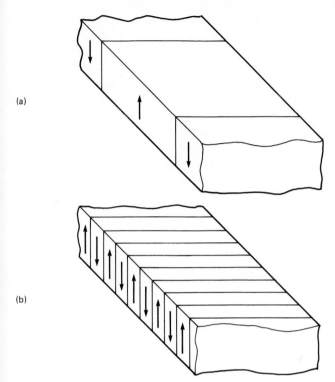

Figure 1.8   Perpendicularly recorded magnetization at (a) a low and (b) a high density.

media made possible by the use of a high-permeability underlayer; and, perhaps most difficult of all, the nonlinear nature and directional properties of the hysteresis loops of the media. The more important micromagnetic factors include the precise structure of the magnetic boundary defining a transition, and the way in which a relatively thick perpendicular medium is magnetically "switched" throughout its depth.

Work is proceeding along these lines using a variety of analytical, computer-modeling, and experimental techniques (e.g., Iwasaki, 1984; Middleton and Wright, 1982; Beardsley, 1982a, 1982b; Bromley, 1983; Weilinga, 1983; Tong et al., 1984). The results are described and discussed extensively in Chap. 2, and specific aspects related to media, heads, recording limitations, and measurements are covered in Chaps. 3 to 6.

1.2.5   Reproduction

Reproduction with a magnetic head is insensitive to the direction of the recorded magnetization. The head reacts to the flux emerging from the surface of the medium, but the vectorial nature of the magnetization that created this flux cannot be deduced from the head response. Thus, for example, a ring head is just as adept at reproducing a perpendicular recording as a longitudinal one. Another simplifying fact is that, unlike the recording process, the reproducing process is linear. Therefore it is possible to deduce the response to complex waveforms from considering sine waves.

For the present purpose, attention is confined to an inductive ring head. If an efficient head of this type makes perfect contact with the medium, and all the available flux ϕ is collected, the voltage induced in a coil of N turns is

$$e = -N\frac{d\phi}{dt} = -NV\frac{d\phi}{dx} \tag{1.5}$$

where the x direction is down the track. If the flux (or a Fourier component of it) is sinusoidal and written as $\phi \cos(2\pi x/\lambda)$, the peak voltage in the coil becomes

$$E = -\frac{2\pi NV\phi}{\lambda} = -2\pi N\phi f \tag{1.6}$$

and is proportional to frequency.

Further information concerning the response of the head can be obtained from a knowledge of the field distribution of the head when energized, by applying the principle of reciprocity (Westmijze, 1953a). One of the key results—and it applies to any type of reproducing head—is that the head output falls off exponentially with the spacing that exists accidentally, or purposely, between the active surfaces of the head and the medium. Thus

$$\frac{e_d}{e_0} = \exp\left(-\frac{2\pi d}{\lambda}\right) \tag{1.7}$$

where the subscripts denote a spacing of d and a spacing of zero. The severity of this spacing loss increases rapidly with decreasing wavelength, and can be conveniently calculated by expressing the loss as $54.6d/\lambda$ in decibels (Wallace, 1951). The spacing loss is always of critical significance in magnetic recording, and often is the dominant cause of loss at high densities.

A further aspect of the spacing loss is that it applies also to the spacing between the head and elementary layers positioned beneath the surface of the medium. As an elementary layer becomes deeper, it becomes less capable of contributing significantly to the short-wavelength, or high-density, output. Consequently, at very high densities, virtually all the output comes from a thin layer near the surface, when the medium has a thickness which is an appreciable fraction of the wavelength (Wallace, 1951).

So far, it has been assumed that the head gap length is infinitesimal. When it is finite, it is the source of another, aperture-type of loss, the gap loss, that produces a null in the reproducing response as the wavelength approaches the gap length (Lübeck, 1937; Westmijze, 1953a). In practice, gap loss is one of the easier losses to control, except when the same head has to be used for recording and reproducing. A similar (usually larger) loss is associated with the finite thickness of the pole in a single-pole head when it is used for reproducing.

1.3 Practical Constraints

1.3.1 Noise and interference

All magnetic recorders produce unwanted signals in the form of noise and interference, and these impose limitations on the achievable performance. The subject is covered extensively in Chap. 5 and receives only brief mention here. Essentially, there are two major sources of noise: the medium and the reproducing head (electronics noise is usually negligible). Medium noise arises from the fact that no medium is magnetically homogeneous. Particulate media are obviously discontinuous, and create noise in accordance with the number, density, size, and spatial distribution of the particles (Mann, 1957; Mallinson, 1969; Daniel, 1972). Deposited-film media are inhomogeneous because they possess a grain structure, or because irregular domains are formed to minimize the energy at transition boundaries (Baugh et al., 1983; Belk et al., 1985). The major head noise arises from the fact that any head possesses an impedance, and the real part of this impedance gives rise to noise of thermal origin. Other forms of head noise are associated with magnetic domain changes (Barkhausen noise) or magnetostriction effects ("rubbing" noise).

Interference—the appearance of signals other than those intended—arises in many ways, such as cross-talk between different elements in a multitrack head; incomplete erasure of a previously recorded signal; track misregistration when a head scans a track on successive occasions; and print-through, the magnetic transfer of signals between the layers in a stored reel of tape. Which effect is the most serious depends on the type of recorder and the application. In most analog recorders, medium noise is dominant, but reproducing head noise will become more significant as signal frequencies go up and track widths diminish. Medium noise is also the dominant form of noise in digital disk drives, but, currently, interference due to track misregistration constitutes a more serious limitation than noise. Print-through is of consequence only in the analog recording of audio signals.

1.3.2   Head-medium interface

Magnetic recording involves mechanical motion between the media and heads, and, as intimated above, the spacing between these components must be made critically small. This combination imposes stringent demands upon the surface characteristics of media and heads: their flatness, smoothness, freedom from asperities, frictional properties, and mutual capability to resist wear.

In the great majority of recorders using flexible media, the heads and media are run nominally in contact in order to reduce spacing losses to the values required for high-density performance at practical head-media speeds. The head material, profile, and surface integrity are critical properties in achieving adequate head life and avoiding undue wear of the medium. The corresponding properties of the medium are, if anything, more critical, because they must be maintained over an enormously larger surface area. Particulate media have the advantage that the magnetic and tribological properties can be controlled independently, through the choice of particle and the choice of plastic binder and additives.

In rigid-disk drives the problem is approached differently. The head is mounted on a slider which flies above the rotating disk, and provides an air bearing which, in principle, avoids wear entirely (Gross et al., 1980). Early air bearings were formed by designing the slider to create hydrodynamic pressure and weighting the head to achieve a flying height of about 20 μm. Modern bearings are designed to be self-adjusting and to fly at a height of 0.5 μm or less. In practice, wear is not entirely eliminated by the use of a flying head. Durability to occasional head-medium contacts at full velocity is required in addition to the slower-speed contact which occurs when the head takes off or lands during starting and stopping the disk rotation. Because of these requirements, the disk magnetic layers are protected by lubricant or protective overcoat layers. Such layers are typically required when metal-film media are used, but their thickness must be small with respect to the flying height in order to avoid excessive spacing loss.

A detailed discussion of the head-medium interface, with particular emphasis on the rigid-disk interface, is given in Chap. 7.

1.4   Recording Technology Emphasis

1.4.1   Areal density

Magnetic recording research and development efforts will continue to be aimed toward achieving higher areal density by a combination of increases in linear and track densities (Mallinson, 1985). Increases in linear density will require improvements in materials and recording techniques, and controlled miniaturization of the key recording components. Increases in track density will require advances in media properties, narrow-track head design, and head-positioning technology. Some of the areas that are expected to receive emphasis are outlined below.

1.4.2 Head and media developments

Sophisticated methods of head fabrication will continue to be developed for ferrite, metal, and composite head structures. In particular, the use of deposition techniques, similar to those used in the semiconductor industry, is an attractive means to achieving some of the miniaturization goals, and applications of film heads can be expected to broaden. This technology has also spurred renewed interest in flux-sensing read heads, particularly those relying upon the use of a thin magnetoresistance element. Such heads give larger reproduce signals than inductive heads, particularly at lower head-to-medium speeds. This higher sensitivity can be used to offset the lower signal flux available when track density is increased.

Media developments cover a broad range of materials and processes. Iron oxide remains the dominant magnetic ingredient of particulate media, but the higher magnetization of metal particles makes them advantageous for certain high-areal-density applications. Deposited-metal-film media have even higher magnetization potential, and a high-output amplitude can be obtained from a very thin layer, which is favorable at high linear and track densities. Metal-film media are receiving rapidly growing attention, and progress is being made in solving some of the problems associated with durability, corrosion, and other defects that inhibited their use in the past.

The dimensions of head magnetic elements and the recorded transitions in media have decreased to the point where micromagnetic structure and switching mechanisms can no longer be ignored. Grain size in ferrite-head poles, domain effects in film-head poles, and micromagnetic irregularities in recorded transition boundaries are becoming of critical concern. Further advances in head and media materials and design will be assisted by a better understanding of, and ability to control, domain-level phenomena.

1.4.3 Recording modes

The dominant mode of recording has been to magnetize the medium predominantly in the longitudinal direction. Early attempts to record in the perpendicular direction failed because the media used would not readily support such magnetization (Hoagland, 1958). The situation was changed in the mid-1970s by the introduction of a cobalt-chromium medium which possesses the requisite properties favoring magnetization in the perpendicular direction (Iwasaki and Ouchi, 1978). Subsequently, the interest in exploring the perpendicular mode for high-density recording increased rapidly, and this area of study continues to attract a significant fraction of magnetic research activities worldwide (Iwasaki, 1980). These activities include work on particulate as well as single- and double-layer metal-film media, and the study of various types of single-pole head with an auxiliary pole on the same or the opposite side.

Despite these research efforts, the practical implementation of perpendicular recording has been slow, and some of the fundamental aspects of this mode of recording are still not fully understood. Once again, a stronger emphasis at the micromagnetic level will be required before the potential of this mode of recording can be fully assessed, and perpendicular media and heads optimized.

1.4.4   Optical recording

In terms of linear density, magnetic recording is ahead of optical recording. Moreover, the linear density of optical recording is limited by the wavelength of the light source. Therefore, short of developing a short-wavelength laser (e.g., by frequency doubling), magnetic recording's lead in linear density will lengthen. With respect to track density, however, optical recording is ahead of magnetic recording. The advantages of optical recording stem from the capability to maintain a signal-to-noise ratio down to very narrow track widths, together with the wide bandwidth with which the beam position can be servoed. The adoption of optical-beam recording and reproducing techniques to magnetic recording is receiving increasing emphasis. It offers an instant large increase in track density and, despite the lower linear density, can result in a net increase in areal density. This advantage will, however, be eroded as the conventional forms of magnetic recording continue their advance.

References

Arnoldussen, T. C., "Thin Film Recording Media," *Proc. IEEE,* **74** (1986).
Bate, G., "Particulate Recording Media," *Proc. IEEE,* **74** (1986).
Baugh, R. A., E. S. Murdock, and B. R. Najarajan, "Measurement of Noise in Magnetic Media," *IEEE Trans. Magn.,* **MAG-19,** 1722 (1983).
Beardsley, I. A., "Effect of Particle Orientation on High Density Recording," *IEEE Trans. Magn.,* **MAG-18,** 1191 (1982*a*).
Beardsley, I. A., "Self-Consistent Recording Model for Perpendicular Recording," *J. Appl. Phys.* **53,** 2582 (1982*b*).
Belk, N. R., P. K. George, and G. S. Mowry, "Noise in High Performance Thin-Film Longitudinal Magnetic Recording Media," *IEEE Trans. Magn.,* **MAG-21,** 1350 (1985).
Bertram, H. N., "Fundamentals of the Magnetic Recording Process," *Proc. IEEE,* **74** (1986).
Bromley, D. J., "A Comparison of Vertical and Longitudinal Magnetic Recording Based on Analytical Models," *IEEE Trans. Magn.,* **MAG-19,** 2239 (1983).

Brown, W. F., *Micromagnetics,* Krieger, New York, 1978.

Camras, M., *Magnetic Tape Recording,* Van Nostrand Reinhold, New York, 1985.

Chikazumi, S., and S. H. Charap, *Physics of Magnetism,* Krieger, New York, 1978.

Daniel, E. D., "Tape Noise in Audio Recording," *J. Audio Eng. Soc.,* **20**, 92 (1972).

Fujiwara, T., "Barium Ferrite Media for Perpendicular Recording," *IEEE Trans. Magn.* **MAG-21,** 1480 (1985).

Gross, W. A., L. Matsch, V. Castelli, A. Eshel, T. Vohr, and M. Wilamann, *Fluid Film Lubrication,* Wiley, New York, 1980.

Hoagland, A. S., "High-Resolution Magnetic Recording Structures," *IBM J. Res. Dev.,* **2**, 91 (1958).

Hoagland, A. S., *Digital Magnetic Recording,* Wiley, New York, 1963.

Iwasaki, S., "Perpendicular Magnetic Recording—Evolution and Future," *IEEE Trans. Magn.,* **MAG-20**, 607 (1984).

Iwasaki, S., and Y. Nakamura, "An Analysis of the Magnetization Mode for High Density Magnetic Recording," *IEEE Trans. Magn.,* **MAG-13,** 1272 (1977).

Iwasaki, S., and K. Ouchi, "Co-Cr Recording Films with Perpendicular Magnetic Anisotropy," *IEEE Trans. Magn.,* **MAG-14,** 849 (1978).

Jeffers, F., "High Density Magnetic Recording Heads," *Proc. IEEE,* **74** (1986).

Jeschke, J. C., East German Patent 8684, 1954.

Jorgensen, F., *The Complete Handbook of Magnetic Recording,* TAB Books, Blue Ridge Summit, Penn., 1980.

Karqvist, O., "Calculation of the Magnetic Field in the Ferromagnetic Layer of a Magnetic Drum," *Trans. R. Inst. Technol. (Stockholm),* **86**, 3 (1954).

Lemke, J. U., "Ultra-High Density Recording with New Heads and Tapes," *IEEE Trans. Magn.,* **MAG-15,** 1561 (1979).

Lubeck, H., "Magnetische Schallaufzeichrung mit Filmen und Ringkopfen," *Akust. Z.,* **2**, 273 (1937).

Mallinson, J. C., "The Maximum Signal-to-Noise Ratio of a Tape Recorder," *IEEE Trans. Magn.,* **MAG-5,** 182 (1969).

Mallinson, J. C., "The Next Decade in Magnetic Recording," *IEEE Trans. Magn.,* **MAG-21,** 1217 (1985).

Mallinson, J. C., and H. N. Bertram. "Theoretical and Experimental Comparison of the Longitudinal and Vertical Modes of Magnetic Recording," *IEEE Trans. Magn.,* **MAG-20,** 461 (1984).

Mann, P. A., "Das Rauschen eines Magnettonbandes," *Arch. Elek. Ubertragung.* **11, 97** (1957).

Masumoto, M., *Magnetic Recording* (in Japanese), Kyoritsu, Tokyo, 1977.

Mee, C. D., *The Physics of Magnetic Recording,* North-Holland, Amsterdam, 1964.

Meiklejohn, W. H. "Magnetooptics, A High Density Magnetic Recording Technology," *Proc. IEEE,* **74** (1986).

Middleton, B. K., and C. D. Wright, "Perpendicular Recording," *IERE Conf. Proc.,* **54,** 181 (1982).

Tjaden, D. L. A., and J. Leyten, "A 5000:1 Scale Model of the Magnetic Recording Process," *Philips Tech. Rev.,* **25**, 319 (1964).

Tong, H. C., R. Ferrier, P. Chang, J. Tzeng, and K. L. Parker, "The Micromagnetics of Thin-Film Disk Recording Tracks," *IEEE Trans. Magn.,* **MAG-20,** 1831 (1984).

Wallace, R. L., "The Reproduction of Magnetically Recorded Signals," *Bell Syst. Tech. J.,* **30**, 1145 (1951).

Weilinga, T., "Investigations on Perpendicular Magnetic Recording," Thesis, Twente Univ. Technol., the Netherlands, 1983.

Westmijze, W. K., "Studies in Magnetic Recording," *Philips Res. Rep.,* **8**, 161 (1953a).

Westmijze, W. K., "Studies in Magnetic Recording," *Philips Res. Rep.,* **8**, 245 (1953b).

White, R. M., *Introduction to Magnetic Recording,* IEEE Press, New York, 1985.

Wohlfarth, E. P., *Ferromagnetic Materials,* North-Holland, Amsterdam, 1980–1982, vols. I–III.

Wood, R., "Magnetic Recording Systems," *Proc. IEEE,* **74** (1986).

2

Recording and Reproducing Processes

Barry K. Middleton

Manchester Polytechnic, Manchester, England

The aim of this chapter is to describe the contributions of the record and reproduce processes to the observed output voltage waveforms. Although the first of these processes is complex in the extreme and defies explanation in anything approaching a rigorous way, much has been learned with simplified models, and the essential quantities controlling the recording process appear to have been identified. The line taken here will be to follow these simpler models and to discuss their limitations and indicate the achievements of the more rigorous works. In contrast, the reproduce process is well understood, and good agreement between theory and experiment has been obtained. This is partly because the reproduce process is easily described mathematically and partly because measurable voltages are generated through the reproduce process and so reveal its character directly.

A number of earlier books have covered the topics of this chapter (Hoagland, 1963; Mee, 1964; Sebestyen, 1973; Jorgensen, 1980), but progress continues and new results need to be presented. Of particular note is the emergence of perpendicular recording and its implications for the record and reproduce processes. Therefore emphasis will be placed upon recent advances in magnetic recording to attempt to complement and extend the available literature rather than just to update the earlier works.

This chapter will deal first with the reproducing process, for the reasons already outlined, and then consider nonlinear and linear, or ac-biased, recording.

2.1   The Reproducing Process

In this section the process by which voltages are generated in the coils of heads of various structures is described. The response to magnetization components directed parallel to the surfaces of the recording media and the recorded tracks, and in directions perpendicular to the surface, will be considered under the broad headings of longitudinal and perpendicular recording.

The treatment here covers both longitudinal and perpendicular recording and makes appropriate comparisons between the two. However, in reference to experimental results, there is substantial information available for longitudinal recording but limited information relating to perpendicular recording, and so a truly balanced approach to the two aspects of the reproduce process is difficult to achieve.

Section 2.1.1 deals with the reciprocity formulas which form the basis of all the calculations of the output voltages generated by recorded magnetization patterns in the media.

2.1.1   The basis of reproduce theory: Reciprocity

Figure 2.1a and b shows the basic geometrical situation as a medium moves past a ring head and a single-pole head, respectively. The diagrams show how the chosen coordinate system applies to both head structures equally well and is potentially suitable for any head structure. Let us begin by assuming the heads produce field components $H_x(x, y, z)$ and $H_y(x, y, z)$ in the x and y directions. Then, for media with x and y components M_x and M_y of magnetization, the corresponding output voltages e_x and e_y at time t can be derived using the principle of reciprocity (Westmijze, 1953; Mee, 1964; Sebestyen, 1973); and this procedure is carried out in Appendix I. The result is

$$e_x(\bar{x}) = -\mu_0 V \int_{-w/2}^{+w/2} dz \int_d^{d+\delta} dy \int_{-\infty}^{+\infty} \frac{dM_x(x - \bar{x})}{d\bar{x}} \frac{H_x(x, y, z)}{i} dx \qquad (2.1a)$$

$$e_y(\bar{x}) = -\mu_0 V \int_{-w/2}^{+w/2} dz \int_d^{d+\delta} dy \int_{-\infty}^{+\infty} \frac{dM_y(x - \bar{x})}{d\bar{x}} \frac{H_y(x, y, z)}{i} dx \qquad (2.1b)$$

where μ_0 = permeability of free space
   $\overline{x}$ = Vt
   V = tape velocity
   i = current in head coil that produces above-mentioned field components
   d = spacing of head from medium of thickness δ
   w = track width

The above formulas may be simplified when the track width w is much larger than other dimensions in the system, whence it may be assumed that there is no variation of magnetization with the z dimension. Thus, Eqs. (2.1a) and (2.1b) may be simplified to

$$e_x(\overline{x}) = -\mu_0 Vw \int_d^{d+\delta} dy \int_{-\infty}^{+\infty} \frac{dM_x(x - \overline{x})}{d\overline{x}} \frac{H_x(x, y)}{i} \, dx \qquad (2.2a)$$

$$e_y(\overline{x}) = -\mu_0 Vw \int_d^{d+\delta} dy \int_{-\infty}^{+\infty} \frac{dM_y(x - \overline{x})}{d\overline{x}} \frac{H_y(x, y)}{i} \, dx \qquad (2.2b)$$

These equations form the basis for the study of the reproduce process and are often referred to simply as the *reciprocity formulas*. Also the occur-

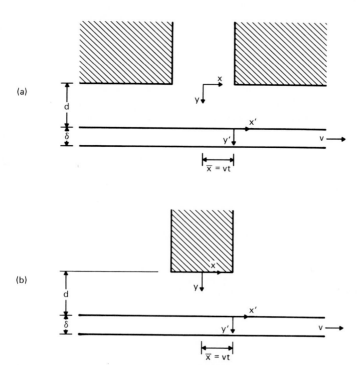

Figure 2.1 Definition of symbols and dimensions employed. The axes z and z' are directed into the paper.

rence in the same equation of head-field distributions means that they are often termed *head sensitivity functions*. However, some discussions of the relevant mathematical forms of magnetization and head fields are necessary to be able to proceed with any calculations, and these will be considered in the following sections.

2.1.2 Reproduction with a ring head

The schematic forms of the head-field distributions H_x and H_y are shown in Fig. 2.2, and a detailed consideration of them is given in Chap. 4. However, space is provided for a brief review of their form in view of their significance to the record process, which will be discussed later, as well as to the reproduce process currently being considered.

For the ring head the field distributions have been determined precisely by a number of workers. For distances from the head of $y > g/2$, the longitudinal and perpendicular field components are sufficiently accurately given by the expressions (Karlqvist, 1954)

$$H_x = \frac{H_g}{\pi}\left(\arctan\frac{g/2 + x}{y} + \arctan\frac{g/2 - x}{y}\right) \tag{2.3a}$$

$$H_y = \frac{-H_g}{2\pi}\ln\frac{(g/2 + x)^2 + y^2}{(g/2 - x)^2 + y^2} \tag{2.3b}$$

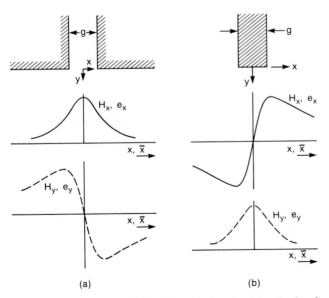

(a)   (b)

Figure 2.2 General form of field distributions for (*a*) a ring head and (*b*) a single-pole head.

where H_g is the field in the gap, and g is the gap length. Figure 2.3 shows the variations of H_x and H_y as functions of spacing as determined from Eqs. (2.3a) and (2.3b). These may be compared with more detailed results from rigorous computations shown in Chap. 4. It can be seen that the precise results differ from those of Eqs. (2.3) at small separations from the head where peaks in H_x appear near to the gap edges. These departures can be expressed analytically by correction terms (Fan, 1961), which simply add to Eqs. (2.3a) and (2.3b). The values of these terms are easily calculable (Fan, 1961; Baird, 1980; Middleton and Davies, 1984) to give precise results if needed.

The variations of the peak amplitudes of the head fields are shown in Fig. 2.4 as functions of reduced head-to-medium separation. The peak field values can be shown using Eqs. (2.3) to be given by

$$H_x(0, y) = \frac{2H_g}{\pi} \arctan \frac{g}{2y} \tag{2.4a}$$

$$H_y(x_0, y) = \frac{-H_g}{2\pi} \ln \frac{(g/2 + x_0)^2 + y^2}{(g/2 - x_0)^2 + y^2} \tag{2.4b}$$

where
$$x_0^2 = \left(\frac{g}{2}\right)^2 + y^2 \tag{2.4c}$$

The distances $\pm x_0$ represent the positions of the maxima and minima of

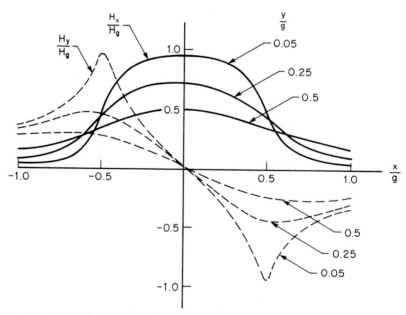

Figure 2.3 Field distributions for a ring head derived using Karlqvist's equations.

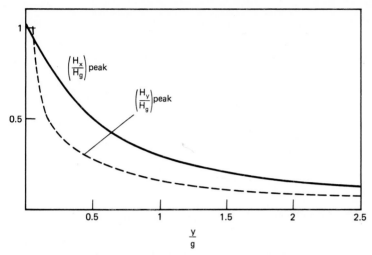

Figure 2.4 Reduced field amplitudes as a function of reduced separation: $(H_x/H_g)_{\text{peak}}$ = longitudinal component; $(H_y/H_g)_{\text{peak}}$ = perpendicular component.

the field distribution H_y, and so $2x_0$ is a measure of its "width." It can be shown that $2x_0$ is also the width of the field distribution H_x at 50 percent of peak amplitude. Calling this p_{50}, for reasons which will become obvious later, leads to

$$p_{50} = 2\left[\left(\frac{g}{2}\right)^2 + y^2\right]^{1/2} \tag{2.5}$$

This quantity is plotted in Fig. 2.5 as a function of reduced spacing y/g. At large separations or, conversely, when head gaps are short, Eqs. (2.3) reduce to the simple forms

$$H_x = \frac{H_g g}{\pi} \frac{y}{x^2 + y^2} \tag{2.6a}$$

$$H_y = \frac{-H_g g}{\pi} \frac{x}{x^2 + y^2} \tag{2.6b}$$

An approximation for the head gap field can be obtained from a simple consideration of the magnetic circuit of the heads, leading to

$$H_g = \frac{ni}{g + lA_g/\mu A_c} \tag{2.7}$$

where n is the number of turns in the head coil, l is the length of the magnetic circuit of the head of relative permeability μ, and A_g and A_c are,

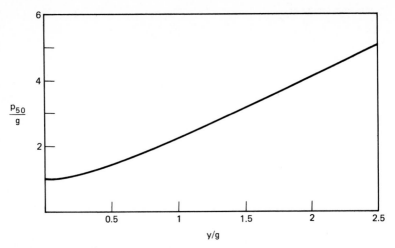

Figure 2.5 Width of field distribution, and also pulse width, as a function of reduced separation.

respectively, the cross-sectional area of the head magnetic circuit at the gap and the average cross-sectional area for the whole magnetic circuit.

With the determination of H_g, the dependence of head field on head gap can be examined. The variation of reduced head field, given by Eqs. (2.4a) and (2.7), is shown plotted against reduced gap in Fig. 2.6 for finite and infinite values of head permeability. The role of head geometry as determined by the quantity $lA_g/\mu A_c$ is an important factor in controlling head-field amplitudes, and obviously its magnitude needs to be kept as small as possible for fields to be highest. The same quantity will, at a later stage, also be shown to have a role in determining reproduced pulse amplitudes.

2.1.3 Output voltage waveforms

2.1.3.1 Special cases. In digital recording the magnetization in the medium is changed in as short a distance as possible along the recorded track to allow the maximum number of transitions per unit distance. In the ideal case, the magnetization would undergo a step change from its maximum value in one sense to its maximum value in the opposite sense. Considering a step change from a value of $-M_r$ to $+M_r$ in the x component of magnetization M_x, where M_r is the remnant magnetization, would cause Eq. (2.2a) to reduce to (Eldridge, 1960; Teer, 1961; Mee, 1964)

$$e_x(\bar{x}) = -2\mu_0 VwM_r \int_d^{d+\delta} \frac{H_x}{i} \, dy \qquad (2.8a)$$

Similarly, a change of the same magnitude in the components of magnetization M_y directed normal to the plane would result in an output voltage easily derived from Eq. (2.2b) as

$$e_y(\bar{x}) = -2\mu_0 VwM_r \int_d^{d+\delta} \frac{H_y}{i}\, dy \qquad (2.8b)$$

When the recording media are thin, the integrations in Eqs. (2.8a) and (2.8b) may be dispensed with and the results become

$$e_x(\bar{x}) = -2\mu_0 VwM_r\delta \frac{H_x(\bar{x}, y)}{i} \qquad (2.9a)$$

$$e_y(\bar{x}) = -2\mu_0 VwM_r\delta \frac{H_y(\bar{x}, y)}{i} \qquad (2.9b)$$

Thus the output voltage waveforms are identical in shape to those of the head-field distributions. Therefore, the shapes of the pulses, their amplitudes, and their widths have already been shown as functions of spacing and gap in Figs. 2.3 to 2.6; and the reason for introducing the quantity p_{50} should now be clear. In all cases it is apparent that spacing is an extremely

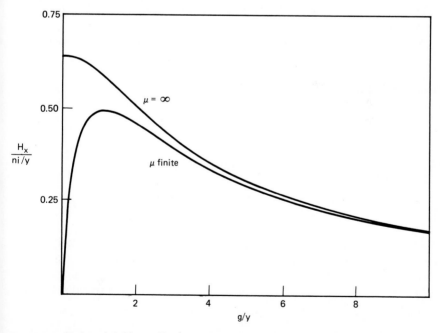

Figure 2.6 Reduced field amplitude as a function of reduced gap for infinite- and finite-permeability heads: $\mu = \infty$, $lA_g/\mu A_h = 0$; μ finite, $lA_g/\mu y A_h = 0.2$.

important parameter and that it must be minimized to maintain pulse amplitudes as high and widths as small as possible.

Another special case is of particular interest. It can be demonstrated by taking Eq. (2.2a) for a thin film and considering the situation when the length of the head-field distribution becomes so short as to approach a delta function. Then

$$e_x(\overline{x}) \propto \frac{dM_x(\overline{x})}{dx} \tag{2.10}$$

This is perfect readback since there are no losses introduced by the reproduce head, and the output is proportional to the magnetization gradient in the recording medium. This prediction is very useful in that it impresses the need for high magnetization gradients, and this implies short magnetization transition lengths between neighboring levels of magnetization. In reality the predicted situation is never quite achieved but is best approached with very short gaps in good contact with very thin recording media.

2.1.3.2 Sine-wave recording.
Many of the features of the reproduce process are most adequately demonstrated by considering the voltage waveforms generated in a reproduce head by a sinusoidal variation of magnetization in the medium. Many of the early works dealing with the reproduce process in longitudinal recording (Westmijze, 1953; Fan, 1961; Wang, 1966) confronted this problem, and the works are in substantial agreement in their predictions.

Consider the magnetization to be given by

$$M_x(x) = M_0 \sin kx \tag{2.11}$$

where k is the wave number, $2\pi/\lambda$. Substitution of this and H_x given by Karlqvist into the reciprocity formula results in

$$e(\overline{x}) = -\mu_0 V w M_0 \frac{H_g g}{i} k\delta \, [e^{-kd}] \left[\frac{1 - e^{-k\delta}}{k\delta} \right] \left[\frac{\sin (kg/2)}{kg/2} \right] \cos k\overline{x} \tag{2.12}$$

The above equation is presented in a form which is intended to highlight the important features of the reproduce process. The terms in the square brackets are all known as *loss terms* and in general have values of less than unity, the smallness being a measure of the loss introduced. These terms will now be considered in detail.

The first term is the spacing loss

$$L_d = e^{-kd} \tag{2.13}$$

and this shows that the output falls exponentially with the ratio of spacing to wavelength (Wallace, 1951). By taking logarithms of both sides of Eq. (2.13), it can be shown that the loss amounts to 54.6 dB per wavelength of spacing between head and medium. Equation (2.13) is plotted in Fig. 2.7 as a function of d/λ.

The second term to be considered is the thickness loss

$$L_\delta = \frac{1 - e^{-k\delta}}{k\delta} \tag{2.14}$$

This is also plotted in Fig. 2.7, where it is seen not to be such a severe source of deterioration of the signal as that due to spacing, but its importance should not be diminished.

The third term is the gap loss term (Lübeck, 1937)

$$L_g = \frac{\sin{(kg/2)}}{kg/2} \tag{2.15}$$

and it is plotted in Fig. 2.8 as a function of g/λ. Its value passes through a null at $\lambda = g$, and at multiples of g thereafter. However, Eq. (2.12) was derived using a head-field distribution which is approximate, and so Eq. (2.15) itself is approximate. Repeating the calculations for an exact head-field distribution leads to the same result as before except that the gap loss term is modified to

$$L_g = \frac{\sin{(kg/2)}}{kg/2} \left[1 + \sum_{n=1}^{\infty} \frac{A_n}{U} \frac{n(-1)^n 4\pi}{4 - (4\pi n/kg)^2} \right] \tag{2.16}$$

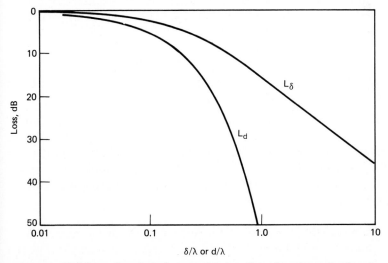

Figure 2.7   Thickness loss L_δ and separation loss L_d as functions of reduced thickness or reduced spacing.

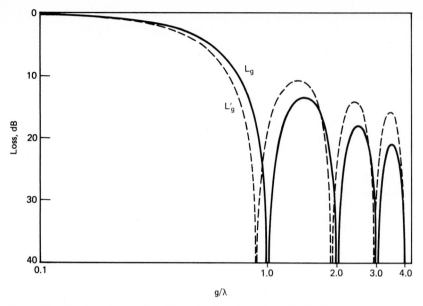

Figure 2.8 Gap loss L_g as a function of reduced gap length g/λ. Curve L_g, approximate expression; curve L'_g, accurate expression.

where the coefficients in the summation were originally given as far as the term $n = 3$ (Fan, 1961) but have more recently been evaluated as far as $n = 6$ (Baird, 1980). The corresponding variation of gap loss with wavelength is also shown in Fig. 2.8, where it is seen that the first null in the output is shifted from $\lambda = g$. The new location can be estimated by taking the first term in the summation (i.e., that involving $A_1/U = -0.08054$) and determining the value of λ/g for which the loss term goes to zero. The null occurs when

$$\lambda = \frac{g}{0.88} \tag{2.17}$$

This result was first produced using calculations based upon a conformal mapping solution of the head-field distribution (Westmijze, 1953). Another way of stating the result is that the "effective" gap length, g_{eff}, is some 14 percent higher than the physical gap length. Replacing g by g_{eff} in Eq. (2.15) gives a simple and accurate approximation to the gap loss in the range $\lambda > g_{eff}$.

2.1.3.3 Finite transition lengths. To investigate the output voltages arising from transitions of finite length, it is necessary to assume that the magnetization distributions take on a particularly convenient mathematical form. For the x and y components of magnetization, these are assumed to be (Miyata and Hartel, 1959)

$$M_x = \frac{2}{\pi} M_0 \arctan \frac{x'}{a_x} \tag{2.18a}$$

$$M_y = \frac{2}{\pi} M_0 \arctan \frac{x'}{a_y} \tag{2.18b}$$

where a_x and a_y are the transition length parameters for longitudinal and perpendicular recording, respectively, and where the magnetization changes from a value $-M_0$ to $+M_0$ over the transition. These distributions are shown in Fig. 2.9, where a tangent through the origin of the curve shows that the transition lengths might reasonably be quoted as πa_x or πa_y for the two different components of magnetization. The output voltage characteristics are now most conveniently considered separately for the hypothetical cases of purely longitudinal and purely perpendicular recording.

Longitudinal recording. The output voltage waveforms in longitudinal recording are obtained by substitution from Eq. (2.18a) into the reciprocity formula and by using Appendix II to obtain

$$e_x(\bar{x}) = \frac{-2\mu_0 VwH_gM_0}{\pi i} \int_0^{d+\delta} \left(\arctan \frac{g/2 + \bar{x}}{a_x + y} \right.$$
$$\left. + \arctan \frac{g/2 - \bar{x}}{a_x + y} \right) dy \tag{2.19}$$

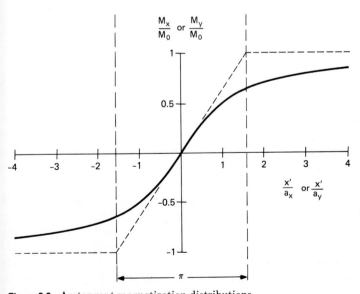

Figure 2.9 Arctangent magnetization distributions.

Upon completion of the y integration, the total output voltage is given by

$$e_x(\bar{x}) = \frac{-\mu_0 VwH_gM_0}{\pi i} \left[2(a_x + d + \delta) \left(\arctan \frac{g/2 + \bar{x}}{a_x + d + \delta} \right. \right.$$
$$\left. + \arctan \frac{g/2 - \bar{x}}{a_x + d + \delta} \right)$$
$$- 2(a_x + d) \left(\arctan \frac{g/2 + \bar{x}}{a_x + d} + \arctan \frac{g/2 - \bar{x}}{a_x + d} \right)$$
$$+ \left(\frac{g}{2} + \bar{x} \right) \ln \frac{(a_x + d + \delta)^2 + (g/2 + \bar{x})^2}{(a_x + d)^2 + (g/2 + \bar{x})^2}$$
$$+ \left. \left(\frac{g}{2} - \bar{x} \right) \ln \frac{(a_x + d + \delta)^2 + (g/2 - \bar{x})^2}{(a_x + d)^2 + (g/2 - \bar{x})^2} \right] \tag{2.20}$$

This equation has been obtained previously (Speliotis and Morrison, 1966a; Potter, 1970; Middleton and Davies, 1984), and it can be used to investigate the factors controlling the amplitudes and widths of pulses. While it is easy to find pulse amplitude by putting $\bar{x} = 0$, it is not possible to find a simple formula for pulse width. Therefore, it is more convenient to consider certain limiting cases which provide a useful insight into the replay process.

For replay from a thin film in which $\delta \ll d$, Eq. (2.19) reduces to

$$e_x(\bar{x}) = \frac{-2\mu_0 VwH_gM_0\delta}{\pi i} \left(\arctan \frac{g/2 + \bar{x}}{a_x + d} + \arctan \frac{g/2 - \bar{x}}{a_x + d} \right) \tag{2.21}$$

By comparing this with the Karlqvist expressions for the head-field distribution, it is apparent that

$$e_x(\bar{x}) = \frac{-2\mu_0 VwM_0\delta}{i} H_x(\bar{x}, a_x + d) \tag{2.22}$$

This states that the output voltage waveform follows the same form as the head-field distribution except that the distance from the head surface y ($= d$) is replaced by $a_x + d$. Therefore, the general characteristics of the output waveform are shown in Figs. 2.3 to 2.6, with the requirement that y is replaced by $a_x + d$.

Equation (2.22) is not limited in validity to the approximate head-field distribution of Eq. (2.3a) but applies for exact head-field distributions. This can be proved by comparing reproduced pulse shapes with detailed computations of exact field distribution (Middleton and Davies, 1984). Figure 2.10a shows the experimental pulse shapes obtained for long-gap heads used in contact recording, while Fig. 2.10b shows the same heads used in conjunction with an increased head-to-tape spacing. In the first case the pulse shows the "double-humped" feature discussed earlier, while at the larger spacing the pulse takes a form satisfactorily covered by the simple Karlqvist formula. The medium used was a chemically deposited cobalt film in which uniformity of magnetization through its thickness is a reasonable assumption. This is in line with the assumption leading up to the derivation of the relevant formulas, and so detailed comparisons between theory and experiment reveal good agreement for a range of $(a_x + d)/g$.

Returning to Eq. (2.21) and putting $\bar{x} = 0$ provides the pulse amplitude $e_x(0)$ as

$$e_x(0) = \frac{-4\mu_0 VwH_g M_0 \delta}{\pi i} \arctan \frac{g/2}{a_x + d} \tag{2.23}$$

and this can be used to provide predictions suitable for comparison with experiment. Figure 2.11 shows experimental results obtained from pulses reproduced from an evaporated cobalt film medium. Output is displayed as a function of spacing d, introduced by adding spacers between head and medium, to which is added an effective spacing d_0 shown to be present in the experiments on contact recording (Bonyhard et al., 1966). Relevant properties of the medium and head, given in the figure captions, were used to obtain the theoretical curves. In all cases it was necessary to multiply the calculated outputs by a factor of 0.6 to obtain the results shown. This quantity is a form of head efficiency, which to some extent explains the losses and leakages of flux in the magnetic circuit of the head. The figures confirm the need for small values of spacing d (and a_x) to maintain highest outputs.

The variation of pulse amplitude with gap length is shown in Fig. 2.12. The results were obtained on the same medium as used earlier and the curve is derived using Eq. (2.23). The latter gives satisfactory results and the overall agreement between theory and experiment is confirmed. The

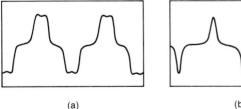

(a)   (b)

Figure 2.10 Observed output pulses for (a) contact reproduction with a 51-μm-gap ring head and (b) reproduction with the same head and a total spacing of 5.8 μm *(Middleton and Davies, 1984).*

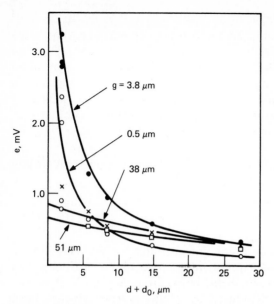

Figure 2.11 Pulse amplitudes as functions of spacing $d + d_o$ for gap lengths of 0.5, 3.8, 38, and 51 μm. The points are experimental and the lines theoretical. The film had $H_c = 41$ kA/m (510 Oe), $\delta = 0.2$ μm, and $M_r = 800$ kA/m (800 emu/cm^3) *(Middleton and Davies, 1984)*.

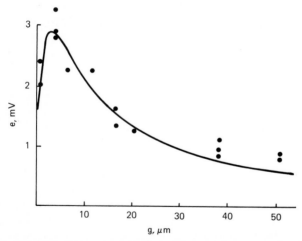

Figure 2.12 Pulse amplitude as a function of gap length. Points are for observations on the medium of Fig. 2.11, and curves are theoretical *(Middleton and Davies, 1984)*.

graph shows the way in which a reduced gap length causes a general rise in pulse amplitude until the losses of the head gap circuit cause a sudden drop of amplitude at small values of gap length.

Pulse widths are also very much of interest, and these can be obtained from Eq. (2.21). For example, solving the equations $e_x(0)/2 = e_x(\bar{x})$ and $e_x(0)/4 = e_x(\bar{x})$ leads to values of $\bar{x}$, which, when doubled, give the pulse widths at 50 and 25 percent of pulse amplitude, respectively. Thus p_{50} and p_{25} are given by

$$p_{50} = 2\left[\left(\frac{g}{2}\right)^2 + (a_x + d)^2\right]^{1/2} \tag{2.24a}$$

and
$$p_{25} = \left[\frac{2(a_x + d)g^2}{[(a_x + d)^2 + (g/2)^2]^{1/2} - (a_x + d)} - (a_x + d)^2 + g^2\right]^{1/2} \tag{2.24b}$$

The variation of p_{25} with total head-to-tape spacing is shown in Fig. 2.13 along with appropriate predictions from Eq. (2.24b). The experimental points are for a thin metallic medium which was separated from the reproducing heads by thin plastic spacers. Figure 2.14 shows the variations of pulse width caused by heads of different gaps. The line is the prediction derived from Eq. (2.24a), agreement being good in all cases. All graphs emphasize the need to keep spacings and gaps to a minimum.

While it has proved possible to test the validity of the thin-film approximations, it has not proved possible to verify the predictions of the full formula of Eq. (2.20). This is because magnetizations are never so simply described as in Eq. (2.18a), and their forms are complicated functions of distance into the media and are not sufficiently well known to be able to calculate expected output voltages. Further there is often a difference

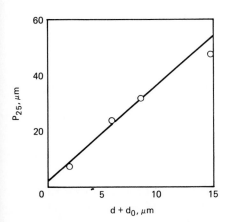

Figure 2.13 Pulse width as a function of total spacing ($d + d_o$). The points are for the medium of Fig. 2.11, and the curve is theoretical (*Middleton and Davies, 1984*).

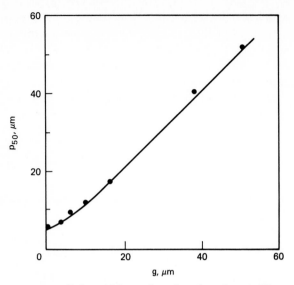

Figure 2.14 Pulse width as a function of gap length. The points are for the medium of Fig. 2.11, and the curve is theoretical *(Middleton and Davies, 1984).*

between the effective depth of recording and the medium thickness, and a detailed consideration of thickness losses is not easily attainable.

The anticipated role of medium thickness is most easily demonstrated by taking the short-gap approximation to the integrand of Eq. (2.19) and then integrating over y to obtain

$$e_x(\bar{x}) = \frac{-\mu_0 VwH_gM_0g}{\pi i} \ln \frac{(a_x + d + \delta)^2 + \bar{x}^2}{(a_x + d)^2 + \bar{x}^2} \tag{2.25}$$

The amplitude of this pulse is obtained by putting $\bar{x} = 0$ in the above equation, while its width at 50 percent of peak amplitude can be derived in the manner described earlier as

$$p_{50} = 2[(a_x + d)(a_x + d + \delta)]^{1/2} \tag{2.26}$$

This equation, when viewed alongside Eq. (2.24a), has led to a suggestion that the following is a suitable approximation for the width of pulses in the general case (Middleton, 1966):

$$p_{50} = 2\left[\left(\frac{g}{2}\right)^2 + (a_x + d)(a_x + d + \delta)\right]^{1/2} \tag{2.27}$$

Figure 2.15a and b shows reduced pulse amplitude, derived from Eq. (2.25), and reduced pulse width, from Eq. (2.26), plotted as functions of reduced medium thickness $\delta/(a_x + d)$. Increasing medium thickness increases pulse amplitude and pulse width; but it should be pointed out that the parameter a_x is also a function of medium thickness, and so its influence is, in reality, more complex than might appear at first sight. Nevertheless it is reasonable to conclude that small thicknesses are needed if pulse widths are to be kept as short as possible.

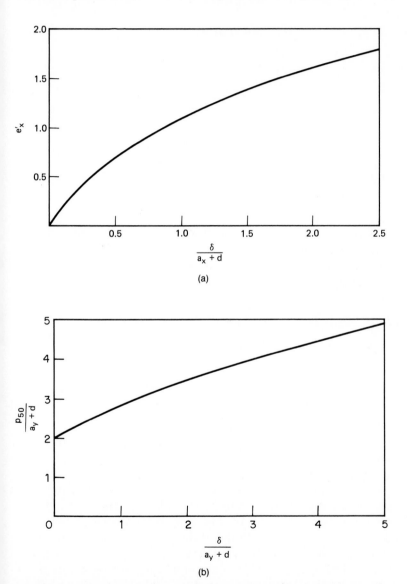

(a)

(b)

Figure 2.15   (a) Reduced output and (b) pulse width for longitudinal recording as functions of reduced medium thickness.

Perpendicular recording.   The calculations in this section follow the form of those in the previous one. Thus to obtain the output voltage in perpendicular recording it is necessary to substitute from Eqs. (2.3b) and (2.18b) into Eq. (2.2b) and evaluate the integrals (see Appendix II). Thus

$$e_y(\bar{x}) = \frac{-\mu_0 V w H_g M_0}{\pi i} \int_d^{d+\delta} \ln \frac{(g/2 + \bar{x})^2 + (a_y + y)^2}{(g/2 - \bar{x})^2 + (a_y + y)^2} \, dy \qquad (2.28)$$

and completion of the y integration gives

$$e_y(\bar{x}) = \frac{-\mu_0 V w H_g M_0}{\pi i} \left[(a_y + d + \delta) \ln \frac{(a_y + d + \delta)^2 + (g/2 + \bar{x})^2}{(a_y + d + \delta)^2 + (g/2 - \bar{x})^2} \right.$$

$$- (a_y + d) \ln \frac{(a_y + d)^2 + (g/2 + \bar{x})^2}{(a_y + d)^2 + (g/2 - \bar{x})^2}$$

$$+ 2\left(\frac{g}{2} + \bar{x}\right)\left(\arctan \frac{a_y + d + \delta}{g/2 + \bar{x}} - \arctan \frac{a_y + d}{g/2 + \bar{x}} \right)$$

$$\left. + 2\left(\frac{g}{2} - \bar{x}\right)\left(\arctan \frac{a_y + d}{g/2 - \bar{x}} - \arctan \frac{a_y + d + \delta}{g/2 - \bar{x}} \right) \right] \qquad (2.29)$$

It is not easy to visualize the shape of this pulse directly from this equation; therefore, as in the case of longitudinal recording, particular and revealing cases are sought.

Consider first the case of thin recording media ($\delta \ll d$) in which the integration over y in Eq. (2.28) can be avoided. The output voltage is given by

$$e_y(\bar{x}) = \frac{\mu_0 V w H_g M_0 \delta}{\pi i} \ln \frac{(g/2 + \bar{x})^2 + (a_y + d)^2}{(g/2 - \bar{x})^2 + (a_y + d)^2} \qquad (2.30)$$

Comparison of this with Eq. (2.3b) shows that

$$e_y(\bar{x}) = \frac{\mu_0 V w M_0 \delta}{\pi i} H_y(\bar{x}, a_y + d) \qquad (2.31)$$

Therefore, in similar vein to the case of longitudinal recording, the output voltage waveform is identical in shape to that of the head field, and is shown in Fig. 2.3 to have what is often termed a *bimodal* form; that is, it is positive- and negative-going at different times. Experimental pulse shapes obtained for reproduction by ring heads from thin films of sputtered Co-Cr are shown in Fig. 2.16. The waveforms are very much of the form indicated in that the amplitudes of the positive- and negative- going parts of the pulses are roughly equal. However, observed waveforms often do not exhibit the same type of symmetry since the positive- and negative-going parts do not have the same amplitudes. The cause of the latter is thought to be considerable departures of the magnetization distributions from that assumed in Eq. (2.18b), as discussed in Sec. 2.2.3 below. This makes it difficult to investigate the replay process in the same detail as was done for longitudinal recording, and the limited availability of experimental results in this area somewhat reflects this difficulty. Returning to Eq. (2.30) it can be shown that the minimum and maximum of the output pulse occur at

$$\bar{x} = \bar{x}_0 = \pm\left[\left(\frac{g}{2}\right)^2 + (a_y + d)^2\right]^{1/2} \tag{2.32}$$

which leads to a *peak-to-peak* distance $p_{p\text{-}p}$ of

$$p_{p\text{-}p} = 2\left[\left(\frac{g}{2}\right)^2 + (a_y + d)^2\right]^{1/2} \tag{2.33}$$

By comparing Eq. (2.33) with Eq. (2.24a) it can be seen that p_{50} and $p_{p\text{-}p}$ coincide and that the graph of p_{50} (Fig. 2.5) applies equally well for $p_{p\text{-}p}$. Therefore pulse widths for longitudinal recording and peak-to-peak distances for perpendicular recording are comparable quantities and both can be termed *pulse widths*.

To investigate the influence of medium thickness, it is convenient to take the narrow-gap approximation to Eq. (2.28) and carry out the y integration. The result is

$$e_y(\bar{x}) = \frac{2\mu_0 VwH_gM_0g}{\pi i}\left(\arctan\frac{\bar{x}}{a_y + d} - \arctan\frac{\bar{x}}{a_y + d + \delta}\right) \tag{2.34}$$

Again this is a bimodal pulse with minimum and maximum at

$$\bar{x} = \bar{x}_0 = \pm[(a_y + d)(a_y + d + \delta)]^{1/2} \tag{2.35}$$

for which the pulse width is

$$p_{50} = 2[(a_y + d)(a_y + d + \delta)]^{1/2} \tag{2.36}$$

This formula for pulse width is identical to that in longitudinal recording.

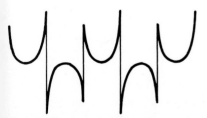

Figure 2.16 The general shape of reproduced pulses obtained in perpendicular recording.

The amplitude of the voltage waveform is obtained by substituting $\bar{x} = \bar{x}_0$ in Eq. (2.34) to obtain

$$
e_y(\bar{x}_0) = \frac{2\mu_0 VwH_gM_0g}{\pi i}\left(\arctan\sqrt{\frac{a_y + d + \delta}{a_y + d}}\right.
$$

$$
\left. - \arctan\sqrt{\frac{a_y + d}{a_y + d + \delta}}\right) \tag{2.37}
$$

The variation of reduced pulse amplitude,

$$
e'_y = \frac{e_y}{2\mu_0 VwH_gM_0g/\pi i}
$$

with reduced medium thickness $\delta/(a_y + d)$, derived from Eq. (2.37), is shown in Fig. 2.17. As expected, increasing medium thickness results in higher output.

The comments made earlier about the experimentally observed pulse shapes indicate the difficulty of comparing the above formulas with observations. Pulse shapes are related to the detail of the magnetization distributions, and so it is necessary to consider the reproducing process along with recording theory; this is done in Sec. 2.2.

Reproduction with a single-pole head can be studied in just the same way as reproduction with a ring head. However, there is no need to work through a complete set of new calculations for the following reason. It was

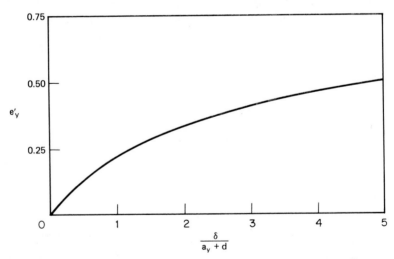

Figure 2.17 Reduced pulse amplitude as a function of reduced medium thickness for perpendicular recording.

shown earlier that there is a general similarity between head-field components produced by ring and single-pole heads. In particular, H_x for the ring head and H_y for the single-pole head are similar, and so it can be expected, in view of Eqs. (2.22) and (2.31), that the output waveforms arising from longitudinal components of magnetization and a ring head are similar to those arising from perpendicular components of magnetization and a single-pole head. Hence all the equations covered under the discussion of longitudinal recording applying to a ring head also apply to perpendicular recording with a single-pole head (Middleton and Wright, 1982). A similar analogy could be made with respect to the reproduction of perpendicular components of magnetization with a ring head and longitudinal components of magnetization and a single-pole head. The only difference to take into account when applying the equations to a single-pole head is to note that H_g in Eq. (2.7) no longer refers to a gap-surface field but is a field at the tip of the single pole.

2.1.4 Recording at high densities: Pulse superposition

When the linear density of information in a recording medium is increased, a number of additional considerations come into play. First, the magnetization transitions in the medium are brought into proximity and interact, and, second, the reproduced pulses overlap. The process by which interactions between neighboring transitions occur in the medium is subject to much discussion; this contrasts with the process of reproduction where pulses from neighboring transitions are simply summed by the reproduce head. As a first approximation to dealing with the pulse crowding situation, it is usually assumed that it is valid simply to superpose and sum isolated pulse waveforms at appropriate spacings to obtain the resultant voltage waveform. This process is known as *linear superposition* (Hoagland, 1963; Mallinson and Steele, 1969; Tjaden, 1973).

In mathematical terminology, pulse superposition simply states that the total voltage $e_t(\overline{x})$ is, for pulses separated by a distance $\lambda/2$, given by

$$e_t(\overline{x}) = \sum_n (-1)^n e\left(\overline{x} + \frac{n\lambda}{2}\right) \tag{2.38}$$

where the values of n employed in the summation determine the recorded pattern. Generally there are two consequences of pulse crowding: the first is that the amplitudes of the pulses are altered; the second is that the positioning of the peaks of the pulses are shifted, leading to a phenomenon known as *peak shift*. Peak shift is a form of *intersymbol interference*, a term often employed in discussions of recording system behavior.

The effects of pulse crowding in longitudinal recording are illustrated in Fig. 2.18. This shows pulse shapes of the type predicted by Eq. (2.25) crowded together in the way suggested by Eq. (2.38). In Fig. 2.18b the pulses are clearly isolated. At the density shown in Fig. 2.18c the pulses are beginning to overlap, and it can be seen that each detracts from its neighbor and causes a reduction of pulse amplitudes. By the time the configuration in Fig. 2.18d is reached, the resultant waveform looks rather like a sine wave of amplitude considerably reduced from that of the isolated pulses. Further increases of density continue to reduce the amplitude of this waveform.

In perpendicular recording, the pulses according to Eq. (2.30) are crowded together in the manner of Eq. (2.38) and are shown in Fig. 2.19. The isolated pulse shape, shown in Fig. 2.19a, has a particularly long "tail" which results in overlapping of pulses even at low densities (Fig.

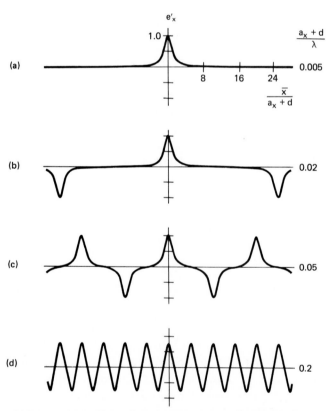

Figure 2.18 Predicted output waveforms for longitudinal recording and various reduced densities $(a_x + d)/\lambda$ *(Middleton and Wright, 1982).*

2.19*b*). In perpendicular recording, the interaction of pulses is initially constructive, and in Fig. 2.19*b*, *c*, and *d* the peak amplitudes rise above those of the isolated pulses. At higher densities than those shown in Fig. 2.19*d*, however, the amplitude of the wave, which now resembles a sine wave, decreases rapidly.

The above results are encapsulated in Fig. 2.20, which shows the variation of output amplitude plotted as a function of reduced linear density. The predictions are given for short-gap heads for a reason to become apparent below. The output from longitudinal recording falls monotonically with increasing density, while that for perpendicular recording rises to a low peak and then falls. The fall of amplitude with density is identical at high densities for both modes of recording, provided the transition lengths are identical.

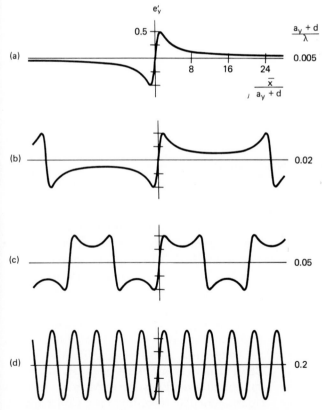

Figure 2.19 Predicted output waveforms in perpendicular recording as a function of reduced densities $(a_y + d)/\lambda$ *(Middleton and Wright, 1982).*

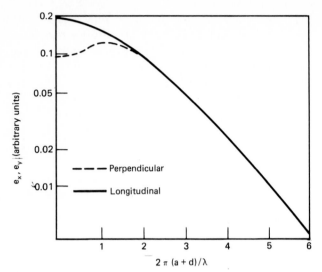

Figure 2.20 Predicted outputs for longitudinal and perpendicular recording as a function of reduced density $(a_x + d)/\lambda$ or $(a_y + d)/\lambda$.

It should be noted that the axes of Fig. 2.20 are log-linear, which implies at high densities

$$e_x(0) \propto \exp \frac{-2\pi(a_x + d)}{\lambda} \tag{2.39a}$$

and $$e_y(0) \propto \exp \frac{-2\pi(a_y + d)}{\lambda} \tag{2.39b}$$

These equations can be obtained by summing the original waveforms using Eq. (2.38) and taking the short-wavelength approximations.

Experimental results which demonstrate the validity of Eq. (2.39a) for longitudinal recording are shown in Fig. 2.21. These results were taken on an instrumentation tape recorder using a high-coercivity recording medium moving with a velocity of 0.76 m/s (Middleton, 1982). At higher frequencies, the straight line satisfies Eq. (2.39a), and measurement of its slope can yield a value for $a_x + d$: the appropriate value is quoted. Graphs such as that in Fig. 2.21 provide a convenient way of investigating values for $a_x + d$ and have been used widely (e.g., Mallinson, 1975; Middleton, 1982; Bertram and Fielder, 1983), although care must be taken to correct for gap losses in any of the experimental readings. An experimental curve of output against density for perpendicular recording on a Co-Cr film is shown in Fig. 2.22 (Hokkyo et al., 1982). Its general form is as expected, with the initial rise of output occurring before the fall as density increases.

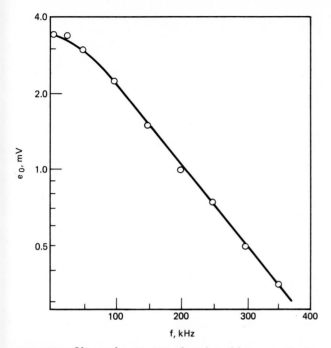

Figure 2.21 Observed output as a function of frequency for longitudinal recording using an instrumentation deck with tape of coercivity 52 kA/m (650 Oe) moving at 0.76 m/s. The value of $a_x + d$, determined from its slope, is 0.90 *(Middleton, 1982)*.

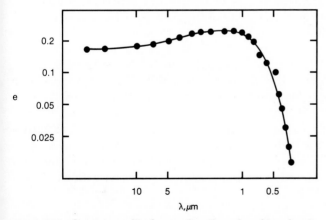

Figure 2.22 Output amplitude as a function of packing density for perpendicular recording with ring heads on a Co-Cr medium (Hokkyo et al., 1982). The properties of the medium were $M_s = 400$ kA/m (400 emu/cm^3), $H_c = 32$kA/m (400 Oe), and $\delta = 0.5$ μm.

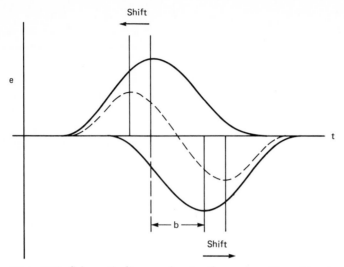

Figure 2.23 Schematic diagram showing the superposition of two isolated pulse shapes to give reduced output and peak shift.

The other consequence of pulse crowding, peak shift, is illustrated in Fig. 2.23. Two pulses, whose shapes are shown as if they are isolated, have their peaks separated by a distance b. When they are summed in the reproduce head, the net output is shown by the broken curve. This summing process, based upon assuming the linear superposition of isolated pulse shapes, results in a waveform with peaks shifted from the positions of the original pulses. The extent of the shift is expected to vary with density as shown in Fig. 2.24. Curves of this form have been observed (e.g., Morrison and Speliotis, 1967; Bertram and Fielder, 1983), although the magnitude of the peak shift tends to be less than that predicted by linear pulse superposition (Bertram and Fielder, 1983). This is because, during the recording process, the demagnetizing field from the previously recorded transition tends to augment the record field, and leads to a transition positioned to give reduced peak shift on reproduction.

In perpendicular recording, the analogous quantity to the peak shift of longitudinal recording is shift in the crossover position. This can be demonstrated graphically in a manner similar to that displayed for peak shift, and graphs of shift as a function of density can be derived which are similar to that in Fig. 2.24.

2.1.5 The reproduce process for different head geometries

The reproduce heads used in practical circumstances do not always conform to a reasonable approximation to the semi-infinite pole head for which the Karlqvist and Fan formulas can be assumed to apply. More often the dimension p of the pole tips shown in Fig. 2.25a has a magnitude comparable to other dimensions in the recording system and so needs to be taken into account. The effect of reducing p is to bring the outer edges of the head near enough to the head gap to influence the fields produced by the head and, consequently, the output voltage waveforms on reproduction.

The x components of fields produced by thin-pole heads have been computed (Westmijze, 1953; Potter et al., 1971) and are shown in Fig. 2.25b. The predicted effect of shortening the pole tips is to shorten the head-field distribution and to cause it to have a negative excursion. In view of the earlier discussions on the reproduce process, it could be expected that the output pulse shapes should show this feature, and indeed this is the case (Valstyn and Shew, 1973). Experimental results are shown in Fig. 2.25c for a long- and a 2-μm-pole head. The corresponding amplitude-versus-density curve is shown in Fig. 2.26. The narrowing of pulses causes a lifting of the high-density outputs over and above what is expected for the long-pole head. This basic result was confirmed by comparing theoretical and experimental curves of output against density for a range of pole dimensions (Kakehi et al., 1982). The results, shown in Fig. 2.27, reveal that the detailed features of the head-field distribution result in undulations in the output-versus-density curve.

Turning now to another pole structure which is under active investigation, consider Fig. 2.28. Here a single main pole for record is placed above the recording medium, while an auxiliary drive pole is placed below the medium. Although a full description of this head structure is given in Chap. 4, a brief description of its operation is given here. The auxiliary

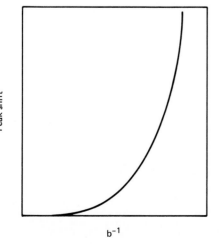

Figure 2.24 Peak shift as a function of packing density for the two-pulse arrangement of Fig. 2.23. Scales on axes are arbitrary.

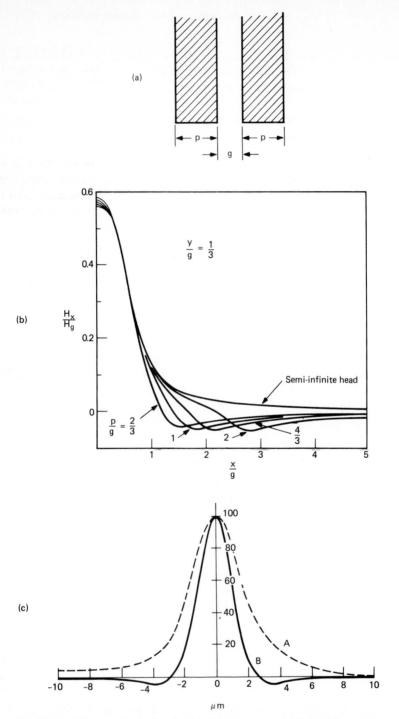

Figure 2.25 Finite-pole-tip heads. (a) Definition of symbols; (b) computed head-field shapes (Potter et al., 1971); (c) observed pulse shapes for (curve A) long- and (curve B) short-pole heads normalized for equal amplitude *(Valstyn and Shew, 1973)*.

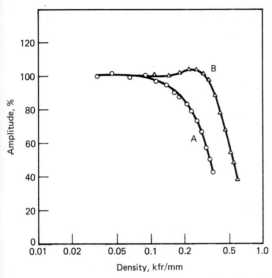

Figure 2.26 Observations of outputs as a function of density for (curve *A*) a long-pole head and (curve *B*) a narrow-pole head *(Valstyn and Shew, 1973).*

pole is magnetized by a current in its coil and produces a low-amplitude but widespread field which penetrates the medium and magnetizes the main pole above the medium. The main pole then produces a localized field of sufficiently high strength to alter the magnetization in the medium. The resulting field distribution is a significant factor in determining output voltage waveforms, particularly when using a film of cobalt-chromium which has a soft nickel-iron backing layer (Iwasaki et al., 1979). In such a case the output voltage pulses should follow the form of the head-field distributions. The waveforms shown in Fig. 2.28*b* were compared with computed field distributions to give good correlation. In addition, the corresponding pulse-crowding curves are shown in Fig. 2.28*c*; the nulls in the output arise from the effects of pole length, as opposed to gap length of a ring head, and were shown to be easily and accurately predicted. Therefore, the predicted similarity between output waveforms for longitudinal recording with a ring head and perpendicular recording with a single-pole head is observed.

2.1.6 Reproduction from double-layer media

Much of the work on perpendicular recording has involved double-layer recording media, in which the high-coercivity layers used for the storage process have thin underlayers of high-permeability material such as Ni-

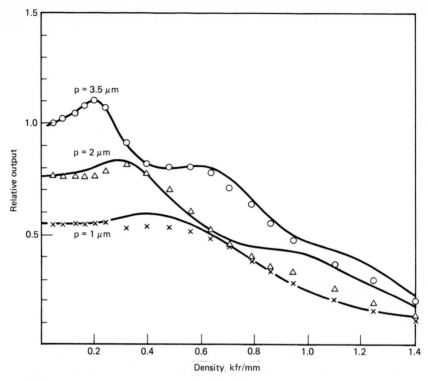

Figure 2.27 The influence of pole width on output versus density. Points are observations, and the lines are the result of calculations *(Kakehi et al., 1982)*.

Fe Permalloy (Iwasaki et al., 1979). The influence of the underlayer is multifold but for simplicity can be categorized in terms of its influence on head-field distributions (Iwasaki, 1980, 1984; Mallinson and Bertram, 1984*b*), the recording process (Iwasaki et al., 1983), and the reproduce process (Yamamori et al., 1981; Nakamura and Iwasaki, 1982; Lopez, 1983; Quak, 1983). Here a few comments are made with respect to the reproduce process.

Figure 2.29*a* shows a magnetization distribution in a single-layer medium and the corresponding outputs at different wavelengths. The voltage waveforms can be considered as arising from the flux induced in the replay head by the magnetic poles on both surfaces of the recording medium. In Fig. 2.29*b* the magnetization distribution occurring in a double-layer medium is shown along with that anticipated for the underlayer. The latter is thought to offer means for the achievement of flux closure at the lower surface, thereby leaving magnetic poles only on the top surface of the medium. Figure 2.30 shows the effect of this on the variation of output with packing density (Yamamori et al., 1981). The output for a

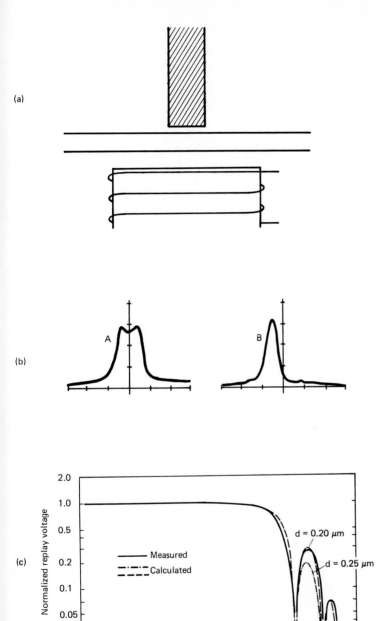

Figure 2.28 Perpendicular recording. (*a*) Main pole and auxiliary pole arrangement for perpendicular recording; (*b*) isolated pulse shapes reproduced by (curve *A*) poles of 2.0-μm and (curve *B*) 0.9-μm widths in contact with double-layer Co-Cr/Ni-Fe media (1.9 μm/per division); (*c*) output amplitude as a function of density *(Iwasaki et al., 1981)*.

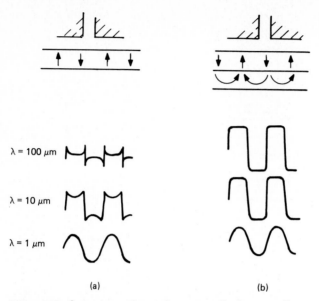

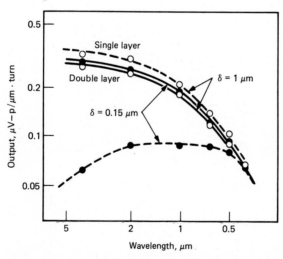

Figure 2.29 Output waveforms for perpendicular recording for (a) single Co-Cr and (b) double-layer media, Co-Cr on Ni-Fe, as a function of wavelength. Recording head—single-pole design; reproducing head—ring design.

Figure 2.30 Output amplitudes as a function of wavelength for perpendicular recording on single- and double-layer (underlayer of 1 μm) Co-Cr media *(Yamamori et al., 1981)*.

single-layer medium is a strong function of medium thickness, whereas the output for the double-layer medium is virtually independent of thickness since the influence of the lower side of the medium has been removed.

2.1.7 Reproduction with magnetoresistive heads

The heads discussed so far have all been of the inductive variety; that is, the output voltages developed in their coils are proportional to the rate of change of the flux induced in the cores. Such heads are not the only types available; there are also flux-sensitive heads which, as their name implies, are sensitive to the flux induced within them rather than its rate of change. Of these, the magnetoresistive head has been most widely studied and is considered fully in Chap. 4. With regard to the operation of a flux-sensitive head, it is sufficient to say at this stage that potential distributions take on the role of head sensitivity functions and replace the field distributions in the reciprocity formulas. Once the potential distributions are known, it is possible to determine output voltages (Cole et al., 1974; Potter, 1974; Davies and Middleton, 1975).

2.2 Nonlinear Recording Processes

The nonlinear recording process is one of great complexity and does not yield easily to rigorous analysis. Nevertheless, much has been learned about the primary factors involved in determining the nature of the process from the study of simplified models. In this section such simplified models will be presented and their success and limitations discussed.

As a lead-in to this topic, consider a medium moving from left to right, first past a record head and then past a reproduce head. Any element of the medium that has passed the record head has been subjected to a complex history of head fields of varying amplitude and direction. The picture is further complicated by the existence of demagnetizing fields within the medium. These fields modify the field experienced by the medium during the record process. They remain present in the medium as it moves away from the record head, causing the written magnetization to change. They are also present as the medium moves under the reproduce head, thus causing a further change. Any precise model of magnetic recording therefore needs, as a whole, to take into account the vector nature of the recording process, the role of demagnetizing fields, and the various stages of the process which take place before a signal is finally produced in the reproduce head.

In most media, the particles are oriented in the direction of motion, thus encouraging longitudinal magnetization. When the medium is thin, the demagnetizing factor of the medium in a direction normal to its plane is large, causing large normal components of the demagnetizing field and discouraging normal components of magnetization. In this situation, it is possible to consider longitudinal recording solely in terms of longitudinal components of head field and magnetization. In cases where the medium is not thin, such an approach is invalid.

2.2.1 Longitudinal recording

Consider the longitudinal recording process where a head field of negative polarity is shown being applied to a medium previously magnetized to a level $+M_r$. The applied longitudinal field has its highest magnitude directly below the head gap, causing the magnetization to take on a value of $-M_r$. This means that to the right of the gap the magnetization undergoes a change from negative values to positive values and therefore a magnetization transition has been recorded.

At the center of the transition, $M_x = 0$, and this state of magnetization is achieved when the field in the medium is equal to the coercivity. Using the short-gap approximation, the locus of the points at which $H_x = H_c$ is found by setting

$$H_c = \frac{H_g g}{\pi} \frac{y}{x^2 + y^2} \tag{2.40}$$

Reordering of this equation leads to

$$x^2 + \left(y - \frac{1}{2} \frac{H_g g}{H_c \pi} \right)^2 = \frac{1}{4} \left(\frac{H_g g}{H_c \pi} \right)^2 \tag{2.41}$$

which is the equation of a circle. This result implies that constant field contours are circular, an example being the field contour for $H_x = H_c$ shown in Fig. 2.31a (Bauer and Mee, 1961). Thus the position of the tran-

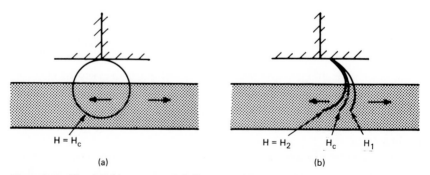

$$H = H_c \qquad\qquad\qquad\qquad H = H_2 \qquad H_c \qquad H_1$$

(a)   (b)

Figure 2.31 Head-field contours. (a) Contour of $H_x = H_c$ below a narrow-gap ring head; (b) contours for $H_x = H_1$ and $H_x = H_2$.

sition center will vary with spacing from the head and, in a thick medium, will be a variable through its depth. In thin media the variations may be considered sufficiently small to be negligible.

The nature of the transition can be studied by assuming that the major change of the magnetization takes place over the field range H_1 to H_2. In a similar fashion to that used for the drawing of the field contours for the coercivity it is possible to obtain the contours for $H_x = H_1$ and $H_x = H_2$. These lines then display the boundaries of the transition and so give a visual picture of the form of the recorded transition as shown in Fig. 2.31b.

Should a more detailed representation of the variation of magnetization within a transition be required, the techniques illustrated in Fig. 2.32 are appropriate. The figure shows the head-field distribution at the time of

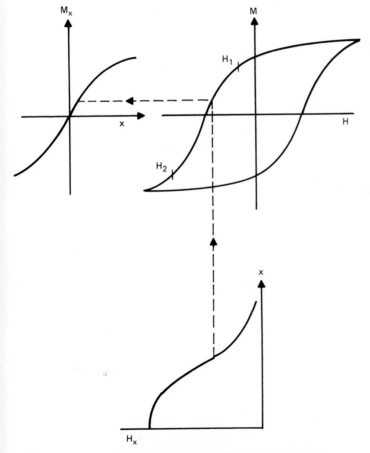

Figure 2.32 Scheme showing how recorded magnetizations can be determined for given media hysteresis and head-field distributions.

recording, the hysteresis loop of the medium, and the means of tracing from the first to the second diagram to find the magnetization variation with position in the medium. This procedure can be applied to different head fields and hysteresis loop shapes. An analytical approach to the same problem can be made by assuming a mathematical form for the hysteresis loop $M(H)$. Then, since the field distribution $H = H_x(x, y)$ is known, the magnetization distribution $M(H)$ in the transition is given by

$$M_x(x) = M(H_x(x, y)) \tag{2.42}$$

which is the mathematical equivalent of the result obtained using the graphical technique. The magnetization gradient, or recorded pulse, is given by

$$\frac{dM_x}{dx} = \frac{dM(H_x)}{dH_x} \frac{dH_x}{dx} \tag{2.43}$$

This formula says that the magnetization gradient is proportional to the product of the slope of the side of the hysteresis loop and the head-field gradient. Hence the requirements for short transition lengths are for rectangular hysteresis loops and high head-field gradients.

To improve the theory and make it more realistic it is necessary to take into account the demagnetizing fields H_d within the recording medium. These add to the applied head field to produce a total field, H_t, and can be taken into account by rewriting Eq. (2.43) as

$$\frac{dM_x}{dx} = \frac{dM(H_t)}{dH_t} \frac{dH_t}{dx} \tag{2.44a}$$

where
$$\frac{dH_t}{dx} = \frac{dH_x}{dx} + \frac{dH_d}{dx} \tag{2.44b}$$

Unfortunately, a complete solution of the equations requires the use of iterative, numerical methods since the magnetization gradient, which we need to calculate, is itself needed for the calculation of H_d. Therefore, to progress with the analytical approach, it is necessary to use an approximate method for solving the equations. The one adopted here is that proposed by Williams and Comstock (1972) but modified and developed in some detail by Maller and Middleton (1973, 1974). This basic method was developed for studies on thin media, which is the focus of the ensuing discussion.

Redrawing the hysteresis loop with more detail, as in Fig. 2.33a, permits a more satisfactory discussion of the recording process. Assuming the medium is initially magnetized, as before, to $+M_r$, the application of a head field of negative sense changes the magnetization according to the

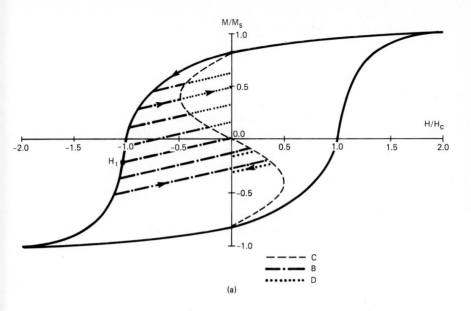

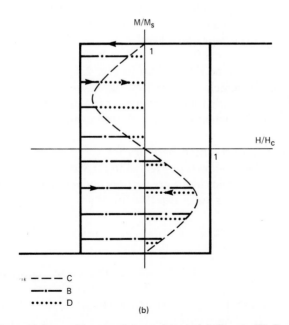

Figure 2.33 Assumed form of hysteresis loop shapes. (a) $S < 1$; (b) $S = 1$. Curve B—demagnetization after removal of applied field; curve C—demagnetizing field as a function of magnetization, and curve D—remagnetization under replay head *(Maller and Middleton, 1973).*

left-hand side of the major hysteresis loop, which is assumed to have the form

$$M(H) = \frac{2M_s}{\pi} \arctan\left(\frac{H + H_c}{H_c} \tan \frac{\pi S}{2}\right) \tag{2.45}$$

where $S = M_r/M_s$ is the remanence squareness. The center of the transition is assumed to be positioned at point H_1 on the hysteresis loop since the removal of the corresponding element of the medium away from the head field leaves it with zero magnetization. This process, termed *relaxation,* will now be described.

When an element of the recorded magnetization distribution leaves the vicinity of the record head field, it relaxes to a new value by moving along one of the curves marked B to end up at some position along the curve C of Fig. 2.33a. The latter is determined by the levels of magnetization allowed by the demagnetizing fields in the medium. When this new distribution is transported under the reproduce head, there is a further readjustment of the magnetization. This arises because the high-permeability reproduce head has induced within it image charges of opposite sign to those in the recording medium. The image charges produce fields in the recording medium that reduce the demagnetizing field therein and so alter the magnetization. This process takes place along the lines marked D, and it is the final magnetization values on these lines that generate the output voltage in the reproduce head.

By way of explanation of curve C of Fig. 2.33, an arctangent magnetization distribution such as that shown in Fig. 2.34a gives rise to a demagnetization field H_d which is plotted in Fig. 2.34b. Each position x' in the medium is associated with unique values of the magnetization and demagnetization fields, and when these are plotted against each other, they give rise to curve C.

The calculations presented here are for the particular case of rectangular hysteresis loops shown in Fig. 2.33b for which $S = 1$. In such cases the sides of the hysteresis loops are vertical and so their slopes are infinite. Substituting $dM/dH = \infty$ in Eq. (2.44) leads to

$$\frac{dH_x}{dx} = -\frac{dH_d}{dx} \tag{2.46}$$

The left-hand side of Eq. (2.46) can be evaluated by taking the short-gap approximation for the head field from Eq. (2.6a) and optimizing H_g to produce the highest field gradient at the center of the transition, that is, at $H_x = H_1 = H_c$. The result is

$$\frac{dH_x}{dx} = -\frac{H_c}{y} \tag{2.47}$$

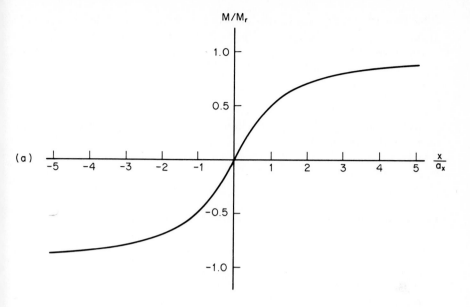

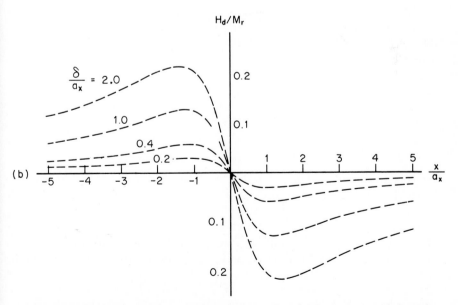

Figure 2.34 Arctangent transition. (*a*) Magnetization; (*b*) demagnetization field along the center plane of media of various thicknesses.

The demagnetizing field in the recorded transition can be calculated only by assuming a form for the recorded magnetization distribution. The form assumed here is

$$M_x = \frac{2}{\pi} M_r \arctan \frac{x}{a_x} \tag{2.48}$$

for which the self-demagnetizing field can be calculated using the results of Appendix III. Thus

$$H_{dx} = -\frac{1}{2\pi} \int \int \frac{\partial M_x(x')}{\partial x'} \frac{x - x'}{(x - x')^2 + (y - y')^2} \, dx' \, dy' \tag{2.49}$$

and substitution from Eq. (2.48) and evaluation results in

$$H_{dx} = -\frac{2}{\pi} M_r \left(\arctan \frac{x}{a_x} - \arctan \frac{x}{a_x + \delta/2} \right) \tag{2.50}$$

which takes the form shown in Fig. 2.34.

The magnetization gradient at the center of the transition and the center plane of the medium ($y = d + \delta/2$) is now determined to be

$$\frac{dH_{dx}}{dx} = -\frac{M_r \delta}{\pi} \frac{1}{a_x(a_x + \delta/2)} \tag{2.51}$$

Substituting this along with dH_x/dx from Eq. (2.47) into Eq. (2.46) and evaluating leads to a transition length of

$$a_x = -\frac{\delta}{4} + \left(\frac{\delta^2}{16} + \frac{M_r \delta y}{\pi H_c} \right)^{1/2} \tag{2.52}$$

which, for thin films, becomes

$$a_x = \left(\frac{M_r \delta d}{\pi H_c} \right)^{1/2} \tag{2.53}$$

If this transition is long, it suffers no further demagnetization when it moves away from the record head since the demagnetization curves B have zero slope (Fig. 2.33b). A similar situation occurs on replay, and a_x given by Eq. (2.53) is the value of the transition parameter that governs the ouput voltage. It should be noted that this is the shortest possible transition since the head field has been optimized to make it so. Should the transition length predicted by Eq. (2.53) be very short, it would result in a very large demagnetizing field. Inspection of the hysteresis loop of Fig. 2.33b reveals that, for a stable magnetization, demagnetizing fields cannot exceed the coercivity (Chapman, 1963; Davies et al., 1965; Bonyhard et al., 1966; Speliotis and Morrison, 1966a). Therefore, according to Eq. (2.50), there will be a minimum value of the transition length allowed by demagnetization, and this can be determined by letting $H_{d,\max} = H_c$. For thin films, this leads to

$$a_d = \frac{M_r\delta}{2\pi H_c} \tag{2.54}$$

The reduced transition lengths caused by writing losses (a_x/δ) and by demagnetization (a_d/δ) are plotted in Fig. 2.35 as functions of $M_r/4\pi H_c$ for different values of y/δ. Two regimes of behavior can be identified corresponding to square root and linear dependency, and these are symptomatic of record demagnetization and self-demagnetization, respectively. For hysteresis loop squareness other than unity, the transition lengths differ at different parts of the record-demagnetization-reproduce cycle. As an example, consider the results for wide-gap heads shown in Fig. 2.36 (Maller and Middleton, 1974). In the figure, a_1/δ represents the transition length under the influence of record head field, a_2/δ the demagnetized transition length, and a_3/δ the remagnetized transition length as it appears under the reproduce head. The implications are that short transition lengths are created under the record head and that these lengthen as they move away from the influence of the head field. When they move under the reproduce head, the reduction of demagnetizing fields results in a shorter transition. Of particular note is that the lengthening of the

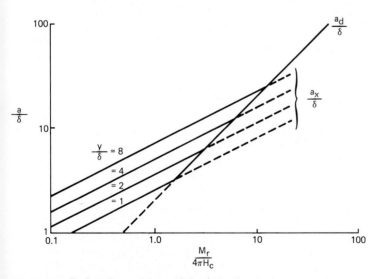

Figure 2.35 Reduced transition width a/δ for longitudinal recording as a function of $M_r/4\pi H_c$. Squareness $S = 1$.

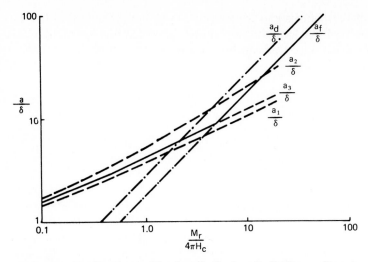

Figure 2.36 Predicted transition lengths for longitudinal recording at the instant of recording (a_1/δ), after relaxation (a_2/δ), and after remagnetization (a_3/δ) as functions of $M_r/4\pi H_c$. Also shown are the transition lengths after demagnetization (a_d/δ) and after remagnetization (a_f/δ) *(Maller and Middleton, 1973)*.

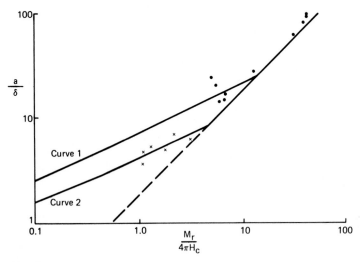

Figure 2.37 Observed transition widths for film media as a function of $M_r/4\pi H_c$. The lines represent predictions *(Maller and Middleton, 1973)*.

transitions due to relaxation and the shortening due to remagnetization almost compensate, making the final length close to the written one. At large values of $M_r/4\pi H_c$, the transition lengths are determined not by the write process but by demagnetization. In this case a_d/δ is the transition length after removal from the vicinity of the record head and a_f/δ is the final value under the reproduce head. Figure 2.36 shows the two basic regimes of record and replay demagnetization, although the picture is slightly more complicated than that shown in Fig. 2.35. These regimes have been confirmed by others employing more precise numerical modeling (Speliotis, 1972; Tjaden and Tercic, 1975).

On the experimental side, much has been done to provide experimental data and to fit theory to it (Davies et al., 1965; Bonyhard et al., 1966; Speliotis and Morrison, 1966a, 1966b; Speliotis, 1972; Tjaden and Tercic, 1975). The selected results shown in Fig. 2.37 are deduced from measurements of pulse lengths from cobalt films, using the replay theory presented here (Maller and Middleton, 1974). The basic features of the theory are confirmed.

In addition to considerations based on bulk properties, it is known that there is fine detail in recorded transitions which may be of importance. This is illustrated diagrammatically in Fig. 2.38a, which shows a transition of arctangent form, and in Fig. 2.38b, which attempts to show that the transition in metallic films takes the form of a domain wall disposed in zig-zag or sawtooth fashion (Dressler and Judy, 1974). The available experimental evidence suggests that the transition length l defined by the

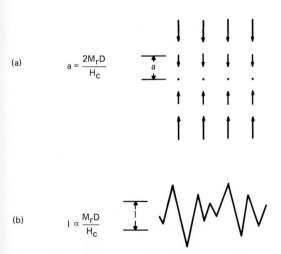

Figure 2.38 Schematic representation of (a) the assumed form of the magnetization at a transition and (b) the observed shape of the transitions.

average sawtooth amplitude in Fig. 2.38*b* varies as (Dressler and Judy, 1974; Yoshida et al., 1983; Gronau et al., 1983)

$$l \propto \frac{M_r \delta}{H_c} \tag{2.55}$$

It is perhaps fortunate that this should have the same form as Eq. (2.54), the formula for transition widths caused by self-demagnetization when the fine detail of the transitions is ignored.

In summarizing the lessons to be learned from the theory, it should be borne in mind that the treatment given here applies solely for thin films and is relevant to thicker, particulate media with less certainty. Nevertheless, for all media, it is necessary to keep M_r/H_c and δ small in order to achieve sharp transitions. This requirement has led to continued developments toward thinner media of higher coercivity. The other major requirement is that hysteresis loop squareness should be high.

2.2.2 Perpendicular recording

In perpendicular recording the medium has a preferred orientation, or easy axis, in a direction normal to its plane so as to encourage components of magnetization in that direction (Iwasaki and Takemura, 1975; Iwasaki and Nakamura, 1977; Iwasaki, 1984). A typical recording medium consists of a sputtered film of cobalt-chromium having high anisotropy (anisotropy field $H_k > M_s$) and easy axis normal to its plane (Iwasaki and Ouchi, 1978). Suppose the head field is magnetizing the medium in an upward direction against a background of downward magnetization. Just as in longitudinal recording, the shape of the head-field distribution is important and needs to be taken into account in considering the recording process.

The model of the recording process to be described is simply an adaptation of that used earlier for longitudinal recording in thin-film media: two variations on the theme exist in the literature (Middleton and Wright, 1982; Bromley, 1983). As in the longitudinal case, the recorded transition lengths are determined by solution of the equation

$$\frac{dM_y}{dx} = \frac{dM(H_t)}{dH_t} \left(\frac{dH_y}{dx} + \frac{dH_d}{dx} \right) \tag{2.56}$$

in which the use of the y subscript denotes the consideration given to perpendicular components. For rectangular hysteresis loops, the slopes of their sides, dM/dH, is infinite, and

$$\frac{dH_y}{dx} = -\frac{dH_d}{dx} \tag{2.57}$$

To aid in obtaining the solution of this equation for a ring head, the maximum field gradient existing at the center of the transition at $H_y = H_c$ can be shown, using the short-gap head-field approximation, to be

$$\frac{dH_y}{dx} = 0.3 \frac{H_c}{y} \tag{2.58}$$

which is less than one-third of that given for the x component of the field. The other quantity needed is the demagnetizing field H_{dy}, which can be evaluated only when the magnetization distribution is known. Assuming an arctangent transition of the type

$$M_y(x) = \frac{2M_0}{\pi} \arctan \frac{x}{a_y} \tag{2.59}$$

where M_0 is the magnitude of the magnetization for each side of the transition, and using this in conjunction with the results of Appendix III leads to an expression for the demagnetizing field in the midplane of the recording medium

$$H_d = - \frac{2M_0}{\pi} \arctan \frac{x}{a_y + \delta/2} \tag{2.60}$$

Differentiation of this gives the field gradient, which, at the center of the transition, is

$$\frac{dH_d}{dx} = - \frac{2M_0}{\pi} \frac{1}{a_y + \delta/2} \tag{2.61}$$

Substitution of this and Eq. (2.58) into Eq. (2.57) leads to an expression for transition width

$$a_y = \frac{2}{0.3\pi} \frac{M_0}{H_c} \left(d + \frac{\delta}{2} \right) - \frac{\delta}{2} \tag{2.62}$$

To complete the equation, it is necessary to consider the value of M_0. In the case when $M_r < H_c$, demagnetization is limited and $M_0 = M_r$. For most practical cases, however, $M_r > H_c$, and consideration needs to be given to the effect of demagnetization. Equation (2.60) shows that at large values of x (distant from the transition), the demagnetizing field takes on a value $-M_0$. However, as in the case of longitudinal recording, a finite magnetization can be maintained only if the demagnetizing field does not exceed H_c. Thus the magnetization is limited to a value of $M_0 = H_c$, and use of this in Eq. (2.62) leads to

$$a_{yr} = \frac{2}{0.3\pi} \left(d + \frac{\delta}{2} \right) - \frac{\delta}{2} \tag{2.63}$$

The subscript r has been added to indicate that this solution applies to a ring head. The message of Eq. (2.63) is that head-to-tape spacing is a primary factor in determining transition length and that medium magnetic properties, such as M_r and H_c, make no contribution.

With regard to recording with a single-pole head, the considerations are the same as those already expounded for a ring head except that the head field, and therefore the head-field gradient, has a different value. Assuming that the y component of the field of a single-pole head takes the same mathematical form as the x component of the field of a ring head, the maximum head-field gradient that can occur at $H = H_c$ is given by

$$\frac{dH_y}{dx} = \frac{H_c}{y} \tag{2.64}$$

Using Eq. (2.64) along with Eq. (2.61) in Eq. (2.67) leads to a predicted transition width of

$$a_{yp} = \frac{2}{\pi}\frac{M_0}{H_c}\left(d + \frac{\delta}{2}\right) - \frac{\delta}{2} \tag{2.65}$$

The use of the extra subscript is to denote single-pole or probe head. Again, unless $M_r < H_c$ it is appropriate to make $M_0 = H_c$, and so

$$a_{yp} = \frac{2}{\pi}\left(d + \frac{\delta}{2}\right) - \frac{\delta}{2} \tag{2.66}$$

Here, once again, neither M_r nor H_c figures in the equation, and the head-to-medium separation is a primary factor.

Transition lengths for longitudinal and perpendicular recording are shown in Fig. 2.39 as a function of medium thickness for various head-to-medium spacings. A number of deductions may be made from this figure. First, perpendicular recording is much more sensitive to variations in head-to-medium spacing than is longitudinal recording, and reduced spacings would lead to shorter transitions regardless of which head was used. Second, perpendicular recording with a single-pole head is much more favorable than with a ring head. Third, longitudinal recording produces short transitions only in very thin media. An alternative is to base the analysis on solving Eq. (2.56) at the surface rather than at the mid-plane of the medium. The corresponding results are plotted in Fig. 2.40. The conclusions are similar to those listed above except that, as expected, the dependence of transition length on medium thickness is much reduced.

On the experimental front no data have been provided on transition lengths. However, with transition lengths independent of M_r and H_c it is to be expected that the output voltage levels according to Eqs. (2.30) and

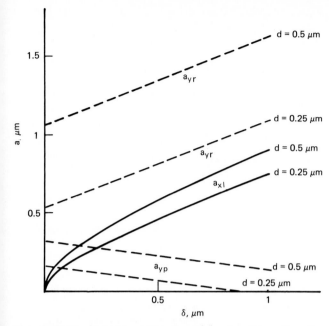

Figure 2.39 Predicted transition lengths as functions of media thickness for different spacing, based on balancing the field gradients at the midplane. *(Middleton and Wright, 1982).*

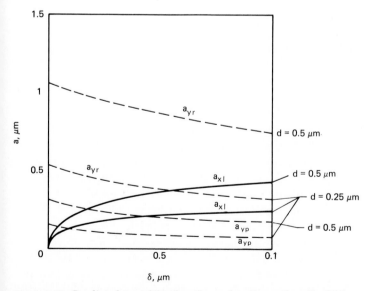

Figure 2.40 Predicted transition lengths as functions of media thickness for different spacing, based on balancing the field gradients at the surface.

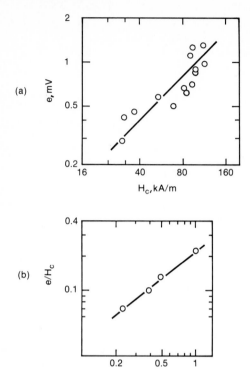

Figure 2.41 Observations of output amplitude as functions of (*a*) H_c (1.2 kfr/mm, δ = 1 μm) and (*b*) δ (67 fr/mm, M_s = 400 kA/m) for perpendicular recording *(Suzuki and Iwasaki, 1982; Hokkyo et al., 1982).*

(2.31) should be proportional to the product $M_0\delta$, that is, $H_c\delta$. Experimental results taken from cobalt-chromium films (Fig. 2.41) reveal the anticipated linear dependencies on H_c and δ (Hokkyo et al., 1982).

2.2.3 Self-consistent modeling

It has been realized for some time that there exist not only longitudinal components of magnetization in longitudinal recording but also perpendicular components (Wallace, 1951; Westmijze, 1953). Evidence for this became available as a result of large-scale physical modeling of the record and replay processes (Tjaden and Leyten, 1964), and more recently as a result of numerical modeling (Potter and Beardsley, 1980). A typical magnetization distribution arising from the latter work is shown in Fig. 2.42*a* and is a result of considering the vector nature of the record process. The modeling employs a vector hysteresis model of recording media (Ortenburger and Potter, 1979) based on an assembly of noninteracting fine particles having their easy axes distributed in three dimensions. The passage of the media through time-varying head fields is considered, along with fully self-consistent, point-by-point determination of the magnetization distribution. These involve calculations of the demagnetizing field for each magnetization distribution, along with procedures to ensure that, in each case, the magnetization and total field within the medium conform to its hysteresis loop. Predictions have been produced not only for longitudinal recording but also for perpendicular recording with ring heads (Fig. 2.42b) and single-pole heads (Fig. 2.42c). While these calculations have exposed the complexity of the true situations, and the limitations of the simple models described earlier, they have not thrown doubt on the basic qualitative predictions of the role of media and head properties. Further, there will remain considerable problems with numerical modeling in that particle-particle interactions in the recording media are not yet taken into account (Ortenburger and Potter, 1979; Potter and Beardsley, 1980) and the modeling process is often sensitive to details of the loop models (Beardsley, 1982a). Nevertheless the numerical computations have provided what are probably the best attempts so far at modeling the recording process.

The complexity of these calculations can be somewhat reduced by neglecting the effects of demagnetizing fields during the record process. The much simplified modeling procedures provide useful insights into the nature of magnetic recording (Bertram, 1984a, 1984b) and are justified by experiments which revealed demagnetization-free recording in situations where head-to-tape spacing is small (Bertram and Niedermeyer, 1982).

In numerical modeling of perpendicularly oriented cobalt-chromium films (Beardsley, 1982b), it was found necessary to restrict the calculations to only uniform magnetization through the media thickness in order to obtain agreement between the calculations and experimental results.

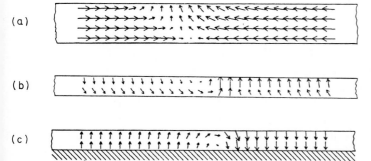

Figure 2.42 Predicted magnetization distributions in (a) longitudinal recording by a ring head of gap 1 μm on a medium of thickness 1 μm, (b) perpendicular recording by a ring head of gap 1 μm on a medium of thickness 0.5 μm, and (c) perpendicular recording with a single-pole head on a medium of thickness 0.5 μm and with a soft magnetic underlayer *(Potter and Beardsley, 1980)*.

This assumption of uniformity has led to the development of simpler analytical modeling procedures as will now be described. However, there is no direct evidence for uniform magnetization.

Consider a medium initially magnetized to saturation in the negative sense by a large field applied in a direction normal to its plane. After the removal of the field, the medium is left with a magnetization of $-H_c$, as already described. The process by which magnetization is recorded into such a medium by the application of perpendicular components of head fields can then be derived with precision subject to a number of initial simplifying assumptions (Wielinga et al., 1983; Middleton and Wright, 1984; Lopez, 1984). The general thrust of these works is contained in the following discussion.

The application of a head field having perpendicular components in the positive sense is assumed to generate only perpendicular components of magnetization, the high anisotropy of the medium discouraging longitudinal components. Provided that the applied field is not so large as to magnetize the medium beyond the vertical side of the hysteresis loop, the total field at any point in the medium must be equal to the coercivity, and therefore

$$H_c = H_y(x, y) + H_d(x, y) \tag{2.67}$$

This equation defines the magnetization distribution since $H_d(x, y)$ is calculated from it. Solution of Eq. (2.67) in closed form is possible for the particular case of uniform magnetization through the depth of the medium. This is a reasonable assumption in strongly coupled media such as metallic films and leads to

$$M_y(x) = H_y(x, d) - H_c \tag{2.68}$$

where the magnetization in the medium is related to the head field along its top surface. Such a result provides a simple picture of the recording process.

By use of the Karlqvist approximation, the distributions recorded by a ring head have been predicted (Wright and Middleton, 1984) and the corresponding outputs calculated. The results are shown in Fig. 2.43a and b. The shapes of these outputs are similar to those observed experimentally and show the correct dependence on record current amplitude. The basic theory has been extended with minor modifications to cope with recording on media having hysteresis loops with sides of finite slope and on double-layer media (Lopez, 1984; Wright and Middleton, 1985).

To progress with the analytical modeling of the generalized recording process, it is necessary to obtain from the detailed numerical predictions, or large-scale modeling (Tjaden and Leyten, 1964; Monson et al., 1981), some pointers as to the essential features of either the record process itself or the recorded distributions. These might allow some simplifications to the modeling procedures which may make analytical approaches tractable. One feature which is already apparent is that the recorded transitions are all shaped round the contour $H = H_c$, where H is the total applied head field (Ortenburger and Potter, 1979; Fayling, 1982). Allying this to a second point that the recorded magnetization follows the vector head-field distribution (Bertram, 1984a) allows some insights into the nature of the processes and opens up avenues of attack for analytical work.

2.2.4   Media orientations

Generally, the recording processes are never purely longitudinal or purely perpendicular but are always a mixture of the two. This applies even when the media are well oriented, but is particularly evident in unoriented

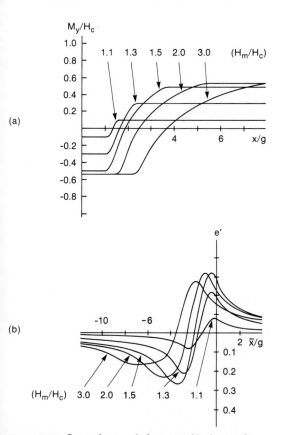

Figure 2.43 Strongly coupled perpendicular media. (a) Magnetization distributions predicted to occur as a result of recording with a ring head. (b) Corresponding reproduced waveforms induced in a ring head. H_m is the peak perpendicular field experienced by the medium *(Wright and Middleton, 1984)*.

media or in media having little or no bulk anisotropy. Media have been produced from particles having multiaxial symmetries, and their hysteresis properties show that they are essentially isotropic, and have a high remanence squareness in all directions (Jeschke, 1954; Krones, 1960). Detailed experimental and theoretical work confirmed suspicions that both longitudinal and perpendicular components of magnetization play a role in determining the outputs and that the latter may be active in keeping outputs high at short wavelengths (Lemke, 1979, 1982).

The process of recording on media of differing orientations has been investigated using self-consistent iterative modeling techniques (Beardsley, 1982a). In these computations a model was developed for a system of particles of differing anisotropies constrained to different orientations. Some of the results are shown in Fig. 2.44, where output voltage is related to packing density for particles which are (1) longitudinally oriented, (2) perpendicularly oriented, and (3) isotropically oriented. The longitudinally oriented medium gives the highest output at long wavelengths; then the output falls off steadily at shorter wavelengths. The perpendicularly oriented medium starts off at long wavelengths with a low output, but the output at short wavelengths is somewhat higher than for the longitudinal medium. The output from the isotropic medium is low at long and intermediate wavelengths, and at short wavelengths its output is intermediate between the outputs for longitudinal and perpendicular recording, as might be expected from intuitive reasoning. In the short-

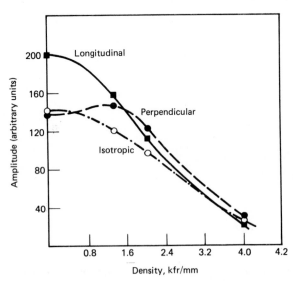

Figure 2.44 Predicted outputs for longitudinal, isotropic, and perpendicular particulate recording media *(after Beardsley, 1982a)*.

wavelength region, the difference between the three types of media does not appear to be large.

2.2.5 Performance of various media and modes of recording

The preceding sections have discussed recording processes and how these are influenced by the magnetic properties of media and the properties of recording heads. Various features of recording processes have been identified, and these have been studied both qualitatively and quantitatively.

In longitudinal recording it has been shown that it is necessary to keep the ratio M_r/H_c of the recording media as low as possible, and to keep the media thin if transition widths are to be narrow and linear densities high. Head-to-tape spacings must also be kept small and, in regard to perpendicular recording, small spacing is the overriding requirement. A comparison of the potential of any media or mode of recording requires that all the variables be defined and the conditions under which the measurements taken or predictions made be specified. However, useful and realistic comparison is obtained by examining the best available experimental data rather than dealing in predictions, however good the theory.

Data have been assembled of the output performance obtained experimentally on various media in laboratories. These results (Mallinson and Bertram, 1984a; Mallinson, 1985), which include only calibrated data, show that there is little difference in attainable output levels for longitudinal recording and for perpendicular recording on metallic media having coercivities in the range 64 to 72 kA/m (800 to 900 Oe). This basic result has been repeated (Stubbs et al., 1985), and the results are shown in Fig. 2.45. Four types of media were investigated with the same short-gap ring head, and there was virtually no difference between recording on a longitudinal Co-Ni medium and recording on Co-Cr with or without a soft magnetic underlayer. The similarity of the film media curves suggests that recording densities are more likely to be limited by a common cause such as head-to-tape spacing than by the properties of the media used. Figure 2.45 shows that signal levels hold up well to extremely high densities and that a considerable potential for achieving high densities exists with all media.

2.3 Linear Recording

So far this chapter has dealt with digital recording wherein the requirement is to be able to detect either of two stable states of magnetization in a recording medium, or the difference between these states. There is no requirement for a linear relationship between the level of magnetization

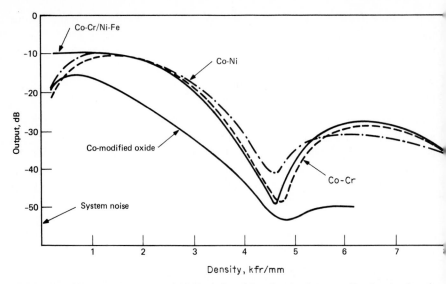

Figure 2.45 Output voltages as a function of packing density for recording longitudinally on Co-Ni-plated media and perpendicularly on single- and double-layer Co-Cr media. The output for a cobalt-modified oxide is also shown *(Stubbs et al., 1985).*

in the recording medium and the recording signal. Linear recording brings with it a need for a linear recording process, and this itself creates a need for recording media which show a linear relationship between their magnetizations and the applied field. The latter conditions are created by the use of what is termed *biased recording,* in which an additional field, termed a *bias field,* is present along with the field intended for the recording of the signal. This section deals with the influence of the bias field on the magnetization and recording processes.

Recording with ac bias is well known as the basic process used in analog audio recording systems, and it also has a range of uses in instrumentation recording. Because it involves essentially linear relationships, the level of understanding of the recording processes involved in this type of recording is relatively high and useful reviews of the physics of ac-bias recording exist in the literature (e.g., Westmijze, 1953; Mee, 1964). Chapters 6 and 7 of Volume II cover the subjects of audio and instrumentation recording.

This section proceeds with a discussion of the basic hysteretic and anhysteretic properties of magnetic recording media. This is followed by a consideration of the behavior of interacting fine-particle arrays partly as a means of explaining the earlier hysteresis properties and partly as a lead-in to the study of the recording process. However, the understanding of particle interactions is still limited. Finally the process of ac-bias recording is discussed in detail.

2.3.1 Hysteresis properties of recording media

It is appropriate to begin with a discussion of the magnetic hysteresis properties of media as they are observed experimentally. This will provide a factual basis before discussion of the less certain area of interpreting these properties in terms of the behavior of individual particles and the nature of particle arrays. This discussion will center on the remanent magnetization curves which contrast with the hysteresis curves previously referenced. In hysteresis curves, it is usual to plot the instantaneous magnetization developed in the presence of an applied field as a function of that field, whereas the remanence curves show the magnetization remaining after removal of the field as a function of that field. There must be a strong connection between the two sets of curves, and indeed it is possible to trace out the remanence curves once a family of hysteresis curves is available.

Consider the application of a unidirectional magnetic field to a sample of medium which had previously been subject to an alternating field of decreasing amplitude so as to leave it in the demagnetized state. The variation of remanent magnetization with applied field is shown in Fig. 2.46. Also shown is the variation of magnetization with field when the specimen had previously been subject to a saturating field in the reverse direction. The curves are modified when an ac-bias field is present in addition to the unidirectional fields (Daniel and Levine, 1960a). The influence of the bias field is greatest when its amplitude is appropriately chosen to diminish the hysteretic processes present in unbiased magnetization and to produce what are termed *anhysteretic magnetization processes*. Figure 2.47 was obtained as a result of the application to a previously demagnetized sample of an alternating field, whose amplitude was reduced to zero prior to the reduction of the unidirectional field ("ideal" anhysteresis). It shows the variation of remanent magnetization with unidirectional field amplitude, with the peak ac-field amplitude as a parameter. Providing that the latter is reduced sufficiently slowly, so that the medium experiences a sufficient number of cycles, the curves are fully reproducible. The curve for zero alternating-field amplitude is identical to that already considered in Fig. 2.46, but the influence of the additional ac fields is drastic. Most notable is the fact that, when the magnitude of the bias field approaches that of the remanence coercivity, this field alone is sufficient to cause the particles in the medium to switch and so little extra unidirectional field is needed to influence the remanent state of magnetization of the particles as the alternating field is reduced. Consequently, the magnetization is a rapidly varying function of applied dc field and is, in fact, a linear function of field up to quite high magnetization values. On the lower parts of these curves the susceptibility of the medium is high. As a result of the linear-

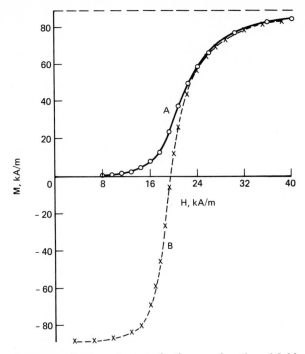

Figure 2.46 Remanent magnetization as a function of field strength for (curve *A*) an ac-demagnetized and (curve *B*) a negatively saturated γ-Fe_2O_3 medium *(Daniel and Levine, 1960a).*

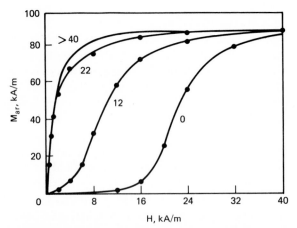

Figure 2.47 Anhysteretic remanent magnetization as a function of dc-field strength for different values of alternating-field amplitude *(Daniel and Levine, 1960a).*

izing process, the curves cease to show hysteresis and so are termed *anhysteretic magnetization curves.*

In the recording situation the linearizing effect of the ac bias, as illustrated by these graphs, is a necessary requirement for linear recording. The experimental procedure adopted does not quite correspond to practical ac-bias recording as will now be described. It is true that when an element of a recording medium passes under a recording head it experiences the combined influence of unidirectional and ac fields, but as it passes away from the recording head, both fields decrease at the same time and at the same rate. Consequently, different curves are required, and these have been plotted under the new conditions of simultaneously reducing fields (Daniel and Levine, 1960*a*). The results are given in Fig. 2.48 and show how the curves steepen to reach a maximum slope as the ac field is increased and thereafter suffer a decreasing slope. This mode of magnetization has been termed "modified" anhysteretic magnetization.

The initial susceptibilities for the ideal and modified anhysteretic processes are shown in Fig. 2.49 as functions of ac-bias-field amplitude. In the ideal anhysteretic process the susceptibility approaches asymptotically a maximum value, whereas in the modified process there is a maximum value at a finite value of bias field. The occurrence of this maximum implies that optimum conditions must exist during the recording process.

2.3.2 Fine-particle assemblies

Particulate recording media consist of assemblies of fine particles of magnetic material which are subject to interactions, and the problems of

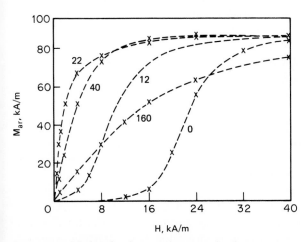

Figure 2.48 Modified anhysteretic magnetization curves as a function of dc-field strength for different values of alternating-field amplitude *(Daniel and Levine, 1960a).*

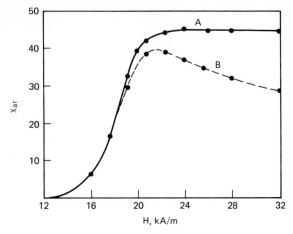

Figure 2.49 Initial susceptibilities for (curve *A*) ideal anhysteretic magnetization and (curve *B*) modified anhysteretic magnetization *(Daniel and Levine, 1960a)*.

understanding and modeling the behavior of particulate media are substantial. These interactions, however, are very important in recording and strongly influence the resulting magnetization achieved in anhysteretic magnetization and in ac-bias recording.

The properties of individual particles when observed in isolation are complex and do not fit easily alongside idealized modes of reversal (Knowles, 1981). In addition, individual particle properties differ considerably from one another, and the properties of assemblies of particles vary considerably from the properties of the individual particles making up the assemblies. These differences arise because of interparticle interactions and also because of variations in particle properties, orientations, and spatial positioning within the recording media. Overall, this represents a difficult, many-bodied problem in solid-state physics, and the attempts that have been made at solving this type of problem are necessarily simplified and somewhat approximate. Nevertheless, some progress has been made using various avenues of approach, and a brief discussion of some aspects of these approaches will now be given.

Beginning with the properties of individual particles, there is considerable background of study of reversal processes in particles having regular geometries. Possible reversal processes such as coherent rotation, curling, buckling, and fanning are discussed in detail in Chap. 3. Simple models of the hysteresis properties of particle assemblies in the form of recording media have been developed by summing the individual loops of particles of varying orientations and switching fields (Ortenburger and Potter, 1979). However, such approaches neglect the effects of interparticle interactions, which are known to be important, and further improvements in modeling are needed in this respect. The potential use of computers for determining the effects of interactions in particle arrays is being studied (e.g., Chantrell et al., 1984).

With regard to anhysteresis, more progress has been made on the modeling in the sense that interactions have been taken into account, albeit in rather simplified arrangements. Generally, modeling has involved the use of Monte Carlo methods to treat systems of interacting fine particles which are disposed on lattices of particular shapes (Bertram, 1971; Chantrell et al., 1984). The particles are assumed to be ellipsoids of revolution arranged in chains with their long axes parallel to one another and pointing along the direction of the applied fields. Each particle is given a switching field according to a Gaussian probability distribution, and placed randomly within the lattice. The lattice is then subject to a large, but decaying, ac field in addition to a dc field. The particle magnetizations are assumed to "freeze" when the total local static fields reduce in value to equal the switching field of the particles. The local fields are given by $H_t = H_{dc} + H_{int}$, where H_{dc} is the externally applied dc field and H_{int} is the interaction field produced by magnetically frozen particles. This modeling process leads to anhysteresis curves of the form shown in Fig. 2.50.

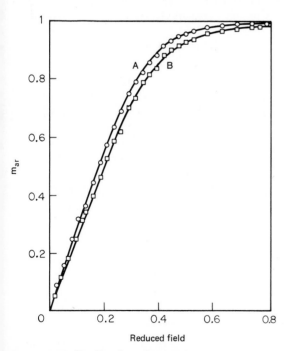

Figure 2.50 Predicted anhysteretic magnetization curves: (curve *A*) particles with identical volume, and (curve *B*) particles with a volume distribution *(Chantrell et al., 1984)*.

Two curves are shown: one for equally sized particles and one for a distribution of particle sizes. The curves are of the right shape, and the susceptibilities, as provided by the slopes of the curves near the origin, are of a reasonable order of magnitude compared with experimental data. While methods such as the above provide a means of modeling particle assemblies, they are not developed to a point where they can be applied to the magnetic recording process. The problems arise with the complexity of the interaction fields, which demand excessive computing capacities.

2.3.3   Interactions and the Preisach diagram

A means of graphic representation of the role of interaction fields is available in the form of the Preisach diagram, which has proved particularly useful (Preisach, 1935). Although this oversimplifies the system, it has provided an insight into the behavior of particle assemblies and its use will now be described.

As a lead-in to this topic, consider two particles subject to interaction fields which are positive and negative, respectively, along their easy axes and which possess rectangular hysteresis loops. Figure 2.51b and c shows how the hysteresis loops are shifted according to the sense of the interactions. Obviously the particles are now much more easily switched by externally applied fields of interaction-field polarity, and so the interactions can be expected to influence measurable parameters.

In particle assemblies, the interaction field experienced by any one particle arises from the effects of all the other particles, and so is a complicated function of particle properties and geometries. The interaction field may be in any direction, and this may alter the mode of magnetization reversal in the particle and therefore its intrinsic switching field. If it is possible to ignore the effects of the vector nature of the interactions, and just consider interaction and switching fields in one dimension, then it is possible to display the effects of the interaction fields in a graphical manner on what is known as the Preisach diagram. In one form of the diagram shown in Fig. 2.52, positive switching fields H_1 are plotted on the horizon-

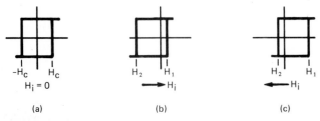

Figure 2.51   Effect of (a) zero, (b) positive, and (c) negative interaction fields on the hysteresis loops of single-domain particles.

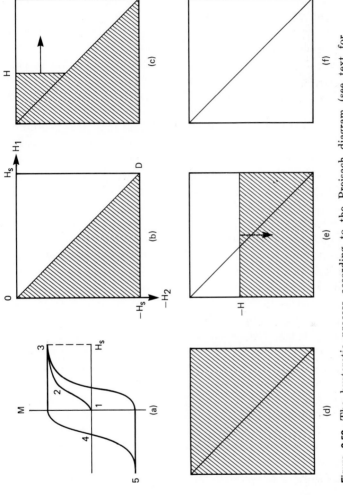

Figure 2.52 The hysteretic process according to the Preisach diagram (see text for explanation).

tal axis while negative switching fields H_2 are plotted on the vertical axis (Néel, 1955; Daniel and Levine, 1960a; Bate, 1962). Thus any particle occupies one point on the diagram and all the particles of an assembly appear on the diagram. The behavior of the assembly corresponding to a wide range of field histories and suitable examples can be portrayed as described below. An alternative form of this diagram plots H_i against H_c (Schwantke, 1958; Woodward and Della Torre, 1960). The two diagrams contain the same basic information, but are rotated through 45° with respect to one another.

Now consider the application of the Preisach diagram to explain the hysteresis process. Positively magnetized particles are indicated on the diagram by shaded areas; unshaded areas are occupied by negatively magnetized particles. The diagrams are shown bounded by $|H_1| = |H_2| = H_s$, where H_s is the maximum switching field (the field required to saturate the assembly). Figure 2.52b shows particles on the Preisach diagram which have been subject to demagnetization by an ac field of reducing amplitude. This has ensured that those particles experiencing positive interaction fields are left in a positive state of magnetization whereas those experiencing a negative interaction field are left in a negative state of magnetization. Thus, provided the distribution of points about the diagonal OD axis is symmetrical, the material has zero net magnetization. The application of a positive field H means that all particles which satisfy the condition $H \geq H_1$ will be switched to the positive sense of magnetization. This process is shown on the Preisach diagram indicated in Fig. 2.52c. All particles to the left of the vertical line $H_1 = H$ are switched to the positive sense of magnetization while those to the right remain magnetized in the negative sense. The total magnetization of the sample depends on the density distribution of the particles on the diagram but at this stage represents the magnetization 2 shown in Fig. 2.52a. Increasing the field moves the switching boundary to the right, as indicated by the arrow, causing more of the medium to be switched until the sample is saturated in the positive sense, Fig. 2.52d, and this state is represented by region 3 on the hysteresis loop. When the applied field is reduced from the value needed for saturation switching, the magnetization is determined by the horizontal line shown in Fig. 2.52e as it sweeps from top to bottom. Figure 2.52f shows the situation corresponding to negative saturation.

The ideal anhysteretic processes can be displayed in a similar way. Figure 2.53a shows the result of the application of a large field in the positive sense followed by a field of slightly lower amplitude in the negative sense. In the presence of a constant dc field H_{dc}, the application of successive fields of positive and negative polarity of decreasing amplitude eventually results in the picture shown in Fig. 2.53b. When the steps in the reduction of the field become infinitesimal, the diagram becomes that shown in Fig.

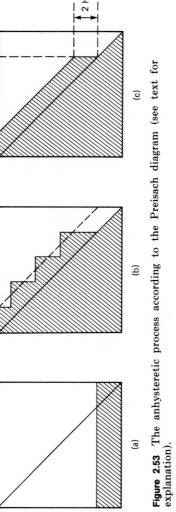

Figure 2.53 The anhysteretic process according to the Preisach diagram (see text for explanation).

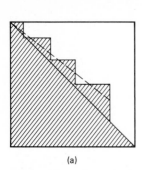
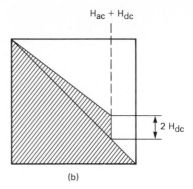

(a)   (b)

Figure 2.54 The modified anhysteretic process according to the Preisach diagram (see text for explanation).

2.54c for which the corresponding magnetization can be obtained if the density distribution of particles is known. In the modified anhysteresis, the diagrams are shown in Fig. 2.54a and b for finite and infinitesimal field decrements.

In all these cases the processes of magnetization can be protrayed and therein lies the value of the Preisach diagram. To enable determination of the magnetization at any stage of any of the processes depicted, it is necessary to have a knowledge of the density distribution of magnetizations of the particles. Sample distributions have been measured (Biorci and Pescetti, 1958; Daniel and Levine, 1960b; Woodward and Della Torre, 1960, 1961; Bate, 1962), and when they conform to simple mathematical forms, the magnetizations can be calculated analytically. Despite such apparent simplicity of use, the diagram may be open to question regarding its stability during the process of magnetization change. For example, the process of reversal of magnetization in one or more particles will cause the interactions experienced by all the other particles in the assembly to change. This would imply that the Preisach diagram should change, although experiments seem to indicate a surprisingly large measure of stability of the diagram (Bate, 1962).

More formal attempts have been made to quantify the effects of interactions, although the cases studied so far fall considerably short of modeling realistic recording media. For example, the hysteresis loops of interacting pairs of particles have been studied in some detail in the case where the interaction fields are in either the positive or negative senses of the particle axes (Néel, 1958; Wohlfarth, 1964). Less restrictive computations have been given involving arbitrary positioning of particles (Bertram and Mallinson, 1969, 1970) and larger arrays of particles (Moskovitz and Della Torre, 1967; Soohoo and Ramachandran, 1974). In the latter work, the spatial positioning of particles was shown to be particularly important in an attempt to characterize the behavior of particle arrays. In view of the haphazard arrangement of particles in a recording medium, it must be conceded that the works mentioned above provide but a starting point for a more accurate modeling of the recording process.

2.3.4 Linear recording theory

The anhysteretic process has been shown to provide linear magnetization curves and so is employed in analog recording, where fidelity of transfer is an essential requirement. In biased recording systems, it is necessary to adjust the amplitude of both bias and signal fields to achieve the required linearity. From the considerations given earlier, it is clear that bias fields must be large and that, to avoid distortions, signal fields must be kept small and on a restricted part of the magnetization curve. When the medium has a rectangular hysteresis loop, the effect of the bias is to alternate the sense of the magnetization of the particles as long as its magnitude is above the coercivity, but when its magnitude falls below the coercivity, the particle magnetizations become frozen into certain directions. The resulting medium magnetization is proportional to the signal-field amplitude, assuming that the signal is of much lower frequency than the bias. This simple description can be envisaged as a basis for the explanation of the linear recording process (Westmijze, 1953; Daniel and Levine, 1960b; Tjaden and Leyten, 1964; Bertram, 1974, 1975).

Consider Fig. 2.55, which shows the contours of constant total field amplitude around a record-head gap. These have been calculated using the Karlqvist approximations and are shown to penetrate a medium moving from left -to -right past the head and in contact with it. If it is supposed that the contour labeled 0.6 is where the total field equals the coercivity, this element of tape will have its magnetization continuously reversed as long as it remains inside the contour, but frozen as it passes

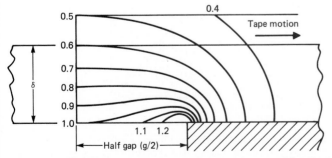

Figure 2.55 Total field contours near the gap of a record head *(Bertram, 1975).*

outside the contour. Once the magnetization is frozen, it will have a value determined by the susceptibility χ_{ar} associated with the anhysteresis curve corresponding to

$$M_x(x) = \chi_{ar}H_x(x) \tag{2.69}$$

It is assumed that the resultant field is effective in the biasing process, but that only the x component of signal field contributes to the output since the susceptibility is large in this, the orientation, direction but is low in a direction normal to it. To investigate the variation of magnetization with depth into the tape, it can be shown that H_x varies almost linearly with y along the coercivity contour and so takes the form

$$H_x(x) = \frac{y}{\delta} H_x(\delta) \tag{2.70}$$

and this result, in combination with Eq. (2.69), leads to a flux in the reproduce head of

$$\phi_0 = \mu_0 w \chi_{ar} \delta H_x(\delta) \frac{1 - e^{-k\delta}(1 + k\delta)}{(k\delta)^2} \tag{2.71}$$

where $k = 2\pi/\lambda$ is the wave number. The variation of this flux with reduced wave number $k\delta$ is shown in Fig. 2.56. For the purpose of comparison, the results obtained by considering only x components of bias field, which lead to x components of magnetization having constant amplitude through the depth of the tape, are also shown. The experimental results follow very closely the predictions obtained using vector fields,

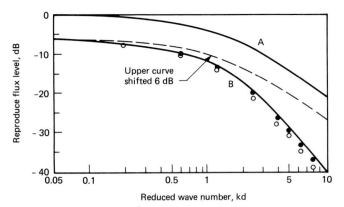

Figure 2.56 Reproduced flux as a function of a reduced wave number for biased sine-wave recording: (curve A) considering only the longitudinal component of bias field; (curve B) considering the resultant bias field *(Bertram, 1975).*

and diverge considerably from those using longitudinal fields. Thus the need for vector modeling is confirmed for linear as well as nonlinear recording processes.

Following on from Eq. (2.71), it is to be anticipated that the output-signal amplitude should be proportional to record-signal amplitude, and this is known to be true when recording is at low levels. Attempts to record large signals on tapes lead, however, to distortions in the output waveform, and so output levels need to be specified along with a measure of nonlinear distortion. The cause of distortion can be appreciated by referring to observations which show that a wide range of magnetic media satisfies a universal anhysteretic magnetization curve, provided the parameters of the curve are properly specified (Köster, 1975). Figure 2.57 shows the reduced anhysteretic magnetization $m_{ar} = M_{ar}/M_r$ as a function of reduced field $h = H/H_{0.5}$, where $H_{0.5}$ is the field required to make $m_{ar} = 0.5$. The experimental points all fall within the shaded area and can be fitted closely by a single empirical curve. The initial part of the curve is linear, with a normalized susceptibility $m_{ar}/h = 0.57$. An empirical expression evaluated for sinusoidal inputs leads to a predicted linearity, equivalent to less than 3 percent third harmonic distortion, being main-

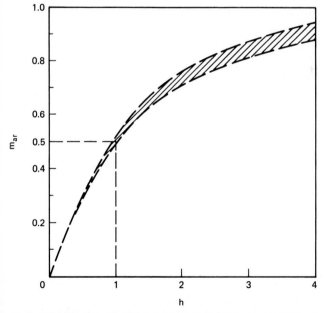

Figure 2.57   Universal anhysteretic magnetization curve derived from experimental results on a range of tapes. The shaded area accommodates all the experimental data from a wide variety of media *(after Köster, 1975)*.

tained up to $m = 0.5$, or 50 percent of saturation. Such predictions are confirmed by experiment (Köster, 1975). The initial susceptibility, defined by $\chi_{ar} = dM_{ar}/dH$, can be shown from Eq. (2.69) to be proportional to M_r. Substitution of this into Eq. (2.71) yields an output voltage, for a specified level of distortion, proportional to $M_r\delta$. This relationship has also been confirmed experimentally (Köster, 1975; van Winsum, 1984).

Regarding the optimization of bias to produce maximum signal amplitude, consider first the dependence of depth of recording into the medium and its relation to head field. In the theory outlined above, the bias field should be large enough to allow the coercivity contour of the head field to penetrate through to the back layer of the medium (Bertram, 1975). When the bias field is larger or smaller, it can be shown to produce lower outputs. Second, the influence of bias field on the initial susceptibility of the modified anhysteretic curves is to cause it to have a maximum at a particular value of bias field near to the coercivity. Thus, on both counts, a direct and linear relationship between optimum bias level and coercivity could be anticipated. Such a relationship has been both predicted and observed (Fujiwara, 1979; van Winsum, 1984).

APPENDIX I   The Reproducing Process: Reciprocity

Consider two coils (1, 2) linked by a mutual inductance L_m. A current i_1 in coil 1 will cause a flux ϕ_2 to thread coil 2 given by

$$\phi_2 = L_m i_1 \tag{2.72a}$$

In a similar way, the flux ϕ_1 threading coil 1 as a result of a current i_2 in coil 2 is given by

$$\phi_1 = L_m i_2 \tag{2.72b}$$

The reciprocal relationships portrayed by the above equations are an example of the principle of reciprocity and are used here as a starting point for the derivation of expressions for the output voltages generated in the coils of reproduce heads.

In the above equations, the mutual inductance is a common factor, and so

$$\frac{\phi_1}{i_2} = \frac{\phi_2}{i_1} \tag{2.73}$$

Now take coil 1 to represent the coil of the reproduce head and coil 2 to carry a solenoidal current representing a magnetized element of the recording medium. Then, considering only x components of the head field $H_x(x, y)$, the flux ϕ_2 is given by

$$\phi_2 = \mu_0 H_x(x, y)\, dy\, dz \tag{2.74}$$

where $dy\, dz$ is the area enclosed by coil 2. The current i_2 represents the x component of the magnetization, at point x' in the medium, via

$$i_2 = M_x(x')\, dx' \tag{2.75a}$$

$$i_2 = M_x(x - \bar{x})\, dx \tag{2.75b}$$

Use of this and Eq. (2.74) in Eq. (2.73) and manipulation of the resulting equation gives the flux in the reproduce-head coil as

$$\phi_1 = \mu_0 M_x(x - \bar{x})\, \frac{H_x(x, y)}{i_1}\, dx\, dy\, dz \tag{2.76}$$

Integration over the total volume of the recording medium leads to the total flux in the head coil

$$\phi_1 = \mu_0 \int_{-\infty}^{+\infty} \int_{d}^{d+\delta} \int_{-\infty}^{+\infty} M_x(x - \bar{x})\, \frac{H_x(x, y)}{i_1}\, dx\, dy\, dz \tag{2.77}$$

Noting that the reproduced voltage e_x is related to the flux by

$$e_x = -\frac{d\phi}{dt} \tag{2.78}$$

leads to an output voltage of

$$e_x(\bar{x}) = -\mu_0 V \int_{-\infty}^{+\infty} \int_{d}^{d+\delta} \int_{-\infty}^{+\infty} \frac{dM_x(x - \bar{x})}{d\bar{x}}\, \frac{H_x(x, y)}{i_1}\, dx\, dy\, dz \tag{2.79}$$

where the quantity $H_x(x, y)/i_1$ is the head field per unit current in the head winding should it be energized.

In a similar way, the output voltage arising from y components of magnetization in the medium is

$$e_y(\bar{x}) = -\mu_0 V \int_{-\infty}^{+\infty} \int_{d}^{d+\delta} \int_{-\infty}^{+\infty} \frac{dM_y(x - \bar{x})}{d\bar{x}}\, \frac{H_y(x, y)}{i_1}\, dx\, dy\, dz \tag{2.80}$$

APPENDIX II Standard Integrals

The following are standard integrals needed to complete certain integrations in the main text:

$$\int_{-\infty}^{+\infty} \frac{1}{c^2 + (\bar{x} - x)^2} \left(\arctan \frac{a + x}{y} + \arctan \frac{a - x}{y} \right) dx$$

$$= \frac{\pi}{c} \left(\arctan \frac{a + \bar{x}}{c + y} + \arctan \frac{a - \bar{x}}{c + y} \right) \tag{2.81}$$

$$\int_{-\infty}^{+\infty} \frac{b^2}{x^2 + b^2} \ln\left[(x - a)^2 + c^2\right] dx = \pi b \ln\left[a^2 + (b + c)^2\right] \tag{2.82}$$

$$\int_{-\infty}^{+\infty} \arctan \frac{x}{y}\, \frac{1}{c^2 + (b - x)^2}\, dx = \frac{\pi}{c} \arctan \frac{b}{c + y} \tag{2.83}$$

APPENDIX III   Demagnetizing Fields

In a magnetic material, there is no divergence of the induction **B**. The latter can be expressed as

$$\mathbf{B} = \mu_0(\mathbf{H} + \mathbf{M}) \tag{2.84}$$

while the condition of zero divergence is signified by

$$\nabla \cdot \mathbf{B} = 0 \tag{2.85}$$

Combining these two equations gives

$$\nabla \cdot \mathbf{H} = -\nabla \cdot \mathbf{M} \tag{2.86}$$

In magnetic field problems it is usual to express the field in terms of the gradient of a potential ϕ; i.e.,

$$\mathbf{H} = -\nabla\phi \tag{2.87}$$

so that introduction of it into Eq. (2.86) produces a version of Poisson's equation, namely,

$$\nabla^2\phi = \nabla \cdot \mathbf{M} \tag{2.88}$$

wherein the quantity $-\nabla \cdot \mathbf{M}$ takes on the significance of a magnetic charge density. Solution of Poisson's equation takes the form

$$\phi(r) = -\frac{1}{4\pi} \int_{V'} \frac{\nabla \cdot \mathbf{M}}{|r - r'|}\, dV' - \frac{1}{4\pi} \int_{A'} \frac{\sigma_m(r')}{|r - r'|}\, dA' \tag{2.89}$$

where the first term is the potential arising from the volume charges within the magnetic material, with the integration taking place over all these charges, while the second term is the potential arising from surface charges $\sigma_m(r')$, with the integration taking place over all the relevant surfaces of the material. The surface charge density can be obtained from

$$\sigma_m(r') = \mathbf{M} \cdot \mathbf{n} \tag{2.90}$$

where **n** is the outward normal to the surface.

Following from Eq. (2.89) the demagnetizing field is

$$H_d(r) = -\frac{1}{4\pi} \int_{V'} \frac{\nabla \cdot \mathbf{M}}{|r - r'|^2}\, dV' + \frac{1}{4\pi} \int_{A'} \frac{\sigma_m(r')}{|r - r'|^2}\, dA' \tag{2.91}$$

Use of the above equation in any generalized way is extremely difficult. However, certain particular cases of interest in this work are now discussed.

First, consider a magnetization which has only x components varying with the x dimension and giving rise to no surface charges. Then

$$\nabla \cdot \mathbf{M} = \frac{dM_x}{dx} \tag{2.92}$$

Use of this in Eq. (2.91) would give rise to a field having, in general, x, y, and z components. Considering only the x components yields

$$H_{dx}(x, y, z) = -\frac{1}{4\pi} \int \int \int \frac{dM_x}{dx'} \frac{(x - x')\, dx'\, dy'\, dz'}{|(x - x')^2 + (y - y')^2 + (z - z')^2|^{3/2}} \tag{2.93}$$

and when the transverse dimension is so large that the integration over z' may take place from $-\infty$ to $+\infty$, Eq. (2.93) becomes

$$H_{dx}(x, y) = -\frac{1}{2\pi} \int \int \frac{dM_x}{dx} \frac{(x - x')\, dx'\, dy'}{(x - x')^2 + (y - y')^2} \tag{2.94}$$

The second case occurs when there are no volume charges and only y components of magnetization. Then $dM_y/dy = 0$, but there are surface charges of magnitude M_y. The y component of magnetic field is then, using Eq. (2.91),

$$H_{dy}(x, y, z) = \frac{1}{4\pi} \int \int M_y \frac{(y - y')\, dx'\, dz'}{|(x - x')^2 + (y - y')^2 + (z - z')^2|^{3/2}} \tag{2.95}$$

Again, when the z dimension is infinite, this becomes

$$H_{dy}(x, y) = \frac{1}{2\pi} \int M_y \frac{(y - y')\, dx'}{(x - x')^2 + (y - y')^2} \tag{2.96}$$

References

Baird, A. W., "An Evaluation and Approximation of the Fan Equations Describing Magnetic Fields Near Recording Heads," *IEEE Trans. Magn.*, **MAG-16**, 1350 (1980).

Barkouki, M. F., and I. Stein, "Theoretical and Experimental Evaluation of RZ and NRZ Record Characteristics," *IEEE Trans. Elec. Comput.*, **EC-12**, 92 (1961).

Bate, G., "Statistical Stability of the Preisach Diagram for Particles of Gamma-Ferric-Oxide," *J. Appl. Phys.*, **33**, 263 (1962).

Bauer, B. B., and C. D. Mee, "A New Model for Magnetic Recording," *IRE Trans. Audio*, **AU-9**, 139 (1961).

Beardsley, I. A., "Effect of Particle Orientation on High Density Recording," *IEEE Trans. Magn.*, **MAG-18**, 1191 (1982a).

Beardsley, I. A., "Self Consistent Recording Model for Perpendicularly Oriented Media, *J. Appl. Phys.*, **53**, 2582 (1982b).

Bertram, H. N., "Monte Carlo Calculation of Magnetic Anhysteresis," *J. Phys.*, **32**, 684 (1971).

Bertram, H. N., "Long Wavelength AC Bias Recording Theory," *IEEE Trans. Magn.*, **MAG-10**, 1039 (1974).

Bertram, H. N., "Wavelength Response in AC Biased Recording," *IEEE Trans. Magn.*, **MAG-11**, 1176 (1975).

Bertram, H. N., "Geometric Effects in the Magnetic Recording Process," *IEEE Trans. Magn.*, **MAG-20**, 468 (1984a).

Bertram, H. N., "The Effect of Angular Dependence of the Particle Nucleation Field on the Magnetic Recording Process," *IEEE Trans. Magn.*, **MAG-20**, 2094 (1984b).

Bertram, H. N., and L. D. Fielder, "Amplitude and Bit Shift Spectra Comparisons in Thin Metallic Media," *IEEE Trans. Magn.*, **MAG-19**, 1605 (1983).

Bertram, H. N., and J. C. Mallinson, "Theoretical Coercivity Field for an Interacting Anisotropic Dipole Pair of Arbitrary Bond Angle," *J. Appl. Phys.*, **40**, 1301 (1969).

Bertram, H. N., and J. C. Mallinson, "Switching Dynamics for an Interacting Anisotropic Dipole Pair of Arbitrary Bond Angle," *J. Appl. Phys.*, **41**, 1102 (1970).

Bertram, H. N., and R. Niedermeyer, "The Effect of Spacing on Demagnetization in Magnetic Recording," *IEEE Trans. Magn.*, **MAG-18,** 1206 (1982).

Biorci, A., and D. Pescetti, "An Analytical Theory of the Behavior of Ferromagnetic Materials," *Il Nuovo Cimento,* **7,** 829 (1958).

Bonyhard, P. I., A. V. Davies, and B. K. Middleton, "A Theory of Digital Magnetic Recording on Metallic Films," *IEEE Trans. Magn.*, **MAG-2,** 1 (1966).

Bromley, D. J., "A Comparison of Vertical and Longitudinal Magnetic Recording Based on Analytical Models," *IEEE Trans. Magn.*, **MAG-19,** 2239 (1983).

Chantrell, R. W., A. Lyberatos, and E. P. Wohlfarth, "Anhysteretic Properties of Interacting Magnetic Tape Particles," *J. Appl. Phys.*, **55,** 2223 (1984).

Chapman, D. W., "Theoretical Limit on Digital Magnetic Recording Density," *Proc. IEEE,* **51,** 394 (1963).

Cole, R. W., R. I. Potter, C. C. Lin, K. L. Deckert, and E. P. Valstyn, "Numerical Analysis of the Shielded Magneto-Resistive Head," *IBM J. Res. Dev.*, **18,** 551 (1974).

Daniel, E. D., and P. E. Axon, "The Reproduction of Signals Recorded on Magnetic Tape," *Proc. IEEE,* **100,** Pt. III, 157 (1956).

Daniel, E. D., and I. Levine, "Experimental and Theoretical Investigation of the Magnetic Properties of Iron Oxide Recording Tape," *J. Acoust. Soc. Am.*, **32,** 1 (1960*a*).

Daniel, E. D., and I. Levine, "Determination of the Recording Performance of a Tape from its Magnetic Properties," *J. Acoust. Soc. Am.*, **32,** 258 (1960*b*).

Davies, A. V., and B. K. Middleton, "The Resolution of Vertical Magneto-Resistive Readout Heads," *IEEE Trans. Magn.*, **MAG-11,** 1689 (1975).

Davies, A. V., B. K. Middleton, and A. C. Tickle, "Digital Recording Properties of Evaporated Cobalt Films," *IEEE Trans. Magn.*, **MAG-1,** 344 (1965).

Dressler, D. D., and J. H. Judy, "A Study of Digitally Recorded Transitions in Thin Magnetic Films," *IEEE Trans. Magn.*, **MAG-10,** 674 (1974).

Eldridge, D. F., "Magnetic Recording and Reproduction of Pulses," *IRE Trans. Audio,* **7,** 141 (1960).

Fan, G. J. "Analysis of a Practical Perpendicular Head for Digital Purposes," *J. Appl. Phys.*, **31,** 402s (1960).

Fan, G. J., "A Study of the Playback Process of a Magnetic Ring Head," *IBM J. Res. Dev.*, **5,** 321 (1961).

Fayling, R. E., "Studies of the 'Magnetization Region' of a Ring-Type Head," *IEEE Trans. Magn.*, **MAG-18,** 1212 (1982).

Fujiwara, T., "Non-Linear Distortion in Long Wavelength AC Bias Recording," *IEEE Trans. Magn.*, **MAG-15,** 894 (1979).

Gronau, M., H. Goeke, D. Schuffler, and S. Sprenger, "Correlation Between Domain Wall Properties and Material Parameters in Amorphous SmCo Films," *IEEE Trans. Magn.*, **MAG-19,** 1653 (1983).

Hoagland, A. S., *Digital Magnetic Recording,* Wiley, New York, 1963.

Hokkyo, J., K. Hayakawa, I. Saito, S. Satake, K. Shirane, N. Honda, T. Shimamura, and T. Saito, "Reproducing Characteristics of Perpendicular Magnetic Recording," *IEEE Trans. Magn.*, **MAG-18,** 1203 (1982).

Iwasaki, S., "Perpendicular Magnetic Recording," *IEEE Trans. Magn.*, **MAG-16,** 71 (1980).

Iwasaki, S., "Perpendicular Magnetic Recording—Evolution and Future," *IEEE Trans. Magn.*, **MAG-20,** 607 (1984).

Iwasaki, S., and Y. Nakamura, "An Analysis for the Magnetization Mode for High Density Magnetic Recording," *IEEE Trans. Magn.*, **MAG-13,** 1272 (1977).

Iwasaki, S., and K. K. Ouchi, "CoCr Recording Films with Perpendicular Magnetic Anisotropy," *IEEE Trans. Magn.*, **MAG-14,** 849 (1978).

Iwasaki, S., and K. Takemura, "An Analysis for the Circular Mode of Magnetization in Short-Wavelength Recording," *IEEE Trans. Magn.*, **MAG-11,** 1173 (1975).

Iwasaki, S., Y. Nakamura, and K. Ouchi, "Perpendicular Magnetic Recording with a Composite Anisotropy Film," *IEEE Trans. Magn.*, **MAG-15,** 1456 (1979).

Iwasaki, S., Y. Nakamura, and H. Muraoka, "Wavelength Response of Perpendicular Magnetic Recording," *IEEE Trans. Magn.*, **MAG-17,** 2535 (1981).

Iwasaki, S., D. E. Speliotis, and Y. Yamamoto, "Head-to-Media Spacing Losses in Perpendicular Recording," *IEEE Trans. Magn.*, **MAG-19**, 1626 (1983).
Jacobs, I. S., and C. P. Bean, "An Approach to Elongated Fine Particle Magnets," *Phys. Rev.*, **100**, 1060 (1955).
Jeschke, J. C., East German Patent 8684, 1954.
Jorgensen, F., *The Complete Handbook of Magnetic Recording*, TAB Books, Blue Ridge Summit, Penn., 1980.
Kakehi, A., M. Oshiki, T. Aikawa, M. Sasaki, and T. Kozai, "A Thin Film Head for High Density Recording," *IEEE Trans. Magn.*, **MAG-18**, 1131 (1982).
Karlqvist, O., "Calculation of the Magnetic Field in the Ferromagnetic Layer of a Magnetic Drum," *Trans. R. Inst. Technol. (Stockholm)*, **86**, 3 (1954).
Knowles, J. E., "Measurements on Single Magnetic Particles," *IEEE Trans. Magn.*, **MAG-14**, 858 (1978).
Knowles, J. E., "Magnetic Properties of Individual Acicular Particles," *IEEE Trans. Magn.*, **MAG-17**, 3008 (1981).
Köster, E., "A Contribution to Anhysteretic Remanence and AC Bias Recording," *IEEE Trans. Magn.*, **MAG-11**, 1185 (1975).
Krones, F., *Technik der Magnetspeicher*, Springer, Berlin, 1960, p. 474.
Lemke, J. U., "Ultra-High Density Recording with New Heads and Tapes," *IEEE Trans. Magn.*, **MAG-15**, 1561 (1979).
Lemke, J. U., "An Isotropic Particulate Medium with Additive Hilbert and Fourier Field Components," *J. Appl. Phys.*, **53**, 2361 (1982).
Lopez, O., "Reproducing Vertically Recorded Information—Double Layer Media," *IEEE Trans. Magn.*, **MAG-19**, 1614 (1983).
Lopez, O., "Analytic Calculation of Write Induced Separation Losses," *IEEE Trans. Magn.*, **MAG-20**, 715 (1984).
Lübeck, H., "Magnetische Schallaufzeichrung mit Filmen und Ringkopfen," *Akust. Z.*, **2**, 273 (1937).
Maller, V. A. J., and B. K. Middleton, "A Simplified Model of the Writing Process in Saturation Magnetic Recording," *IERE Conf. Proc.*, **26**, 137 (1973).
Maller, V. A. J., and B. K. Middleton, "A Simplified Model of the Writing Process in Saturation Magnetic Recording," *Radio Electron. Eng.*, **44**, 281 (1974).
Mallinson, J. C., and C. W. Steele "Theory of Linear Superposition in Tape Recording," *IEEE Trans. Magn.*, **MAG-5**, 886 (1969).
Mallinson, J. C., "A Unified View of High Density Digital Recording Theory," *IEEE Trans. Magn.*, **MAG-11**, 1166 (1975).
Mallinson, J. C., "The Next Decade in Magnetic Recording," *IEEE Trans. Magn.*, **MAG-21**, 1217 (1985).
Mallinson, J. C., and H. N. Bertram, "Theoretical and Experimental Comparison of the Longitudinal and Vertical Modes of Magnetic Recording," *IEEE Trans. Magn.*, **MAG-20**, 461 (1984a).
Mallinson, J. C., and H. N. Bertram, "On the Characteristics of Pole-Keeper Head Fields," *IEEE Trans. Magn.*, **MAG-20**, 721 (1984b).
Mee, C. D., *The Physics of Magnetic Recording*, North-Holland, Amsterdam, 1964.
Middleton, B. K., "The Dependence of Recording Characteristics of Thin Metal Tapes on Their Magnetic Properties and on the Replay Head," *IEEE Trans. Magn.*, **MAG-2**, 225 (1966).
Middleton, B. K., "Performance of a Recording Channel," *IERE Conf. Proc.*, **54**, 137 (1982).
Middleton, B. K., and A. V. Davies, "Gap Effects in Head Field Distributions and the Replay Process in Longitudinal Recording," *IERE Conf. Proc.*, **59**, 27 (1984).
Middleton, B. K., and P. L. Wisely, "The Development and Application of a Simple Model of Digital Magnetic Recording to Thick Oxide Media," *IERE Conf. Proc.*, **35**, 33 (1976).
Middleton, B. K., and P. L. Wisely, "Pulse Superposition and High Density Recording," *IEEE Trans. Magn.*, **MAG-14**, 1043 (1978).
Middleton, B. K., and C. D. Wright, "Perpendicular Recording," *IERE Conf. Proc.*, **54**, 181 (1982).

Middleton, B. K., and C. D. Wright, "The Perpendicular Recording Process," *IEEE Trans. Magn.*, **MAG-20**, 458 (1984).

Middleton, B. K., and C. D. Wright, "Perpendicular Recording: The Replay Process," to be published (1986).

Miyata, J. J., and R. R. Hartel, "The Recording and Reproduction of Signals on Magnetic Medium Using Saturation-type Recording," *IRE Trans. Elect. Comp.*, **EC-8**, 159 (1959).

Monson, J. E., R. Fung, and A. S. Hoagland, "Large Scale Model Studies of Vertical Recording," *IEEE Trans. Magn.*, **MAG-17**, 2541 (1981).

Morrison, J. R., and D. E. Speliotis, "Study of Peak Shift in Thin Recording Surfaces," *IEEE Trans. Magn.*, **MAG-3**, 208 (1967).

Moskovitz, R., and E. Della Torre, "Hysteretic Magnetic Dipole Interaction Model," *IEEE Trans. Magn.*, **MAG-3**, 579 (1967).

Nakamura, Y., and S. Iwasaki, "Reproducing Characteristics of Perpendicular Magnetic Head," *IEEE Trans. Magn.*, **MAG-18**, 1167 (1982).

Néel, L., "Some Theoretical Aspects of Rock Magnetism," *Philos. Mag. Suppl. (Adv. Phys.)* **4**, 191 (1955).

Néel, L., "Sur les Éffets de Couplage entre Grains Ferromagnetique Doués Hysteresis," *C. R. Acad. Sci. (Paris)*, **246**, 2313 (1958).

Nishikawa, M., "Characteristics of the Readback Signal in Digital Magnetic Recording," *IEEE Trans. Magn.*, **MAG-6**, 811 (1970).

Ortenburger, I. B., and R. I. Potter, "A Self-Consistent Calculation of the Transition Zone in Thick Particulate Media," *J. Appl. Phys.*, **50**, 2393 (1979).

Potter, R. I., "Analysis of Saturation Magnetic Recording Based on Arctangent Magnetization Transitions," *J. Appl. Phys.*, **41**, 1647 (1970).

Potter, R. I., "Digital Magnetic Recording Theory," *IEEE Trans. Magn.*, **MAG-10**, 502 (1974).

Potter, R. I., and I. A. Beardsley, "Self-Consistent Computer Calculations for Perpendicular Magnetic Recording," *IEEE Trans. Magn.*, **MAG-16**, 967 (1980).

Potter, R. I., R. J. Schmulian, and K. Hartmann, "Fringe Field and Readback Voltage Computations for Finite Pole-Tip Length Recording Heads," *IEEE Trans. Magn.*, **MAG-7**, 689 (1971).

Preisach, F., "Magnetic After-Effect," *Z. Phys.*, **94**, 277 (1935).

Quak, D., "Influence of the Layer Thickness of a Double Layer Medium on the Reproduced Signal in Perpendicular Recording," *IEEE Trans. Magn.*, **MAG-19**, 1502 (1983).

Satake, S., K. Hayakawa, J. Hokkyo, and T. Simamura, "Field Theory of Twin Pole Head and a Computer Simulation Model for a Perpendicular Recording," *IECE Tech. Group Meeting Magn. Rec.*, **MR77-26**, 33 (1977).

Schwantke, G., "Der Aufsprechvongang beim Magnetton ver fahren in Preisach-Darstellung," *Frequenz*, **12**, 383 (1958).

Schwantke, G., "The Magnetic Tape Recording Process in Terms of the Preisach Representation," *J. Audio Eng. Soc.*, **9**, 37 (1961).

Sebestyen, L. E., *Digital Magnetic Tape Recording for Computer Applications*, Chapman and Hall, London, 1973.

Soohoo, R. F., and K. Ramachandran, "Switching Dynamics of an Assembly of Interacting Anisotropic Ferromagnetic Particles," *AIP Conf. Proc.*, **18**, 1098 (1974).

Speliotis, D. E., "Digital Recording Theory," *Ann. N.Y. Acad. Sci.*, **189**, 21 (1972).

Speliotis, D. E., and J. Morrison, "A Theoretical Analysis of Saturation Magnetic Recording," *IBM J. Res. Dev.*, **10**, 233 (1966a).

Speliotis, D. E., and J. R. Morrison, "Correlation between Magnetic and Recording Properties of Thin Surfaces," *IEEE Trans. Magn.*, **MAG-2**, 208 (1966b).

Stoner, E. C., and E. P. Wohlfarth, "Mechanism of Magnetic Hysteresis in Heterogeneous Alloys," *Proc. R. Soc. (London)*, **A240**, 599 (1948).

Stubbs, D. P., J. W. Whisler, C. D. Moe, and J. Skorjanec, "Ring Head Recording on Perpendicular Media: Output Spectra for CoCr and CoCr/CoNi Media," *J. Appl. Phys.*, **57**, 3970 (1985).

Suzuki, T., and S. Iwasaki, "Magnetization Transitions in Perpendicular Magnetic Recording," *IEEE Trans. Magn.*, **MAG-18**, 769 (1982).

Talke, F., and R. C. Tseng, "Effect of Submicrometer Transducer Spacing on the Readback Signal in Saturation Recording," *IBM J. Res. Dev.*, **19**, 591 (1975).

Teer, K., "Investigation of the Magnetic Recording Process with Step Functions," *Philips Res. Rep.*, **16**, 469 (1961).

Tjaden, D. L. A., "Some Notes on Superposition in Digital Magnetic Recording," *IEEE Trans. Magn.*, **MAG-9**, 331 (1973).

Tjaden, D. L. A., and J. Leyten, "A 5000:1 Scale Model of the Magnetic Recording Process," *Philips Tech. Rev.*, **25**, 319 (1964).

Tjaden, D. L. A., and E. J. Tercic, "Theoretical and Experimental Investigations of Digital Magnetic Recording on Thin Media," *Philips Res. Rep.*, **30**, 120 (1975).

Valstyn, E. P., and L. F. Shew, "Performance of Single-Turn Film Heads," *IEEE Trans. Magn.*, **MAG-9**, 317 (1973).

van Winsum, J. A., "Effect of Orientation Ratio and Powder Properties on Anhysteretic Linearity of Magnetic Tape Coatings," *IEEE Trans. Magn.*, **MAG-20**, 87 (1984).

Wallace, R. L., "The Reproduction of Magnetically Recorded Signals," *Bell Syst. Tech. J.*, **30**, 1145 (1951).

Walther, E. L., "Digital Signal Response of Magnetic Recording Heads," *Philips Res. Rep.*, **19**, 281 (1964).

Wang, H. S. C., "Gap Loss Function of Certain Critical Parameters in Magnetic Data Recording Instruments and Storage Systems," *Rev. Sci. Inst.*, **37**, 1124 (1966).

Westmijze, W. K., "Studies on Magnetic Recording," *Philips Res. Rep.*, **8**, 161 (1953).

Wielinga, T., J. H. J. Fluitman, and J. C. Lodder, "Perpendicular Standstill Recording in CoCr Films," *IEEE Trans. Magn.*, **MAG-19**, 94 (1983).

Williams, M. L., and R. L. Comstock, "An Analytical Model of the Write Process in Digital Magnetic Recording," *A.I.P. Conf. Proc.*, **5**, 738 (1972).

Wohlfarth, E. P., "A Review of the Problem of Fine Particle Interactions with Special Reference to Magnetic Recording," *J. Appl. Phys.*, **35**, 783 (1976).

Woodward, J. G., and E. Della Torre, "Particle Interaction in Magnetic Recording Tapes," *J. Appl. Phys.*, **31**, 56 (1960).

Woodward, J. G., and E. Della Torre, "Particle Interaction in Magnetic Recording Tapes," *J. Appl. Phys.*, **32**, 126 (1961).

Wright, C. D., and B. K. Middleton, "The Perpendicular Record and Replay Processes," *IERE Conf. Proc.*, **59**, 9 (1984).

Wright, C. D., and B. K. Middleton, "Analytical Modeling of Perpendicular Recording," *IEEE Trans. Magn.*, **MAG-21**, 1398 (1985).

Yamamori, K., R. Nishikawa, T. Asano, and T. Fujiwara, "Perpendicular Magnetic Recording Performance of Double Layer Media," *IEEE Trans. Magn.*, **MAG-17**, 2538 (1981).

Yoshida, K., T. Okuwaki, N. Osakake, H. Tanabe, Y. H. Onuchi, T. Matsuda, K. Shinagawa, A. Tonomura, and H. Fujiwara, "Observation of Recorded Magnetization Patterns by Electron Holography," *IEEE Trans. Magn.*, **MAG-19**, 1600 (1983).

Chapter

3

Recording Media

Eberhard Köster*

BASF, Aktiengesellschaft, Ludwigshafen, West Germany

Thomas C. Arnoldussen†

IBM Corporation, San Jose, California

3.1 Types of Recording Media

Historically, magnetic recording media evolved from metal wires and tapes to magnetic oxide particles held in an organic binder and carried by a polymer tape substrate. This design of recording medium has the advantage that the substrate and the magnetic layer can be independently optimized. Furthermore, the magnetic layer itself can be tailored for optimum magnetic and mechanical properties by adjustment of the magnetic particles and the binder system. Particulate recording tape continues to be the dominant tape medium, with countless refinements introduced over the last several decades. As computer use grew, a new need arose for rapid access to stored data. The sequential access of information stored on tape was too slow for many applications. Plated metal films on spinning drums were used for a time. A read-write head could quickly be moved to any

---

*Sections on particulate recording media.

†Sections on film recording media.

spot on the drum, providing improved access speed at the expense of storage capacity. The use of rotating rigid disks replaced drum storage since they could be more easily coated, and disks can be stacked and recorded on two sides, thus increasing the volume storage capacity compared with drums. Although metal and oxide films were considered for disk media in the early 1960s, particulate media dominate the disk as well as the tape industry. With the appearance of mini- and microcomputers, a third form of magnetic storage grew to major proportions, namely the flexible (floppy) disk. This serves the role of providing rugged, low-cost, transportable storage with relatively high capacity and rapid access. Again particulate technology is dominant.

Although particulate magnetic coatings have been the mainstay of media technology, it is generally recognized that the highest storage densities can be achieved by using (probably metal) films. The broad application of film media remains in the future, however, because particulate media have not reached their ultimate potential, and because some of the problem areas of thin-film technology have yet to see economical solutions.

3.1.1 Storage formats

Magnetic media of the three general types (tape, rigid disks, and flexible disks) have played major roles in the storage of information. Applications using magnetic stripes and cards also continue in special areas. Tape is the oldest form and remains an important medium today. Its main use in computer applications is for archival storage and mass memory systems. Tape media are, however, central to most audio, video, and instrumentation recording applications. These three recording applications have historically used linear or frequency-modulation techniques for recording analog information on a tape, in contrast to the digital information stored by a computer tape. Audio, video, and instrumentation recorders are, however, evolving toward the use of digital techniques, so that computer and noncomputer requirements are converging.

The second major class of media, the rigid disk, is designed to provide rapid access to data files which need to be called up frequently. Rigid-disk drives are designed to operate at high rotational speeds, typically in the range of 3000 to 5000 rpm, to minimize the "latency time" which elapses before a piece of data passes the head again. These drives are also operated continuously to eliminate the time for getting up to speed. Rigid substrates are needed for such high-speed operation to avoid hazardous mechanical resonances as well as mechanical distortion of the disk. Because of the high speed and rigidity of the disk, the head is designed to "fly" above the disk surface (on the order of 0.30 μm in advanced drives). Flying the head minimizes disk and head wear, which would severely shorten the life of the file if continuous head-disk contact were permitted. Tapes and flexible disks can tolerate such contact because of the slower speeds, mechanical compliance of the media, and lower duty cycles.

The flexible disk, like the rigid disk, permits direct access to data files. The substrate material is essentially the same as is used for tapes, only thicker. A magnetic coating is applied to both sides of a polyester base out of which disks are stamped and then packaged in a protective envelope. Flexible disks are not capable of as high a track density as rigid media, yet they fill a very significant need for easily transportable, direct-access storage of moderate capacity and low cost. The primary market niche is for personal computers, where rapid access is important but instant access unnecessary. The flexible disk also serves as a very convenient means of distributing programs and data.

3.1.2 Substrates

Standard substrate materials for recording media are biaxially oriented polyethylene terephthalate for flexible media and high-purity aluminum-(4–5%) magnesium alloys for rigid disks. Other substrate materials being explored include polyimides and polyimide-polyamide copolymers for flexible media, and glass and ceramics for rigid disks. New flexible media substrate development stems from the need for improved dimensional and thermal stability. Ultraviolet and electron-beam cross-polymerization of substrate materials is being explored, as well as laminated structures. The efforts to develop new rigid substrate materials originate in the need for smoother, more uniform surfaces to allow closer head-to-media spacings, as well as reduced substrate-induced noise and dropouts. Glass and ceramics have been considered for many years but have yet to appear in any product. Besides cost, the main concern has been possible brittle fracture during file assembly or operation. Nevertheless, the ability to obtain extremely fine surface finishes on such substrates gives them continued appeal.

3.1.3 Particulate media

The most commonly used magnetic media are particles of gamma ferric oxide. These are normally acicular, with a saturation magnetization of about 350 kA/m (emu/cm^3). When diluted by an organic binder to a volume fraction of 20 to 45 percent, the coating magnetization is accordingly smaller. The coercivity of a coating is in the range 23 to 32 kA/m (290 to 400 Oe) if the acicular particles are aligned by a magnetic field during the coating process, and smaller if the particles are randomly oriented. Tape and rigid-disk applications generally call for particle alignment to improve the recording properties. Flexible disks, on the other hand, are obliged to use random orientation, so that large web substrates can be coated, from which the disks are subsequently stamped.

Other particles that are being used for more advanced applications include cobalt-modified γ-Fe_2O_3, chromium dioxide, and metal particles. The properties which are sought in these particles are higher coercivity, smaller particle size, or higher magnetization. The chief advantage of γ-Fe_2O_3 is its proven long-term stability. Often the alternative particles are accompanied by thermal or chemical instabilities which can compromise their advantages and limit their application.

In addition to magnetic particles, nonmagnetic particles such as Al_2O_3 are often added to the coating to improve the tape or disk abrasion resistance. Their use can, however, lead to noise and magnetic or mechanical defects. Liquid or solid lubricants are also commonly used to reduce friction and wear.

3.1.4 Film media

Thin magnetic films which have been explored for recording media are usually cobalt-based alloys, sputtered iron oxide being the principal exception. The cobalt alloys often possess a hexagonal crystal structure in which magnetocrystalline anisotropy aids in achieving a high coercivity. These films have been electroplated, chemically plated, evaporated, and sputtered. Most of the past work has focused on longitudinal media in which the easy axis of magnetization lies in the plane of the film. Such films range in thickness from 25 to 200 nm. Cobalt-chromium sputtered alloys have more recently gained attention for perpendicular media in which the easy axis is normal to the film plane.

The saturation magnetization of deposited films can range from as low as 240 kA/m (emu/cm^3) for sputtered γ-Fe_2O_3 to over 1000 kA/m for cobalt-iron alloys. The coercivity can range from under 32 to 160 kA/m (400 to 2000 Oe), depending on the deposition process and the material. The chief advantage of deposited films is the higher magnetization, which allows the use of thinner recording layers while maintaining signal amplitude. Thin recording layers lead to a better-defined magnetization reversal and consequently to higher recording densities. Film technology also makes possible the tailoring of magnetic properties to meet specific design requirements. Unlike a particulate coating, whose magnetic properties are largely determined by the type of particle used, a film of a given composition can be made to be isotropic or anisotropic, of high or low coercivity, of in-plane or perpendicular easy axis, by modifying the deposition process or substrate preparation.

Film media, because they are thin, require nonmagnetic processing or coating layers which are unnecessary with particulate media. In rigid-disk media, for example, an underlayer is usually deposited on the substrate first to improve the impact resistance of the total structure, and an over-coat is applied after the magnetic layer to provide abrasion resistance. Nonmagnetic wear or load-bearing particles cannot be used as with particulate coatings. Similarly, lubricant cannot be retained in a thin solid film as it can in an organic resin particulate coating. Therefore, new concepts in lubrication or elimination of lubricant are necessary for film media.

3.1.5 Requirements of magnetic recording media

Regardless of the application (audio, video, or data) and the type of recording (analog or digital pulse), the direction of progress is toward higher information density. In some system designs, linear density (along the recording track) may be favored for optimization, while in others, track density (number of tracks per width of tape or band of disk) may be favored. However, at a more global level it is higher areal density or even volume density which is the goal.

Digital pulse recording is a convenient format for evaluating the potential storage density of a medium. In this mode of recording, information is written as magnetization reversals along a track, which are read back as electrical pulses in the windings of an inductive head. The sharper the reversals, the slimmer the pulses and the greater the linear density capability. As the reversals are written closer, they destructively interfere with each other, causing the readback pulse to diminish in amplitude and shift in position. At sufficiently close spacing the distinction between sinusoidal and pulse recording vanishes. Hence pulse width is a good measure of linear information density capability, for pulse or sinusoidal recording.

The sharpness of a magnetic transition in a longitudinal medium is found to be proportional to an expression of the form $(M_r\delta/H_c)^n$, where M_r is the remanent magnetization, δ is the magnetic coating thickness, H_c is the coercivity, and n is between 0.5 and 1 (see Chap. 2), In any event, small pulse lengths and high linear density call for small $M_r\delta$ and/or high H_c, but the former diminishes the signal amplitude, while the latter limits the ability of the head to write on the medium. The choice of magnetization, thickness, and coercivity therefore depends on the linear density design goals and cannot be made without considering the head design and practical head-to-medium spacing.

Designing for high linear density cannot be the sole objective. Track density and adequate signal must also be considered. Signal amplitude is proportional to the track width as well as $M_r\delta$. If the linear density is increased by reducing $M_r\delta$, the track width needs to be increased (and track density decreased) to maintain a given signal level. In this sense, linear density and track density are traded against each other, unless the minimum required signal can be lowered by reducing noise. On the other hand, raising the linear density by increasing H_c does not suffer this trade-off. Similarly, decreasing the head-to-medium spacing allows simultaneous increases in linear and track densities.

Over the years, most of the areal density gains have been achieved by a combination of lowering system noise, decreasing coating thickness, and reducing flying height (for rigid disks). This is because for $\gamma\text{-Fe}_2\text{O}_3$, the primary recording medium, M_r and H_c are substantially fixed by materials properties and pigment-loading constraints. For this reason, as practical limits of coating thickness, flying height, and noise reduction are approached, it becomes more important to explore the use of alternative particles and deposited films with higher coercivities and magnetizations.

Besides the basic magnetic requirements outlined above, the most important properties needed in advanced media are of a mechanical nature. Substrate and coating surfaces must be smooth to reduce noise in all media and to reduce head-to-medium spacing on rigid disks. Mechanical wear resistance is highly important for all types of media, but especially so for thin films or thin particulate coatings. Magnetic uniformity in film media may be governed by compositional or grain size uniformity, which is the counterpart to pigment dispersion uniformity in particulate coatings; in both cases inhomogeneities result in noise. Gains in these mechanical areas can be ranked equal in importance to gains in magnetic properties.

3.2   Particulate Media

3.2.1   Magnetic anisotropy and the magnetization process of fine particles

3.2.1.1   Magnetization.   The spontaneous magnetization of ferromagnetic or ferrimagnetic materials is one of the two fundamentals of any magnetic material used for magnetic recording. The other is the magnetic anisotropy, which allows some of the magnetization acquired in an applied field to remain even after the field is removed. The remanent magnetization, or remanence, constitutes the basis for a magnetic storage device.

As an intrinsic bulk property, the spontaneous magnetization is the parallel alignment of the uncompensated magnetic spin moments of ferromagnetic elements such as iron, cobalt, and nickel, or of the metal ions in ferromagnetic compounds such as chromium dioxide or europium sulfide. The ferrimagnetic oxides have a more complicated magnetic structure. For many ferrites the crystal structure is inverse spinel, which can be described by a unit cell with a cubic close packing of 32 oxygen ions in which di- and trivalent metal ions occupy certain interstices. Each unit cell contains eight [MFe_2O_4], where M is a divalent metal ion. The trivalent iron ions occupy eight tetrahedral and eight octahedral sites with an antiferromagnetic alignment, that is, with opposite direction of their spin moments. Their resultant magnetic moment is zero, and only the divalent ions which are located on eight additional octahedral sites contribute to the spontaneous magnetization. Thus, the magnetic moment decreases in steps of one unit from five to one Bohr magnetons in the order of Mn, Fe, Co, Ni, and Cu as divalent ions in the spinel ferrites. Another group of ferrites has a hexagonal lattice of oxygen ions with some ions substituted by Ba, Sr, Pb, or Ca ions; they may also contain interstitial divalent ions.

The spontaneous alignment of the spin moments is due to quantum-mechanical exchange forces (Smit and Wijn, 1959; Kneller, 1962; Chikazumi, 1964; Tebble and Craig, 1969). Any deviation from this alignment gives rise to an increase of the free exchange-energy density

$$F_A = A \left(\frac{d\phi}{dx}\right)^2 \tag{3.1}$$

where A is the exchange-energy constant, which is of the order 10^{-11} J/m

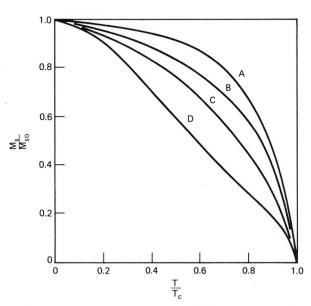

Figure 3.1 Normalized spontaneous magnetization M_s/M_0, where M_0 is the spontaneous magnetization 0°K, versus normalized temperature T/T_c, where T_c is the Curie temperature, for (curve A) ferromagnetic metals, (curve B) Fe_3O_4, (curve C) $CoFe_2O_3$, and (curve D) $BaFe_{12}O_{19}$.

$(10^{-6}$ erg/cm), and $d\phi/dx$ is the gradient of the angle ϕ between the individual spins. At zero absolute temperature, the spontaneous magnetization is given by the number of uncompensated Bohr magnetons. At higher temperatures the parallel alignment of the spins is increasingly disturbed by thermal fluctuations. Thus, the spontaneous magnetization, given by the mean value of the spin moments in zero field, decreases monotonically with increasing temperature until it becomes zero at the Curie temperature. This is shown in Fig. 3.1 for various magnetic materials. In mixed ferrites, the different sublattices may have different rates of fall of the spontaneous magnetization, leading to an intermediate compensation point with zero magnetization. In Table 3.1, the spontaneous magnetization of various bulk materials is listed, together with other intrinsic room temperature properties of interest.

In slowly varying magnetic fields, no further basic differences between ferro- and ferrimagnetic materials are observed. In ac fields, the electric conductivity of the ferrites, which is several orders of magnitude lower, can be significant. However, the importance of conductivity is diminished when the material is in the form of fine particles well dispersed in a dielectric binder with few electrically conducting bridges between them.

The magnetization M is expressed in magnetic moment per unit volume, kA/m (emu/cm^3), or more conveniently in the case of porous material or powders, in magnetic moment per unit mass σ, A·m^2/kg (emu/g). In the latter case, or when the powder is dispersed in a liquid binder which is later solidified, as is the case with particulate recording media, the sat-

TABLE 3.1   Spontaneous Magnetization M_{sb}, Curie temperature T_c, Constants of Magnetocrystalline Anisotropy K_1 and K_2, Magnetocrystalline Anisotropy Field H_k, and Saturation Magnetorestriction λ_s of Various Magnetic Materials at Room Temperature

Material	M_{sb}, kA/m (emu/cm^3)	T_c, °K	K_1, 10^3 J/m^3 (10^4 erg/cm^3)	K_2, 10^3 J/m^3 (10^4 erg/cm^3)	H_k, kA/m (4π Oe)	λ_s 10^{-6}
Fe	1710	1043	45	20	42	−4
Fe 30 at % Co)	1900	1223	30	18	25	45
Co	1430	1393	430	120	479	−100
Fe$_4$N	1385	761				
Ni	483	658	−4.5	2.3	4.6	−36
CrO$_2$	480	387	22	4	73	3
Fe$_3$O$_4$	480	858	−11	−2.8	20	40
γ-Fe$_2$O$_3$	350	870	−4.6		21	−5
CoFe$_2$O$_4$	425	793	200		749	−110
BaFe$_{12}$O$_{19}$	380	728	330		1380	10

uration magnetization M_s and the magnetization M are given, respectively, by

$$M_s = \frac{M_{sb}}{v_{tot}} \sum_i v_i = pM_{sb} \tag{3.2}$$

$$M = \frac{M_{sb}}{v_{tot}} \sum_i v_i \cos \theta_i = pM_{sb} \langle \cos \theta \rangle_{av} \tag{3.3}$$

where v_i = volume of the ith particle
θ = angle between magnetization and field directions
v_{tot} = total volume of sample
p = volumetric packing fraction of the magnetic particles

The saturation magnetization of the bulk material M_{sb} is introduced in distinction to the mean value M_s of systems of fine particles. It is assumed in Eq. (3.3) that the magnetization is uniform in each particle, an assumption which is discussed in Sec. 3.2.1.4. The saturation magnetization, as usually measured in high magnetic fields, is equal to the spontaneous magnetization only at temperatures low compared to the Curie temperature. Near the Curie point, the spins are partially reoriented in the magnetic field, thus leading to a saturation magnetization which is higher than the spontaneous magnetization.

If the particle dimensions are decreased to the order of the lattice parameter, the spontaneous magnetization may be particle-size-dependent. From an extensive discussion of theoretical and experimental results (Jacobs and Bean, 1963) it appears, however, that for particle diameters down to 2 nm, the spontaneous magnetization does not significantly differ from the bulk value. Smaller particles are not important for consideration for magnetic recording media.

3.2.1.2 Local anisotropies. The spin-orbit coupling of the electrons leads to energetically preferred orientations of the magnetic moments in the crystal lattice, and the symmetry of these orientations is reflected in the resulting magnetocrystalline anisotropy. While the spontaneous magnetization is a cooperative phenomenon, the magnetocrystalline anisotropy is a local effect, where each spin senses the local crystal-lattice symmetry (Birss, 1965). Consequently, lattice defects which introduce local variations of the crystal lattice symmetry can lead to an induced anisotropy. Interstitials, vacancies, or foreign atoms may diffuse under the influence of a magnetization distribution which is aligned by an applied field or stress, particularly at elevated temperatures. Such diffusion leads to a regular arrangement of the defects, resulting in an extrinsic form of uniaxial magnetocrystalline anisotropy.

The dependence of the free magnetocrystalline energy density F_k on the direction of the magnetization can be expressed in appropriate coordinates. The direction cosines α_1, α_2, α_3 are suitable for multiaxial cubic symmetry:

$$F_k = K_1(\alpha_1^2\alpha_2^2 + \alpha_2^2\alpha_3^2 + \alpha_1^2\alpha_3^2) + K_2(\alpha_1^2\alpha_2^2\alpha_3^2) + \cdots \tag{3.4}$$

For the uniaxial anisotropy found with hexagonal and tetragonal lattice structures, F_k is expressed in terms of the angle ϕ of rotation from the preferred axis:

$$F_k = K_1 \sin^2 \phi + K_2 \sin^4 \phi \tag{3.5}$$

The magnetocrystalline energy constants K_1 and K_2 are expressed as energy densities in joules per cubic meter (J/m^3). They usually change much more rapidly with increasing temperature than the spontaneous magnetization.

The minima of Eqs. (3.4) and (3.5) represent the preferred, or easy, directions of magnetization. The maximum field necessary to rotate the magnetization from one easy direction into another is the anisotropy field H_k. This may be looked at as the magnetic field needed to produce a small initial rotation from the equilibrium direction (Chikazumi, 1964):

$$H_k = \frac{d^2 F(\phi)}{\mu_0 M_{sb} \, d\phi^2} \tag{3.6}$$

(Here, as with other energy equations in SI units, $\mu_0 M_{sb}$ has to be replaced by M_{sb} for use with cgs units.) The easy directions and the anisotropy fields of the most important configurations are shown in Table 3.2. The initial magnetic susceptibility due to rotation of the magnetization is inversely proportional to H_k.

A deformation of the crystal lattice changes its symmetry. This introduces an additional anisotropy, with a consequent change in the direction of the magnetization. Conversely, a change of the magnetization is accompanied by a fractional change in length, $\lambda = \Delta l/l$, which is known as *magnetostriction*. The magnetostriction is directionally dependent in a similar way to the magnetocrystalline anisotropy. With an isotropic orientation distribution of crystal grains or particles, the free anisotropy energy density under the influence of an uniaxial stress σ has the form

$$F_\sigma = \tfrac{3}{2}\lambda_s\sigma \sin^2 \phi \tag{3.7}$$

in which λ_s is the saturation magnetostriction. This description of the uniaxial stress anisotropy is a sufficient approximation for many problems. More details on magnetostriction are given in review publications (Lee, 1955; Birss, 1959).

TABLE 3.2   Easy Directions and Anisotropy Fields H_k for Various Combinations of the Constants of Magnetocrystalline Anisotropy K_1 and K_2

K_1	K_2	Easy directions	$\mu_0 H_k M_{sb}$
	(a) Cubic Anisotropy		
>0	$-\infty < K_2 < -9K_1$	$\langle 111 \rangle$	$-\frac{4}{3} K_1 - \frac{4}{9} K_2$
>0	$-9K_1 < K_2 < \infty$	$\langle 100 \rangle$	$2 K_1$
<0	$-\infty < K_2 < \frac{9}{4}|K_1|$	$\langle 111 \rangle$	$-\frac{4}{3}K_1 - \frac{4}{9}K_2$
<0	$\frac{9}{4}|K_1| < K_2 < 9|K_1|$	$\langle 110 \rangle$	$K_1 + \frac{1}{2}K_2$
<0	$9|K_1| < K_2 < \infty$	$\langle 110 \rangle$	$-2K_1$
	(b) Uniaxial Anisotropy		
	$K_1 + K_2 > 0$	$\langle 001 \rangle$	$2K_1 + 4K_2$
	$K_1 + K_2 < 0$	Basal plane	0

SOURCE: Bozorth (1936).

Another severe change of symmetry occurs at internal or free surfaces of the magnetic material, where the environment of the surface atoms has a lower symmetry than that of the crystal lattice. This may become of importance if the particles are small enough to have a comparable number of atoms in the volume and at the surface. For instance, in a cube of a face-centered cubic (FCC) crystal having a length of 10 lattice parameters, about one-third of the atoms are surface atoms. This fraction rapidly decreases with increasing dimensions. The surface atoms sense a uniaxial anisotropy perpendicular to the surface of the order of 0.1 to 1.0 J/m^2 (0.1 to 1.0 $\times$ 10^3 erg/cm^2) (Néel, 1954; Jacobs and Bean, 1963). This surface anisotropy has been used to explain the increase of coercivity of gamma ferric oxide by cobalt surface adsorption (Tokuoka et al., 1977), and may give rise to an anomalous increase in coercivity of iron oxide particles coated with sodium polyphosphate (Itoh et al., 1977). The experimental evidence of this anisotropy is still very much under discussion (Morrish and Haneda, 1983).

Exchange anisotropy is another kind of surface anisotropy which is located, for example, at the phase boundary between a ferro- or ferrimagnet and an antiferromagnet. It has been discovered on cobalt particles with an antiferromagnetic cobalt monoxide surface layer (Meiklejohn and Bean, 1956). The exchange coupling of the last plane of the magnetically fixed antiferromagnetic lattice to the first ferro- or ferrimagnetic lattice plane leads to unidirectional or vector anisotropy. This acts like a dc-bias field that displaces the hysteresis loop (Meiklejohn and Bean, 1957) and leads to a finite anhysteretic magnetization in zero external field (Köster and Steck, 1976).

3.2.1.3 Magnetostatic anisotropies. The change of the normal component of the magnetization at the surface of a particle leads to an external as well as to an internal field with an associated energy density. The internal field is oppositely directed to the originating magnetization M, and hence is called the *demagnetizing field* H_d. For convenience, demagnetization is described by a geometrical shape factor N:

$$H_d = -NM \tag{3.8}$$

For the general ellipsoid, with major semiaxes a, b, and c, N can be directly calculated (Osborn, 1945). In the SI unit system, the sum of the demagnetization factors of the three major axes is $N_a + N_b + N_c = 1$, whereas it is $N_a + N_b + N_c = 4\pi$ in the cgs unit system. Of particular interest is the rotational ellipsoid with $a > b = c$, which is the most convenient mathematical model to describe the elongated particles used in most magnetic recording media. With parallel rotation of the magnetization, the difference in demagnetizing field energy perpendicular and longitudinal to the particle axis constitutes the uniaxial shape-anisotropy energy density F_m. The easy axis coincides with the particle axis. The energy density varies with the angle ϕ between the magnetization and the particle axis as follows:

$$F_m = \tfrac{1}{2}\mu_0 M_{sb}^2 (N_b - N_a) \sin^2 \phi \tag{3.9}$$

The corresponding anisotropy field is

$$H_m = (N_b - N_a)M_{sb} \tag{3.10}$$

The difference between the demagnetization factors perpendicular and longitudinal to the particle axis, $N_b - N_a$, increases steeply with the axial ratio a/b of the particles, as shown in Fig. 3.2. Consequently, in almost all cases of fine-particle magnetism, shape anisotropy is highly significant. In fact, up to recently it has been the intention of research and application to rely predominantly on shape anisotropy. An advantage is that H_m is directly proportional to M_{sb}, which shows little temperature dependence in the vicinity of room temperature for the materials in use. In contrast, the magnetocrystalline anisotropy field of most of the materials available is strongly dependent on temperature. Further, the frequently used demagnetizing criterion M_r/H_c (remanence divided by coercivity) is independent of temperature for shape anisotropy.

Another form of magnetostatic anisotropy is the interaction anisotropy which occurs with a linear chain of isotropic magnetic spheres (Jacobs and Bean, 1955). This leads to a preferred direction of magnetization along the chain axis. The maximum anisotropy field corresponds to uniform parallel rotation of the magnetization in an infinite chain of contacting

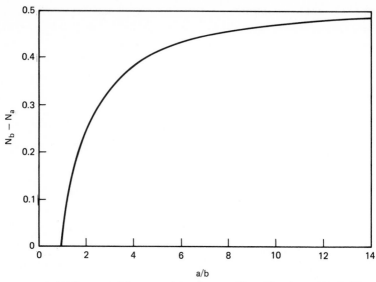

Figure 3.2 Difference of demagnetizing factors, $N_b - N_a$, of rotational ellipsoids with major semiaxes $a > b = c$ as a function of the axial ratio a/b.

spheres. It amounts to $0.3M_{sb}$ ($1.2\pi M_{sb}$ in cgs units), and is smaller than the maximum shape-anisotropy field of $0.5M_{sb}$ ($2\pi M_{sb}$) for infinite cylinders.

3.2.1.4 Single-domain particles. Consider a large ferromagnetic crystal which is magnetized to saturation in a strong magnetic field H (positive direction). When the field is reversed continuously, the magnetization M is also reversed. It is usually not zero at zero field but has a finite value, the saturation remanence $M_r(\infty)$ and changes sign in a field $-H_c$, where H_c is the coercivity. For moderate crystalline anisotropies, H_c is found to be much smaller than the magnetocrystalline anisotropy field H_k. This is because the magnetization is not rotated spatially in unison in the volume of the crystal. Instead, it is rotated locally in thin sheets which are moved through the crystal, the domain walls. These walls separate domains with uniform magnetization oriented along the preferred axes of magnetization, but with different directions in neighboring domains. The magnetocrystalline anisotropy energy to be overcome for magnetization reversal is reduced to that in the domain walls. This reduction is at the expense of exchange energy, which enters through the angles between the spin vectors in the domain wall. The total energy of the crystal is further minimized by the reduction of the external field energy through the formation of the alternately magnetized domains. The overall minimization of external field, magnetocrystalline, and exchange energies leads to a certain geometric configuration and number of domains. The thickness of the domain wall, satisfying equilibrium between exchange and magnetocrystalline anisotropies, is of the order of

$$\delta = 5 \sqrt{\frac{A}{K}} \tag{3.11}$$

With $A \approx 10^{-11}$ J/m (10^{-6} erg/cm) and $K_1 = -4.6 \times 10^3$ J/m^3 (-4.6×10^4 erg/cm^3) for gamma ferric oxide, the domain wall thickness is 200 nm. For iron, the thickness is 70 nm. The domain structure of large crystals, various types of domains and domain walls, wall displacement, and the physical mechanisms of "wall friction," which control the coercivity, are fairly well understood (Kneller, 1962; Chikazumi, 1964; Craik and Tebble, 1965).

If the size of the crystal becomes sufficiently small, there is a critical particle size below which nonuniform magnetization configurations will no longer be stable, at least in zero external magnetic field. From discussions of critical particle dimensions for this to happen (Wohlfarth, 1963; Mee, 1964; Kneller, 1969), it appears that Eq. (3.11) represents a reasonable limit for predominant magnetocrystalline anisotropy. Throughout this chapter we use the definition of a single-domain particle in the following sense: nonuniform modes of magnetization reversal are admitted provided the magnetization is uniform through the volume of the particles in zero external field. This definition encompasses the situation where this condition is slightly violated by the fields from surface irregularities, which are deviations from the ideal ellipsoid. Such surface irregularities result in a mean magnetization in the easy direction, at zero external field, which is smaller than M_{sb} and must be taken into account in considerations of the remanence of particle systems.

3.2.1.5 Magnetization processes. The basic mode of magnetization reversal for single-domain particles is uniform rotation of the magnetization (Stoner and Wohlfarth, 1948). The magnetization curves for particles with uniaxial anisotropy are shown in Fig. 3.3 for various angles θ between the easy direction of magnetization (the particle major axes in the case of shape anisotropy), and the applied field direction. The strong influence of particle orientation on the shape of the magnetization curve is evident. For $\theta = 0$, the normalized coercivity $h_c = H_c/H_a$ equals unity, where H_a stands for any of the uniaxial anisotropy fields described earlier. The coercivity decreases rapidly with increasing θ. At angles less than 45° it is controlled by irreversible and reversible magnetization changes but, at angles greater than 45°, by reversible rotation only. Here, the reduced coercivity is given by the equation

$$h_c = \frac{1}{2} \sin 2\theta \tag{3.12}$$

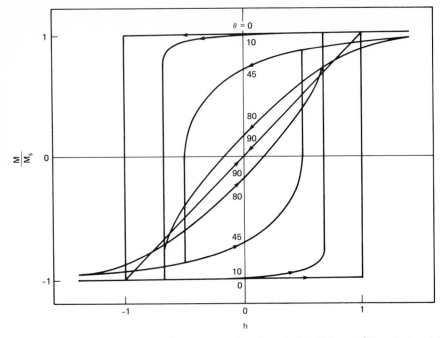

Figure 3.3 Magnetization M/M_s versus reduced applied field $h = H/H_a$, calculated for a uniaxial particle reversing its magnetization by uniform rotation. The parameter θ is the angle between field and particle axes *(Stoner and Wohlfarth, 1948)*.

In order to distinguish between the processes, it is useful to define the switching field H_0, or reduced switching field $h_0 = H_0/H_a$, for irreversible magnetization reversal. In the range of angles $\theta \leq 45°$, $h_c = h_0$. At angles $\theta > 45°$, h_c continues to decrease with increasing θ, while h_0 starts to increase again. At $\theta = 90°$, $h_c = 0$ and $h_0 = 1$. Owing to the angular variation of h_c and h_0, an assembly of identical particles with a distribution in particle-axis orientations has a distribution of both quantities. A random distribution of orientations leads to a reduced coercivity of $h_c = 0.479$, and to a reduced remanence coercivity of $h_r = H_r/H_a = 0.529$. The remanence coercivity H_r is a measure of the mean switching field, and is defined as the reverse field applied after saturation which reduces the remanent magnetization to zero. The reduced saturation remanence, called *remanence squareness* $S = M_r(\infty)/M_s$, of the random assembly becomes $S = \langle \cos \theta \rangle_{av} = 0.5$.

Magnetization curves $M(H)$ of assemblies of typical elongated single-domain, gamma ferric oxide particles are shown in Fig. 3.4. They demonstrate the characteristic hysteresis of magnetically hard materials. The curve for partially aligned particles has a higher saturation remanence $M_r(\infty)$, and a steeper slope at coercivity H_c, than the curve for randomly oriented particles. The distribution of switching fields can be characterized by the field $H_{0.25}$, where 25 percent of the particles have their magnetization reversed after saturation $[M_r(H_{0.25})/M_r(\infty) = 0.5]$, and $H_{0.75}$, where 75 percent of the particles are reversed $[M_r(H_{0.75})/M_r(\infty) = -0.5]$. The switching-field distribution is then expressed by the normalized quantity:

$$\Delta h_r = \frac{H_{0.75} - H_{0.25}}{H_r} \tag{3.13}$$

This type of measurement is shown in Fig. 3.4, where $M_d(H)$ is the remanence acquired after saturation in the positive field direction and successive application of a dc field in the negative direction. For random distribution of particle axes of identical single-domain particles with uniaxial anisotropy and with uniform magnetization reversal, Δh_r amounts to 0.13. In summary, any particle orientation distribution deviating from the mathematically parallel alignment usually leads to a coercivity smaller than the anisotropy field H_a, a distribution in switching fields, and a sat-

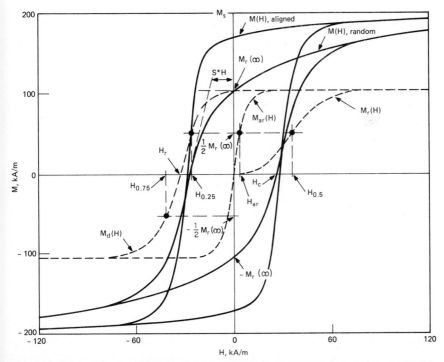

Figure 3.4 Magnetization $M(H)$, remanence $M_r(H)$, dc demagnetizing remanence $M_d(H)$, and anhysteretic remanence $M_{ar}(H)$ of a sample of randomly oriented γ-Fe_2O_3 particles. For the same particles, partially aligned, $M(H)$ is also included.

uration remanence smaller than M_s. In addition, particles with a distribution of the anisotropy field H_a further broaden the switching-field distribution. This may be due either to a distribution of the axial ratio a/b, and consequently of the shape-anisotropy field, or to a distribution of the magnetocrystalline anisotropy of the individual particles. The latter may be caused by unevenly distributed dopants that strongly influence the magnetocrystalline anisotropy, such as cobalt in iron oxide or cobalt and titanium in barium ferrite particles. For practical purposes, another parameter for characterizing the switching-field distribution, the *coercivity squareness S**, has been introduced. As shown in Fig. 3.4, S^* can be conveniently obtained from the intersection of the slope of the magnetization curve $M(H)$ at the coercivity point $M(H_c) = 0$,

$$\frac{dM(H)}{dH} = \frac{M_r(\infty)}{H_c(1 - S^*)} \tag{3.14}$$

with the line $M = M_r(\infty)$. The quantity $1 - S^*$ almost equals Δh_r, since the contributions by reversal magnetization processes to $1 - S^*$ cancel each other. Other characterizations of the switching-field distribution, like the half-pulse width of the differentiated magnetization curve, or H_r/H_c, very much depend on reversible magnetization processes and may be used only if the angular distribution of the particle axes is kept constant (Köster, 1984).

The theoretical predictions for magnetization reversal by uniform rotation have been experimentally verified in cases with predominant magnetocrystalline anisotropy $H_k \gg M_{sb}$. Single-domain barium ferrite particles in the size range of 0.1 to 0.3 μm can be prepared by a flux method. These exhibit as a random assembly the theoretical values of H_r = $0.529(H_k + H_m)$ = 530 kA/m (6660 Oe), and Δh_r = 0.13, within the uncertainty of the data. Other examples are chromium dioxide doped with iridium (Kullmann et al., 1984b), and cobalt-doped gamma ferric oxide (Köster, 1972), both in the remanence coercivity range of 300 kA/m (3800 Oe). However, for predominant shape anisotropy, $H_k \ll M_{sb}$, much lower coercivities and a much broader switching-field distribution are obtained than expected for uniform rotation. For gamma ferric oxide, which is the most widely used material for magnetic recording media, the largest switching field H_0 found along the particle axis in a single particle measurement has been 84 kA/m (1050 Oe) (Knowles, 1980). The theoretical value is 156 kA/m (1960 Oe) for shape anisotropy with $a/b = 7$.

In fact, uniform rotation almost never occurs in particles with predominant shape anisotropy. Instead, the magnetization is reversed in a nonuniform fashion by trading long-range magnetostatic field energy against short-range exchange energy. This necessitates that the particle dimensions are large enough to accommodate incremental angles between the spins without introducing too high an exchange energy. Consequently, this reversal mode is strongly size-dependent, and there is indeed a critical particle diameter

$$2b_0 = 2 \left(\frac{4\pi A}{\mu_0 M_{sb}^2} \right)^{1/2}$$ (3.15)

above which nonuniform rotation is energetically favorable. With $A \approx 10^{11}$ J/m (10^{-6} erg/cm) and $M_{sb} = 350$ kA/m (emu/cm^3) for gamma ferric oxide, $2b_0 \approx 60$ nm. For iron, $2b_0 \approx 12$ nm. Experimentally, however, the critical diameter of gamma ferric oxide seems to be about 15 nm, which is considerably smaller than expected (Eagle and Mallinson, 1967). The nonuniform reversal mode which takes place above this critical diameter can be considered as a form of nonuniform magnetization reversal intermediate between uniform rotation in very small particles, and local nonuniform rotation in domain walls of very large particles.

Three models of nonuniform rotation modes have been investigated intensively (Wohlfarth, 1963; Kneller, 1969). The first mode is buckling, which is a periodic fluctuation of the magnetization direction from the particle axis. This mode involves magnetostatic field and exchange energy. The second mode is curling, which is best described by a bundle of twisted wires along the particle axis. It avoids any magnetostatic field energy but involves exchange energy. The third mode is fanning, described by a chain of spheres in which the angles between magnetization and chain axis for neighboring spheres are of opposite sign. This mode relies on magnetostatic energy and excludes exchange energy. With the reduced particle radius $b/b_0 > 0$, the switching field in the case of buckling in an infinite circular cylinder becomes

$$H_0 = 0.645 M_{sb} \left(\frac{b}{b_0} \right)^{-2/3}$$ (3.16)

while for curling in rotational ellipsoids,

$$H_0 = \left[0.5k \left(\frac{b}{b_0} \right)^{-2} - N_a \right] M_{sb}$$ (3.17)

Here k is a numerical factor, which increases monotonically with decreasing axial ratio a/b, from $k = 1.08$ for $a/b = \infty$ (infinite cylinder) to $k = 1.39$ for $a/b = 1$ (sphere). In a formal sense, N_b, the demagnetization factor perpendicular to the particle axis in Eq. (3.10), is reduced with increasing particle diameter. In the case of fanning, which is independent of particle diameter, the maximum nucleation field of an infinite chain of contacting spheres is reduced from $0.3 M_{sb}$ for uniform rotation to values between $0.08 M_{sb}$ and $0.125 M_{sb}$. The reversal mode for a given cylinder radius is always the one with the least switching field H_0. For the magnetic particles used in magnetic recording media, b/b_0 has values between 2 and 4. Consequently, curling or fanning are the most likely modes to occur. Fanning may apply only if the particles are distinctively subdivided in crystallites, separated by boundaries, with little or no exchange coupling.

In reality, none of the above models seems to apply strictly, since the particle shape in general deviates considerably from the ideal ellipsoid, and offers additional sites for nucleation of magnetization reversal. Experimentally, the coercivity tends to be proportional to the inverse particle diameter rather than the inverse squared diameter as predicted for the curling mode. On the other hand, the coercivity is not independent of particle diameter as expected for fanning. This is demonstrated in Fig. 3.5, where the coercivity of gamma ferric oxide and chromium dioxide powder samples is plotted versus the inverse mean particle diameter. The two sets of data are not directly comparable because $2b$ has been determined from electron micrographs for the $\gamma\text{-Fe}_2\text{O}_3$ particles, and from x-ray analysis for the CrO_2 particles. Apparently some nonuniform mode of magnetization reversal takes place which involves less exchange energy, and consequently is less dependent on particle diameter, than curling. But the reversal mechanism is not solely dependent on magnetic interactions, and therefore independent of particle diameter, as with fanning. Recently, a mode of pseudo-curling or buckling has been proposed (Knowles, 1984). Instead of occurring simultaneously everywhere in the particle volume, as is the case with the above models, the switching is thought to start at one

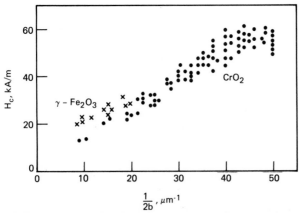

Figure 3.5 Coercivity H_c as a function of the inverse of the mean particle diameter $2b$ of isotropic powder samples of gamma ferric iron oxide and chromium dioxide *(Steck, 1985)*.

tip of an elongated particle and move through the particle in the fashion of falling dominoes.

Actual recording media always have a distribution of the direction of the particle axes, and the angular dependence of the coercivity must be considered. Further interest arises from the fact that the magnetic head field close to the head surface rotates from a longitudinal direction in the gap zone to a perpendicular direction over the pole piece. This process becomes of increasing interest in short-wavelength recording. Both the reduced coercivity h_c, and the switching field h_0, of infinitely long cylinders are plotted in Fig. 3.6 for the curling model as a function of the angle θ between the particle axis and the applied field direction. Similar curves are expected for the buckling or fanning modes. The parameter is the reduced cylinder radius b/b_0. The magnetocrystalline anisotropy is assumed to be negligible. As has already been discussed for the mode of uniform rotation (curve for $b/b_0 \to 0$ in Fig. 3.6), h_c and h_0 at low and

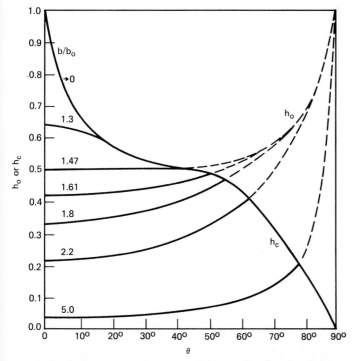

Figure 3.6 Reduced coercivity $h_c = H_c/H_a$, and reduced switching field $h_0 = H_0/H_a$, of infinite cylinders ($H_a = 0.5M_s$) as a function of the angle θ between the field and the cylinder axis. The parameter is the reduced cylinder radius b/b_0. *(After Shtrikman and Treves, 1959.)*

medium angles coincide until a critical angle is reached. Above this angle, h_c is controlled by reversible magnetization rotation according to Eq. (3.12) and becomes zero for $\theta = 90°$. The switching field increases further and is controlled by uniform rotation when the respective curves in Fig. 3.6 join the curve for uniform rotation. Experimental data are shown in Fig. 3.7. Here, the coercivity H_c and the remanence coercivity H_r of samples of gamma ferric oxide, chromium dioxide, and iron particles with different degrees of particle alignment, measured longitudinal and trans-

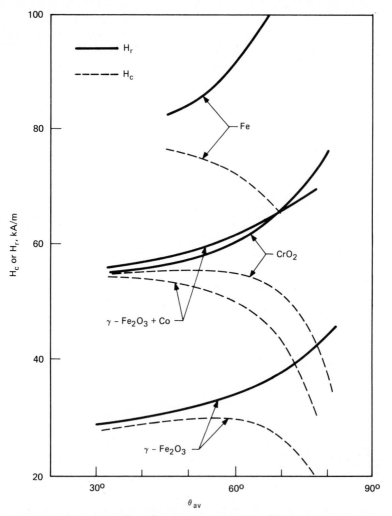

Figure 3.7 Coercivity H_c and remanence coercivity H_r versus the average angle, $<\theta>_{av} = \arccos S$, of particle alignment $[S = M_r(\infty)/M_s]$, for iron, gamma ferric iron oxide, and chromium dioxide particles.

verse to the direction of alignment, are plotted as a function of the mean angle of particle orientation $\langle \theta \rangle_{av}$ = arccos S. At low values of $\langle \theta \rangle_{av}$, H_r ≈ H_c, as expected. At larger $\langle \theta \rangle_{av}$, $H_r > H_c$. The rate of increase of H_r at large values of $\langle \theta \rangle_{av}$ is far less than expected from the curling model, while H_c indeed tends to zero according to Eq. (3.12). Measurements of H_c and H_r on samples of aligned particles as a function of sample rotation (Bate, 1961) more prominently exhibit the maximum of H_c in Fig. 3.7.

The switching-field distribution of an isotropic particle assembly of identical particles is expected to increase with increasing b/b_0, from Δh_r = 0.13 for uniform rotation to Δh_r = 0.56 for b/b_0 = 3 of the curling mode, as shown in Fig. 3.8. Here, Δh_r is plotted versus b/b_0 for particle assemblies with various degrees of particle alignment. An azimuth-invariant distribution function for θ is assumed of the form exp $(\alpha \cos^2 \theta)\sin \theta$, where α is the orientation parameter (Bertram, 1976). A lower value of

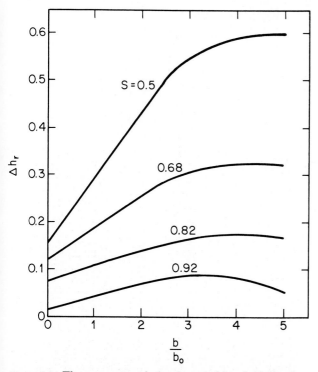

Figure 3.8 The parameter Δh_r for the switching-field distribution, calculated for various degrees of particle orientation, that is, for various values of the normalized saturation remanence S as a function of the reduced cylinder radius b/b_0. An azimuth invariant distribution function of the orientation angle θ of infinite cylinders has been used of the form exp $(\alpha \cos^2 \theta)\sin \theta$ used *(Bertram, 1976)*.

b/b_0 means a higher coercivity, and also a lower switching-field distribution as shown in Fig. 3.9. Here, Δh_r of numerous isotropic powder samples of gamma ferric oxide, chromium dioxide, and iron particles is plotted as a function of the remanence coercivity H_r. The curves are shifted in proportion to the saturation magnetization of the respective materials and, in the case of chromium dioxide, in addition by the magnetocrystalline anisotropy field. With increasing alignment of identical particles, that is, with increasing S, Δh_r is expected to decrease to zero for $S \rightarrow 1$, as is indicated in Fig. 3.10. Here, Δh_r is shown as a function of the squareness S for the reduced radii $b/b_0 = 2$ and $b/b_0 = 4$, together with measurements on samples of particles with different degrees of alignment made from a number of magnetic materials. Contrary to theory, the experimental data tend to a finite value of Δh_r for $S \rightarrow 1$. This is due to a distribution of particle diameters and a consequent distribution of switching fields, even for mathematically aligned particles. Assuming a triangular distribution for b/b_0, the experimental data in Fig. 3.10 can be quite well described by the curling model (Bertram, 1976).

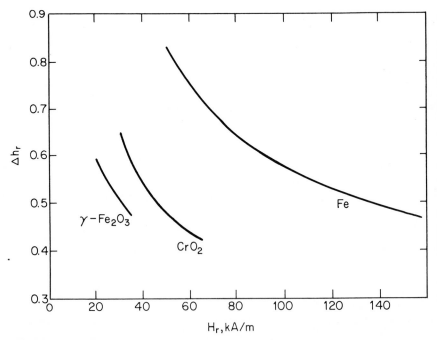

Figure 3.9 Switching-field distribution Δh_r as a function of remanence coercivity H_r for isotropic powder samples with a volumetric packing fraction of $p = 0.25$ of gamma ferric oxide, chromium dioxide, and iron particles. The curves represent the mean value of over 20 samples of each material with a standard deviation of 4 to 7 percent.

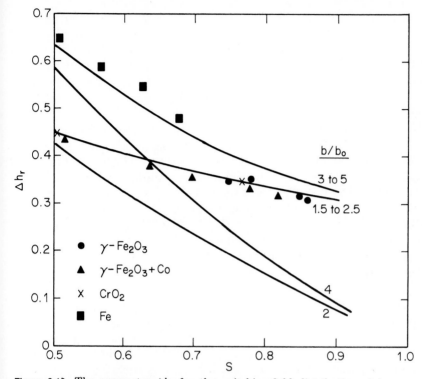

Figure 3.10 The parameter Δh_r for the switching-field distribution of iron, gamma ferric iron oxide, cobalt-modified gamma ferric iron oxide, and chromium dioxide particles of different degrees of particle alignment as a function of the remanence squareness S. The solid curves represent theoretical values for infinite cylinders of different reduced radii b/b_0, either constant in value or having a triangular distribution, calculated under the same conditions as in Fig. 3.8.

3.2.1.6 Mixed anisotropies and saturation remanence. So far, only one type of magnetic anisotropy has been considered at a time. In the case of collinear uniaxial shape and magnetocrystalline anisotropies, Eq. (3.17) extends to

$$H_0 = H_k + \left[\frac{k}{2} \left(\frac{b}{b_0} \right)^{-2} - N_a \right] M_{sb} \tag{3.18}$$

where the two anisotropies are simply additive. The magnetization curves of particles with cubic anisotropy are more complicated than those for uniaxial anisotropy (Johnson and Brown, 1961). Thus, the coercivity value

$$H_c = 0.32 H_k \tag{3.19}$$

as given by Néel (1947) for an isotropic assembly of cubic particles may not apply in all cases, although it serves well for predominant magnetocrystalline anisotropy (Köster, 1972). Magnetization curves of an elongated single-domain particle, with a ⟨111⟩ fiber texture parallel to the long axis, and with ⟨111⟩ easy directions of magnetocrystalline anisotropy, have also been considered (Johnson and Brown, 1959). It has been found that the existence of collinear shape and magnetocrystalline easy axes does not raise the coercivity under all circumstances. In general, the situation of mixed anisotropies is little understood.

In the case of multiaxial anisotropies, the reduced saturation remanence, or squareness S, for isotropic particle assemblies increases monotonically with the number n of easy axes, from $S = 0.5$ for $n = 1$ (uniaxial anisotropy) to $S = 0.83$ for $n = 3$ (magnetocrystalline anisotropy with $K_1 > 0$), and to $S = 0.87$ for $n = 4$ (magnetocrystalline anisotropy with $K_1 < 0$) (Gans, 1932; Wohlfarth and Tonge, 1957; Tonge and Wohlfarth, 1958). The transition from $n = 1$ to $n = 3$ can be directly observed by measuring the temperature dependence of S for elongated cobalt-doped gamma ferric oxide particles (Sec. 3.2.2).

During the manufacturing process of particulate media, a step of compression such as calendering is usually involved, which may introduce a more planar distribution of the particle axes. It is therefore of interest to note that for a planar isotropic distribution of uniaxial single-domain particles, $S = 2/\pi$, a value larger than $S = 0.5$ which results from an isotropic spatial distribution. Attempts have been made to differentiate between a spatial and a planar distribution of particle axes in magnetic tape samples (Bate and Williams, 1978; Bertram, 1978). It appears that the angular distribution is close to cylindrical around the direction of alignment, although some results are difficult to interpret because the remanence squareness of the individual particles along their axes is not necessarily unity. Their irregular shape, or spin canting at the surface (Morrish and Haneda, 1983), may cause local deviations from the otherwise uniform spin alignment and consequently a squareness smaller than unity. Thus, for isotropic powder samples, S is often below 0.5, and typically has a value of 0.47. For this reason the orientation ratio M_{rx}/M_{rz} is often used, instead of the squareness S, to characterize the degree of particle alignment. Here M_{rx} is the saturation remanence in the direction of particle alignment and M_{rz} is the saturation remanence transverse to the particle alignment. The orientation ratio is independent of the effect of nonuniform spin alignment which enters equally in M_{rx} and M_{rz}.

3.2.1.7 Magnetostatic particle interactions.

In the previous sections, it has been assumed that the particles in an assembly are independent. In practice, magnetostatic interactions will mutually influence their magnetization reversal. The maximum average interaction-field energy density is of the order of $0.25p\mu_0 M^2$, where $p = M_s/M_{sb}$ is the volumetric packing fraction of the particles. If shape anisotropy dominates ($K_1 \ll \mu_0 M_{sb}^2$), the energy density of the particles in the magnetostatic interaction field may have the same order of magnitude as the maximum shape-anisotropy energy density $F_m = 0.25 \, \mu_0 M_{sb}^2$. Only for predominant magnetocrystal-line anisotropy ($K_1 \gg \mu_0 M_{sb}^2$), the magnetostatic interactions may be negligible. The anhysteretic remanence $M_{ar}(H)$, which is acquired after application of a dc field H simultaneously with a slowly decreasing ac field H' of sufficiently high initial amplitude, $H' > H_0$, is directly controlled by the interaction fields. Here, the ac field serves to overcome the energy barriers that prevent irreversible magnetization changes from ocurring in H alone, and $M_{ar}(H) = M_r(\infty)$ for $H > 0$ for independent particles if thermal fluctuation effects can be neglected. In practice, the presence of magnetostatic interaction effects dictates that $M_{ar}(H)$ has a finite slope; that is, the initial anhysteretic susceptibility $\chi_{ar} = dM_{ar}(H)/dH$ is finite. The anhysteretic remanence is a single-valued, odd function of the applied dc field (Fig. 3.4), and is thus used as the basic process for linearization of the recording channel in ac-bias recording (Chap. 2). Apart from this, the anhysteretic remanence is a useful tool for the study of interaction effects.

Another interesting measure of the interaction fields can be drawn from remanence curves. The static remanence curve $M_r(H)$ can be measured after ac demagnetization and successive application of a dc field H, with $m_r(H) = M_r(H)/M_r(\infty)$. The dc demagnetizing remanence $M_d(H)$ is acquired after saturation in one direction and successive application of a dc field in the opposite direction with $m_d(H) = M_d(H)/M_r(\infty)$. For independent uniaxial particles, the two remanence curves are interrelated as

$$m_d(H) = 1 - 2m_r(H) \tag{3.20}$$

When particles interact, this condition is not obeyed. According to Eq. (3.20), the field $H_{0.5}$ for irreversibly reversing 50 percent of the particles [$m_r(H_{0.5}) = 0.5$] and the remanence coercivity H_r [$m_d(H_r) = 0$] should be identical. In practice, $H_r < H_{0.5}$ (Fig. 3.4). For oriented gamma ferric oxide media, this difference amounts to 2 to 4 kA/m (25 to 50 Oe) or about 10 percent of H_r.

Different theoretical models for the interaction problems have been investigated. They are based on cooperative phenomena in Ewing-type models, on statistical considerations of pair models and their extension, the Preisach diagram, on statistical thermodynamical considerations, and on mean-field concepts. Most of these theories have not been advanced enough to give direct relations between the static and the anhysteretic magnetic magnetization parameters. The Preisach diagram (Chap. 2) is a powerful tool in describing the interaction-field distribution and its influence on various remanence curves. Also, a dynamic mean-field theory (Kneller, 1968, 1969, 1980) leads to a relationship between the initial anhysteretic susceptibility and the spontaneous magnetization M_{sb}, the volumetric packing fraction p, the remanence squareness S, and the remanence coercivity H_r:

$$\chi_{ar} = \frac{0.5p(1 - p)M_{sb}^2}{H_r^2(1 - S^2)} \tag{3.21}$$

The distribution of switching fields is not incorporated but, by empirical replacement of the numerical factor 0.5 by 0.27, good agreement with experiment is obtained. This theoretical approach can be extended to thermoremanent magnetization, the magnetization acquired after cooling in a dc field from temperatures above, to temperatures below, the Curie temperature.

The dc field can be normalized by writing $h = H/H_{ar}$, where H_{ar} is the field for which the anhysteretic remanence equals half the saturation remanence. The reduced anhysteretic remanence m_{ar} is then experimentally found to follow an almost universal curve versus h for tape samples with partially aligned particles, $0.70 \leq S \leq 0.87$, of the form (Köster, 1975)

$$m_{ar} = 0.569h - 0.0756h^3 + 0.0065h^5 \tag{3.22}$$

for $h < 2$. Consequently, up to 50 percent of $M_r(\infty)$ can be utilized in ac-bias recording with less than 3 percent third harmonic distortion, independent of the magnetic material used or of the magnetic media preparation. The third-order factor in Eq. (3.22) varies from -0.095 for $S = 0.6$ to -0.063 for $S = 0.9$ in a systematic magnetic tape series, thus indicating the role of the distribution of the switching fields in the degree of linearity of the anhysteretic magnetization curve. The mean-field approach predicts that $H_{ar} = H_{0.5} - H_r$; this difference is directly related to χ_{ar} according to Eq. (3.22) with $h \rightarrow 0$:

$$(H_{0.5} - H_r) = \frac{0.569M_r(\infty)}{\chi_{ar}} \tag{3.23}$$

The particle interactions not only cause a shift of $H_{0.5}$ and H_r but also reduce these properties with increasing p. There have been many attempts to calculate this dependence rigorously. They lead to equations with various exponents of p near unity, which are almost impossible to discriminate experimentally. In addition, if shape anisotropy is dominant, there may be an increase of the critical radius b_0 for uniform rotation which is expected to be proportional to $b_0(0)(1 - p)^{-1/2}$. This would reduce the rate of decrease of H_r with p (Kneller, 1969). With these questions open to discussion, it seems still legitimate to use Néel's formula

$$H_r(p) = H_{r,\text{cryst}} + H_{r,\text{shape}}(0)(1 - p) \tag{3.24}$$

for which experimental examples are given in Fig. 3.11. Here, H_c and H_r are plotted for increasingly compacted iron, gamma ferric oxide, and chromium dioxide powders as a function of p. Equation (3.24) holds likewise for H_r and H_c. It appears, although still questioned (Huisman, 1982), that the contribution by the magnetocrystalline anisotropy, as a local effect to H_c and H_r, is independent of the long-range interaction fields and hence of the packing fraction p (Morrish and Yu, 1955). In fair agreement with the expected values, the experimental curves of Fig. 3.11 extrapolate to coercivities of 7 kA/m for iron, 6 kA/m for gamma ferric oxide, and 27 kA/m for chromium dioxide (88, 75, and 340 Oe, respectively). Thus, the magnetocrystalline contribution has been added in Eq. (3.24) to the original equation by Néel (Wohlfarth, 1963; Kneller, 1969).

The magnetostatic interactions also change the saturation remanence. A simple model of mean interaction fields predicts an increase of the remanence squareness S with increasing packing fraction p (Wohlfarth, 1955; Köster, 1970; Bertram and Bhatia, 1973). This is shown in Fig. 3.12, where the squareness of an isotropic sample of gamma ferric oxide particles, with a mean value of the axial ratio $a/b = 7$, is plotted as a function of p together with the theoretical curve (see also Smaller and Newman, 1970). The initial value of $S < 0.5$ at low p is due to particle imperfections as discussed before. Contradicting results of S decreasing with increasing densification of gamma ferric oxide powders may be due to the formation of flux-closing clusters. The increase of S with p, which is more pronounced for smaller values of a/b, also serves to explain the high remanence squareness of cobalt-phosphorus thin films. Such films are thought to consist of highly packed cobalt crystallites, more or less isolated from each other with respect to exchange forces, by phosphorus segregation in the grain boundaries. The parameter Δh_r, as defined earlier, is also plotted in Fig. 3.12. It is found to decrease with increasing p, in corroboration with the rather low value of the highly packed cobalt-phosphorus film. Generally, the magnetostatic interaction fields tend to level out all effects which rely on magnetostatic energy.

Finally, the initial reversible susceptibility χ_0 of an ac-demagnetized isotropic powder sample can be shown to increase owing to particle interactions (Köster, 1970):

$$\chi_0(p) = \frac{2pM_{sb}}{3H_a\left[1 - pN_b(\tfrac{2}{3}M_{sb}/H_a)\right]} \tag{3.25}$$

This equation allows the field H_a to be determined independently of the ambiguity of the nucleation of the magnetization reversal, provided the demagnetization factor N_b perpendicular to the particle axis is known:

$$H_a = \frac{2}{3}pM_{sb}\left(\frac{1}{\chi_0} + N_b\right) \tag{3.26}$$

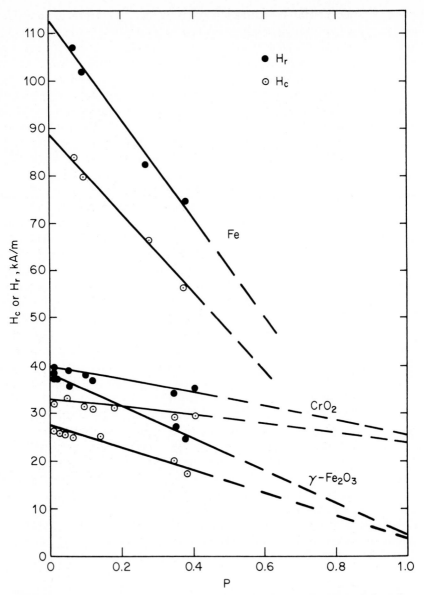

Figure 3.11 Coercivity H_c and remanence coercivity H_r as a function of the volumetric packing fraction p for isotropic assemblies of iron, gamma ferric iron oxide, and chromium dioxide particles.

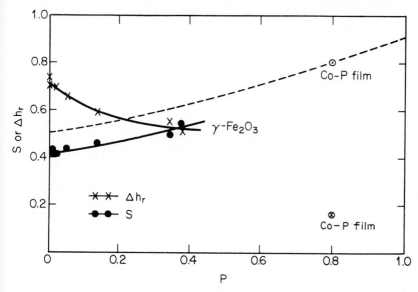

Figure 3.12 Remanence squareness S and the parameter Δh_r for the switching-field distribution versus the volumetric packing fraction p, for an isotropic assembly of gamma ferric iron oxide particles. The dashed curve has been calculated for infinite cylinders. Data for a cobalt-phosphorus film are included.

Equation (3.26) may be written as

$$\frac{2}{3}\frac{1}{\chi_0} = \frac{H_a}{pM_{sb}} - \frac{2}{3}N_b \tag{3.27}$$

giving the opportunity to determine H_a/M_{sb} and N_b from plotting $1/\chi_0$ as a function of $1/p$ for a series of differently compacted powder samples. In addition, the temperature dependence of χ_0 indicates whether shape anisotropy alone, or shape plus magnetocrystalline anisotropy, is present since, according to Eq. (3.25), χ_0 is independent of temperature for shape anisotropy alone ($H_a \propto M_{sb}$). This analysis may fail, as is the case with chromium dioxide, if the magnetocrystalline anisotropy field H_k happens to have the same temperature dependence as the spontaneous magnetization, instead of decreasing as usual more rapidly with temperature (Köster, 1970; Flanders, 1983).

3.2.1.8 Thermal fluctuations. In analogy to Brownian movement, thermal energy causes fluctuations in the magnetization of a particle as a whole. The magnetization may be reversed statistically, as a time and temperature effect, if the total anisotropy energy of the particle, vK, is of the order of the thermal energy kT. Here, v is the volume of the particle, k is the Boltzmann constant, T is the absolute temperature, and K stands for any of the anisotropy energy density constants. This behavior is called *superparamagnetism* because the complete magnetic moments of the particles act like the individual spin moments of paramagnetic materials. There is a critical volume v_p given by

$$v_p = \ln (2tf_0) \frac{2kT}{\mu_0 M_{sb} H_a} \tag{3.28}$$

below which superparamagnetism exists (Wohlfarth, 1963). Here t is the time period of observation (duration of measurement), $f_0 = 10^9 \text{ s}^{-1}$ is the Larmor frequency, and H_a is the total anisotropy field. The latter may have to be replaced by the switching field H_0 of the particles or, consequently, by the remanence coercivity H_r, as the mean switching field of the sample under investigation. This has been done in order to calculate an upper limit of v_p for $t = 100$ s and $t = 300°\text{K}$ of gamma ferric oxide, of chromium dioxide, and of iron particles. The results are listed in Table 3.3 together with typical mean particle volumes of these materials. The implication of this effect for magnetic recording is that, even for $v > v_p$, the remanence may not be stable with time and temperature since

$$M_r \left(t, \frac{v}{v_p} \right) = M_r(0) \exp \left[-(2tf_0)^{1-v/v_p} \right] \tag{3.29}$$

and $\qquad H_c \left(\frac{v}{v_p} \right) = H_c(0) \left[1 - \left(\frac{v_p}{v} \right)^{1/2} \right] \tag{3.30}$

Equation (3.29) is illustrated in Fig. 3.13 for $t = 100$ s (static measurement) and for spherical Co-Fe particles. The remanence increases very rapidly with increasing particle dimensions, from zero at $v = v_p$ to its maximum value $M_r(0)$ which is practically stable at $v > 4v_p$. The influence of thermal fluctuations on H_c according to Eq. (3.30) is also shown in Fig. 3.13. They lead to a parabolic increase of H_c, from $H_c = 0$ at v_p to its maximum value $H_c(0)$ at relatively large volumes $v > 100v_p$. Before this happens, however, nonuniform magnetization reversal, or even multidomain behavior, sets in, which reduces H_c with increasing particle size.

TABLE 3.3 Critical Volume v_p and Critical Particle Diameter $2b_p$ for Superparamagnetic Behavior, Average Particle Volume $\langle v \rangle_{av}$, and Average Particle Diameter $\langle 2b \rangle_{av}$

Material	v_p, $10^{-4} \mu m^3$	$2b_p$, nm	$\langle v \rangle_{av}$, $10^{-4} \mu m^3$	$\langle 2b \rangle_{av}$, nm
γ-Fe$_2$O$_3$	0.2	15	2–25	30–70
CrO$_2$	0.09	11	1–10	25–52
Fe†	0.01	6	0.6–1	20–25

†Iron particles with $a/b = 7$.

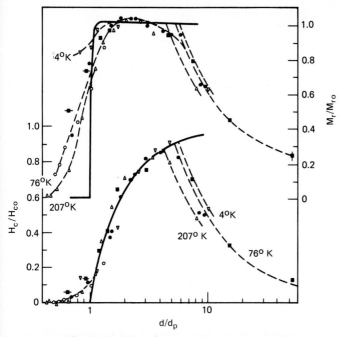

Figure 3.13 Coercivity H_c and saturation remanence $M_r(\infty)$ of iso-tropic assemblies of essentially spherical Co-Fe particles as a function of the reduced particle diameter d/d_p, where d_p is the critical diameter for superparamagnetic behavior. The parameter is the temperature of measurements. The solid curves are calculated from Eqs. (3.29) and (3.30) *(Kneller and Luborsky, 1963).*

For optimum coercivity, a particle size must be chosen which is just before the peak on the right-hand side of the coercivity–versus–particle-size curve. This implies that the coercivity always depends on the speed of the magnetic field change during measurement, because v_p depends on t according to Eq. (3.28). In Table 3.4, the critical factor $\ln (2tf_0)$ is listed for various conditions of magnetic media application. It varies by a factor of less than 10 for a factor of 10^{17} in t. Although this reduces the impact of the time factor on the value of the coercivity, it has long been observed that H_c of gamma ferric oxide, for instance, as measured in a 50- or 60-Hz loop tracer is 5 to 10 percent higher than H_c as measured statically in a vibrating sample magnetometer (see also Sharrock, 1984). For the smaller particle size of chromium dioxide particles, this difference can be as high as 20 percent. Consequently, the static coercivity of chromium dioxide magnetic tapes usually is 10 to 20 percent lower than that of cobalt-modified gamma ferric oxide tapes designed for the same application.

Under the bias of a small applied field, $H \ll H_a$, the magnetization of particles which are only slightly larger than v_p can be reversed, and a

TABLE 3.4   Time Factor ln $(2tf_0)$ in Eq. (3.28) under
Various Conditions of Magnetic Media Application

Application	t	$\ln (2tf_0)$
Archival storage	100 years	43
Static measurement	100 s	26
50- or 60-Hz measurement	20 ms	17
Audio ac-bias recording	10 μs	10
Video recording	100 ns	5

remanence is acquired which depends on time and temperature. This effect is known in analog audio recording as print-through from one layer to another in a roll of tape (Sec. 3.2.4).

3.2.2 Preparation and properties of magnetic particles

3.2.2.1 Preparation of magnetic particles. Ideally, to assure good magnetic and dispersibility properties, fine magnetic particles should be prepared by a direct precipitation process. This is in order to obtain isolated particles with smooth surfaces. In addition, they should resemble the ideal ellipsoid on which all theoretical considerations regarding the magnetization reversal processes are based. Unfortunately, gamma ferric oxide, γ-Fe_2O_3 (synthetic maghemite), the most widely used magnetic recording material, is metastable and can be prepared only by oxidation of ferrous oxide, Fe_3O_4 (synthetic magnetite). To make things even more complicated, Fe_3O_4, because of its cubic lattice structure, cannot be directly prepared as elongated particles. Magnetite can be directly precipitated only as cubes, which allows a maximum coercivity of an isotropic powder sample, after oxidation to γ-Fe_2O_3, of about 15 kA/m (190 Oe). The coercivity is controlled by magnetocrystalline and shape anisotropy in about equal parts.

The preparation steps of various magnetic particle materials (Bate, 1980) are shown in Fig. 3.14 in terms of the achieved bulk density. First, needle-shaped iron hydroxide, FeOOH, is grown on precipitated seeds from a solution of iron salts and then dehydrated to the nonmagnetic modification α-Fe_2O_3 (synthetic hematite), which in turn is being reduced to Fe_3O_4. The hydroxide has three different acicular modifications which lead to γ-Fe_2O_3 particles of different morphological properties. Their principal methods of preparation and the reactions leading to γ-Fe_2O_3 are shown in Fig. 3.15. The orthorhombic α-FeOOH (synthetic goethite) has been mostly used so far. It consists of more or less well-defined particles which often have a fibrous structure (Fig. 3.16a). The tetragonal β-FeOOH (synthetic akageneite) can be made in well-defined needles (Fig.

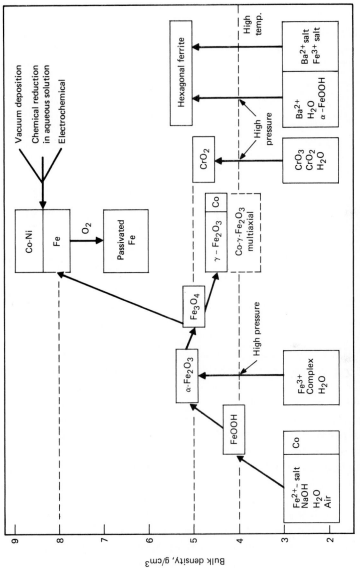

Figure 3.14 Stages of preparation and bulk density of magnetic particles.

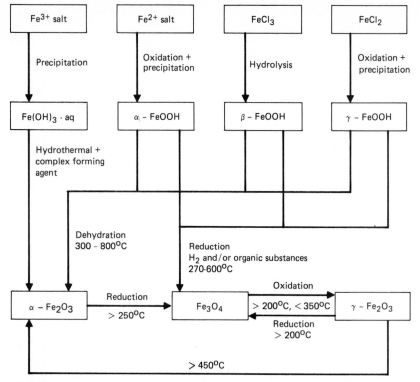

Figure 3.15   Principal methods of preparation and reactions for gamma ferric oxide particles.

3.17c) which, however, consist of hollow tubes which collapse in the dehydration and reduction processes to particles with ill-defined shape. Thus, β-FeOOH is not in commercial use for the preparation of fine magnetic particles. Synthetic lepidocrocite (γ-FeOOH) again is orthorhombic and forms preferably bundles of particles which are conserved through all stages of the preparation process (Fig. 3.16c and d).

The large material transport and the density change that take place in the solid-state reactions of the dehydration and reduction processes lead to Fe_3O_4 and consequently to γ-Fe_2O_3 particles with a rather irregular shape containing many pores (Fig. 3.16b and d). These deficiencies can be reduced to some extent by a direct hydrothermal synthesis of α-Fe_2O_3 from trivalent iron complexes under high pressure (Matsumoto, 1980). The directly precipitated α-Fe_2O_3 has a very dense structure and a well-defined shape (see Fig. 3.17a for particles made by a similar method). The final γ-Fe_2O_3, however (Fig. 3.17b), has some pores and irregularities at the surface. This is because, although the dehydration step has been eliminated, there is still material transport connected with the ensuing reduction and oxidation processes.

In order to assure well-crystallized particles with large coercivities, the dehydration and reduction processes need temperatures up to 700°C, which may cause the particles to form aggregates by sintering. Since this must be avoided at all costs, the FeOOH, or the directly precipitated α-Fe_2O_3 particles, are coated with organic or inorganic substances. Sometimes, elements such as Pr, Ni, Zn, or Sn are added in the preparation of FeOOH for this purpose, or for controlling the particle geometry. Indeed, most of the art of influencing the size and axial ratio of the magnetic particles takes place in the stage of seed formation and precipitation of the nonmagnetic precursor FeOOH.

The reduction process can be performed with hydrogen or organic substances such as oils of various kinds. The final oxidation should take place at temperatures below 350°C, because the ferrimagnetic γ-Fe_2O_3 will increasingly be transformed to the weakly magnetic α-Fe_2O_3 at higher temperatures. The pseudomorphic transformation from orthorhombic FeOOH to the inverse spinel structure of Fe_3O_4, and finally to the defect spinel structure of γ-Fe_2O_3, leads to polycrystalline particles with a predominantly $\langle 110 \rangle$ fiber axis parallel to the long axis of the particles. The intermediate product of partially oxidized Fe_3O_4, $(\gamma$-$Fe_2O_3)_x(Fe_3O_4)_{1-x}$

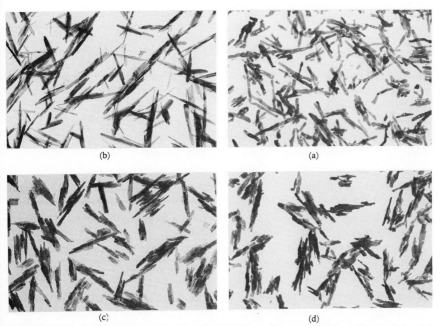

(b)   (a)

(c)   (d)

Figure 3.16   Electron micrographs of iron hydroxide and gamma ferric oxide particles: (a) α-FeOOH; (b) γ-Fe_2O_3 made from (a); (c) γ-FeOOH; (d) γ-Fe_2O_3 made from (c).

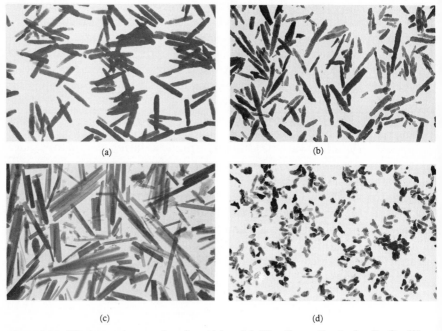

Figure 3.17 Electron micrographs of particles. (*a*) Directly synthesized α-Fe$_2$O$_3$; (*b*) γ-Fe$_2$O$_3$ made from (*a*); (*c*) β-FeOOH; (*d*) *Co-doped* γ-Fe$_2$O$_3$ *with multiaxial magnetic anisotropy.*

(synthetic berthollide), which has a slightly higher coercivity than γ-Fe$_2$O$_3$, is simply produced by interrupting the oxidation process at the desired value of x.

After the oxidation process, iron oxide powder batches are usually mechanically treated by tumbling in large drums or pressing between steel rolls, or by other means. Apart from the improved handling of powders with a higher apparent density, the mechanical treatment or densification also improves the dispersibility of the particles, as indicated in Fig. 3.18, possibly by a breaking up of agglomerates. Here the orientation ratio, the gloss of the as-coated and dried dispersion, and its coercivity are plotted as a function of the apparent density. Gloss is the reflectivity of the coating which is measured in this instance under an angle of incidence of 85°. It represents a convenient figure of merit for the homogeneity of particle dispersion. While the orientation ratio and the gloss indicate a maximum of dispersibility at a density of about 0.7 g/cm^3, the decreasing coercivity signals an increasing mechanical damage of the particles at higher degrees of densification. Sometimes, the particles are simultaneously treated with dispersants during this mechanical process to promote better dispersibility.

The substitution of a small amount of cobalt ions for iron ions causes a large increase of the magnetocrystalline anisotropy, as was initially shown for Fe_3O_4 (Bickford et al., 1956). Similarly, the magnetocrystalline anisotropy of γ-Fe_2O_3 can be increased by cobalt substitution, with a corresponding increase of the coercivity of the γ-Fe_2O_3 particles. Gamma ferric oxide with cobalt ions evenly distributed in the volume of the particles is made either by adding cobalt salts in the solution before the precipitation of the FeOOH, or by precipitation of cobalt hydroxide on the already formed FeOOH particles. The ensuing reduction and oxidation processes secure the homogeneous diffusion of the cobalt ions into the volume of the

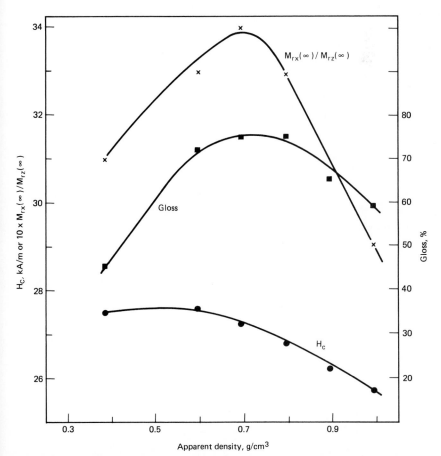

Figure 3.18 Orientation factor $M_{rx}(\infty)/M_{rz}(\infty)$, coercivity H_c, and gloss of as-coated and dried dispersion, versus the apparent density of γ-Fe_2O_3 particles after different degrees of densification *(Jakusch, 1984)*.

particles. The Co substitution causes a multiaxial magnetic anisotropy of the particles which can be used for the preparation of isotropic recording media. Such media may be used for a combined longitudinal and perpendicular recording (Lemke, 1982). These particles either are cubic in shape, or have a low axial ratio a/b of about 2 (Fig. 3.17d). The latter are a compromise between a dominant multiaxial anisotropy with a pronounced temperature dependence of coercivity and remanence, and a dominant uniaxial anisotropy with a less pronounced temperature dependence of the magnetic properties. In the preparation process of this type of particle, the content of Fe^{2+} ions must be kept as low as possible because, in

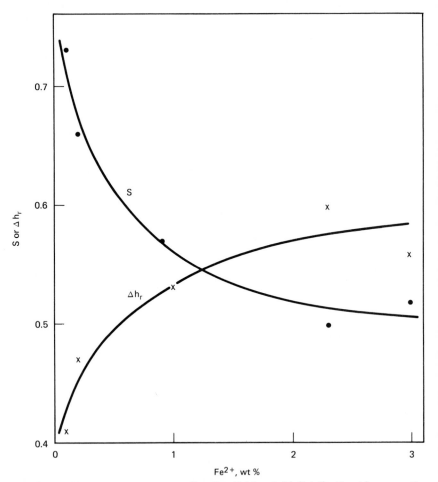

Figure 3.19 Remanence squareness S and switching-field distribution Δh_r versus the amount of Fe^{2+} ions present in volume-doped γ-Fe_2O_3 particles with an axial ratio of $a/b = 2$ *(Steck, 1985)*.

the presence of Fe^{2+} ions, an induced uniaxial anisotropy tends to be developed which reduce the squareness. This is shown in Fig. 3.19, where S is plotted versus the amount of Fe^{2+} present, for a constant amount of 5.5 wt % Co. It is of interest to note that, simultaneously with a decreasing S, the switching-field distribution Δh_r becomes larger, another reason to keep the Fe^{2+} content as low as possible.

The modification of γ-Fe_2O_3 with Co exclusively at the particle surface almost completely conserves the uniaxial magnetic anisotropy of γ-Fe_2O_3. This approach is normally used in order to increase the coercivity of gamma ferric oxide particles beyond the value of 32 kA/m (400 Oe) achievable for pure gamma ferric oxide. These surface-modified particles are prepared by precipitation of cobalt hydroxide, with or without the presence of Fe^{2+} ions from an aqueous solution, and by a subsequent annealing of the dry powder (Schönfinger et al., 1972; Umeki et al., 1974; Shimizu et al., 1975). This annealing step must be controlled carefully in order to avoid diffusion of the Co or Fe^{2+} ions far into the volume of the particles (Witherell, 1984). An interesting dry process of preparing Co-modified γ-Fe_2O_3 is the pyrolytic decomposition of Co organometallic vapor on the surface of Fe_3O_4 particles in a fluid-bed reactor (Monteil and Dougier, 1980).

Carrying the reduction of FeOOH beyond Fe_3O_4 leads to a further increase of the bulk density to that of the pure metal (Fig. 3.14). This step is even more sensitive to sintering effects, making a protective surface treatment of the FeOOH precursor mandatory in order to obtain well-defined particles like those shown in Fig. 3.20a. Sintering is aggravated because the FeOOH particles must be chosen to be even smaller than those for the preparation of γ-Fe_2O_3. For instance, the specific surface area may be up to 100 m^2/g instead of up to 60 m^2/g. This is in order to obtain reasonable values of coercivity for iron or iron-cobalt particles, since the critical radius b_0 of iron for uniform magnetization reversal is about five times smaller than that of γ-Fe_2O_3 (Sec. 3.2.1). With pure iron, coercivities of up to about 140 kA/m (1760 Oe) have been obtained. The high specific surface area of the iron particles makes them pyrophoric, and they must be passivated by a controlled oxidation of their surface. In practice, about 50 percent of a particle's volume is oxidized with a kernel of pure iron. Consequently, the bulk density is lowered as indicated in Fig. 3.14.

The methods for a direct synthesis of metal particles, such as evaporation of the metal in an inert gas atmosphere (Tasaki et al., 1979), chemical reduction with borohydride (Oppegard, 1961), and electrochemical deposition (Luborsky, 1961), are economically of minor importance. More recently, there has been some interest in metal particles with intermediate coercivities which may be obtained by reducing the spontaneous magnetization. For this purpose, the iron can be alloyed to some extent with

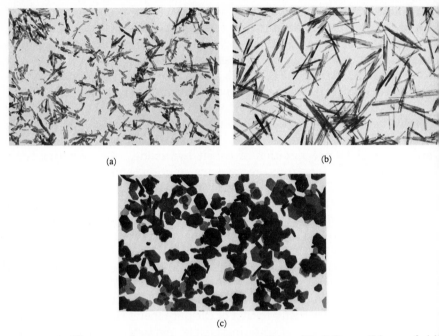

(a)   (b)

(c)

Figure 3.20 Electron micrographs of (a) iron particles, (b) CrO_2 particles, and (c) $BaFe_{12}O_{19}$ platelets.

nickel by coprecipitation from Fe and Ni salts when starting from FeOOH. An easier method is by the precipitation of nickel hydroxide onto the already formed FeOOH particles. An evaporation method can also be used. Another way of reducing the spontaneous magnetization is the formation of Fe_4N by heating Fe particles in a mixture of hydrogen and ammonia (Tasaki et al., 1981). Iron carbonitrides ($Fe_4N_xC_{1-x}$), prepared by a similar process, have been reported to be more stable to oxidation than Fe_4N (Andriamandroso et al., 1984). Another route for reducing the coercivity is the reduction of the axial ratio of the particles. An increase of particle diameter at constant axial ratio is not feasible, since the increased particle volume leads to prohibitive noise characteristics.

Chromium dioxide can be directly synthesized in a hydrothermal process at pressures between 3 and 7×10^6 Pa (4 to 10×10^6 lb/in^2) by reduction of CrO_3, or by oxidation of Cr_2O_3. In both cases, the particle size and its axial ratio are controlled by adding small amounts of Fe, Te, and Sb as doping elements. This process leads to perfectly crystallized, elongated particles (Fig. 3.20b). The addition of Fe and Ir increases the magnetocrystalline anisotropy of CrO_2. The reduction in size by the addition of antimony is illustrated in Fig. 3.21, where lines of constant specific surface area are drawn in a plot of Sb versus Fe content. Here Sb effectively increases the specific surface area; that is, it reduces the particle size and consequently its radius. The radius controls the contribution of shape anisotropy to the overall coercivity. The addition of Fe increases the coercivity through an increase of magnetocrystalline anisotropy as shown by Fig. 3.22, where lines of constant coercivity are drawn in the same diagram of Sb versus Fe content (Chen et al., 1984).

Since CrO_2 can be a chemically oxidizing substance, the particles are made to have a thin protective layer, for example, of orthorhombic CrOOH. This layer reduces the specific saturation magnetization but not the saturation magnetization which controls the coercivity. In large-scale production, CrO_2 with a coercivity of up to about 55 kA/m (690 Oe) can be prepared.

The hexagonal ferrites, in particular barium ferrite, are of interest for perpendicular magnetic recording because they can be precipitated by

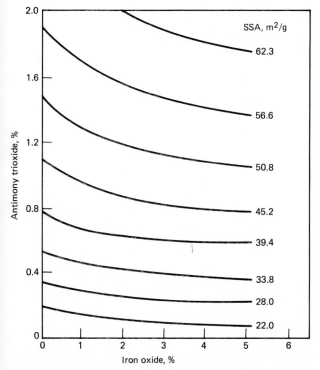

Figure 3.21 The specific surface area (SSA) that results from varying the percentages of antimony trioxide and iron oxide in the preparation of chromium dioxide particles *(Chen et al., 1984).*

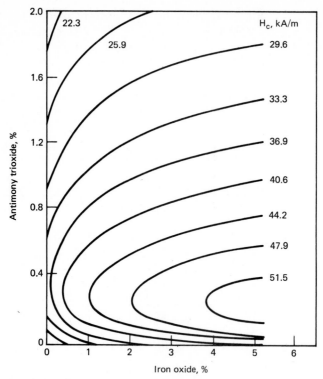

Figure 3.22 The coercivity H_c that results by varying the percentages of antimony trioxide and iron oxide in the preparation of chromium dioxide particles *(Chen et al., 1984).*

various processes as hexagonal platelets (Fig. 3.20c). Their easy axis of magnetization is parallel to the c axis, which is perpendicular to the plane of the platelets. The possible preparation methods are precipitation in a NaCl-KCl flux, hydrothermal synthesis from $Ba(OH)_2$ with α-FeOOH, or from $Ba(NO_3)_2$ with $Fe(NO_3)_3$ and NaOH, and quenching of a BaO + B_2O_3 + Fe_2O_3 melt to an amorphous glass with subsequent cystallization by annealing. Since the coercivity of barium ferrite can be as high as 480 kA/m (6000 Oe), it is usually reduced by adding Co, Ni, Mn, Zn, or Ti in order to be used in magnetic recording (Smit and Wijn, 1959; Kubo et al., 1982). Needle-shaped barium ferrite can also be made by a pseudomorphic reaction of elongated FeOOH particles as used for the preparation of γ-Fe_2O_3 particles with Na_2CO_3 (Takada et al., 1970). The c axis as the easy direction of magnetocrystalline anisotropy is perpendicular to the particle axis. A review of preparation methods is given by Hibst (1982).

The magnetic powder properties for the materials discussed in this section are given in Table 3.5, together with their bulk density, range of specific surface area, and particle volumes.

3.2.2.2 Iron oxides. The first experimental magnetic tapes were made with carbonyl-iron particles of more than 1 μm in diameter, and the first commercial audio tapes were produced with equant Fe_3O_4 and, later, with equant γ-Fe_2O_3 particles. Elongated γ-Fe_2O_3 particles were introduced in 1954 and are still most widely used in magnetic recording. Gamma ferric oxide combines the advantages of being chemically and structurally stable, and has a magnetic anisotropy which is dominated by shape. The particles tend to consist of individual crystallites with a $\langle 110 \rangle$ fiber axis parallel to the particle axis with other orientations occurring as well (Gustard and Vriend, 1969). Both magnetocrystalline and shape anisotropy seem to be additive, with a contribution of the magnetocrystalline anisotropy to the coercivity of about 6 kA/m (75 Oe) (Eagle and Mallinson, 1967; Köster, 1970). The crystallite size plays an important role because it degrades the magnetic properties of the particles if it becomes smaller than the particle diameter. This is demonstrated to the extreme by the data in Fig. 3.23 (Berkowitz et al., 1968). Here, the specific saturation magnetization σ_s and the coercivity H_c are plotted as a function of the crystallite size. With decreasing crystallite size, σ_s is reduced by nonmagnetic grain boundaries approximately 0.6 nm thick. Simultaneously, the crystallites become more sensitive to thermal fluctuations. The coercivity has a typical maximum as a function of crystallite size as discussed earlier. Consequently, the print-through effect in analog audio recording increases with decreasing crystallite size (Tochihara et al., 1970).

The particle size, or more precisely its volume v, has been reduced considerably over the years from over 30×10^{-4} to about 2×10^{-4} μm^3 in order to reduce the noise of the recording media. The corresponding reduction of the particle diameter results in a more uniform reversal of magnetization. This leads to an increase of coercivity from below 20 to

TABLE 3.5   Specific Saturation Magnetization σ_s, Bulk Density ρ, Coercivity H_c, Mean Particle Volume $\langle v \rangle_{av}$, and Specific Surface Area (SSA) of Isotropic Powder Samples†

Material	σ_s, A·m²/kg (emu/g)	ρ, g/cm³	H_c, kA/m (4π Oe)	$\langle v \rangle_{av}$, 10^{-4} μm^3	SSA, m²/g
Fe	125–170	5.8	75–130	< 1	35–50
Fe(70 at %) Co (30 at %)	140–190	5.8	90–160	< 1	35–50
Fe_4N, Fe-Ni	110–120	5.8	55–90	< 1	35–50
CrO_2	76–84	4.8	30–50	1–10	18–36
γ-Fe_2O_3	73–75	4.8	20–32	1–25	12–35
γ-Fe_2O_3 + Co	70–75	4.8	30–70	1–10	16–33
$BaFe_{12-2x}Co_xTi_xO_{19}$					
$x = 0.7$	58	5.3	95	< 2	20–35
$x = 0.8$	58	5.3	56	< 2	20–35

†With a volumetric packing fraction of $p = 0.25$.

more than 32 kA/m (250 to 400 Oe), and a decrease of the switching-field distribution Δh_r from 0.6 to 0.5. These data are for an isotropic powder sample with a volumetric packing fraction of $p = 0.25$. Unfortunately, the effect of thermal fluctuations, and consequently print-through, become more prominent with smaller particle volumes. A decrease in the distribution of particle volumes and larger crystallite sizes have helped to keep the print-through signals at reasonable values.

The use of γ-FeOOH as a precursor for γ-Fe$_2$O$_3$ was introduced commercially in 1970. The advantages are a better degree of orientation and a higher volumetric packing compared to particles made from the α-FeOOH used previously (Gustard and Wright, 1972). These improvements have been attributed to bundlelike aggregates of parallel particles. The remanence squareness can be increased from 0.72 to 0.85, and the volumetric packing fraction from 0.35 to 0.45. Consequently, the saturation remanence of audio tapes can be improved from about 95 to 135 kA/m (emu/cm^3). Meanwhile, the particles made from α-FeOOH were improved by carefully avoiding any sintering effects. Recently, particles based on a direct hydrothermal precipitation of α-Fe$_2$O$_3$ (Matsumoto, 1980) were introduced which are similar to the particles made from α-FeOOH, but which lead to about 10 percent higher saturation remanence under the same preparation conditions.

Magnetite Fe$_3$O$_4$ has the intrinsic advantage of a 37 percent higher

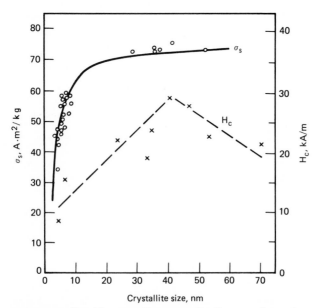

Figure 3.23 Specific saturation magnetization σ_s and coercivity H_c as a function of the crystallite size of elongated γ-Fe$_2$O$_3$ particles *(Berkowitz et al., 1968).*

spontaneous magnetization than γ-Fe_2O_3. It has, however, been little used in magnetic recording media owing to relatively high print-through signals and the precautions necessary to guard against partial oxidation during the medium manufacturing process. The intermediate state, $(\gamma$-$Fe_2O_3)_x(Fe_3O_4)_{1-x}$, has a maximum in coercivity at $x = 0.85$, which can be as much as 30 percent larger than that of the corresponding γ-Fe_2O_3 particles. However, these nonstoichiometric iron oxides suffer from prohibitive aging effects, even at room temperature, due to a directional pair-ordering effect of divalent iron ions (Imaoka, 1965; Chikazumi, 1974). Whether this divalent pair-ordering effect, or vacancy ordering, or even stress anisotropy, is the origin of the increased coercivity is still under discussion (Kishimoto, 1979; Kaneko, 1980; Kojima and Hanada, 1980).

3.2.2.3 Cobalt-modified iron oxides.

It has been known for some time that the coercivity of γ-Fe_2O_3 is increased by introducing cobalt ions which replace trivalent iron ions on octahedral sites and simultaneously fill octahedral site vacancies (Jeschke, 1954; Khalafalla and Morrish, 1972). Since the cobalt ions have a large positive magnetocrystalline anisotropy, the $\langle 100 \rangle$ axes are easy directions of magnetization. Two of them are at 45° and one at 90° to the $\langle 110 \rangle$ fiber axis, which is the easy direction of the shape anisotropy. Consequently, for an isotropic powder sample, the remanence squareness increases from 0.5 for unmodified γ-Fe_2O_3, to 0.83 for dominating magnetocrystalline anisotropy. In close accordance with theoretical considerations (Tonge and Wohlfarth, 1958), the transition from dominating shape anisotropy to dominating multiaxial magnetocrystalline anisotropy occurs at $H_k/H_m = 1.16$ (Köster, 1972). The strong temperature dependence of the magnetocrystalline aniostropy thus leads to a temperature dependence of S (Speliotis et al., 1964). This is demonstrated in Fig. 3.24, where remanence squareness S and remanence coercivity H_r are plotted as a function of temperature. While γ-Fe_2O_3 (sample A) shows almost no variation of S and H_r in the temperature range from 140 to 420°K, Sample C, with 2.5 wt % Co in the particle volume, has a sharp increase of both quantities below 370°K. For particles (sample D) which have an axial ratio of 2 instead of the usual value of 7, S is much higher at room temperature, and H_r has a much more pronounced temperature dependence. The continuous increase of H_r in the curves of Fig. 3.24 indicates that shape and magnetocrystalline anisotropy add constructively to H_r, although their easy axes do not coincide. The particles of sample D behave more like multiaxial particles and are candidates for use in isotropic recording media (Lemke, 1982). However, such particles show strong temperature dependence of their magnetic properties. Also, reductions of recorded signals may occur after repeated use because of the high magnetostriction effects associated with cobalt ions. These defects may limit extensive commercial use of these particles.

Although it has raised interest since the early days of magnetic recording (Krones, 1960), cobalt-substituted Fe_3O_4 or $(\gamma\text{-}Fe_2O_3)_x(Fe_3O_4)_{1-x}$ has been little used in magnetic recording. Such materials suffer from the same defects described in the previous section for the unmodified materials. Furthermore, Co in $(\gamma\text{-}Fe_2O_3)_x(Fe_3O_4)_{1-x}$ develops large induced anisotropies after aging in external fields (Kamiya et al., 1980). This leads to recording signals which cannot be erased after prolonged storage, because they are permanently "recorded" by locally induced anisotropies.

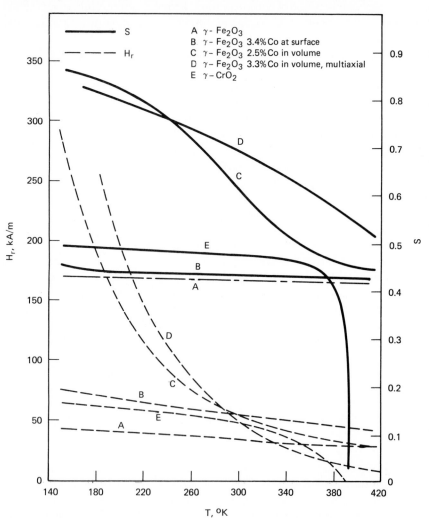

Figure 3.24 Remanence squareness S and remanence coercivity H_r as a function of temperature T for isotropic samples of (curve A) elongated $\gamma\text{-}Fe_2O_3$ particles, (curve B) the same with 3.4% Co concentrated at the particle surface, (curve C) the same with 2.5% Co evenly distributed in the particle volume, (curve D) almost equant $\gamma\text{-}Fe_2O_3$ particles with 3.3% Co evenly distributed in the particle volume, and (curve E) elongated CrO_2 particles.

Because of the pronounced temperature dependence of their magnetic parameters, little commercial use has been seen for γ-Fe$_2$O$_3$ with Co ions diffused throughout the particle volume. A breakthough occurred, however, with the idea to adsorb cobalt ions at the particle surface (Schönfinger et al., 1972; Umeki et al., 1974). A strong anisotropy energy of the order of 1.5×10^5 J/m^3 (1.5×10^6 erg/cm^3) is reported to be found at the particle surface which inhibits the nucleation of magnetization reversal at any irregularities (Tokuoka et al., 1977). Thus, the shape anisotropy as given by Eq. (3.16) can be more fully utilized, and the coercivity is raised to about 42 kA/m (530 Oe) without an increase of its temperature dependence. For this to happen, the additional anisotropy at the particle surface need only be above a certain threshold of the order of the magnetostatic energy $0.25 \, \mu_0 M_{sb}^2 = 3.8 \times 10^4$ J/m^3 ($\pi M_{sb}^2 = 3.8 \times 10^5$ erg/cm^3) in order to prevent local demagnetizing fields from nucleating magnetization reversal at the surface. The coercivity increases proportionally to the amount of Co adsorbed until about 2 wt % of Co. Above this point, H_c stays at a constant value and the specific saturation magnetization starts to decrease (Tokuoka et al., 1977). The annealing step involved in the preparation of these particles makes it likely that, up to 2% Co concentration, the Co ions are incorporated into the iron oxide lattice at the particle surface. Thus, a thin layer of cobalt-modified iron oxide may be formed which has a high anisotropy energy density. Consequently, attempts were made to increase the thickness of this layer by depositing onto the particles Co^{2+} and Fe^{2+} ions simultaneously (Shimizu et al., 1975). A similar increase of coercivity, proportional to the amount of cobalt added, has been found without an increase of its temperature dependence. Thus, sample B in Fig. 3.24 exhibits little more temperature dependence of S and H_r than γ-Fe$_2$O$_3$ itself, although it has the same room temperature value of $H_r = 55$ kA/m (690 Oe) as sample C with the Co evenly distributed in the particle volume. The coercivity and specific saturation magnetization increase continuously with the amount of added Co and Fe ions in contrast with the Co-adsorbed type. This and the observed increase in particle size suggests the formation of a shell of cobalt-modified iron oxide around the γ-Fe$_2$O$_3$ particle. Beyond a coercivity of about 60 kA/m (750 Oe), the particles increasingly lose their uniaxial anisotropy, and, consequently, their magnetic parameters become increasingly temperature-dependent (Kishimoto et al., 1981). This indicates that beyond this coercivity, cobalt-modified iron oxide, with its cubic magnetocrystalline anisotropy, dominates the nucleation of magnetization reversal, an effect which seems to be retarded by an induced anisotropy with the easy axis of magnetization parallel to the particle axis. It is of great importance at this point that no nonstoichiometric iron oxide compounds are created. These will lead to additional induced anisotropy phenomena as has been discussed before in the context of cobalt substitution in the volume of the particles.

3.2.2.4 Chromium dioxide. Chromium dioxide (CrO_2) particles are single-crystal needles with a tetragonal rutile structure. The easy direction of magnetocrystalline anisotropy has been reported either as parallel to the c axis, which is parallel to the particle axis (Rodbell, 1966), or inclined by 40° to the c axis (Cloud et al., 1962). The remanence coercivity H_r of a spatially isotropic particle assembly is expected to be $H_r = 0.529H_k = 38.6$ kA/m (485 Oe) for uniaxial magnetocrystalline anisotropy, and $H_r = 0.529(N_b - N_a)M_{sb} = 0.23M_{sb} = 110.4$ kA/m (1387 Oe) for shape anisotropy ($a/b = 6$) and uniform magnetization reversal. If the axes of easy magnetization coincide, both contributions add to give $H_r = 149$ kA/m (1872 Oe). In the case of CrO_2 made with 0.15 wt % Sb and 0.25 wt % Fe, a value of $H_r = 54.8$ kA/m (689 Oe) has been measured, extrapolated to zero volumetric packing fraction. This value can be separated into a contribution by magnetocrystalline anisotropy of 21.2 kA/m $= 0.29H_k$, and a contribution by shape anisotropy of 33.6 kA/m $= 0.07M_{sb}$ (Köster, 1973). The value for shape anisotropy may be attributed to a typical reduced particle radius of $b/b_0 = 3$. There is, however, no obvious explanation for the rather small contribution of magnetocrystalline anisotropy, unless the easy axis of magnetocrystalline anisotropy is indeed assumed to be inclined with respect to the c axis. In the case of 40° of inclination, a contribution of $0.3H_k$ can be calculated, together with an equilibrium position of the magnetization direction in zero field of about 10° off the c axis. Whether this is evidence enough against collinear magnetocrystalline and shape anisotropy in CrO_2 particles is still open to discussion.

The temperature dependence of the magnetocrystalline anisotropy energy density of CrO_2 is about equal to that of M^2_{sb} over a wide range; therefore H_a/M_s is almost independent of temperature up to about 80°C. Beyond this temperature it drops at an increasing rate to zero at the Curie temperature, typically 125°C (Rodbell, 1966; Köster, 1973). This is equally true for the inverse demagnetization criterion H_c/M_r and for the remanence squareness (curve E in Fig. 3.24), thus making CrO_2 a viable material for magnetic recording in spite of its relatively low Curie temperature.

The coercivity of CrO_2 can be controlled not only by the particle radius (already shown in Fig. 3.5) but also by its magnetocrystalline anisotropy. In Fig. 3.25, the magnetocrystalline anisotropy field H_k, and Curie temperature T_c, are plotted as a function of the amount of Fe added. Values as high as $H_k = 160$ kA/M (2000 Oe) have been achieved. An even more dramatic increase of H_k is found by using Ir additives in the hydrothermal synthesis of CrO_2 which increase the spin-orbit coupling constants (Maestro et al., 1982). Values of H_k up to 580 kA/m (7300 Oe) are possible, and the remanence coercivity H_r can almost equal $0.529H_k$, the value expected for magnetization reversal by uniform rotation (Kullmann et al., 1984b). Apart from this rather expensive possibility, it has been possible to obtain coercivities of up to 55 kA/m (690 Oe) in production quantities.

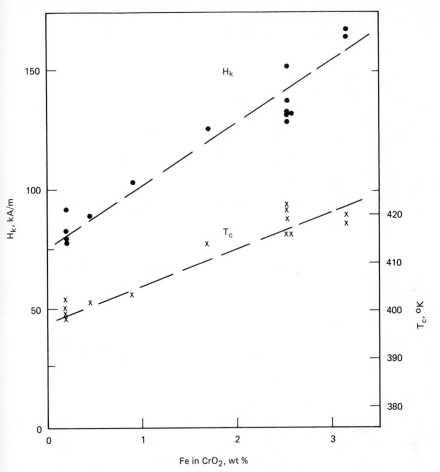

Figure 3.25   Magnetocrystalline anisotropy field H_k and Curie temperature T_c versus the amount of iron in chromium dioxide particles.

Chromium dioxide particles can be made in very small sizes, with mean particle volumes of the order of 10^{-4} μm^3, using the direct hydrothermal precipitation process. This partially explains a marked difference between coercivities measured statically, and the coercivity in the rapidly changing fields as experienced in magnetic recording. Thus, depending on the type of magnetic recording application, the 55 kA/m cited above is equivalent to about 62 to 70 kA/m (780 to 880 Oe) in cobalt-modified iron particles of larger particle size.

The almost perfect shape of CrO_2 particles enables them to be well aligned, with a reduced saturation remanence as high as $S = 0.88$. The more irregularly shaped cobalt-modified iron oxide particles have a lesser degree of particle alignment. They may, however, have about the same value of S due to some multiaxial anisotropy contribution of the cobalt ions.

3.2.2.5 Metal particles. Iron particles, as prepared by reduction of the various precursors of gamma ferric iron oxide particles, consist of a kernel of 45 to 55 vol % metal with a shell of ill-defined magnetic iron oxides. The magnetostatic energy and nucleation fields for magnetization in this situation can be analyzed theoretically, based on a model of coaxial prolate spheroids (Stavn and Morrish, 1979). It was found that the coercivity is almost unaltered by the magnetic oxide shell, and determined by the shape anisotropy of the iron kernel only. Thus, the maximum possible coercivity of a powder sample of infinite cylinders with a volumetric packing fraction of 0.25 is 307 kA/m (3860 Oe). So far, values have been obtained of up to 140 kA/m (1760 Oe). The contribution of magnetocrystalline anisotropy may be neglected, since its contribution to H_c amounts to only 13 kA/m (163 Oe) (Fig. 3.11). The specific saturation magnetization σ_s of the particles is reduced to 125 to 170 A·m²/kg (125 to 170 emu/g) and the bulk density to 5.4. to 6.0 g/cm³ because of the surface oxide shell. Consequently, magnetic recording media made of iron particles have a saturation remanence of about 240 kA/m (emu/cm³).

Apart from the application in high-coercivity recording media, attempts have been made to utilize the higher saturation magnetization in existing audio and video recording systems which are designed for coercivitites between 45 and 60 kA/m (560 and 750 Oe). In order to achieve these low coercivities, however, the particle diameter, which controls H_c, and consequently the particle volume, becomes too large for reasonable noise values. In addition, the switching-field distribution Δh_r increases with decreasing H_c due to a less uniform magnetization reversal mode. Consequently, a compromise solution has been evolved by keeping the particle dimensions constant and reducing the saturation magnetization. This is achieved by alloying with Ni or by forming Fe_4N.

The metal particles obtained by condensation of vapor in an inert gas atmosphere (Tasaki et al., 1979) are spherical, and form linear chains similar to those obtained by the reduction of metal salts with borohydride (Oppegard et al., 1961), or the decomposition of metal carbonyls (Thomas, 1966). Using the evaporation method, a coercivity range of 55 to 180 kA/m (700 to 2300 Oe) has been achieved by alloying with Ni or Co, and keeping the particle diameter constant at approximately 20 nm (Tasaki et al., 1983). The specific saturation magnetization has been varied correspondingly, from 110 to 200 A·m²/kg (110 to 200 emu/g).

3.2.2.6 Barium ferrite. Barium ferrite ($BaFe_{12}O_{19}$), as such, is of little use for magnetic recording since its intrinsic anisotropy field $H_a = (2K_1/\mu_0 M_{sb}) - M_{sb} = 1000$ kA/m (13,000 Oe) is extremely high. Shape and

magnetocrystalline anisotropy are orthogonal, and contribute with opposite sign to the anisotropy field. This is because the easy direction of magnetocrystalline anisotropy is perpendicular to the plain of the thin hexagonal barium ferrite platelets. The resulting remanence coercivity for
uniform magnetization reversal is given by $H_r = 0.529 H_a$, or 530 kA/m
(6660 Oe). This value is indeed found with particles in the size range of
0.1 to 0.3 μm, as precipitated in a NaCl flux. Increased interest has been
raised for this type of material through the possible use in recording
media for perpendicular recording. The idea is to orient the platelets flat
in the plane of the recording media, thus having an easy direction of magnetization perpendicular to the plane. The magnetocrystalline anisotropy
can be lowered through the substitution of the Ba^{2+} or Fe^{3+} ions by di- or
trivalent ions, together with a reduction in M_{sb} and T_c. A more drastic
reduction in anisotropy is produced using Co with Ti, or Zn with Ti, in
$BaMe_x^{2+} Me_x^{4+} Fe_{12-2x}O_{19}$, as a combined substitution of the trivalent iron
ions. Here Me^{2+} and Me^{4+} stand for the di- and tetravalent substitution
ions, respectively. This type of substitution dramatically reduces H_k, and
leads to an easy plane of magnetization perpendicular to the c axis, for x
> 1.2 and $x > 0.3$ for the Co-Ti and Zn-Ti substitution, respectively.
Thus, little saturation magnetization is lost in varying H_c over the whole
range from almost zero to the maximum possible value. On the other
hand, particles which are small enough to provide good noise characteristics (particle volumes of the order of $2 \times 10^{-4} \mu m^3$) are difficult to prepare in a way such that they are easily dispersed and aligned in a magnetic
field.

3.2.2.7 Morphological and physicochemical aspects.
So far, only the
mean particle dimensions have been addressed. Particle length and particle diameter are both subject to a distribution function, as illustrated in
Fig. 3.26. Here, the normalized particle count is plotted versus particle
diameter $2b$, using a Gaussian distribution as an approximation for the
measured histograms. Representative samples of γ-Fe_2O_3, CrO_2, and Fe
particles have been chosen. Similar distributions exist for the particle
length. The characteristic feature of these distribution functions is an
increase in the standard deviation, as well as a shift of the distribution
curve, with increasing mean particle dimensions. The most conveniently
measured quantity for characterization of the mean particle size is its specific surface area, and this is widely used. It scales quite well with the
particle size and varies from 16 m^2/g for sample A of Fig. 3.26 to 48 m^2/g
for sample F. Some caution has to be applied when using the specific surface area as a measure for particle volume. Measured areas are influenced
by the axial ratio a/b, by the bulk density, and in particular by pores in
the particles. Such factors can lead to misinterpretations of the specific
surface area data, particularly for metal particles.

The typical shapes of various representatives of magnetic fine particles

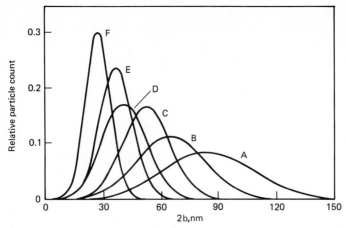

Figure 3.26 Normalized particle count versus particle diameter $2b$ for (curve A) γ-Fe_2O_3 particles (SSA = 16 m^2/g) as used in computer tapes or floppy disks, (curve B) Fe_2O_3 particles (SSA = 20 m^2/g), (curve C) γ-Fe_2O_3 particles (SSA = 24 m^2/g), (curve D) CrO_2 particles (SSA = 27 m^2/g) as used in audio cassette tapes (curve E) Co-modified γ-Fe_2O_3 (SSA = 36 m^2/g) and (curve F) Fe particles (SSA = 48 m^2/g) as used for video tapes *(Jakusch, 1984).*

have been shown in Figs. 3.16, 3.17, and 3.20. It is common experience that particles with irregularly shaped surfaces yield less perfect degrees of alignment than those with smooth surfaces, such as CrO_2. This is due to a steric, frictionlike interference between the particles. The magnetic torque of the field applied for alignment increases proportionally to the volume v of the particles. Also, the friction between the particles increases with particle volume at a rate between $v^{1/3}$ and $v^{2/3}$. For these reasons, larger particles usually yield a better degree of alignment than smaller particles. Thus γ-Fe_2O_3 made from γ-FeOOH, which exists in the form of bundles of particles that rotate as a single-particle unit, gives a much better degree of alignment than the more individual particles made from α-FeOOH. In addition, large particles lend themselves to a higher volumetric packing fraction than small particles, an effect which again is connected with the surface-to-volume ratio (Patton, 1979).

The mechanical energy needed to disperse the particles in a binder solution, apart from depending on the degree of agglomeration produced in the synthesis of the particles, increases with decreasing particle size. This is demonstrated in Fig. 3.27, where the gloss of the as-coated and dried dispersion is plotted versus the milling time for the same particles used for Fig. 3.26. With increasing specific surface area, a longer milling time is needed to obtain the optimum gloss value (Patton, 1979). On the other hand, smaller particles yield a smoother final dispersion, as is indicated by the larger final gloss value. Exceptions from this general trend are γ-Fe_2O_3 particles made from γ-FeOOH (curve B), and CrO_2 particles

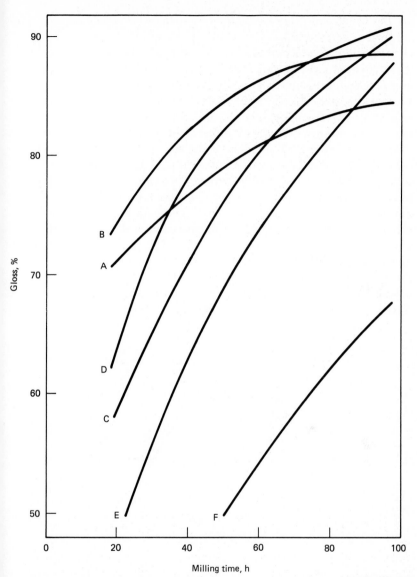

Figure 3.27 Gloss versus milling time for the same powder samples as in Fig. 3.26: (curve A) through (curve C) γ-Fe$_2$O$_3$, (curve D) CrO$_2$, (curve E) Co-modified γ-Fe$_2$O$_3$, and (curve F) Fe particles *(Jakusch, 1984)*.

(curve *D*), which are more easily dispersed than other particles of comparable size (curves *A* and *C*).

When considering the concentration of particles in a magnetic coating, one has to distinguish between the pigment-volume concentration and the packing fraction introduced earlier. The former is the volume occupied by the particles as related to the nonvolatile components of a dispersion. The latter is related to the nonmagnetic volume of the coating (Patton, 1979). At low concentrations, both are identical since the particles are embedded in a continuous matrix of binder and additives. At higher pigment-volume concentrations, typically above a critical value of 0.3, the packing fraction increases more slowly and finally stays at a maximum value independent of the pigment-volume concentration. The closest packing of the particles is approached, and the magnetic coating increasingly contains pores (Huisman and Rasenberg, 1984). This situation is illustrated schematically in Fig. 3.28, where the packing fraction of the particles is shown versus the pigment-volume concentration. Usually, a pigment-volume concentration slightly above the critical value is chosen to exploit the compressive effect of the calendering and to obtain the highest possible saturation remanence (Rasenberg an Huisman, 1984). For iron oxide and

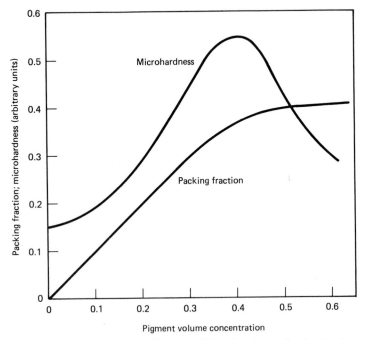

Figure 3.28 Schematic variation of packing fraction and microhardness with pigment-volume concentration of an as-dried coating.

chromium dioxide pigments, this amounts to a pigment-volume concentration of 0.45 to 0.55, and for metal pigments of 0.35 to 0.45. In comparison, the maximum values of the packing fraction are about 0.47 and 0.37, respectively. The lower values for metal pigments are accounted for by their smaller particle volume; that is, the critical volume concentration and the packing fraction usually decrease with decreasing particle volume. On the other hand, both can be considerably influenced by the dispersant used. An upper usable limit for the pigment-volume concentration is governed by the weakening of the bonds between the particles in the magnetic coating caused by increasing porosity. Below the critical particle-volume concentration, the particles act like a filler that leads to increased mechanical wear resistance. Above the critical value, the mechanical strength is weakened by the porosity of the coating (Dasgupta, 1984; Huisman and Rasenberg, 1984). This effect is shown in Fig. 3.28 in terms of the microhardness of the coating.

The effect of the dispersants in the milling process depends on the chemical nature of the surface of the particles (Sugihara et al., 1980), the dispersant, and the binder system used. The chemical nature of the surface of the particles can be investigated by measuring the wetting energy of test substances such as ethyl alcohol as a proton donor, dioxane as a proton acceptor, and nitromethane as a molecule with a dipole moment. The differences in wetting energies of the three test substances using typical γ-Fe_2O_3 and Co-modified γ-Fe_2O_3 particles are listed in Table 3.6. These data confirm the general experience that unmodified γ-Fe_2O_3 has an acidic surface character and polar characteristics which are reduced by the Co-modification process taking place in an alkaline solution. Surprisingly, CrO_2 exhibits a strong proton-donor adsorption which is not fully understood. The wetting energy can also be used to select suitable dispersants, and the isothermal curve of adsorption can be used to determine the amount of dispersant needed. A combination of both measurements can be performed in a flow calorimeter with the chosen solvent. The results obtained, however, have to be confirmed, or be adjusted, by tests with the binder system in the real formulation. The reason for this is a possible competitive adsorption of the binder and the dispersant on the

TABLE 3.6   Wetting Energy of Ethanol as a Proton Donor, Dioxane as a Proton Acceptor, and Nitromethane with a Dipole Moment, for γ-Fe_2O_3, Co-Modified γ-Fe_2O_3, and CrO_2 Particles

Material	Ethyl alcohol, mJ/m^2	Dioxane, mJ/m^2	Nitromethane, mJ/m^2
γ-Fe_2O_3	260	350	460
γ-Fe_2O_3 + Co	390	340	310
CrO_2	670	350	310

SOURCE: Jakusch (1984).

surface of the particles (Jaycock, 1981). The adsorption characteristics of various binder systems have been studied, for example, on Co-modified γ-Fe$_2$O$_3$ particles, by a method of magnetic rotation of the particles in the binder solution (Sumiya et al., 1984). Another measure for the efficiency of a dispersant is the amount of binder solution containing the dispersant needed to just wet a given amount of powder, a method similar to the oil adsorption test. In Fig. 3.29, the critical amount of the solution is plotted versus the amount of dispersants for a γ-Fe$_2$O$_3$ powder and various dispersants. The less binder dispersant solution needed, the better the effi-

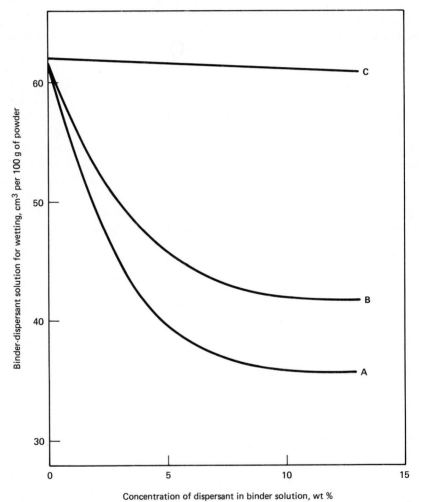

Concentration of dispersant in binder solution, wt %

Figure 3.29   Amount of binder-dispersant solution needed for the wetting of γ-Fe$_2$O$_3$ powder as a function of concentration of dispersant in the binder solution for (curve *A*) phosphate ester, (curve *B*) lecithin, and (curve *C*) silicone oil *(Jakusch, 1984)*.

ciency of the dispersant. The optimum amount of dispersant is indicated by the transition from a rapid decrease to a constant value of binder-dispersant solution needed. The lubricant included in Fig. 3.29 (curve C) is not supposed to be adsorbed at the surface of the particles and, hence, has no influence on the dispersibility properties. The intersection of the curves with the ordinate in Fig. 3.29 is equivalent to the oil absorption, a widely used test for the degree of densification of the powder (Patton, 1979).

3.2.3 Magnetic media preparation

3.2.3.1 Binder, additives, and dispersion. In order to form a cohesive and durable magnetic coating of defined thickness and surface structure, the particles are well dispersed in a polymeric binder, together with solvents, dispersants, lubricants, antistatic agents, and, if necessary, fillers for improved mechanical wear resistance or light absorption. The principles and operations are similar to those used in the paint industry. The dispersion is then coated onto the substrate, dried, and given a surface finish which, for flexible media, comprises a compression by calendering. If applicable, the particles are oriented after coating by a strong magnetic field, applied while the coating is still wet.

The binder is expected to lend itself to a high volumetric packing fraction of the particles and, in combination with the particles as a filler, must have high flexibility, high elasticity, high wear resistance, and low friction on metal or plastic surfaces. These partially contradictory requirements often make it necessary to use mixtures of compatible polymers and plasticizers. For instance, polymers with a lower molecular weight lead to smoother dispersions, with a higher volumetric packing fraction of the particles, but are less wear-resistant and show higher friction values. Thus, cross-linking or radiation curing after dispersion and coating is often used to improve these properties. For flexible recording media including nonmagnetic backcoats, the following groups of organic polymers may be used:

Copolymers of vinyl chloride

Copolymers of vinylidene chloride

Polyvinyl formal

Polyvinyl acetate resins

Acrylate and methacrylate resins

Combinations of polyether with OH groups with polyesters and polyisocyanates

Soluble polyurethane elastomers

Modified cellulose derivatives

Epoxy and phenoxy resins

Polyamides

In recent years, polyurethane binder systems have received the greatest interest (Mihalik, 1983; Williams and Markusch, 1983). For the cross-linking of polymers, polyfunctional isocyanates, melamine-formaldehyde or urea-formaldehyde condensation products of low molecular weight may be used. More recently, electron-beam curing has come into use (Brown et al., 1983). For better adhesion of the coating, the substrate is often precoated with polyvinylidene chloride copolymerizates or polyester.

The solvent, or solvent system, selected for a chosen binder system is one that most economically holds the solubles in solution and maintains the magnetic particle dispersion. It should be reasonably fast-drying, although too rapid an evaporation rate can cause problems in the coating and the drying process. Thus, combinations of solvents having different boiling points are often used. Typical solvents are as follows:

Methyl ethyl ketone

Methyl isobutyl ketone

Cyclohexanone

Tetrahydrofuran

Dioxane

N-methyl pyrrolidone

N-dimethyl formamide

Toluene as an inexpensive diluent

Rigid disks, with thermostable metal substrates, are usually coated with a reactive binder system which cures at temperatures up to 250°C. These are phenol-formaldehyde resins, urea-formaldehyde resins, epoxy resins sometimes together with amine or acidic anhydride hardeners, or polyvinyl acetate resins and silicone resins. These well-hardening resins are essential for the polishing and lapping process which is used to finish the coated disks with respect to coating thickness and surface flatness.

The main steps of preparing a dispersion are wetting and disagglomeration of the particles, stabilization, and final letdown with binder solution to the appropriate viscosity for the coating operation. The most important properties of the liquid for the wetting process are a low contact angle between liquid and solid and a low viscosity. It can thus be advantageous to perform the wetting with the solvent alone. In addition, the low-viscosity solvent may break up bonds between the particles through the Rehbinder effect (Parfitt, 1981).

In contrast to many paint manufacturing procedures, the particles must not be damaged during the dispersion process when magnetic pigments are used. The agglomerates need mechanical energy to break them up. This is introduced by the impact and shear forces in ball mills, sand mills, or kneaders, but care must be taken to maintain the mechanical integrity of the particles. This is done by controlling the milling charge and the time, temperature, and viscosity of the slurry. Usually a high viscosity in the range of 0.2 to 0.6 Pa·s (200 to 600 cP) is used to make use of the shear forces. This is achieved by adding only a part of the binder solution of the final formulation. The effect of milling time is shown in Fig. 3.30, in terms of the gloss of the as-coated, calendered tape surface, the saturation remanence, the coercivity, and the ac-bias noise as well as the print-through of the calendered tape. The detrimental effect of overmilling is evident and, in this particular case of γ-Fe_2O_3 made from γ-FeOOH, a milling time of 40 h in a ball mill is sufficient. In general, the milling time in ball mills can be between 20 and 200 h, depending on the magnetic pigment, the molecular weight of the binder, the dispersant, the viscosity of the slurry, the size and type of the balls, and the equipment used. Nowadays, continously working sand mills are most widely used (Wheeler, 1981; Missenbach, 1983).

In order to achieve a stable dispersion, it is necessary to prevent flocculation and sedimentation. To this end, the magnetic particles must be enveloped by a stabilizing agent that prevents further intimate contact between them. This is partially accomplished by the organic polymer molecules of the binder system. These have polar groups and form an adsorbed layer on the particles with a tail or loop configuration. The stabilizing action is given by entropic effects which depend on the thickness of the adsorbed layer, the structure of the molecules, and the configuration of the polar groups. In addition, active dispersants of low molecular weight with hydrophobic and hydrophilic groups are used. The hydrophobic groups usually consist of a hydrocarbon chain with 6 to 30 C atoms. The hydrophilic groups, which react with the particle surface, are either ionic or have a dipole moment. Ionic dispersants may be anionic, such as carboxylates, sulfonates, and phosphates, or cationic, such as quaternary ammonium salts, alcanol ammonium salts, or imidazolinium salts. Amphoteric dispersants are betaines, amino oxides, and aminocarbon acids. Nonionic dispersants typically consist of arylpolyglycol ether, polyglycol ester, polyglycol amides, or polyalcohols. All dispersants must be carefully selected and their amount chosen with respect to the surface chemistry of the particles and the binder system used.

After dispersion, the remaining binder solution and solvent of the formulation are added, together with lubricants that assure good tape life. Extreme uses include still-framing in video where the same area of tape is scanned some 3600 times/per minute, or in computer tape application

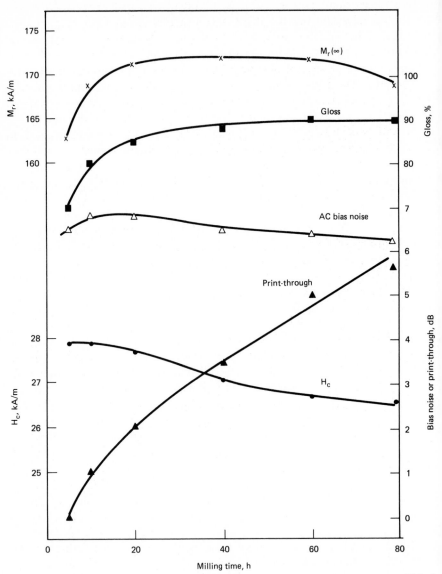

Figure 3.30 Gloss of the as-coated and dried dispersion, saturation remanence $M_r(\infty)$, coercivity H_c, ac-bias noise, and print-through signal of the calendered coating as a function of milling time for γ-Fe_2O_3 particles. For noise and print-through a relative scale has been used. The difference of 1 dB is indicated in the drawing *(Jakusch, 1984)*.

where the tape is subjected to repeated start and stop applications. The basic idea is that the lubricant forms a monomolecular layer on the surface of the magnetic coating, and is continuously replenished by diffusion from within the magnetic layer after wearing off during use. This necessitates a delicate balance between wear at too low a concentration, and deposits on the heads at too high a concentration of the lubricant. Silicone oils, substituted silicone oils, fatty acids, fatty acid esters, fluor polyether stearates, and natural or synthetic hydrocarbons in amounts of 1 to 6 wt % of the nonvolatile substances are used. Sometimes solid lubricants such as carbon black or polyfluor hydrocarbon are added. In addition, antistatic agents such as polyalkylene ether or quarternary ammonium salts are added to keep the resistivity of the magnetic coating in an acceptable level below 10^{11} Ω per square. Often carbon black is added as a conductive filler for this purpose as well (Burgess, 1983).

Nonmagnetic fillers like silicic acid, carborundum, alumina, chromium trioxide, or alpha ferric oxide are often added in small amounts in order to improve the wear resistance and frictional properties of the magnetic coating, as well as to obtain a cleaning action of the head surface by the surface of the magnetic coating. These fillers typically have a size range from 0.2 to 1 μm in diameter. For back coats with a defined surface roughness, which often are necessary for good tape winding characteristics in professional applications, carbon black and silicic acid below a size of 0.1 μm in diameter are most frequently used.

3.2.3.2 Tape and disk substrates. The substrates for particulate magnetic recording media are almost exclusively films of polyethylene terephthalate (polyester) for flexible media, and aluminum disks for rigid media. The polyester films are produced by extrusion and subsequent stretching to the necessary film thickness. According to the ratio of the degree of stretching in the longitudinal to that in the transverse direction, a distinction is drawn between balanced films for which the above ratio is about unity, and tensilized films which have a ratio of about 3. The latter have a tensile strength at an elongation of 5 percent of the order of 150 N/mm^2 (20×10^3 lb/in^2), which is about twice that of balanced films. Tensilized film is usually used in the thickness range below 10 μm. A final annealing step is employed at temperatures between 200 and 250°C, but some shrinkage may still take place at elevated temperatures. This leads to dimensional instabilities for flexible-disk applications. After exposure to 72°C, an initially circular track can become elliptic at room temperature, with a deformation of about 0.3 percent or 180 μm on a track radius of 60 mm (Greenberg et al., 1977). This behavior severely limits the application for high-density recording. Another example is helical-scan video recording, where a shrinkage after recording of 0.1 percent may occur at elevated temperatures. This causes a distortion of the video picture due to a mismatch of the line synchronization. Additional annealing steps after calendering are often added to reduce these possible distortions. Alternatively, other substrate materials, such as polyimides and polyimide-polyamide copolymers, have been explored for better dimensional stability.

Other important properties of the polyester films are their surface smoothness and their resistance to wear. Both are related to avoiding defects in the final magnetic coating which may cause local signal reductions (dropouts). For good winding and handling properties, films below 20-μm thickness must have a certain surface roughness which is produced by inorganic and, more recently, by organic fillers. Agglomerates of filler particles can lead to asperities of 5 to 30 μm in diameter and up to 2 μm in height. Thus, the control of size distribution of the filler is of great importance and determines the quality of the film for high-density recording. Another solution is to coextrude one formulation with, and one without, a filler. This leads to a film which is almost optically flat on one side but has a defined structure on the other side (Holloway, 1983). The high electrical surface resistance of about 10^{16} Ω per square leads to electrostatic surface charges which can cause difficulties in handling the films. In many applications it is therefore necessary to include a conductive back coating on magnetic recording tapes.

For rigid disks, aluminum disks of 1.3- and 1.9-mm thickness and 95- to 356-mm diameter are used. Glass, ceramics, and plastic substrates have often been discussed but so far have not been used in products. Recently, a combination of flexible- and rigid-disk technology has been suggested. It consists of an injection-molded composite plastic substrate with raised edges at the inner and outer diameters of the disk (Fig. 3.31). A web-coated, flexible media is isotropically stretched and bonded to the raised edges. This results in a compliant surface with a 250-μm gap between the substrate and the media (Knudsen, 1985). For aluminum disks, raw sheet made of an aluminum-magnesium alloy is punched into disks with an inner hold. Apart from an annealing step for improved flatness, the finishing processes include diamond turning, polishing, and lapping in order to meet the necessary dimensional and surface specifications. Typical

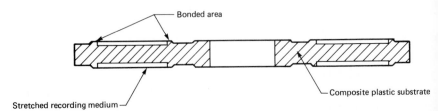

Figure 3.31 Schematic cross section of a rigid disk in combination with a stretched, web-coated medium.

specifications are an axial runout below 75 μm, a waviness below 0.5 μm, and a surface roughness below 0.03 μm. For improved adhesion of the magnetic coating and corrosion protection, the surface may be treated by a chromate process.

3.2.3.3 Coating and orienting.

In the production of flexible recording media, the coating takes place between an unwind and a rewind station that must have precise tension controls in order to handle substrate films as thin as 6 μm, and as wide as 1.5 m, at coating speeds as high as 5 m/s. The master rolls of lengths of the order of 5000 m are usually changed without interruption of the coating process. The coating head applies a coating, about 4 to 10 times as thick as the dry coating, uniformly over the width of the substrate web with tolerances of generally ± 5 percent. There are a number of different types of coating systems (Landskroener, 1983). One is gravure coating (Fig. 3.32a), where small grooves in a roll pick up the dispersion and transfer it to the substrate film in contact. This has the advantage of giving a very uniform average coating thickness, but the disadvantage of needing a postsmoothing of the discontinuous coating pattern. Gravure coating also has the disadvantages of a restricted viscosity range and a fixed coating thickness for a given gravure roll. Knife coating (Fig. 3.32b) is much more flexible. Here the substrate web runs on a backing roll under a knife edge set at a defined distance from the film, and this determines the thickness of the coating. Modern extrusion knife coaters are closed units which are pressure-fed. Their advantages are their low cost and their ability to coat wide thickness ranges. The disadvantage of knife coaters is the critical adjustment of the straightness of the knife edge for a uniform coating thickness across the web. In a reverse roll coater (Fig. 3.32c), the dispersion is applied in excess by a nozzle onto a roll, the thickness being determined by a metering roll, and either transferred directly from that roll, or via a second coating roll, onto the film substrate, which is held by a compliant backing roll. The coating roll rotates in opposite direction to the movement of the film substrate. Application and metering can also be performed with a knife coater. The reverse roll coater is very demanding as to the precision of the rolls and their respective relative velocities. Other coaters are based on the extrusion of the dispersion through the slotted opening of an extrusion die onto the film substrate. Before coating, the dispersion is carefully filtered with filter systems that effectively will remove particles as small as 3 to 5 μm. In case of cross-linked magnetic coatings, the reactive agent is mixed into the dispersion shortly before coating.

Directly after the coating operation, unless a magnetically isotropic medium is desired as is the case for flexible disks, a field parallel to the web transport direction is applied for the alignment of the particles. The preferred configuration of the magnetic pole pieces of permanent-magnet

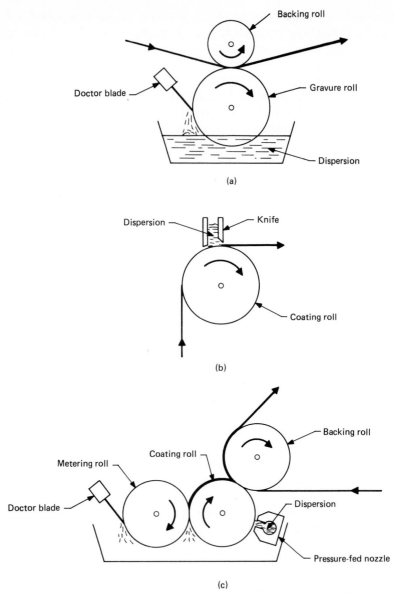

Figure 3.32 Coating systems: (a) gravure, (b) knife, and (c) reverse roll coating.

or electromagnet circuits is one of opposing poles on both sides of the web, which have no perpendicular field component in the center plane. Naturally, attracting poles are used if an orientation perpendicular to the recording media is desired, for which a special ac-field method has been reported (Ohtsubo et al., 1984). It is of importance that the alignment, or orienting, field is large enough to reverse the magnetization of the particles such that they need only to be rotated by angles less than 90°. In Fig. 3.33, the degree of particle alignment as measured by the orientation factor $M_{rx}(\infty)/M_{rz}(\infty)$ is plotted versus the applied field. Here, $M_{rx}(\infty)$ is the reduced saturation remanence parallel, and $M_{rz}(\infty)$ the saturation remanence transverse, to the direction of alignment. Alignment takes place only when the field approaches the remanence coercivity H_r of the particles. At fields higher than $3 \times H_r$ no further improvement in alignment is obtained. Instead, the dragging force of the magnetic field on the particles leads to disturbances of the magnetic coating. It is of great importance that no reverse field exists at the end of the magnet configu-

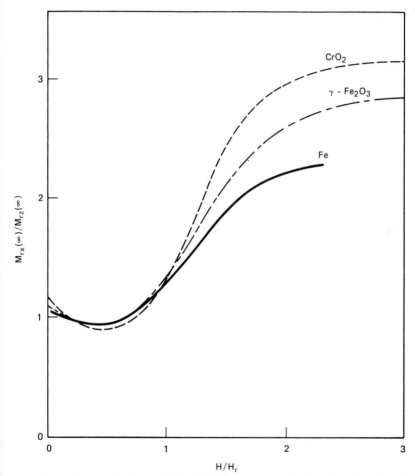

Figure 3.33 Orientation factor $M_{rx}(\infty)/M_{rz}(\infty)$ of gamma ferric oxide, chromium dioxide, and iron particles as a function of the reduced orientation field H/H_r. The quantity H_r is the remanence coercivity of the powder at a volumetric packing fraction of $p = 0.25$.

ration, since field strengths as low as a few kiloamperes per meter (tens of oersteds) can destroy the state of particle alignment. Special designs use antennae of high permeability (Fig. 3.34a), or a low-field solenoid at the end of the magnet configuration (Fig. 3.34b), in order to compensate any field opposing the direction of the field of alignment. Opposing electromagnets have been designed with two air gaps that produce two consecutive orienting fields of equal magnitude, but of opposite sign, thus eliminating any negative fields outside the magnetic configuration (Fig. 3.34c). The duration of the aligning field may become of importance if it is less than 50 ms (Newman, 1978).

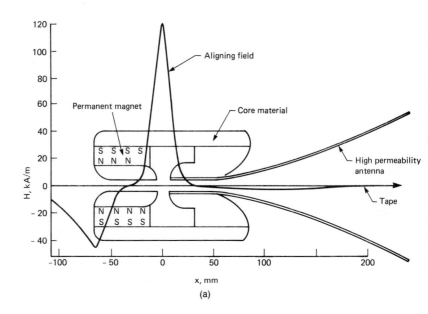

(a)

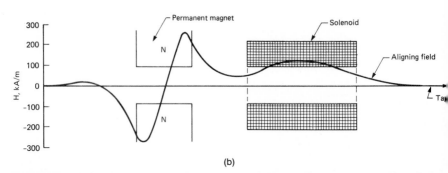

(b)

Figure 3.34 Orienting systems. (a) High-permeability antennae for the reduction of reverse field *(Bate and Dunn, 1980)*, (b) opposing poles with a solenoid for the compensation of reverse field *(Sumiya et al., 1983)*, and (c) double-gap electromagnet *(Köster et al., 1983)*.

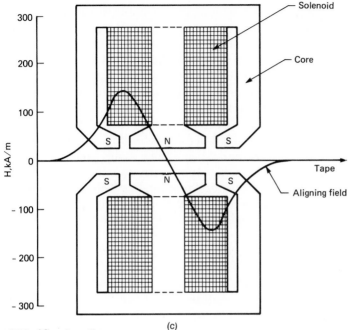

Figure 3.34 (*Continued*)

The drying of the coating takes place in a closed unit at temperatures between 60 and 100°C. For environmental protection, the solvent vapor is either burned or recycled by adsorption on active charcoal (Landskroener, 1983). Sequential coaters with several coating and drying stations are in use for the application of precoatings, multiple recording layers, and back coatings.

In the production of rigid disks, the aluminum disk can be coated in various ways. However, of all methods tried so far, such as spray coating, dip coating, and fluid-bed coating, spin coating has emerged as the dominant method. After the substrate is rinsed with a solvent at low rotational speeds, the dispersion is applied and distributed over the substrate surface in ample quantity. The substrate is then accelerated to high rotational speeds for the spin-off of the surplus dispersion. The viscosity and flow characteristics of the dispersion determine the thickness and thickness profile of the final coating. The latter usually is an increase of thickness going from inside to outside of the substrate. The alignment of the particles in a circumferential direction takes place at low rotational speeds shortly before drying sets (Meijers, 1983).

3.2.3.4 Surface finish and slitting. The dried magnetic coating usually is not completely solid, but has a compressible spongelike structure. Consequently, in flexible recording media, the dried web is calendered between a highly polished steel roll, and a compliant roll having a paper, fabric, or plastic surface. The linear pressure can be as high as 3×10^5 N/m, and the steel roll temperature as high as 90°C. Usually several alternating steel and compliant rolls are combined in a row for high-speed operation. While the dispersion techniques and their theoretical background are based on the experience of the paint industry, the calendering process has been borrowed from the paper industry (Sharpe et al., 1978). The most important effect of calendering is to smooth the surface of the magnetic coating which, for high-density recording, should have a peak-to-valley roughness of less than 0.1 μm. Such highly smooth surfaces are not without problems. In conjunction with equally smooth recording heads, guiding elements, or drums, "blocking" or "slip-stick" phenomena can occur which can cause severe functional problems. Here, a careful compromise between short-wavelength response and frictional properties must be maintained.

Flexible disks, after calendering and punching, are submitted to a burnishing process where surface asperities are removed with ceramic rolls or with polishing tape. Rigid disks cannot be calendered. The curing of the thermosetting resins at temperatures up to 250°C, however, leads to a contraction of the coating and a corresponding densification. The surface finish of rigid disks is achieved by polishing and lapping the hardened magnetic coating. In a final step, rigid disks, and sometimes flexible disks, are lubricated by applying perfluoroalkyl polyether diluted in a fluorocarbon solvent. It is thought to be adsorbed in the pores of the magnetic coating, thus securing a thin surface layer of the lubricant even after repeated use (Lindner and Mee, 1982).

The slitting of magnetic tape is extremely critical. The abrasive nature of the magnetic particles and the tough characteristics of the polyester substrate make it difficult to obtain sharply cut edges. The use of hardened rotary-shear blades and a constant monitoring of the slitting process is important, as well as a careful and precise guidance and tensioning of slit tapes.

3.2.3.5 Surface defects. Any deviation from an ideal flat surface of the magnetic recording layer may be seen as a surface defect. Long-wavelength undulations of the order of 0.1 to 10 mm in length usually contribute to multiplicative (modulation) noise. They may be caused by structures originating in the coating process that depend on the rheological properties of the dispersion, such as striations in the case of knife and reverse roll coating, or replicas of the structure of the gravure roll. Other long-wavelength variations may develop during the calendering process. Here, the compliant rolls must be as smooth as possible, and the line pressure and temperature of the rolls must be kept at a level just necessary to obtain the desired compression. Homogeneous rolls, cast from elastomeric compounds, can give better results with respect to modulation noise than composite compliant rolls made from paper or felt sheets (Mills et al., 1975).

Asperities or holes several tens of micrometers in dimension give rise to a signal reduction (dropout) or an extra signal (drop-in) of short duration. In tape or flexible-disk applications, asperities are the most common type of defect because they cause a spacing between the head and the recording surface that extends far beyond the lateral size of the defect, due to a tent effect (Alstad and Haynes, 1978). Asperities generally cause dropouts unless the asperity is magnetic and causes a spike signal (drop-in) when it passes the reproduce gap (Lee and Papin, 1982). Additives in the film substrate, and shedded film or foreign particles on the substrate, can cause surface defects. In a magnetic coating, agglomerates of magnetic or foreign particles may give rise to asperities. After the coating and drying process, shedded film or foreign particles may be firmly pressed onto the magnetic coating during calendering. If these particles break off again at a later stage, sharp indentations occur which may cause additional signal losses and drop-ins. Scratches are another source of defects that occur if hard foreign particles are caught by the recording head, or by any guiding element, and damage the recording surface. A typical distribution of dropouts in home video recording is shown in Fig. 3.35 as a function of the amount of dropout depth and length. The distribution increases markedly toward small dropouts in depth and length.

With rigid disks, scratches and holes in the aluminum substrate caused by handling are one possible source of defects. Holes in the magnetic coating may be due to wetting problems, or to agglomerates, as well as to foreign particles that have been broken away in the final polishing process. Microcracks can occur during the spin coating and particle alignment process as a result of unfavorable rheological properties of the dispersion. Finally, radial scratches on the polished coating may lead to dropout and drop-in signals.

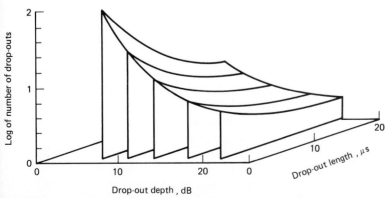

Figure 3.35 Number of dropouts of a home video tape as a function of dropout depth and length (recorded wavelength 1 μm, track width 22 μm).

3.2.4 Magnetic media and recording stability

3.2.4.1 Mechanical stability. The wear of the magnetic recording layer, or that of the recording head, is not easily amenable to precise theoretical or experimental analysis. It depends on the ill-defined geometrical configuration of the contact area. Additional factors are the intrinsic material properties such as yield stress or hardness, the microstructural properties such as porosity or grain structure, the environmental conditions such as temperature and humidity, and, last but not least, the frictional coefficient as influenced by the lubricant employed. Basically, the shear stress must not be allowed to exceed the yield stress, a condition which is usually obeyed. However, because of asperities or local sharp edges of the head, as well as magnetic or additive particles sticking out of the magnetic recording layer, much higher stresses are present locally which cause mutual erosion to take place, aided by increased local friction and heating effects.

Early in the development of rigid disks, alumina particles in the micrometer size range were incorporated in the magnetic coating as a supporting protection against the flying head damaging the magnetic layer in the event of head-disk interferences. Similarly, the trend to increasingly smaller particles in audio and video recording tapes makes it necessary to incorporate larger particles, often of nonmagnetic ferrite, into the magnetic layer. These serve as supporting posts for the head, thus reducing the shear forces on the magnetic coating. The same route has been taken with flexible disks. Obviously, a reasonable compromise between the mutual wear of head and coating must be found. This means making a compromise between wear of the head and a cleaning action of the tape that prevents clogging of the head. While head wear usually occurs gradually, the wear of the magnetic coating often starts with a loss of lubricant on its surface, followed by a cohesive failure of the coating. Then comes an avalanche effect of cohesive and adhesive failure caused by the debris of the previous erosion. Foreign materials like dust particles may cause the same effect. The cohesive strength of the coating depends not only on the yield stress of the binder system but also on the strength of the interaction betwen the binder and the particles. Finally, the adhesive strength of the coating on the substrate is of equal importance. With these properties carefully adjusted, modern magnetic recording media show extremely high wear resistance. For instance, over 100 million revolutions are achieved in flexible-disk applications, corresponding to 200 years of average use; and over 1 h of still-frame video application is possible, with rotating heads continuously running on the same track with a head-to-medium velocity of several meters per second. The wear rate of γ-Fe_2O_3 and CrO_2 particles on head materials of various hardness follows the known linear decrease of abrasivity with increasing relative hardness of the test body, and the steep increase of the wear rate at relative humidities above 70 percent (Mayer, 1974). In addition, the relative wear rates of γ-Fe_2O_3 and CrO_2 tapes depend on the binder system, the degree of cross-linking, the lubricant, and the size of the particles.

Another mechanical effect that can reduce the reliability of recording media is stick-slip and, in its worst manifestation, blocking. Stick-slip is usually due to a high coefficient of friction and leads to an irregular transport of the magnetic medium, with consequent frequency modulation of the recorded signal. Blocking can cause the complete standstill of sensitive tape transports, or leads to heads adhering immovably to a disk surface. A similar phenomenon of occasional blocking is called *stiction,* and this can be the initial step of the wear process of recording media (Chin and Mee, 1983). These effects may occur if the surfaces of the head or guide elements and of the recording layer are too smoothly polished, and is aggravated by high humidity and temperature. To avoid stick-slip or blocking, the frictional properties must be controlled by introducing and maintaining a defined microtexture of the surfaces involved. This is in conflict with the desired close proximity between head and recording layer. Further, the lubricant and adsorbents play important roles in controlling the coefficient of friction, which tends to be smaller at high pigment-volume concentrations (Miyoshi and Buckley, 1984).

The relaxation of a pack of tape wound on a reel is of concern in the archival storage of computer tapes, and in the shipment and use of tape, such as packs of audio cassette tape intended for use in high-speed duplication. The prolonged storage of a reel of tape leads to a relaxation of the stress in the tape pack caused by creep of the substrate film. This can lead to a destructive tape slippage (cinching) during wind or rewind, or to complete disintegration of a tape pack. Further, an increase of temperature above the winding temperature increases the stress and the resulting pressure in the tape pack. These effects occur because of the differential expansion of pack and hub and the anisotropic thermal expansion coefficient of the polyester substrate film. The stress relaxation of the polyester film and plastic flow of the magnetic coating lead to an increased tendency for pack disintegration once the tape has been returned to the operating environment. A decrease of humidity leads to similar effects due to a hygroscopic contraction of the pack relative to the hub (Bertram and Eshel, 1979).

3.2.4.2 Magnetic stability.
The printing of recorded signals from one layer of a wound reel of magnetic recording tape to another has already been mentioned in context with thermal fluctuation effects. It is commonly known as print-through, and is caused by reversals of magnetization, in the presence of the signal field from an adjacent layer, of particles

whose volume is only slightly larger than the critical volume v_p for super-paramagnetic behavior. The range of volume v of the particles affected is given by Eq. (3.30) if the apparent coercivity is set equal to the signal field. For a given signal field, this range extends from v_p to a fixed multiple of v_p. Hence, if v_p increases with the elapsed time of observation t, or with the temperature T according to Eq. (3.28), the volume of the affected particles increases as well. If we assume a triangular distribution function of particle volumes, or just a linearized part of a real distribution function, the total volume of all particles involved, and consequently the printed signal, is proportional to v_p. Thus, following Eq. (3.28), the printed magnetization and its related print-through signal are proportional to the logarithm of the time of exposure to the printing field, and proportional to the storage temperature. The strength of the printing field for a given recorded magnetic flux is wavelength-dependent, and passes through a maximum whose location depends on the thickness of the base film δ_f, and the thickness of the magnetic coating δ, as

$$\lambda_{\max} = 2\pi(\delta + \delta_f) \tag{3.31}$$

where $\lambda_{\max}$ is the wavelength of maximum print-through signal at a given recorded magnetic flux. The corresponding print-through signal is proportional to $\delta/(\delta + \delta_f)$. The details of these experimentally established print-through effects are fairly well understood (Tochihara et al., 1970; Bertram et al., 1980). The printed magnetization can be erased in low magnetic fields and is usually reduced by stress effects; that is, its final level depends to some extent on the mechanical construction of the recorder.

As mentioned earlier, an induced anisotropy is possible in the system $(\gamma\text{-}Fe_2O_3)_x(Fe_3O_4)_{1-x}$, with or without Co-modification, which leads to irreversible changes of the magnetic properties with time and temperature. This effect is also a cause for print-through signals and may, in addition, have serious consequences with respect to the ease of erasure of recorded tapes (Salmon et al., 1979). It is thought to be due to a migration and ordering of cation vacancies, and divalent iron ions, which lower the free-energy density of the magnetization state of each particle according to the magnetization distribution given by the recorded signal. Hence, the recorded magnetization pattern is "frozen" into the recording medium and cannot be erased even with the highest dc field or in a bulk eraser. Therefore, this behavior is also called "memory" effect. It is demonstrated in Fig. 3.36 for a series of audio recording tapes made from $\gamma\text{-}Fe_2O_3$ particles with a surface Co modification using different amounts of Fe^{2+} ions. Immediately after recording, all tapes have a posterasure signal of about -70 dB, independent of the Fe^{2+} or Co^{2+} content, and independent of the coercivity. After storage of the recorded tapes for 4 days at 54°C, the posterasure signal increases as much as 20 dB for $x = 0.3$ in $(\gamma\text{-Fe}_2\text{O}_3)_x(\text{Fe}_3\text{O}_4)_{1-x}$. In addition, the increase of print-through signal with Fe^{2+} content is shown in Fig. 3.36 for this tape series. The memory effect can be simulated in a vibrating-sample magnetometer by applying an ac field which continuously decreases to zero from an initial value of 133 kA/m (1670 Oe) (Fig. 3.36). The samples were stored in the state of saturation remanence for 24 h at 65°C and subsequently erased. The particles containing no Co ions show a maximum in the posterasure signal at an x value of about 0.7 (Salmon et al., 1979), while the particles containing

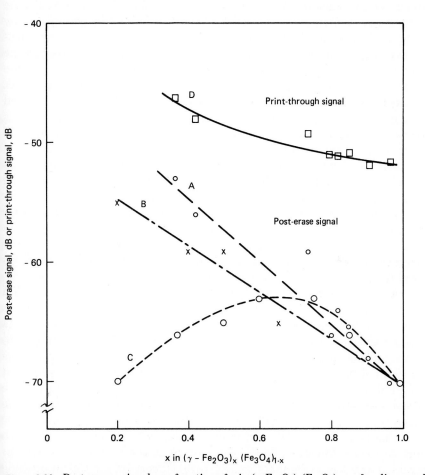

Figure 3.36 Posterasure signal as a function of x in $(\gamma\text{-Fe}_2\text{O}_3)_x(\text{Fe}_3\text{O}_4)_{1-x}$ of audio recording tapes made from surface Co-modified iron oxide particles. Measurements were made after storing a signal recorded to saturation remanence at 65°C for various times, then erasing in an ac field of initial amplitude 133 kA/m (1670 Oe). (Curve A) 1.5 wt % Co, 3 days' storage; (curve B) 1.5 wt % Co, 24 h storage; (curve C) particles without Co, 24 h storage. The print-through signal of tape A is shown with curve D.

1.5 wt % Co have a linear increase of the posterasure signal with increasing Fe^{2+} content. Similar measurement indicate that tapes made from CrO_2 particles can be more easily erased than those made from γ-Fe_2O_3 and even more so than those made from Co-modified γ-Fe_2O_3 particles (Manly, 1976). The above magnetometer measurements suggest the existence of an exchange anisotropy of the gamma ferric oxide kernel coupled to a cobalt-ferrite shell with a magnetocrystalline anisotropy field that can be as high as 749 kA/m (9400 Oe), and whose magnetization can be reversed only in fields much larger than the erase field applied in the experiments. The exchange anisotropy may lead to an anhysteretic magnetization in an ac-erase field, caused by the biasing effect of the exchange anisotropy. A similar effect occurs in the Co-CoO exchange anisotropy system (Köster and Steck, 1976). For a memory effect to occur, the direction of magnetization of the iron oxide kernel must be transferred to the cobalt-ferrite shell. This may occur at elevated temperatures where, due to its small volume and the reduced magnetocrystalline anisotropy field, thermal fluctuation effects cause the cobalt-ferrite shell to be magnetized according to the magnetization of the iron oxide kernel. However, this possible effect, which may contribute to print-through signals as well, needs further investigation. The chance of an induced anisotropy being produced by the mechanical rearrangement of particles at elevated temperatures in the plastic binder as a possible source for a nonerasable signal is minute, although it cannot be totally excluded (Manly, 1976).

Magnetostriction effects can lead to a reduction in signal level after repeated playback of a recorded tape, or after a head-disk interference in rigid-disk application. The reduction of signal level is proportional to the logarithm of the number of passes, to the tension of the tape, to the reciprocal of the wavelength of the recording, and to the constant of magnetostriction (Woodward, 1982). The loss is caused by the combined action of the bending stresses around tape guide elements, or to the shear stresses in disk application, together with the demagnetizing fields of the magnetization transitions (Izawa, 1984). In addition, there is a correlation in magnitude to the print-through effect which suggests that the same particles of low magnetic anisotropy energy are involved. The significant contribution of the demagnetizing fields is shown in Fig. 3.37, where the signal loss after twenty passes on an audio recorder is shown to be inversely proportional to the recorded wavelength. The figure also shows that the signal loss increases in the sequence CrO_2, γ-Fe_2O_3, γ-Fe_2O_3 with surface Co modification, in the same order as the increase in magnetostriction constant (Flanders et al., 1979). There exists the interesting possibility of a zero magnetostriction composition in the system $(\gamma$-$Fe_2O_3)_x(Fe_3O_4)_{1-x}$, as a result of the compensation of the negative magnetostriction of γ-Fe_2O_3 or of Co-modified γ-Fe_2O_3, and the positive magnetostriction of Fe_3O_4 (Flanders, 1979).

3.2.4.3 Chemical stability. Many of the polymeric binders used for magnetic recording media are susceptible to a degradation by hydrolysis, which can lead to the tapes becoming sticky or even shedding gummy and tacky materials. This effect is accelerated by an increase in temperature and humidity (Bertram and Chuddihy, 1982). For archival storage, the conditions of 18°C and 40 percent relative humidity have been recommended, with tight limits on fluctuations in order to keep the stress relaxation as low as possible.

Corrosion may play a role in the use of aluminum-disk substrates. With chromate treatment of the substrate surface, however, no problems are encountered in practical use. Similarly, the stabilization of metal or chromium dioxide particles effectively prevents any sizable signal degradation under reasonable storage conditions.

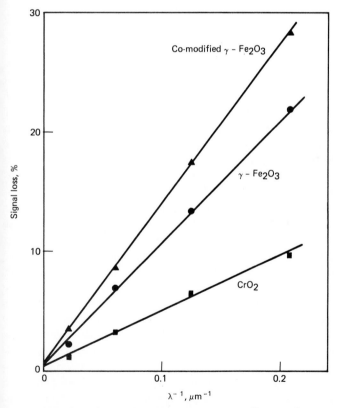

Figure 3.37 Signal loss after 20 passes on an audio recorder as a function of the recorded wavelength λ for tapes made with (1) CrO_2, (2) γ-Fe_2O_3, and (3) Co-modified γ-Fe_2O_3 particles.

3.3 Film Media

3.3.1 Magnetic properties of deposited films

The magnetic films discussed here fall into the category of hard magnetic materials. Such materials require relatively large applied fields to become magnetized or to have their magnetization direction reversed. Once magnetized by an applied field, these materials retain a substantial fraction of their saturation magnetization after the applied field is removed. These are the properties required of a recording medium which will not easily lose written information as a result of stray external magnetic fields or internal demagnetizing fields. Such films are in contrast with soft magnetic films used in recording heads, where low coercivity, low magnetic remanence, and high permeability are required.

Much of the discussion earlier in this chapter of the magnetic properties of particles (such as ferromagnetic exchange coupling, crystalline anisotropy, shape anisotropy, magnetostriction) applies also to films and bulk materials. The reader is referred to the earlier sections of this chapter for these defining concepts and equations, and this section will focus on how these properties are manifested in films.

3.3.1.1 Magnetization. Saturation magnetization is usually considered to be an intrinsic property of the material. For a given composition of a metal alloy, the saturation magnetization M_s will be nearly the same for a film alloy as for a bulk alloy. Provided that the film and bulk materials have the same local atomic ordering and compositional homogeneity, deviations of film magnetization from that of the bulk should occur only when the several atomic layers associated with the surface (or interface between films) constitute a substantial fraction of the total film thickness. Such effects are due to changes in local atomic ordering at the surface and consequently changes in the ferromagnetic coupling. For very thin films, thermal fluctuations can also reduce the magnetization.

More commonly, deviations of measured film from bulk magnetization are observed even for films many atomic layers thick. In reality, what is measured is an effective, or average, magnetization. While true surface atomic ordering effects may be negligible, interfacial composition variations may be significant. Oxidation of a surface or interdiffusion at an interface can cause appreciable changes in the measured effective saturation magnetization. Similarly, deposited films often possess nonequilibrium, microcrystalline structures and solute distributions which can yield magnetizations somewhat different from bulk materials of the same nominal composition. For example, certain stainless-steel alloys are weakly magnetic in bulk form, but films sputtered from such bulk materials may be strongly ferromagnetic. The opposite can also occur. Finally, films have varying degrees of surface roughness associated with them, which may be large compared with the film nominal thickness (for example, 10 to 20 nm roughness on a film of thickness 50 to 100 nm). Such roughness can lead to lower effective magnetization values for films, because of the inability to measure the film volume accurately.

Saturation magnetization may be considered an intrinsic property of a material, with due attention paid to the qualifications mentioned above. Remanent magnetization M_r is an extrinsic property. The magnetization remaining after a saturating field is removed depends very much on the microstructure, film thickness, the nature of the substrate on which the film is deposited, and the conditions of the film deposition process itself. In continuous films, irreversible magnetization changes may occur by domain nucleation and wall motion. In addition, when the magnetic anisotropies of crystallites in a film are not uniformly oriented in the applied field direction, saturation magnetization is achieved by a final reversible rotation of magnetization in the individual grains. This phenomenon occurs at applied fields ranging from about 20 percent greater than the coercivity to several times the coercivity. When the field is removed, the magnetization relaxes by rotation to a lower value determined largely by domain wall energetics and magnetostatic coupling between individual grains. The ratio $S = M_r/M_s$ is called the *remanence squareness*.

3.3.1.2 Magnetic anisotropies.
The mechanisms of magnetic anisotropy described earlier in this chapter are equally operative in deposited films. However, it is much more difficult, in practice, to isolate the individual contributions of magnetocrystalline, magnetostrictive, and magnetostatic anisotropies for polycrystalline films than for particulate recording media.

Magnetocrystalline anisotropy, which arises from spin-orbit coupling, is of prime importance in determining the magnetic behavior of a single (hypothetically isolated) grain, and may thus be a starting point in characterizing a polycrystalline film. However, in real polycrystalline films, this anisotropy may not be the sole mechanism governing magnetization direction and reversal phenomena. It may not even be the most important mechanism. Most metal-film media consist of cobalt-based alloys, and consequently possess a magnetocrystalline anisotropy derived from the hexagonal structure of cobalt. In some preparations of longitudinal media, an effort is made to grow a cobalt-based film with the hexagonal c axis oriented in the plane of the film, inasmuch as the c axis corresponds to the magnetocrystalline easy axis. In perpendicular Co-Cr recording media, a c axis perpendicular to the film plane is sought.

Individual magnetic grains are always in close enough proximity to experience strong magnetostatic coupling, provided that grain boundaries disrupt stronger exchange coupling. To the extent that exchange coupling across grain boundaries exists, free magnetic charge and magnetostatic coupling are diminished. Either of these coupling mechanisms has the effect of causing groups of neighboring grains to act, more or less, in unison. If magnetocrystalline anisotropies of individual grains are nearly aligned, magnetostatic coupling will tend to enhance this predominant anisotropy.

When exchange coupling is disrupted across a grain boundary, a single grain can be approximated as a magnetic dipole, whose field is given by the following expression:

$$H = \frac{m}{4\pi L} \left[\frac{(x - L/2)\mathbf{x} + y\mathbf{y}}{[(x - L/2)^2 + y^2]^{3/2}} - \frac{(x + L/2)\mathbf{x} + y\mathbf{y}}{[(x + L/2)^2 + y^2]^{3/2}} \right] \tag{3.32}$$

Here m is the dipole magnetic moment and L is the dipole length. The dipole is shown oriented in the x direction in Fig. 3.38, where $\mathbf{x}$ and $\mathbf{y}$ are unit vectors. This equation permits a first-order estimation of magnetostatic coupling between grains. For a disk-shaped grain with in-plane magnetization, L may be taken to be approximately the grain diameter and m is roughly $(\pi L^2 \delta/4) M_s$, where δ is the film thickness.

Setting $y = 0$ in Eq. (3.32) gives the magnetostatic field collinear with the dipole, which would tend to align an adjacent dipole. Setting $x = 0$ gives the field antiparallel to the dipole, favoring antiparallel alignment of adjacent grains. The collinear alignment field falls off slightly faster than the antiparallel alignment field with increasing spacing, but the collinear field is always larger (5:1 for adjacent grains and 2:1 for large separations). Thus for two crystallites which have no strong easy axes of their own, alignment of the grains is favored. However, for a two-dimensional uniform array of grains, the multiple magnetostatic interaction fields do not lead to an overall alignment of magnetization. Instead, local fluctuations

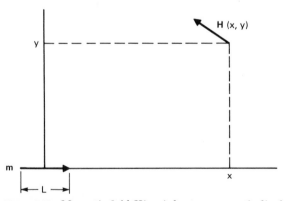

Figure 3.38 Magnetic field $H(x, y)$ due to a magnetic dipole of magnitude m and length L, oriented in the x direction, and positioned at the origin.

lead neighboring grains to form a cluster with a net alignment, due to minimization of magnetostatic energy. The cluster-to-cluster alignments remain random. Factors which can initiate or dominate local cluster alignment include statistical fluctuations of the following: anisotropy energies of individual grains, geometric local arrangement of grains, grain size, and orientation. This is further discussed in Section 3.3.1.5 on magnetization reversal.

In addition to coupling magnetic grains, magnetostatic energy can itself introduce strong anisotropies. Such anisotropy is usually referred to as *local shape anisotropy*. In films, individual grains have anisotropies derived from their shapes, quite analogous to those described earlier for particles. The shape of a grain can be the result of the natural growth habit of a particular crystal structure, but it is more often controlled by the specific conditions under which the film is deposited. In vacuum evaporation, or low-pressure sputtering, the angle at which incident atoms impinge on the substrate can strongly influence the growth morphology of the crystallites and their shape anisotropy. However conditions such as deposition rate, substrate temperature, alloy composition, substrate material, and surface conditions can also affect such grain shape anisotropies. For example, oblique incidence evaporation of metal films can produce strong shape anisotropy, with the easy axis parallel or perpendicular to the incident vapor stream, depending on the angle of incidence. This effect is enhanced by increased deposition rates, but it is somewhat diminished by increased substrate temperature, because the adatom surface mobility and the deposition rate are competing effects.

Local shape anisotropy may arise from the local arrangement of crystallites as well as the morphology of individual grains. Grains which, themselves, might be isotropic, yield very strong anisotropies when arranged in a linear chain during the growth process. This condition can also occur under certain conditions in oblique-angle evaporation. In films where the deposition process has induced the formation of such chains in a preferred direction, the center-to-center spacing of grains within a chain is essentially one grain diameter, while parallel chains tend to have separations which are about 10 percent larger. When magnetostatic interaction energies are computed, assuming that the center-to-center grain spacing between chains is only 5 to 10 percent larger than the center-to-center grain spacing within a chain, a strong anisotropy is produced with magnetization aligned in the direction of the chains.

In addition to magnetocrystalline and magnetostatic (shape) anisotropies, magnetostriction is a third major source of anisotropy. Magnetostriction refers to the interaction of magnetic and elastic properties of a material. The mechanical nature of conventional recording systems makes magnetostrictive properties of a medium generally undesirable, because unpredictable mechanical stresses occur during operation, and these can degrade the recorded signal integrity. So while magnetostriction can be used intentionally to induce magnetic anisotropy, it is difficult to control and seldom used. More will be said about this subject in the next section, where it is treated as a problem to control rather than a useful property in media design.

Before leaving the topic of anisotropy, mention should be made of an induced anisotropy which does not fit neatly into one of the above categories. This induced anisotropy results from applying a magnetic field during the growth of a film, or applying a field during a postdeposition anneal. The applied field affects the motion of atoms during growth or anneal. This atomic migration can result in a local ordering of magnetic species, vacancies, or other point defects, akin to magnetocrystalline anisotropy but not necessarily accompained by long-range crystallographic order. Thus it is a spin-orbit coupling effect like crystalline anisotropy and not really a different physical mechanism. Atomic migration under an applied field, say during an anneal, can also induce grain growth, growth of precipitates, or defect ordering in a preferred direction. Again, this is not a distinct physical mechanism, but rather a technique to produce or enhance shape or magnetocrystalline anisotropies.

3.3.1.3 Magnetostriction. Magnetostriction, described earlier in this chapter, can be even more critical in deposited films than in particulate media. The mechanical compliance of an organic binder in particulate coatings tends to mitigate stress effects on magnetic particles, whereas continuous films experience the full effect of any stresses. While accurate descriptions and measurements of magnetostriction are possible in single-crystal materials, the situation is more ambiguous for polycrystalline film structures.

Magnetostriction in a single crystal can involve one or more anisotropy directions. In a polycrystalline film, such anisotropies become averaged over the various crystallographic orientations taken by the grains. These orientations may be random or preferred, depending on the materials and the deposition conditions. Moreover, the finer the grain structure, the greater will be the effect of grain boundaries in determining the composite magnetostriction coefficients of a sample. Simple schemes for averaging the magnetostriction coefficients over all crystallite orientations in a film have been used, but they generally invoke assumptions of uniform stress or uniform strain and ignore grain boundary effects. The complexity of real film structures makes such exercises of little practical value. Besides the polycrystalline nature of film recording media, the fact that they are always multilayered structures makes a general formulation of the magnetostriction problem intractable. Each layer contributes to the overall stress of the structure, and the stress may vary spatially.

Despite the difficulty in calculating magnetostrictive effects in polycrystalline multilayered media, the phenomenon is extremely important in media design. When theoretical calculation is impractical, empirical correlation with process parameters can be very useful. In deposition processes like sputtering, the built-in stresses depend on sufficiently many variables (rate, substrate temperature, sputtering pressure, substrate bias, and film thickness, to name some of the more important ones) that the final stresses, and consequent magnetic characteristics, will show some degree of random fluctuation. It is best to choose materials whose measured single-crystal magnetostriction coefficients are as low as possible, since this will minimize process control problems. Magnetostriction and its composition dependence are further discussed under magnetic and mechanical instabilities later in this chapter (Sec. 3.3.3).

3.3.1.4 Film media classification.

Film recording media are usually discussed in terms of the process used to deposit the magnetic layer, for example, plated, evaporated, or sputtered. This is a limiting perspective, since the fabrication of a multilayered film disk or tape often involves different processes for undercoats or overcoats from those used for the magnetic layer, such as a plated magnetic film and a sputtered overcoat.

The approach used here will be to classify and discuss the various types of film media according to more generic properties, which reflect their functional application and probable micromagnetic characteristics. The broadest division is between longitudinal recording media, where the magnetization vectors lie predominantly in the plane of the film, and perpendicular media, where the magnetization is predominantly perpendicular to the film plane. Most of the recording industry's attention in the past has been directed at materials and processes for longitudinally recorded media, and these will appropriately receive the most attention here. Beginning in the late 1970s, increased attention was given to perpendicular recording, with the principal media using sputtered Co-Cr alloys.

There is little need for defining subclassifications of perpendicular media. However, the richly varied history of longitudinal media warrants further categorization. In this area the broadest division is between metal and nonmetal (chiefly iron oxide) magnetic films. The second level of classification for longitudinal media will be according to whether the macroscopic magnetic properties are isotropic or anisotropic (that is, possess a preferred orientation) in the plane of the film. With only minor possible exceptions, all magnetically hard films used for recording media possess one or more types of anisotropy at the level of a single grain, local chains of grains, or clusters of grains. However, at the macroscopic level (about 1 μm and larger) films may appear to be isotropic or anisotropic, depending on the manner in which they are deposited. Except for what may be termed *accidental anisotropies,* plated films and most sputtered films fall in the isotropic category. The most notable anisotropic films are produced by oblique-incidence evaporation of metals or by annealing in the presence of an applied field.

3.3.1.5 Magnetization reversal and noise. Control of the magnetization reversal process in a medium is of central importance, because information is recorded by inducing total or partial reversals via the recording head field. Coercivity determines a film's ability to withstand demagnetizing fields. It is a key magnetic parameter in determining linear information density in longitudinally magnetized films, and signal amplitude in perpendicularly magnetized films. But coercivity only describes the average behavior of all the crystallites in a film during the reversal process. Beyond controlling the average field at which magnetization switches, the film media designer must be concerned with the microscopic configuration in which the grains of the film are magnetized after a reversal is written. This micromagnetic structure is a key determinant of media noise.

Magnetization reversal in thin films may occur in three conceptually distinct ways: (1) individual grains undergo reversal independently; (2) local groups of grains, which are magnetically coupled to one another, undergo reversal in unison but independently of other clusters; (3) grains which are coupled over long range form magnetic domains, and reversal occurs by movement of domain walls. In real films more than one of these modes may be operative. In mode 1 coercivity is determined by the magnetocrystalline anisotropy in a grain, the grain shape, and the stress state. In mode 2 the factors in mode 1 apply, as well as the type of grain coupling within a cluster. In mode 3 wall energy controls coercivity and is affected by magnetocrystalline anisotropy, magnetostatic energy at the wall, stress variations, film roughness, and compositional irregularities. Numerous models have been formulated on the basis of these three modes of reversal, and applied to real materials with semiquantitative success at best. We will not attempt to elaborate the details of these models, but will focus on the implications of these different reversal modes.

The single most important reason for considering film media in preference to particulate coatings is the possibility of achieving higher effective magnetization. This can yield higher readback signals, or allow thinner films with the attendant recording benefits of lower thickness losses. Higher signals can increase the signal-to-noise ratio if media noise is insignificant compared with other noise sources in a recording system. If, however, head and electronic noise is sufficiently low, media noise sources must be considered when evaluating the relative merit of film and particulate media.

Longitudinal recording media. In particulate media the intrinsic (ac-erased) signal-to-noise power ratio is proportional to the particle density, if the particles are well dispersed and act independently. Following this line of thought, if the grains of continuous films were uncoupled, such films might be expected to show higher signal-to-noise ratios since they have smaller and more densely packed grains. Longitudinally recorded metal films are indeed low in noise at low recording densities, but the noise increases with write frequency to levels comparable with particulate media (Baugh et al., 1983). This noise is related to the micromagnetic structure of the written magnetization reversals, and intergranular coupling is a key factor.

The remanent magnetization of metal films ranges from 400 to 1000 kA/m (emu/cm^3) compared with about 300 kA/m for magnetic oxide particles, and the grains in a metal film are far more densely packed than the particles in a particulate coating. Thus, while magnetic interactions between particles are significant and affect the switching properties of a particulate coating, even stronger interactions occur in metal-film layers. In a continuous metal film, high magnetization grains should be coupled to one another by strong exchange forces or strong magnetostatic forces, or both together. This strong coupling may give rise to the formation of domains and domain walls in continuous film media.

The existence of domains may be expected to alter the nature of magnetization reversal and therefore the recording process, although the full ramification of these effects on recording characteristics is not yet fully assessed. In particulate coatings, the major questions of magnetization reversal relate to the physics of relatively isolated particle reversal. In continuous films, the nature of isolated grain magnetization reversal is only one aspect of the process. Beyond the individual grain, account must be taken of the intergranular coupling and the energetics and dynamics of domain nucleation and wall motion. These factors are influenced by the interaction of the magnetic layer with the substrate, roughness in various layers, defects, and so on. While insight can be gained by analyzing certain idealized models, the problem of describing domain behavior in film media is extremely complex.

To make an analysis of domain walls tractable, it is necessary to make some simplifying assumptions about which energy terms are important in a particular film. For example, there might be reason to expect that terms involving exchange energy, magnetocrystalline anisotropy energy, and magnetostatic energy are all important. Approximate calculations can be made for simple domain wall configurations, assuming these are the key energies. Energy terms which are more difficult to treat include inhomogeneous compositions and topographies, random defects, and stress fluctuations.

To illustrate an approach to analyzing domain walls, the Néel calculation for a 180° wall is given here in brief form (Néel, 1955). The wall is approximated as an elliptical cylinder. The exchange-energy density may be written as $e_e = A(\pi/D)^2$, where A is the exchange-energy constant in joules per meter and D is the wall width. The uniaxial crystalline anisotropy energy density associated with a 180° wall is $e_a = K_u/2$, where K_u is the anisotropy energy density constant and the anisotropy is assumed uniformly oriented in the film. The demagnetization energy for such a wall is given by $e_d = 2\pi\mu_0 N_d \langle M^2 \rangle = \pi\mu_0 N_d M_s^2$, where N_d is the demagnetization factor, and the mean squared magnetization $\langle M^2 \rangle$ in the wall is $M_s^2/2$. The total wall energy is then

$$E_t = (e_e + e_a + e_d)\left(\frac{\pi L D \delta}{4}\right) \tag{3.33}$$

where the sum of the energy densities has been multiplied by the wall volume. L is the wall length, assumed linear, and δ is the film thickness. The standard procedure is then to minimize the total energy with respect to D to obtain the domain wall width, and thereby the wall energy per unit area, $E_t/L\delta$. To do this one must express the demagnetization factor characteristic of the wall, N_d. In the Néel approximation, $N_d = D/(D + \delta)$ for a Bloch wall and $N_d = \delta/(D + \delta)$ for a Néel wall. Calculated wall dimensions and energies are shown in Figs. 3.39 and 3.40 for Permalloy, including the wall energy for cross-tie walls which are discussed in most magnetism texts (Middelhoek, 1963).

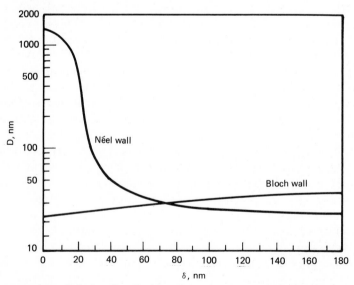

Figure 3.39 Calculated wall width D as a function of Permalloy film thickness T, for a Néel and a Bloch wall. The Néel wall broadens rapidly with thickness, resulting in a lower wall energy, as seen in Fig. 3.40 *(Middelhoek, 1963)*.

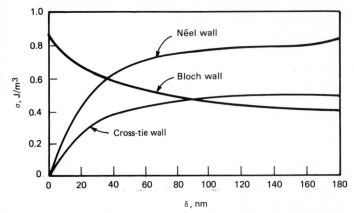

Figure 3.40 Calculated wall energy densities for Néel, Bloch, and cross-tie walls in Permalloy. Néel walls are more stable than Bloch walls in very thin films, but cross-tie walls are the most stable according to the calculation *(Middelhoek, 1963)*.

The treatment above assumes that the wall is parallel to the domain magnetizations on either side, which is a stable configuration. However, a 180° wall normal to the domain magnetization is the situation of interest, since in longitudinal recording on film media, head-on domains are formed in the process of writing data. Instead of straight walls normal to the magnetization vectors, irregular zigzag walls are observed at recorded magnetization reversals (Daval and Randet, 1970; Dressler and Judy, 1974; Chen, 1981; Tong et al., 1984). Figure 3.41 shows a Lorentz microscopy image of such reversal boundaries, written on highly oriented evaporated Fe-Co-Cr film. Isotropic films also show irregular boundaries, but the irregularities seem to be delineations of grain clusters rather than well-defined zigzag domain walls (see Chap. 6).

Various analyses have been made of zigzag domain walls, none of which accounts for the complexity of the irregular walls often observed on recorded tracks. One model has been relatively successful in predicting zigzag wall angles in anisotropic films, though less accurate in predicting zigzag amplitude (Freiser, 1979). The salient features of this model will be discussed here to gain insight into the physical phenomenon rather than to evaluate the merit of this theory or other theories.

In reference to Fig. 3.42, the magnetization within a zigzag region is assumed to converge toward the peaks in the wall pattern. This sort of behavior was observed on evaporated gadolinium-cobalt films, using ferrofluid decoration. Formation of the zigzag reduces the magnetic charge density on the domain wall itself by spreading it over a longer wall length. The magnetization convergence further distributes the free magnetic charges throughout the zigzag region rather than concentrate it near the domain wall. Together, these effects reduce the magnetostatic energy,

2 μm

Figure 3.41 A portion of a recorded track showing two magnetization reversals written by a film head on an obliquely evaporated Fe-Co-Cr film medium, which is highly oriented in the direction of the track. For the medium, $H_c = 40$ kA/m (500 Oe), $M_r = 900$ kA/m (emu/cm^3), and film thickness is about 35 nm. The texture lines, running from top to bottom in the track direction, are from the substrate. *(Photo courtesy of H. C. Tong.)*

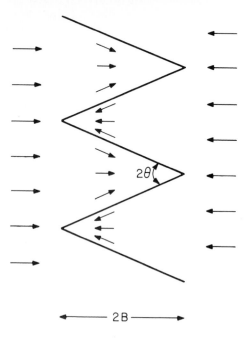

2θ

$2B$

Figure 3.42 Schematic model of a zigzag wall, showing the magnetization vectors converging toward the peaks of the zigzag *(Freiser, 1979)*.

which drives the formation of this type of wall. For simplicity, it is assumed that the magnetic charge is distributed uniformly throughout a region having zigzags of amplitude $2B$. The magnetostatic energy density is then proportional to M_s^2 times the ratio δ/B, where δ is the film thickness. The zigzag wall is assumed to be an uncharged 180° Néel wall, achieved by virtue of magnetization convergence as shown in Fig. 3.42. The Néel wall energy can be calculated as before. The wall energy density per zigzag period varies inversely with the sine of the wall angle θ, reflecting the proportionality of the total Néel wall energy to the total zigzag wall length.

The magnetization convergence in the zigzag region creates a distributed anisotropy energy in addition to the distributed magnetic charge. Assuming a uniaxial anisotropy constant K_u, the anisotropy energy density at any point is proportional to the square of the sine of the angle which the magnetization makes with respect to the easy axis. This energy density, when averaged over the entire zigzag region, is proportional to the square of the sine of the zigzag angle θ. When the sum of these three energies (wall, magnetostatic, and uniaxial anisotropy) is minimized with respect to zigzag angle and amplitude (using a small-angle approximation), the characteristic zigzag half angle is given by

$$\theta = \left(\frac{\pi}{2}\right)^2 \frac{(A/\mu_0)^{1/2}}{M_s \delta} \tag{3.34}$$

and the zigzag half amplitude is given by

$$B = \frac{48}{\pi^5} \frac{\mu_0^2 M_s^4 \delta^3}{K_u A} \tag{3.35}$$

These results may be useful in estimating head-on domain wall characteristics, especially the wall angle. Other models proceed along similar lines, balancing various energy terms. While such models describe important aspects of the physics of such walls, they are too simplistic to describe the magnetic structure of polycrystalline media. They fail to account adequately for zigzag amplitudes, and for statistical variations in zigzag amplitude and angle. The wall in Fig. 3.41 shows extreme irregularity compared with those reported by Freiser and others to which these energy-balance models have been applied. This may be due to the substrate texture used for the film shown in the figure. Also, zigzag wall amplitudes have been observed to depend on the gradient of the field creating the head-on domain wall (Kullmann et al., 1984a), as well as the number of repeated field pulses applied (Hsieh et al., 1974).

For magnetic recording, models like the arctangent models (Williams and Comstock, 1971), which assume smooth and homogeneous magnetization changes, seem to be superior in predicting an effective transition length than the Freiser-type micromagnetic models, even though they ignore completely the wall structure. In large part, these arctangent models are successful because they take account of the effect of applied field gradients.

Magnetization reversal, which is the essence of the recording process, may proceed by domain nucleation and wall motion or by irreversible rotation. When the latter occurs, it probably proceeds in a cooperative manner, where clusters of grains undergo simultaneous rotation due to magnetic coupling. This has been demonstrated by means of a large-scale mechanical model, in which an array of pivoting magnets was shown to reverse its orientation in clusters (Reimer, 1964). A more recent computer simulation considers the magnetostatic coupling of randomly oriented magnetic grains which are isolated from each other by nonmagnetic grain boundaries (Hughes, 1983). The width of the grain boundaries determines the degree of magnetostatic coupling. The simulation demonstrates that greater coupling produces higher remanent and coercive squareness ratios, but lower coercivities. The computer simulation even generated irregular domain walls which, in this case, were more associated with the boundary of magnetostatically coupled grain clusters (Fig. 3.43) than with dispersion of magnetic charge as suggested in the analytical model discussed above. The Hughes model is appropriate to isotropic (randomly oriented polycrystalline grain) films which exhibit softly contoured reversal boundaries, while the Freiser model applies to uniaxial anisotropic films which exhibit sharp zigzag reversals.

When cooperative phenomena like wall motion or cluster reversal are present, squarer hysteresis loops are expected. Indeed this is usually seen in film media. Figure 3.44 compares typical hysteresis loops of longitudinal particulate and film media. In both cases reversible rotation regions exist at the beginning of the demagnetization legs of the loop and near the

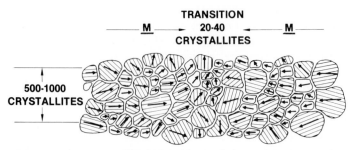

Figure 3.43 Schematic illustration of recorded transition on a polycrystalline film *(Hughes, 1983).*

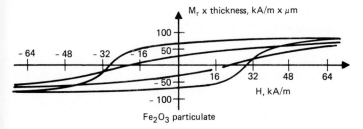

M_r x thickness, kA/m x μm

Fe$_2$O$_3$ particulate

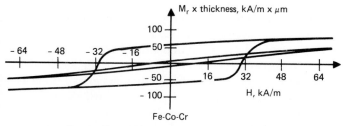

M_r x thickness, kA/m x μm

Fe-Co-Cr

Figure 3.44 Comparison of hysteresis loops of a typical iron oxide, oriented acicular particle coating (top), and those of a highly oriented, obliquely evaporated Fe-Co-Cr film (bottom). In both cases the larger, squarer loop is for the easy axis and the narrower loop is for the transverse direction *(Rossi et al., 1984)*.

saturation region. Between these regions irreversible magnetization changes take place. For particulate coatings, irreversible rotation occurs, but since the particles are varied in size, shape, and degree of misorientation, and because they are only weakly coupled, they switch at slightly different fields. As a result, the coercivity squareness tends to be limited to $S^* = 0.7$ to 0.8. Film media may also have grains which are misoriented (in fact they are often randomly oriented) and vary in size and shape, but because of their strong coupling, they reverse magnetization cooperatively. Consequently $S^* > 0.9$ is not uncommon. Similar considerations also lead to a larger remanence squareness for films.

The densely packed, fine-grain structure of films leads to high signal and low noise on dc-erased, media, as well as to cooperative magnetization reversal, which imparts high hysteresis squareness. But, as a counterpoint to these advantages, the cooperative behavior gives rise to the irregular magnetization reversal boundaries in longitudinal film media and noise originating in these reversals. Granularity noise estimates for particulate media usually assume that the particles act independently. Similarly, granularity noise estimates for films might be based on the smallest magnetic unit that can act somewhat independently, which could be a domain or a grain cluster rather than a crystallite. A film medium has 10 to 50 times as many grains in a recorded bit compared with a particulate medium. However, if clusters of 5 to 10 grains act as magnetic units, the film signal-to-noise ratio advantage may be expected to be in the range of 1 to 10 rather than 10 to 50.

The smallest magnetic unit affects the shape of the magnetization reversal boundary at which information is recorded and at which longitudinal film media noise is manifested. Ideally a magnetization reversal on a recorded track would be a straight boundary traversing the width of the track. In practice, this boundary cannot be abrupt because of the demagnetization effects of adjacent oppositely magnetized regions. If the medium consisted of arbitrarily small independent grains, the reversal would take on a smooth appearance of continuously varying local average magnetization. This cannot occur in longitudinally magnetized films with strongly coupled grains. Instead, a domain wall forms with zigzags across the track width, serving to spread the demagnetization fields over a larger area. The total energy is thereby decreased, resulting in a finite transition length.

In digital saturation recording, the noise cannot be due to the zigzag walls as such, but rather is due to the irregularity of the wall or magnetic inhomogeneities. The walls would produce no noise in real films if the governing energies were as homogeneous and well defined as Freiser's calculation suggests. No deterministic model can explain noise, because the theoretical zigzag angles and amplitudes should, in principle, be quite predictable. Instead, wall energies and coercivities are influenced, if not dominated, by substrate roughness, defects, compositional variations, stress, and the like. These random effects cause the tortuous wall contours seen in Fig. 3.41, and are the logical source of the noise. Unfortunately some of these inhomogeneities are precisely the factors often used to produce a high coercivity. For example, a film roughness model for explaining the origin of coercivity in various cobalt films is in reasonable agreement with experiment (Soohoo, 1981). If roughness is a major coercivity mechanism in a film medium, its random nature may also produce the erratic walls and, hence, noise.

The cooperative behavior of grains in longitudinal thin films can thus be a two-edged sword. On the one hand, it produces square hysteresis which promotes sharper switching and better writability. On the other hand, a noise source is introduced which has its basis in a different type of granularity, namely, the domain or grain cluster and the inhomogeneities governing them.

Perpendicular recording media. Continuous film perpendicular recording media might also be expected to exhibit cooperative magnetic behavior, but whether domain formation and wall motion occur seems to depend on the film preparation conditions. There are two principal models evolving to describe the nature of film perpendicular media, mainly focusing on Co-Cr alloys. One model treats the film as a continuum, forming stripe domains similar to those in bubble materials. The other treats the columnar grains (or multigrains) isolated by nonmagnetic grain boundaries, which have been observed in Co-Cr alloys, as relatively independent particles. The stripe domain model may be applicable for low-Cr-content films. However, under conditions leading to nonmagnetic grain boundary formation, such as increased chromium or oxygen, the particle model may be more realistic, because nonmagnetic grain boundaries can degrade or disrupt any exchange coupling between grains.

With regard to determining linear recording density, it probably makes little difference whether a perpendicular anisotropy film undergoes magnetization reversal by either particlelike switching or reverse domain nucleation and wall motion. However, the micromagnetic structure of a recorded pattern, and therefore media noise, may depend on the mode of reversal. Experimental evidence supporting both models suggests that the mode of magnetization reversal depends on the detailed film deposition conditions.

Perpendicular recording media are generally recognized as supporting more stable high-density recording patterns than longitudinal media, because magnetostatic demagnetizing fields decrease as perpendicular reversals are spaced closer together, whereas they increase for closely spaced longitudinal reversals. This same phenomenon favors smaller micromagnetic granularity in perpendicular media, and therefore potentially lower noise. Figure 3.45a depicts a longitudinally oriented film uniformly magnetized in the positive direction along a recording track, except for a small reversely oriented region. The long regions to either side of the reversed region exert a strong demagnetizing field on the reversed region, favoring the annihilation of the reversed portion. If exchange coupling is operative across grain boundaries, extra energy is required to support the reversed domain because of the 180° reversal of spin direction across the wall. These facts mean that the growth of longer domains at the expense of shorter ones is thermodynamically favored in longitudinal media.

In Fig. 3.45b, a perpendicularly oriented film is shown uniformly magnetized upward, except for a small reversed region. As with the longitudinal film, exchange coupling across grain boundaries means that energy must be added to the system to create the reversed domain, because of the creation of 180° domain walls. However, magnetostatic energy is low-

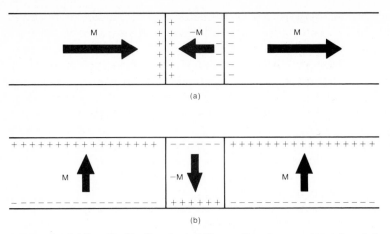

Figure 3.45 (*a*) Longitudinally oriented film, uniformly magnetized from left to right except for a small reversed region. Demagnetization field due to adjacent regions makes the reversed region thermodynamically unstable. (*b*) Perpendicularly oriented film, uniformly magnetized from bottom to top except for a small reversed region. Magnetostatic fields due to adjacent regions tend to stabilize the reversed region.

ered by creation of the reversed region. This then favors the formation of short domains at the expense of long domains along the track direction, leading to increased granularity. If the domain model applies, the smallest magnetic unit is a domain size determined by the balance between the magnetostatic energy decrease and the increase in wall energy due to exchange energy. If the particle model applies, the smallest magnetic unit is the grain or multigrain constituting the quasiparticle. At any rate, exchange forces drive toward the formation of shorter domain wall length and larger domains for both longitudinal and perpendicular films. However, magnetostatic forces drive toward shorter domains in perpendicular media and longer domains in longitudinal media. This suggests the possibility of lower media noise for perpendicular media, particularly for fine-grained media exhibiting particlelike behavior rather than domain behavior.

The shape of the hysteresis loop can give an important clue to the microscopic nature of perpendicular recording media. The hysteresis loops for Co-Cr alloys, measured perpendicular to the film plane, usually resemble rectangular loops which have been sheared into parallelogram shapes, as illustrated in Fig. 3.46. This shearing effect is directly related to micromagnetic structure, and will be addressed in the following discussion of magnetic reversal.

We will first consider coercivities and loop shapes expected for simple coherent rotation of magnetization. For the sake of argument, assume that the magnetic film grows such that it has a uniaxial magnetocrystalline anisotropy, with the easy axis perpendicular to the film plane. This anisotropy is characterized by an energy constant $K_\perp$. In the absence of demagnetization, the film would have a uniaxial anisotropy characterized by the energy constant $K_u = K_\perp$. The magnetocrystalline energy is a function of the angle ϕ between the magnetization vector and the perpendicular easy axis, namely $K_\perp(\sin \phi)^2$. The coercivity would be given by $H_c = H_k = 2K_u/\mu_0 M_s$, where H_k is the anisotropy field. For a Co-(23.5 at %) Cr alloy $K_\perp$ has been reported to be 9.74×10^4 J/m^3 (9.74×10^5 erg/cm^3) with $M_s = 300$ kA/m (emu/cm^3) (Fisher et al., 1984). This would give $H_c = 517$ kA/m (6460 Oe). The hysteresis loop should be perfectly square. If the grains of the film had a finite range of anisotropy constants, that is, a finite switching-field distribution, the hysteresis loop would not be perfectly square, but would have minor shearing due to these grain-to-grain variations.

An infinite-area film with its saturation magnetization perpendicular to the plane has a demagnetizing field of $-M_s$ ($-4\pi M_s$ in cgs units). When the magnetization is canted at an angle ϕ from the normal, the demagnetization energy is given by $\frac{1}{2}\mu_0 M_s^2(\cos \phi)^2$ $[2\pi M_s^2(\cos \phi)^2]$. If the inequality $K_\perp > \frac{1}{2}\mu_0 M_s^2$ holds, the uniaxial anisotropy energy constant would be

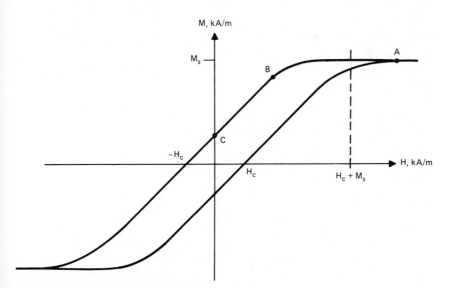

Figure 3.46 Hysteresis loop typical of a perpendicular medium, skewed by the aggregate perpendicular demagnetization field.

$K_u = K_\perp - \frac{1}{2}\mu_0 M_s^2$ ($K_u = K_\perp - 2\pi M_s^2$). The coercivity for uniform rotation should be $H_c = H_k$ as before, but now $H_k = 2K_u/\mu_0 M_s = 2K_\perp/\mu_0 M_s - M_s$. The hysteresis loop again will be perfectly square, aside from the small effects of a finite switching-field distribution, but demagnetization lowers the coercivity by $-M_s$. For the Co-Cr parameters given above, this coercivity is 217 kA/m (2710 Oe). Both the calculations above predict nearly square hysteresis loops and coercivities far higher than the 67.2 kA/m (840 Oe) observed for this composition film (Fisher et al., 1984). In fact, coercivities calculated assuming uniform rotation are almost always higher than measured values, which are typically less than about 100 kA/m (1250 Oe) for Co-Cr alloys.

Coherent rotation is therefore unlikely to be the actual reversal mechanism, and other mechanisms, which more closely predict coercivity, will be described shortly. Nevertheless, under the assumption that a grain or grain cluster being switched sees only its local self-demagnetization field, and not demagnetization from surrounding regions, most reversal mechanisms would still predict either square hysteresis loops or no perpendicular remanence at all. If $H_{c\perp}$ denotes the intrinsic coercivity in the absence of demagnetization, then square hysteresis results when $H_{c\perp} > M_s$, with a coercivity $H_{c\perp} - M_s$. No hysteresis is expected and no perpendicular remanence can be sustained when $H_{c\perp} \leq M_s$.

The shearing of hysteresis loops which is usually observed (Fig. 3.46) is best explained by a variable aggregate demagnetization field. For a film consisting of many small, semi-independent particles (or domains), an individual particle may react to the aggregate demagnetization field of many nearby particles rather than to its own self-demagnetization only. This aggregate demagnetization field, H_d, is approximately equal to $-M$, where M is the average state of magnetization of particles in a given region. We will assume that this M is equal to the macroscopic average which is measured.

In reference to Fig. 3.46, when a positive external field is applied which is large enough to overcome the coercivity and maximum demagnetization field ($H_c + M_s$), all the particles have their magnetization aligned in the direction of the applied field (point A). The aggregate magnetization is M_s, with the demagnetization field equal to $-M_s$ ($-4\pi M_s$ in cgs units). If the applied field is decreased to just under $M_s - H_{c1}$, where H_{c1} is the demagnetization-free coercivity of the most easily switched particles, the easiest-to-switch particles reverse (point B). This lowers the average magnetization as well as the magnitude of the aggregate demagnetization. If the magnetization changes by ΔM, the demagnetization changes by $\Delta H_d \approx -\Delta M$ and the external field required to maintain that state changes by $\Delta H = -\Delta H_d = \Delta M$. As the applied field is further decreased, what would be a nearly vertical side of a hysteresis loop without aggregate variable demagnetization is mapped onto a sloped side with $dM/dH \approx 1$.

When the applied field becomes zero, some remanent aggregate magnetization remains (point C), corresponding to the difference between the unswitched particles and those which have switched. As the applied field is reversed in polarity, the net magnetization becomes zero at $H = - H_c$, which is approximately the average switching field of the individual particles. The description above applies to materials whose coercivity is less than M_s ($4\pi M_s$ in cgs units), which is true for most practical perpendicular media. Under these conditions the remanent magnetization for an infinite-area film is equal to H_c ($H_c/4\pi$). If the coercivity is greater than M_s, shearing of the hysteresis loop still occurs, but the remanent magnetization is nearly equal to that for the demagnetization-free case, rather than being equal to H_c.

Both the particle and stripe-domain models are capable of explaining the sheared hysteresis loops observed, provided that the particles or domains are small enough to produce an aggregate demagnetization effect, and provided that there is a finite distribution of switching fields. For the particle model this means the "particles," or columnar grains, must be noninteracting or weakly interacting, and must have grain diameters sufficiently small so that the aggregate demagnetizing field from surrounding grains overrides an individual grain's self-demagnetizing field. For the stripe-domain model, the balance between wall energy and demagnetization energy must be favorable to the formation of a high density of small domains.

The conclusion reached above, that $dM/dH = 1$ along the side of the hysteresis loop, may be violated in certain cases. Namely, this slope may be greater than unity. This will occur in the particle model when the film thickness is small compared with the grain diameter, so that an individual grain's self-demagnetizing field is large compared with the composite effect of surrounding grains, or when the magnetization is small enough compared with the coercivity that shearing becomes a weak effect. In the domain model, a slope higher than unity can occur when the wall energy is large or the magnetization small, such that the formation of many small domains, with increased total wall length, becomes unfavorable. Decreased film thickness favors decreased wall length and therefore fewer, larger domains. Again, the effect of an aggregate demagnetizing field becomes less important and less loop shearing occurs. Although sheared hysteresis is not sought for its own sake, the fact that it represents a finer grain or domain structure (and therefore lower media noise) suggests that it may be desirable for perpendicular recording media to have loop shearing and a slope dM/dH close to unity.

Both the particle and domain models are able to account qualitatively for the relatively low coercivities observed in perpendicular film media like Co-Cr films. However, the quantitative prediction of coercivities has not been successful with any model. The particle models invoke more or less independent particle switching by a curling or buckling mechanism (Ohkoshi et al., 1983). Coercivity seems to be determined by a combination of mostly magnetocrystalline anisotropy and some shape anisotropy (Wuori and Judy, 1984, 1985). Ample evidence of columnar magnetic grains, isolated by nonmagnetic grain boundaries, in Co-Cr films makes this particle interpretation plausible. The domain models assume that coercivity is governed by wall motion or reverse domain nucleation. As in domain wall models for longitudinal film media, it is common to assume that the coercivity is controlled by wall energy variations due to film thickness irregularity (Wielinga, 1983). Some parameters, such as hysteresis loop shearing slope, tend to favor the wall motion model of coercivity. Microstructure (Chen and Charlan, 1983; Grundy et al., 1984) and temperature dependence of coercivity and anisotropy constant (Wuori and Judy, 1985) tend to favor the particle model.

3.3.2 Film materials and processes

Many of the techniques used to deposit films for magnetic recording media are similar to those used in other film technologies. The reader will be referred to literature from these other fields for detailed treatments of standard techniques and processes whenever appropriate. Attention here will be concentrated on features which are unique or especially important to film recording media.

3.3.2.1 Fabrication overview. Unless otherwise indicated, the term *media* refers to the entire recording structure, not simply the magnetic layer. This broader use of the term is especially important for film media, where the substrate and any film layers beneath or on top of the magnetic layer can profoundly affect the recording properties and performance. Figure 3.47 shows the general structure of thin-film recording media, consisting of a substrate, an undercoat, a magnetic layer, and an overcoat. In some cases there is an additional layer between the magnetic layer and the undercoat for the purpose of controlling the magnetic properties of the recording layer. Similarly, the overcoat has been shown as a single layer, but may consist of multiple layers and perhaps a lubricant on the surface.

The substrate may be rigid or flexible, depending on the final application. Rigid substrates are used for high-density, rapid direct-access disk files. Aluminum-magnesium alloy substrates have been universally used for rigid disks, whether particulate or film. Rigid aluminum alloy substrate disks require a hard undercoat when used for film media, because the Al-Mg alloy itself is too soft to provide adequate impact resistance. Substrates used for tape and flexible-disk applications are generally polyethylene terephthalate (PET), with some surface treatment or thin adhesion-promoting layer (instead of a thick undercoat) required to promote strong bonding between the organic substrate and inorganic film structure.

Most magnetic films are cobalt-based metal alloys. Co-P and Co-Ni-P are the most frequently used plated alloys for longitudinal media, the phosphorus being incorporated from hypophosphite ions in the plating bath. Evaporated metal films have generally been elemental, such as cobalt or iron, or alloys whose constituents have similar vapor pressures, such as Co-Ni. Use of corrosion-retarding additives such as chromium or tungsten require more sophisticated deposition rate controls to cope with their vastly different vapor pressures, but such alloys as Fe-Co-Cr have been successfully evaporated (Rossi et al., 1984; Arnoldussen et al., 1984). The list of metal alloys which have been rf- or dc-sputtered for use as recording films is by far the most extensive, because of the broad compositional flexibility available through this deposition technique. Cobalt, Co-Cr, Co-Ni-W, and Co-Re are but a few of the alloys which have been sputtered for film media. Sputtering is also the preferred technique for producing nonmetal magnetic films, such as γ-Fe_2O_3. Different magnetic films usually require different substrate preparations, as is discussed below.

Finally, some means of alleviating mechanical abrasion of the magnetic film is required, because of the mechanical nature of disk and tape magnetic recording. The use of wear particles, such as Al_2O_3 in particulate media coatings, is obviously not possible for continuous films. Instead, a wear-resistant overcoat is used for virtually all forms of thin-film media.

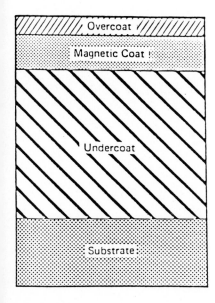

Figure 3.47   General structure of a film medium.

It is necessary to ensure strong bonding between layers, because mechanical wear of thin-film structures often occurs by adhesion failure. For this reason an adhesion-promoting interlayer is sometimes used between the magnetic coating and the overcoat, or between the substrate and first deposited layer.

The film deposition processes generally in use are electroplating, electroless (autocatalytic) plating, evaporation (normal and oblique incidence), and sputtering (rf, dc, and magnetron enhancements of rf and dc). Plating and evaporation techniques are described in greater detail in a later section. Sputter deposition is depicted in Figs. 3.48 and 3.49. Sputtering is performed in a vacuum chamber which has been pumped down to a pressure of 10^{-4} to 10^{-5} Pa (0.75×10^{-6} to 0.75×10^{-7} torr), most often by means of diffusion or cryopumping. The chamber is then backfilled with a sputtering gas to a pressure of 1.3×10^{-1} to 10 Pa (10^{-3} to 7.5×10^{-2} torr). In simple dc sputtering (Fig. 3.48), a dc voltage up to several thousand volts is applied between the substrate and a target of source material to be deposited, the target being biased negatively with respect to the substrate. The voltage used depends on the pressure and

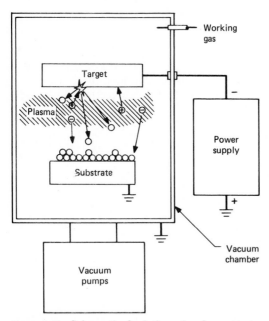

Figure 3.48 Schematic depiction of a dc-sputtering system. An evacuated chamber is backfilled with a working gas (usually argon). Biasing the target negatively with respect to the substrate and chamber initiates ionization of the gas. The physical arrangement of an rf-sputtering system is similar, but rf power is used instead of direct current, and the negative bias on the target is achieved by the system geometry.

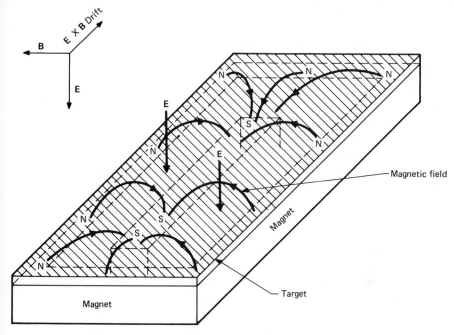

Figure 3.49 One type of magnetron sputtering target. The magnetic field from the permanent (or electro-) magnet behind the target penetrates the target material plate, confining most of the electrons from the plasma to a region near the target and increasing the ionization efficiency.

system geometry, but must be large enough to strike and maintain a plasma. The positively charged, sputtering-gas ions are accelerated toward the target and, upon impact, knock off atoms of target material by mechanical recoil. These atoms, on the average, move toward the substrate by their acquired momentum, and are deposited there. At higher pressures, the sputtered atoms suffer multiple collisions with the sputtering gas, arriving at the substrate with relatively low energy. At lower pressures they may reach the substrate without collision, thereby showing much adatom surface mobility due to retained energies of tens of electronvolts. Argon is generally used as the primary sputtering gas, because its mass is suitable for efficient momentum transfer and because it is chemically inert.

One disadvantage of simple dc sputtering is that fairly high pressures are needed to maintain a plasma, which, in turn, limits process versatility and the energy of deposited atoms. In addition, electrons in the plasma are accelerated toward the substrate with few collisions, thereby heating the substrate by high-energy bombardment. This makes control of substrate temperature a problem. Figure 3.49 illustrates magnetron-enhanced dc sputtering, which alleviates these problems. Electromagnets, or permanent magnets, are placed behind the target, so that the magnetic field penetrating the target causes the electrons to follow helical paths near the target surface. This accomplishes several things. First, the electron path is much longer, resulting in more ionizing collisions with gas atoms. This allows operation at lower pressures and lower voltages, while achieving high deposition rates. Lower pressures, in turn, permit higher adatom energies with consequently better adhesion. Moreover, the magnetic field near the target prevents direct acceleration of electrons toward the substrate, thereby reducing bombardment heating. Magnetic materials pose a limitation to magnetron enhancement because the target tends to shunt the field of the target magnets. This restricts the target permeability and/or thickness, or calls for high-field electromagnets or special target designs.

Both dc and dc-magnetron sputtering require use of conductive target materials. To sputter insulating materials, a radio-frequency (rf) voltage is applied. Geometric asymmetry of the target, substrate holder, and chamber shielding allows a net removal of material from the target and deposition on the substrate, despite the ac field. Otherwise rf sputtering is conceptually similar to the dc methods above. This technique is used for sputtering iron oxides from an oxide target. It also permits reactive sputtering from a metal target. For example, if a partial pressure of oxygen is included in the sputtering gas while sputtering an iron target, an iron oxide film can be deposited. Oxygen can combine chemically with the iron atoms at the target surface, at the substrate, or in transit (to varying degrees). If dc sputtering were used, an insulating oxide layer could build up on the target surface, slowing or stopping the sputtering process. This is not a problem in rf sputtering, which is also a candidate for magnetron enhancement.

It may be advantageous to use different processes for different layers in the multifilm structure. For instance, an electroless-plated Ni-P undercoat is appropriate on an Al-Mg alloy disk substrate where a thickness of many micrometers is required. An obliquely evaporated magnetic film achieves an oriented shape anisotropy, and a sputtered overcoat is favored for strong adhesion and wide materials choice. There is little to be gained by forcing the use of one film deposition process for all layers. In fact, for sputtered γ-Fe_2O_3 films on Al-Mg disk substrates, the undercoat used is generally not deposited at all, but rather it is an anodized Al_2O_3 layer grown from the Al-Mg substrate itself. Furthermore, SiO_2 overcoats have been deposited by spinning on a hydroxysilane solution and subsequently laser curing. For more information on film deposition processes, excellent texts are available (Chopra, 1969; Maissel and Glang, 1970; Vossen and Kern, 1978).

3.3.2.2 Substrate and undercoat preparation.
The basic requirements of the starting substrate are similar for film and particulate media. Substrate

cleanliness and materials compatibility must be carefully maintained, because the mechanical nature of these storage systems demands strong adhesion between the films and between the film structure and the substrate. Film media tend to have more problems in this area than particulate media, in part because they are multilayered structures, having two to four interfaces at which adhesion failure may occur. Many of the standard thin-film techniques for controlling adhesion are used in film media (Chopra, 1969; Maissel and Glang, 1970; Vossen and Kern, 1978).

For rigid-disk systems, substrate hardness is also an important issue for film media. Although the head is usually in flight above the disk surface, it does come into contact with the disk during the starting and stopping of the file, as well as during operation when momentary contacts are possible. The disk should be able to withstand a certain level of such interaction without disturbing the hydrodynamic stability of the head, which can cause the head to crash into the disk. Films used for recording are generally a few tens to a few hundred nanometers thick, and do not substantially increase the composite film substrate hardness, even if the microhardness of the films themselves is high. Since the only rigid substrate in commercial use is the aluminum-(4–5 wt %) magnesium alloy, which is too soft for film rigid-disk requirements, an undercoat is typically used to increase the composite hardness. The most commonly used undercoat is an electroless-plated layer of nickel and phosphorus, which is amorphous and contains sufficient phosphorus to make it nonmagnetic (8–12 at % P). Thicknesses from 15 to 25 μm are typical. Other undercoats that have been used include a 4-μm sputtered layer of Haynes alloy (Rossi et al., 1984), and a 3-μm layer of Al_2O_3, formed by anodization, which is used mainly with sputtered iron oxide and some perpendicular Co-Cr media. Table 3.7 shows the effect of undercoat thickness on the

TABLE 3.7   Knoop Hardness Values in kg/mm^2†

	Load, g		
Surface tested	25	50	200
Bare Al-Mg substrate	. . . .	92	
4-μm Haynes 188 Alloy	370	225	110
10-μm Haynes	963	. . . .	249
20-μm Haynes	939	. . . .	481
Bulk Haynes	401	. . . .	297
15-μm Ni-P	612	602	370
21-μm Ni-P	620	614	478
35-μm Ni-P	642	634	529
45-μm Ni-P	646	643	532
56-μm Ni-P	654	648	532

†For aluminum-(4%) magnesium substrates with and without various undercoats.
SOURCE: S. Doss, Private Communication (1985).

composite hardness of the undercoat and an Al-Mg substrate using electroless-plated Ni-P and a harder sputtered Haynes 188 stainless alloy.

Extremely fine surface finishes are required on substrates, especially for
film media. Since read-write heads fly over the surfaces of rigid disks, the
surface microtopography must be very uniform. Smoothness is not necessarily required or even desired. Controlled texturing to minimize stiction (sticking of the head to the disk) may be advantageous, although this
adds to head-to-medium spacing losses. The disk must, however, be free
of the occasional asperities which may disturb the head flight. Since films
tend to replicate the topography on which they are deposited, elimination
of asperities at the substrate level is crucial. Depressions, or pits, in the
substrate are also troublesome in that a film will follow the contour of the
depression. This represents a variation in the head-to-medium spacing
and, consequently, a modulation of the signal amplitude such as a dropout, as illustrated in Fig. 3.50. Particulate coatings in polymer binders,
depending on their thickness, are somewhat more tolerant owing to the
tendency of the liquid polymer to level by flow prior to cure.

The use of undercoats, in addition to their hardening role, can be helpful in masking such substrate defects. To do this, the undercoat must be
deposited in a sufficiently thick layer that some fraction of it can be polished away. The Al-Mg substrate is normally diamond-turned on a lathe

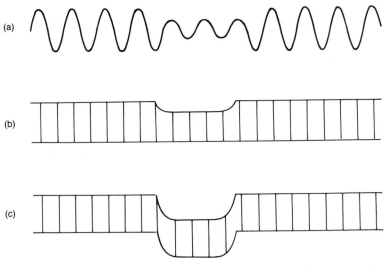

Figure 3.50   (*a*) A reproduced signal with a string of missing pulses, along with
two types of film defects which can produce the signal decrease. (*b*) A portion
of the film is missing. This could be due to a local thickness change across the
entire track as depicted, or the entire magnetic film could be missing, but only
from a portion of the track (e.g., delamination). (*c*) The magnetic film is present across the entire track, with no thickness variation, but the topography of
a substrate pit is replicated by the film, causing a head-to-medium spacing
variation.

to produce a flat, nearly mirror finish with a peak-to-valley surface finish of 50 to 100 nm. While such finishes, and even finer ones, can be achieved by lathe finishing, the diamond-cutting process tends to pull out inter-metallic inclusions, which are present in most Al-Mg alloys, leaving pits in the surface. If a sufficiently thick undercoat is applied to the substrate, it can then be polished back several micrometers to diminish the effect of pits or other irregularities. For such a technique to be effective, the under-coat must itself be polishable, without inclusions or nodules which might produce additional defects. While the use of film and other highly sophis-ticated coating techniques is often the focus of attention for advanced media, mechanical polishing and surface-finishing techniques, such as those shown in Fig. 3.51, are one of the most important factors in produc-ing high-performance recording media.

Figure 3.52 shows a relative comparison of extra and missing pulse defects on an oxide particulate and a metal-plated disk medium. Extra pulses are measured by first dc-erasing the disk. Since any data are thus erased, the head operating in the read mode should ideally record no volt-age pulses. Any pulses whose amplitudes are equal to or greater than a chosen percentage of the average written pulse amplitude are recorded as extra pulses at that threshold. Missing pulses are measured by first writ-ing a string of data pulses on the disk track. The data are then read back, and any data pulses whose amplitudes are equal to or less than a chosen percentage of the average pulse amplitude are recorded as missing pulses at that threshold.

An absence of magnetic coating can be the cause of a missing pulse because a portion of the track width is removed; or it can be the cause of an extra pulse because the coating continuity is broken, allowing flux to fringe from the surface of a dc-erased medium. Large nonmagnetic oxide wear particles in particulate media, or areas of delaminated coating in film media, can thus cause both extra and missing pulses. Substrate pits in continuously coated magnetic film media (Fig. 3.50c), produce missing pulses rather than extra pulses, because the coating is not broken. Thus extra pulses in film media reflect delamination-type defects and are usu-ally far lower in number than extra pulses in particulate media. Missing pulses for film and particulate media tend to be comparable in number.

Flexible-media substrates are not inherently as prone to pits and asper-ities as rigid disks, which suffer from second-phase intermetallic inclu-sions. Moreover, since the head is usually in constant contact with the medium at slower speeds on flexible media, the catastrophic head crash is not a problem. Nevertheless, any topographic variations which do exist in the substrate, or are created in the coating process, can introduce signal amplitude variations and errors. An asperity can move a large portion of the tape or floppy disk sufficiently far away from the head to cause a sig-nificant signal decrease. A pit or depression will cause a smaller signal modulation, since it affects only a region of the physical size of the defect.

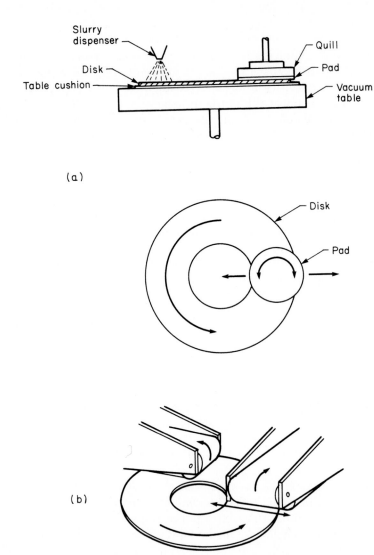

Slurry
dispenser

Quill

Disk

Pad

Table cushion

Vacuum
table

(a)

Disk

Pad

(b)

Figure 3.51 (*a*) Side and top view of a free-abrasive polishing process for rigid-disk preparation. This is a stock removal process used on the substrate and/or the undercoat intended to produce a smooth surface for the magnetic layer. (*b*) Fixed-abrasive buffing of a disk, used on the substrate or undercoat to produce a desired uniform texture. A similar process may be used on the overcoat, with different pressure and buffing tape, to remove loose particles or asperities formed in the film deposition processes *(Rossi et al., 1984).*

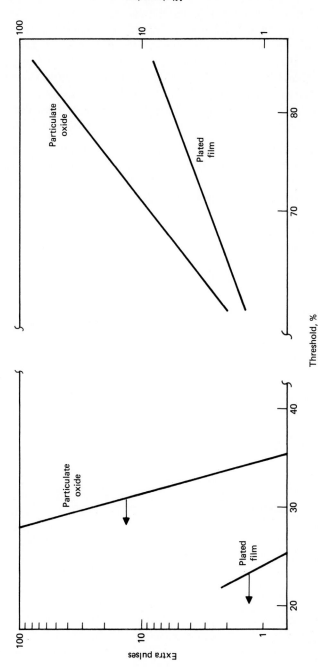

Figure 3.52 Extra and missing pulses as a function of threshold level (percent of average written pulse amplitude), comparing relative numbers of such defects on particulate oxide and plated-film media coercivity H_c = 48 kA/m (600 Oe).

3.3.2.3 Magnetic film processes and properties. Earlier mention was made of the importance of increasing the recording layer magnetization and particle density, which allows the magnetic coating thickness to be decreased in order to achieve high-density recording with equal or better signal-to-noise ratios. This is the primary advantage of continuous magnetic films for recording media. To be useful, the films should show square hysteresis loops, with M_r at least several hundred kA/m (emu/cm^3), and H_c greater than about 40 kA/m (500 Oe). As with particles, film media should be stable against stress, temperature, and corrosive effects.

Metal in-plane isotropic films. Plating was the first process developed for producing commercial magnetic film media, and remains today an important approach for rigid disks. Cobalt and cobalt-nickel alloys have been electroplated from solutions containing salts of Co and Ni and hypophosphite salts (e.g., NaH_2PO_2), along with buffers to maintain the solution pH and improve film uniformity (Bonn and Wendell, 1953). Such plated films have several weight percent P incorporated in the alloy. The phosphorus is key to controlling film coercivity, although plating conditions such as pH can also dramatically affect the magnetic properties (Moradzadeh, 1965).

A similar plating bath allows autocatalytic, or electroless, deposition onto surfaces which have been prepared to initiate the reaction (Brenner and Riddell, 1946). Techniques for electroless plating onto plastic sub-

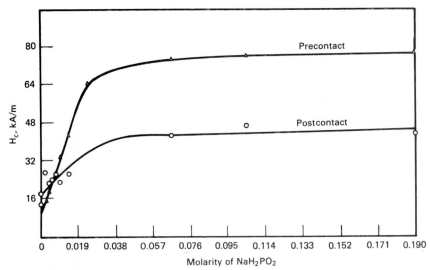

Figure 3.53 Coercivity of electrodeposited Co-Ni-P is seen to increase with concentration of NaH_2PO_2 in the plating bath, which also reflects the phosphorus concentration in the film. The two curves represent two different plating currents *(Judge et al., 1966)*.

strates, such as polyester tapes, involve soaking the substrate in hot chromic-sulfuric acid, followed by hot sodium hydroxide to render them hydrophilic (Fisher and Chilton, 1962). The substrate is then sensitized with a $SnCl_2$ solution which is thought to cause Sn^{2+} to adsorb on the substrate. Following this, immersion in a palladium chloride solution allows Sn^{2+} to reduce Pd^{2+} ions, which displace the resultant Sn^{4+} from the surface. The palladium thus deposited forms catalytic nucleation sites. After the Co(-Ni)-P plating reaction is begun, the plated transition metal itself acts as the catalyst to continue the process. The hypophosphite ion is the reducing agent for cobalt and nickel ions in the presence of a catalyst, according to the probable reaction (Pourbaix, 1974)

$$Me^{2+} + H_2PO_2^- + H_2O \rightarrow Me + H_2PO_3^- + 2H^+$$

where Me represents either Co or Ni.

To a lesser extent, phosphorus can undergo a reduction reaction. As with electroplating, the incorporation of phosphorus in the film is important to controlling coercivity. Figures 3.53 and 3.54 show coercivity dependence on phosphorus content and on film thickness. Complexing agents such as citrates, tartrates, or malonates are generally included in the plating bath to tie up some of the metal ions and release them at a controlled rate by dissociation, thereby regulating the plating rate. Solutions employing malonates codeposit virtually no phosphorus, while citrate solutions may result in as much as 6 wt % P. Films of Co-P and Co-Ni-P

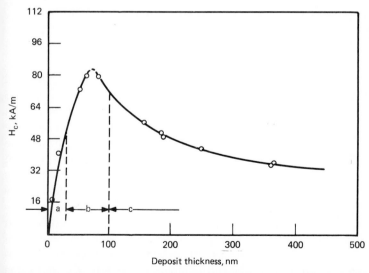

Figure 3.54 Coercivity of electroless-deposited Co-P versus film thickness, showing a maximum which characteristically occurs below 100 nm thickness *(Judge et al., 1965b).*

can be plated with coercivities from about 15 to over 130 kA/m (188 to 1625 Oe). The low-coercivity range corresponds to Co films with negligible P content, while the highest coercivities include high P content and a Co/ Ni ratio of about 4:1 (Judge et al., 1965). In these films high coercivity is believed to result from (1) high (in-plane) crystalline anisotropy grains, of random orientation, decoupled from one another by nonmagnetic phosphorus-rich channels, and (2) high stacking-fault concentration resulting from a mixture of cobalt-rich hexagonal close-packed (hcp) structure and nickel-rich face-centered cubic (fcc) structure, which impedes domain nucleation and wall motion.

Phosphorus is not unique in segregating at the grain boundaries of cobalt alloys. Other additives can also decouple the magnetic grains by formation of nonmagnetic grain boundaries. When Co-X and Co-Ni-X films were plated, where X is either one of the group VA elements P, As, Sb, Bi or one of the group VIB elements W, Mo, Cr, the group VIB sequence of elements produced effects on the magnetic properties similar to the group VA sequence (Luborsky, 1970). In Luborsky's work, however, phosphorus still produced the highest coercivities. Tungsten was found to improve the mechanical hardness and chemical stability, suggesting it would be an advantage to incorporate W or both P and W in plated films, with some compromise on the magnetics. The most significant aspect of this work is that it links plated media to sputtered and evaporated metal films, where Cr or W are commonly used as additives or undercoats. There appears to be a great deal of commonality among the metal in-plane isotropic cobalt-based films with regard to coercivity and magnetization reversal mechanisms, whether they are plated, sputtered, or evaporated. Indeed, rf-sputtered Co-Ni-W has shown magnetic properties basically similar to the plated Co-Ni-X and Co-X alloys (Fisher et al., 1981).

Besides compositional control of magnetic properties in Co-Ni-X and Co-X films, undercoat enhancement layers such as chromium and tungsten have proved useful in vacuum deposition. For example, using normal incidence evaporation, the coercivity of pure Co films was observed to increase with increasing thickness of chromium undercoats (Daval and Randet, 1970). This effect saturates at about 400-nm Cr thickness. The observed coercivity was attributed to magnetocrystalline anisotropy at the granular level. The Cr underlayer was thought to foster epitaxial growth of hcp Co with its c axis in the plane of the film. As the Cr layer thickness was increased, its grain size increased and produced larger cobalt grains by epitaxy. Although this work indicated that Cr underlayers may enhance the in-plane c axis orientation of cobalt, this orientation is often found in thin cobalt-based films (under 100 nm thick) even without chromium underlayers. In fact special care must be taken to prevent this orientation from forming during the initial stages of growth of Co-Cr films intended for perpendicular recording media. The more significant finding was that the Cr underlayer acted to control the size and uniformity of crystallites in the cobalt film deposited on it.

While underlayers can certainly affect the grain size and orientation of magnetic films, as described above, they can also affect the film roughness and the number of wall-pinning defects. Vacuum-deposited films, especially Cr and W, tend to show increased roughness as they grow thicker, saturating at a thickness of several hundred nanometers (Thornton, 1977). Grain size increases with temperature, and so does its associated surface texture. It follows that those conditions favoring epitaxy and grain growth also favor increased roughness, which can raise coercivity by impeding wall motion (Lloyd and Smith, 1959). For roughness to have a strong influence on coercivity, it should be of the same order of magnitude as film thickness, grain size, or domain wall width, all of which are comparable in film media (10 to 100 nm). Various calculations of roughness effects in thin magnetic films have been made (Néel, 1956; Soohoo, 1981). Although these calculations tend to be oversimplified, they serve to demonstrate the pronounced effect film roughness can have on coercivity.

In addition to the use of composition and enhancement underlayers to control coercivity, magnetic film thickness and grain size have generally been found to have a strong effect on coercivity and squareness. Many factors can influence grain size, but substrate temperature and deposition rate are two of the major process parameters. For sputtered or evaporated films, increased temperature or decreased deposition rate usually produces a larger grain structure. It is often found that the coercivity increases with grain size, up to a point where multidomain grains form. Similarly, and perhaps not independently, coercivity often increases with the film thickness up to a point, and then decreases with further increase in the magnetic film thickness (Fig. 3.54). To some extent this may be related to grain size effects, inasmuch as grains tend to grow and coalesce as the films grow thicker. However, superparamagnetic effects, magnetostatic coupling, and film roughness are also capable of producing this behavior. The point at which coercivity begins to decrease with thickness depends, of course, on the coercivity-controlling mechanism in a particular film, but that point is usually in the 20- to 40-nm range for sputtered or evaporated films and 40 to 80 nm for plated films.

The following is a summary of some of the key factors which can control the coercivity of in-plane isotropic film media.

1. *Composition of the magnetic film.* In Co(-Ni)-X films, where X is a group VA element such as P or a group VIB element such as Cr or W, coercivity tends to increase with X to a maximum depending on the species X. This is thought to be due in large part to the formation of nonmagnetic X-rich grain boundaries, which magnetically decouple the grains. Increasing the Ni content in such films increases the coercivity up

to about 20 to 25% Ni, which has been attributed to the mixture of hcp Co-rich phases and fcc Ni-rich phases, causing stacking faults which impede wall motion.

2. *Enhancement underlayers.* Underlayers such as Cr, deposited just prior to the magnetic layer, can increase the coercivity. Coercivity increases with underlayer thickness up to about 400 nm and then saturates. Grain size and orientation are possible mechanisms, although roughness and interdiffusion may be equally important.

3. *Magnetic film thickness.* Over most thicknesses of practical interest, coercivity decreases with increasing magnetic layer thickness. An initial rise produces a maximum, but this maximum often occurs at thicknesses where the film is still discontinuous.

4. *Temperature.* Coercivity generally increases with substrate temperature, which promotes larger grain size. Epitaxially oriented growth on suitable underlayers, interdiffusion, and roughness have also been cited as mechanisms.

5. *Deposition rate.* Low deposition rates produce effects similar to elevated temperatures, with the same probable mechanisms.

Many other materials have been deposited in isotropic film form, producing similar results to the Co(-Ni)-X and the Co on Cr described above. Some of these others are listed in Table 3.8. One material in this classification which deserves mention is Co-Pt, with Pt in the 10 to 25 at % range, because of the very high coercivities achieved in this material. Sputtered films of this alloy have exhibited coercivities peaking at about 25% Pt with values as high as 160 kA/m (2000 Oe). Coercivity in this system is extremely thickness-dependent, so that separate control of H_c and $M_r\delta$ is difficult (Aboaf et al., 1983). Magnetic hardness is correlated with formation of an fcc-hcp phase mixture. Presumably stacking fault impediments to domain wall motion, similar to one hypothesized mechanism in Co-Ni-P films, is the origin. Low-Pt-content alloys are very susceptible to corrosion, but 28% Pt films and 20% Pt–10% Ni films have been reported to show reasonable corrosion resistance (Yanagisawa et al., 1983).

Metal in-plane anisotropic films. Magnetic anisotropy can be induced in evaporated ferromagnetic films by controlling the angle of incidence of the vapor flux with respect to the substrate normal (Smith et al., 1960). The magnetic easy axis lies in the film plane and is perpendicular to the plane of incidence, for angles of incidence from 0° (normal) to about 60°, and this anisotropy increases with angle. Beyond 60° the easy axis rapidly becomes parallel to the plane of incidence. This behavior has been attributed primarily to shape anisotropy induced during the film growth. For angles less than 60° self-shadowing of grains occurs, resulting in formation of chains of microcrystallites perpendicular to the plane of incidence.

Beyond 60° the tendency for crystallites to elongate in the direction of the incoming vapor stream induces shape anisotropy with the easy axis parallel to the plane of incidence. Substrate temperature, magnetostrictive effects, and film composition also play important roles in oblique deposition.

Oblique incidence effects such as these are not limited to evaporation, but could be produced by ion-beam sputtering, ordinary dc or rf sputtering, or ion plating. Sputtering processes employing planar targets parallel to the substrate are often termed "normal incidence" deposition, but the sputtered atoms impinge on the substrate at a distribution of angles, which depends on the pressure and target-to-substrate separation. The resultant films may be isotropic, as discussed in the previous section, due to the averaging effect of all incident angles and adatom mobility on the surface, or they may be inadvertently anisotropic due to a predominance of certain incident angles. Low sputtering pressures can produce this if the sputtered atom mean free path is comparable to the target-to-substrate spacing. Rotation, or other complex motion, of the substrate through the sputtering flux is sometimes employed to ensure thickness uniformity and averaging of incident angles when isotropic films are required.

Early studies of oblique evaporation of Permalloy magnetic films (Smith et al., 1960) were followed by a number of similar studies of evaporated Fe, Co, Ni, and some of their alloys (Schuele, 1964; Speliotis et al., 1965) for possible recording media. Most of these showed similar results to the earlier Permalloy work. Coercivities over 80 kA/m (1000 Oe) for iron and cobalt, parallel to the plane of incidence, have been reported. Remanence squareness has also been found to be a function of deposition angle.

Full-scale processes for fabricating tapes and disks were reported in the early 1980s (Nakamura et al., 1982; Feuerstein and Mayr, 1984; Rossi et al., 1984). Manufacturing systems for oblique deposition onto a continuous PET web recording tape substrate make use of the effect of a high angle of incidence ($> 60°$). The web is transported via rollers and drums past one or more deposition stations, as illustrated in Fig. 3.55. As the tape moves around the drum, it passes by an aperture mask which controls the range of vapor-stream incident angles seen by the tape. If the tape is moved in the direction of decreasing incident angles, higher coercivities and squareness ratios result than if the tape is run in the opposite direction. The conditions of film nucleation are apparently critical to the canted columnar-growth morphology and resulting magnetic properties. Additional stations could be included within the same vacuum system for substrate cleaning, adhesion layer precoating, and postdeposition glow discharge stress relief. Alloys of Co-Ni (20 to 30%) have been used to achieve coercivities as high as 64 kA/m (800 Oe) and squareness ratios

TABLE 3.8 Typical Properties of Magnetic Film Media

Material	Deposition process[a]	Orientation[b]	Dominant crystal structure[c]	M_s, kA/m	H_c, kA/m	S, S^*	K_u, 10^5 J/m^3
Co	OIE	IPA	hcp	1100–1400	60–120	. . .	4 (bulk)
	NIE, SP	IPI	hcp	1100–1400	30–60		
Fe	OIE	IPA	bcc	1600	60–90	. . .	0.3–3.0
	SP	IPI	bcc	1600	10		
Ni	OIE	IPA	fcc	400	20–28		
	SP	IPI	fcc	400			
Co-Ni	OIE	IPA	hcp-fcc	800–1200	30–70		
Co-Fe	OIE	IPA	bcc	1400–1600	60–120		
Co-Sm	NIE (e)	IPA	Amorphous	500–1000	33–55	0.9, 0.9	
Co-P	EL, EP	IPI	hcp	800–1100	36–96	1, 1	
Co-Re	SP	IPI	hcp-fcc	500–750	18–58	0.9, 0.9	
Co-Pt	SP	IPI	hcp-fcc	800–1400	60–140	0.9, 0.9	

Co-Ni-P	EL, EP	IPI	hcp-fcc	600–1000	40–120	0.8, 0.8	
Co-(30 at %) Ni:N$_2$	SP	IPI	hcp-fcc	650	80	0.95	
Co-Ni:O$_2$	OIE	IPA	hcp-fcc	300–400	80	0.7–0.8	
Co-Ni-Pt	SP	IPI	hcp-fcc	800–900	60–70	0.9, 0.97	
Co-Ni-W	SP	IPI	hcp-fcc	450	30–50	0.8, 0.8	
Fe$_3$O$_4$	NIE, SP	IPI	I.S.	400	17–32		
γ-Fe$_2$O$_3$:Co	SP	IPI	I.S.	220–250	40–100	0.8, 0.8	
γ-Fe$_2$O$_3$:Os	SP	IPI	I.S.	240	160	0.8, 0.8	
Co-(18 at %) Cr	SP	⊥	hcp	300–550	80–100 (⊥)	.....	−1.0
Co-(20 at %) Cr	SP	⊥	hcp	400	65–95 (⊥)	.....	0.15
Co-(22 at %) Cr	SP	⊥	hcp	300–340	80–105 (⊥)	.....	0.4

[a] OIE = oblique incidence evaporation; NIE = normal incidence evaporation; SP = sputtered; EL = electroless-plated; EP = electroplated.
[b] IPA = in-plane anisotropic; IPI = in-plane isotropic; ⊥ = perpendicular.
[c] hcp = hexagonal close-packed; fcc = face-centered cubic; bcc = body-centered cubic; I.S. = inverse spinel.
[d] Coercivities for Co films deposited on Cr or W underlayers.
[e] Co-Sm values given for films deposited in the presence of an orienting field.

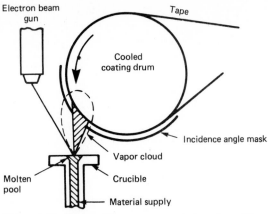

Figure 3.55 Schematic arrangement for depositing obliquely evaporated metal films onto a continuous web for video tape applications. The rod-fed Co-Ni source is evaporated by an electron beam. The incidence angle mask location and aperture are critical in controlling the angles at which nucleation and growth occur as the web passes around the drum *(Feuerstein and Mayr, 1984).*

> 0.8. Composition control is not a serious problem because cobalt and nickel have comparable vapor pressures. The process was developed mainly for video and audio recording media, but could also be used for computer tape.

Oblique evaporation has also been used for rigid-disk fabrication (Rossi et al., 1984; Arnoldussen et al., 1984). In this process a magnetic layer of Fe(42.5 at %)-Co(42.5 at %)-Cr(15 at %) was evaporated at a 60° angle of incidence, while the substrate was constantly rotated to ensure circumferential uniformity (Fig. 3.56). A strong shape anisotropy was produced, with the easy axis in the plane of the film but at right angles to the plane of incidence. The self-shadowing, grain-chaining mechanism is the presumed origin of the anisotropy. Sophisticated rate control was necessary to maintain a constant composition, because the vapor pressure of chromium is significantly higher than that of iron or cobalt. A 4-μm Haynes 188 stainless alloy undercoat was first dc-magnetron-sputtered onto an Al-Mg substrate. This layer was polished with a free abrasive to remove irregularities, and then buffed with a fixed abrasive to produce a controlled circumferential texture, as described in Fig. 3.51. The texturing reduced the sticking of head to disk by reducing the contact area, but also enhanced the strong circumferential orientation of the easy axis of the magnetic layer. By controlling the deposition parameters, including the oxygen partial pressure, coercivities from 24 to over 80 kA/m (300 to 1000 Oe), remanent moments from 600 to 1000 kA/m (emu/cm^3), and squareness ratios S and S^* as high as 0.95 were reported. Chromium was used in the magnetic alloy to retard corrosion (Phipps et al., 1984) rather than for coercivity control as in Luborsky's plated films (Luborsky, 1970).

Magnetic field–induced anisotropy offers another approach to producing in-plane oriented films. Amorphous Co_xSm_{100-x} with $75 \leq x \leq 90$,

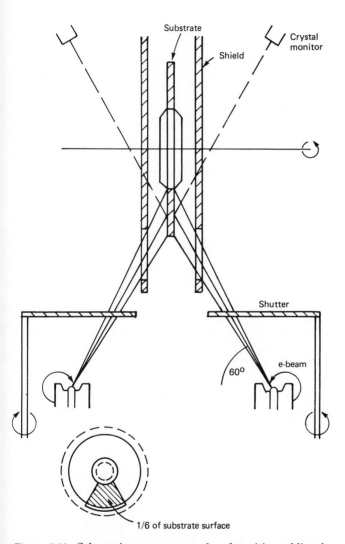

Figure 3.56 Schematic arrangement for depositing obliquely evaporated metal films onto a rotating Al-Mg disk substrate, from oppositely placed rod-fed, electron-beam-heated sources. The evaporant passes through an aperture mask, shown in the lower part of the figure, which controls incidence angle as well as thickness distribution *(Rossi et al., 1984)*.

when evaporated at normal incidence, but in the presence of a 20-kA/m (250-Oe) orienting field, shows extremely high orientation (Kullmann et al., 1984a). The coercivity increases from 20 to 88 kA/m (250 to 1100 Oe) with increasing samarium content. Magnetization varies with Sm in the opposite sense from 1000 to 350 kA/m (emu/cm^3). Both S and S^* were indistinguishable from unity. The anisotropy and hysteresis behavior in these films may be related to local ordering of the spin-orbit coupling between Co-Sm and Sm-Sm atom pairs.

Nonmetal in-plane isotropic films. The primary nonmetal-film media have been iron oxide–based and isotropic. This is a natural outgrowth of the predominance of iron oxide particulate media and concern about corrosion of metal films. The inverse spinel magnetite (Fe_3O_4) films were the first candidates investigated, in large part because they were relatively easy to prepare. In one method, pure Fe films were evaporated, and then oxidized at 450 to 500°C to nonmagnetic α-Fe_2O_3. A second layer of Fe was then deposited, which reduced the α-Fe_2O_3 to Fe_3O_4 upon heating at 350 to 400°C. Excess iron was then etched away in dilute nitric acid (Feng et al., 1972).

In another approach Fe_3O_4 was formed in situ by reactive rf sputtering from an iron target while cycling the oxygen background from high to low partial pressure. In this way a laminated structure of fully oxidized layers and iron-rich layers was formed at low substrate temperatures, while at temperatures of 200 to 250°C, homogeneous Fe_3O_4 is formed by interdiffusion (Heller, 1976).

Concern over the instability of Fe_3O_4 as a recording medium (a disputed issue) drove oxide film work toward techniques for producing γ-Fe_2O_3, as used in particulate media. Borrelli et al. (1972) chemically deposited α-Fe_2O_3 and showed that it could be completely converted to Fe_3O_4 by reduction in an H_2-H_2O atmosphere at 500°C. They then showed that the Fe_3O_4 could be converted into a solid solution of Fe_3O_4 and γ-Fe_2O_3 by heating in air or oxygen at 300°C for periods up to 1.5 h. Gamma-Fe_2O_3 is itself a defect spinel, similar in structure to Fe_3O_4; so the solid solution is more accurately written $Fe^{3+}[(Fe^{3+})_{1+2x}(Fe^{2+})_{1-3x}V_x]O_4$, where x is the defect content in the range $0 \leq x \leq \frac{1}{3}$, and V indicates a vacancy. They observed a peak coercivity of about 52 kA/m (650 Oe) for $x \approx 0.2$, attributing this to Fe^{2+}-V interactions, producing a preferred alignment of the vacancy on an octahedral site. This, in turn, was thought to yield additional magnetic anisotropy superimposed on the cubic magnetic anisotropy.

Refinements to the basic process described above have been made by adding Co (2 to 3%) to increase the coercivity and comparable amounts of Ti and Cu to control grain size and the effective range of heat treatment temperatures (Terada et al., 1983). Two processes are under development for oxide films. One process involves reactive rf sputtering α-Fe_2O_3, which is then reduced to Fe_3O_4 in wet H_2 at about 300°C. The Fe_3O_4 is then reoxidized in air or oxygen at about 300°C to obtain γ-Fe_2O_3 (Hattori et al., 1979). In a second, similar process, Fe_3O_4 is directly sputtered, thus eliminating the reduction step (Ishii et al., 1980). This shorter process would have a manufacturing advantage. However, achieving the desired Fe_3O_4 stoichiometry and structure depends on delicate control of sputtering rate, substrate temperature, and oxygen partial pressure. Cobalt-doped Fe_3O_4 with approximately 2 at % cobalt showed M_s = 263 kA/m (emu/cm^3), H_c = 38 kA/m (470 Oe), S = 0.64, and S^* = 0.54 (Ishii et al., 1980). After oxidization to Co-modified γ-Fe_2O_3, M_s remained the same but coercivity and squareness rose to H_c = 48 kA/m (600 Oe) and S = S^* = 0.85.

A much more effective dopant has been found to be osmium (Ishii and Hatakeyama, 1984). Coercivity and S^* increase sharply with increasing Os content up to 2.6 at %. At that level H_c = 154 kA/m (1920 Oe) and S^* = 0.8. Annealing such a film in a 560-kA/m (7000-Oe) field further increases H_c in the direction of the applied field by about 10 percent, producing a uniaxial anisotropy constant of 4×10^6 J/m^3. More significantly, the magnetic anneal raises the S^* to 0.95 or higher. In previous Co-doped oxide films, Ti dopant served to increase S^* and Cu decreased grain size (decreasing noise), whereas Os seemed to produce the combined effects of Co, Ti, and Cu and with superior results. The exact function of Os in enhancing the magnetic properties is not clearly understood, but it is likely that the similarity between the Fe $3d$ and the Os $5d$ electron occupancy plays a role.

Saturation magnetization in all the γ-Fe_2O_3 films is in the 240- to 260-kA/m (emu/cm^3) range, which is only about two-thirds that of the bulk value for γ-Fe_2O_3 particles. This suggests that a significant fraction of the film is nonmagnetic α-Fe_2O_3 (Kay et al., 1985). While this magnetization is about four times higher than particulate iron oxide media (20% volume packing fraction), it is only one-half to one-fourth the magnetization of metal-film media. Thus proportionately thicker oxide films must be used, compared to metals, for comparable signal amplitudes. This is a disadvantage because it represents a thickness loss of resolution. In early work this was considered to be acceptable because the γ-Fe_2O_3 films were thought not to need a wear-resistant overcoat as do metal films. However, a later study (Terada et al., 1983) showed that significant signal loss occurred when the disk was subjected to pressure and friction, such as might be exerted by a start-stop head contact. This signal loss was attributed to removal of small crystallites from the surface of the film. With such a problem, overcoats are likely required, but the oxide media remain attractive for corrosion resistance, and can give lower noise level than longitudinal metal films (Baugh et al., 1983).

Perpendicular anisotropy films. All the media discussed so far have been designed for longitudinal recording, in which the magnetization vector lies predominantly in the plane of the coating and is recorded by forming magnetization reversals along the line of a recording track. A limitation of this mode of recording is demagnetization, which broadens the transitions. For perpendicular recording, the storage layer is designed so that the magnetization is oriented normal to the plane of the coating. In some designs a soft magnetic underlayer is also required.

An infinite-area thin film with saturation magnetization normal to the plane has a demagnetization field of M_s ($4\pi M_s$ in cgs units), with associated shape anisotropy energy $\frac{1}{2}\mu_0 M_s^2$ ($2\pi M_s^2$). This shape anisotropy will force the magnetization into the plane of the film unless the film has additional anisotropy (e.g., crystalline) perpendicular to the plane with energy $K_\perp > \frac{1}{2}\mu_0 M_s^2$ ($2\pi M_s^2$). Cobalt can be made to grow with its c axis (and accompanying uniaxial anisotropy) normal to the film plane. However, $K_\perp$ for Co is only 6×10^5 J/m^3 (6×10^6 erg/cm^3), while $\frac{1}{2}\mu_0 M_s^2$ ($2\pi M_s^2$) is 12.95×10^5 J/m^3 (12.95×10^6 erg/cm^3). A solution is to dilute the Co moment with Cr, a nearly linear relation, until the required inequality is satisfied ($K_\perp$ decreasing at a less than quadratic rate) (Iwasaki and Nakamura, 1977). For an alloy with 20.5 at % Cr, $K_\perp \approx 1.15 \times 10^5$ J/m^3 (1.15 $\times 10^6$ erg/cm^3), and $M_s \approx 450$ kA/m (emu/cm^3), so that the demagnetization energy is $\frac{1}{2}\mu_0 M_s^2 = 10^5$ J/m^3 (10^6 erg/cm^3) (Fisher et al., 1984). The energy requirement defined above is a necessary condition to maintain a perpendicular saturation magnetization in a uniformly magnetized infinite-area film which reverses by uniform rotation. The calculation indicates that this is satisfied for Cr content greater than about 20 at %. However, since most real films are better characterized by nonuniform reversal and aggregate demagnetizing fields, as discussed earlier, the above inequality need not be strictly met. The more appropriate inequality uses the remanent magnetization of the sheared hysteresis loop, rather than M_s, in calculating the demagnetization energy. Alloys with as low as 15 at % Cr have been used.

Without a sufficiently high coercivity, stray fields can cause data erasure even when the energy inequality above is met. Moreover, because of the loop shearing caused by demagnetization, the remanent magnetization is roughly equal to the coercivity ($H_c/4\pi$ in cgs units). Therefore the coercivity is key to achieving sufficient signal amplitude. This contrasts with longitudinal media, where M_s is the primary determinant of signal amplitude and coercivity contributes to amplitude only secondarily through transition slimming.

Dispersion of the uniaxial perpendicular orientation can cause loss of coercivity, but this cannot account for the large discrepancy between uniform rotation calculations and measured coercivities. Magnetization reversal has been attributed to single-particle (columnar grain) behavior with a curling- or buckling-type reversal (Ohkoshi et al., 1983; Wuori and Judy, 1985). This is a plausible explanation for the coercivity generally being less than 100 kA/m (1250 Oe).

Most of the experimental and theoretical investigations of perpendicular recording have focused on rf-sputtered Co-Cr alloys with Cr in the 15 to 22 at % range. Moderate substrate temperatures up to 300°C are used to facilitate growth of the proper crystal and grain structure. The desired structure is hexagonal close-packed with the c axis normal to the plane, since that is the direction of the easy axis. Moreover, these films generally grow with a columnar grain texture, due partly to the hcp crystal structure and partly to the thickness of the films (0.5 to 2.0 μm). While magnetocrystalline anisotropy is the dominant mechanism, the grain texture orientation is also important for two reasons. First, it superimposes some degree of shape anisotropy. Second, the columnar grains are often separated by nonmagnetic grain boundaries which diminish or eliminate the exchange coupling between grains (Chen and Charlan, 1983; Grundy et al., 1984). The growth of perpendicular anisotropy columnar grains depends on the substrate (or underlayer) and the process conditions. In some cases a transition layer (10 to 100 nm) forms at the early stages of film growth and has poor orientation (Wuori and Judy, 1984). Beyond this point oriented columnar growth proceeds. Films which possess such a transition layer show greater c-axis dispersion and poorer recording properties than those with little or no transition layer.

Additives other than Cr, and processes other than rf sputtering, have been explored for perpendicular media, but not as extensively. For example, substituting up to 6 at % Rh for Co in a sputtered $Co_{78-x}Cr_{22}Rh_x$ film, has been shown to improve the c-axis orientation normal to the surface without altering H_c and M_s (Kobayashi and Ishida, 1981). Chemically deposited or plated cobalt-based films (Chen and Charlan, 1983) have also been shown to produce acceptable perpendicular orientations. Sputtered barium ferrite films have been prepared with perpendicular anisotropy (Matsuoka et al., 1985); however, because of the high substrate temperatures required (400 to 650°C), they have not been widely considered for practical application.

For perpendicular recording, a single-pole head is most suitable for achieving high resolution during the writing process. But a second, return pole is required on the opposite side of the medium, and two-sided recording has practical difficulties. Some head designs have been proposed to simulate the performance of a single-pole head, without the need for a physical pole on the opposite surface, by employing a soft magnetic underlayer beneath the Co-Cr to form a magnetic image below the medium. Permalloy is usually used for such an underlayer. However, in order for a true image pole to be formed, much thicker soft underlayers are required than is practical. Instead of using a single-pole head, newer designs employ a thin primary pole for high resolution in conjunction with one or more large auxiliary poles on the same side of the recording surface (Hokkyo et al., 1984). This achieves a higher field strength and requires a soft magnetic underlayer, not to form a true image pole but simply to act as a flux return path to the large auxiliary pole. Another design uses a thin primary pole near the recording surface, and a large auxiliary pole on the opposite side of the medium, again requiring a soft magnetic underlayer to concentrate the field lines during the write process. Suitable recording head designs which can take advantage of the promised density advantage of perpendicular media are the greatest obstacle to practical use of this mode of recording. By comparison, the medium materials and process development are far more advanced.

Because most head designs require a soft magnetic underlayer, the head and medium must be considered as an integral system, more so than for longitudinal media. Not only is the permeability of the soft underlayer important, but the optimum thickness of the recording layer and the underlayer depend on the head design. For a single, thin primary pole and a large, opposite-surface auxiliary pole, the underlayer enhances write resolution and read sensitivity up to thicknesses of 0.5 to 1.0 μm. Little is gained by making it thicker. The recording layer also has an optimum thickness. Below a few hundred nanometers, read signal amplitude suffers, whereas thicker recording layers cause the soft underlayer to be spaced too far from the primary pole to achieve high resolution. This conflicts with early claimed advantages of perpendicular recording that high resolution and high signal amplitude could be simultaneously realized by using relatively thick films of 1 μm or more. Also sensitivity to substrate defects and surface abrasion were thought to be minimized by the use of thick films. If a practical recording system limits the perpendicular recording layer thickness to a few hundred nanometers, the advantage of perpendicular over longitudinal recording must be diminished.

In addition to the considerations above, the use of a soft magnetic underlayer like Permalloy poses other problems. The Permalloy underlayer has been reported to degrade the Co-Cr orientation, though the specific deposition processes used are probably as important as the materials themselves. This ranks as a practical engineering complication more than an intrinsic or insurmountable obstacle. The misorientation can be partially alleviated by inserting a Ti film about 50 nm thick to nucleate properly oriented grain growth (Kobayashi et al., 1983). Another potential problem is the formation of a corrosion couple. Corrosion has indeed been seen to increase for such a double-layer structure (Dubin and Winn, 1983). A third problem area is noise in the readback signal due to the presence of the soft magnetic underlayer (Uesaka et al., 1984). By its nature, a soft film contains domain walls which are relatively easy to move. Such walls are generally pinned, or at least their movement is impeded, by defects and inhomogeneities of any sort. As a result, walls do not all move smoothly. Rather they tend to move in jumps between pinning sites (Barkhausen noise), and local hysteresis is exhibited. It is not clear whether domain walls actually move during the read process, although this is conceivable, especially at low head-to-media spacings. Nevertheless, since the soft underlayer domains possess their own equilibrium configurations, they may randomly shift the written bit pattern to produce a modulation noise. Figure 3.57 shows the type of complex domain configuration which can occur in a soft Ni-Fe underlayer. This was observed by differential phase-contrast Lorentz microscopy of a cross-sectioned perpendicular medium.

3.3.2.4 Overcoats.

Small head-to-medium spacings are critical to reaching higher densities on rigid disks. This raises the risk of head-disk interactions, which can cause disk and head wear, or even a catastrophic head crash. Wear can also occur when the drive starts and stops, because the head contacts the disk at relatively high speeds. With flexible media the head is in contact continuously during read, write, and accessing operations. Particulate media have so far coped with this problem by the use of hard-wear particles like Al_2O_3 or SiO_2 incorporated in the magnetic coating itself, along with the use of lubricants.

Film media require a different approach, namely, the use of wear-resistant overcoats. Listing the desirable qualities of an overcoat is not difficult; specifying the appropriate materials properties needed to achieve those qualities is impossible to do with any confidence. The film should be as thin as possible (compared with the desired head-to-medium spacing), have no asperities, resist wear by the head, and, at the same time, not inflict wear on the head. Also the film should have virtually no static or dynamic friction with the head and, as much as possible, protect against corrosion. Intuition and empiricism suggest that overcoat materials should be relatively hard and chemically inert, and should bond well to the magnetic layer but not to the head material. They should probably have high tensile strength and not be prone to brittle fracture.

Silicon dioxide, SiO_2, has been used with comparative success, even though it falls in the brittle category and has surface properties which are humidity-sensitive. A spin-coating technique has been used to deposit SiO_2 on plated Co-Ni-P media (Suganuma et al., 1982). A tetrahydroxysilane and alcohol solution is spun on the disk, leaving a layer of $Si(OH)_4$ on the surface. This is then laser-irradiated, driving off water and leaving a nearly stoichiometric SiO_2 about 80 nm thick. By itself, this overcoat has shown high friction and stiction (head sticking to the disk), which varies with exposure to the atmosphere. Suganuma et al. applied an unspecified polar solid lubricant to the surface to minimize these problems.

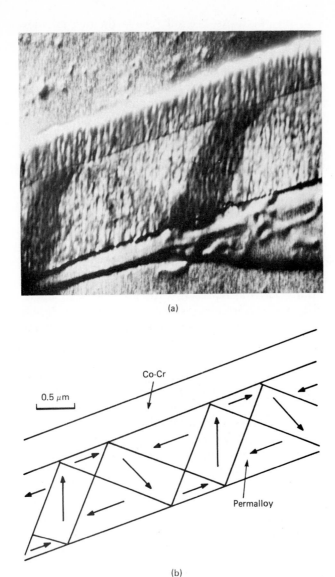

(a)

0.5 μm

Co-Cr

Permalloy

(b)

Figure 3.57 (*a*) Domain structure, observed by differential phase-contrast Lorentz microscopy, of a cross section of perpendicularly oriented Co-Cr on a soft Permalloy underlayer. *(Courtesy of J. Chapman.)* (*b*) Arrows indicate the direction of magnetization in the soft underlayer domains seen in (*a*).

A plasma-polymerized fluorocarbon film has been described in a U.S. patent in which the fluorine content was varied from low values at the metal magnetic film surface to high values at the free surface (Arai and Nahara, 1983). By this, it was claimed, the smooth, lubricating properties of a fluorocarbon are achieved at the free surface with improved adhesion at the metal. This may be considered a combined overcoat and solid bonded lubricant.

A sputtered rhodium overcoat (75 nm) over a Cr adhesion layer (10 nm) has also been used, in combination with a 5-nm perfluorinated polyether lubricant applied by a spray-wipe process (Rossi et al., 1984). Although rhodium is a tough metal, it is not an adequate overcoat without a lubricant on its surface. A solid lubricant is not readily bonded to metals like rhodium, and liquid lubricants tend to spin off a rotating disk. However, when a slight texture is imparted to the disk at the undercoat level, one to several molecular layers of liquid lubricant can be retained by the rhodium surface, giving adequate wear properties. This comes at the expense of added spacing losses in the readback signal due to the texturing.

One of the more promising materials currently being explored is a hard carbon which is often sputtered, but could be deposited by other techniques, such as glow discharge decomposition of hydrocarbon gases (King, 1981; Kolecki, 1982). Carbon films deposited by low-energy processes, such as conventional diode sputtering, tend to form local atomic bonding similar to that in graphite, rather than diamond, but can nevertheless be very hard and wear-resistant. These films should not be confused with another class of hard-carbon film called I-carbon, or amorphous diamond-like carbon, which was first studied for wear-resistant and optical coatings outside the recording industry. These terms strictly refer to films in which there is substantial tetrahedral local bonding as in crystalline diamond. Such films require high-energy ion bombardment during growth to achieve this coordination.

Other friction-, wear-, and corrosion-resistant coatings being explored for various recording and nonrecording applications have been reviewed by Hintermann (1984). This excellent review includes TiC, TiN, SiC, Cr_2C_3, and Al_2O_3, together with the deposition processes used.

In addition to the deposition of a wear-resistant overcoat, some final mechanical surface finishing is generally required, especially for rigid disks where asperities can cause catastrophic head crashes. As smooth and uniform as the disk may have been at the substrate or undercoat level, asperities can form during the multiple film depositions that follow. Non-uniform nucleation and growth, spitting during evaporation, and sparking electrical discharges in sputtering are some of the possible causes of asperity formation during film deposition. Thicker films which show columnar growth, such as perpendicular magnetic films, are most prone to growth-induced roughness.

Whatever the source of process-induced asperities, a final surface buff may be used to produce a reliable head-medium interface. One type of buffing process is illustrated in Fig. 3.51. A disk is shown being buffed by a light-pressure, oscillating, fixed-abrasive polishing tape. The idea is neither to cause any stock removal nor to produce a texture as is done in the undercoat polishing process, but merely to knock off asperities. An alternative method, when few asperities are present, is to use a special burnishing head which is designed to fly lower than a recording head, clipping off asperities.

Surface lubrication may be used, though the extremely smooth surfaces and nonporous structures of film media make liquid lubricants generally undesirable. Without porosity to hold the lubricant, it is restricted to the surface and tends to spin off a disk during operation. Moreover, when the head is at rest in contact with a disk, liquid lubricant can wick under the head, causing the head to stick to the surface. Texturing the undercoat (Rossi et al., 1984) mitigates both the spin-off and stiction problems, but at the expense of added spacing loss. Other approaches have employed solid lubricants with some success, but it is considered preferable to design film media which do not require lubricants for reliable performance.

3.3.3 Magnetic and mechanical stability

Practical commercial recording systems must show long-term stability of all sorts. The subject of device reliability is only sparsely treated in the literature because it falls more in the category of product engineering rather than research. Often information of a proprietary nature is involved, and long-term statistical testing of specific commercial designs is required. For such reasons, it is difficult to make strong generalizations about media stability and reliability. Nevertheless, the key reliability factors can be outlined and some media comparisons made, at a qualitative level. It should also be kept in mind that film properties depend so much on the specific processes used that data on a particular type of film may vary significantly from one lab to another, even when the same nominal deposition process is used.

3.3.3.1 Thermal effects. Two distinct thermal effects are of interest: reversible and irreversible. Temperature dependence of magnetization and of magnetocrystalline anisotropy constant are usually of the reversible type, at least in a temperature range below a stabilizing anneal temperature. Such dependencies stem from the energetics of ferromagnetic exchange or spin-orbit coupling. For example, saturation magnetization can be characterized by the material's Curie temperature according to

$$M_s(T) = M_{s0} \tanh \frac{M_s \theta_c}{M_{s0} T} \tag{3.36}$$

where θ_c is the Curie temperature in degrees Kelvin and M_{s0} is the saturation magnetization at $0°K$.

Reversible thermal effects in films generally are related to those in bulk materials of similar composition, but the correspondence is seldom exact. Most often quantities like the Curie temperature depend on prior heat treatment of a thin film, which can affect the microstructure and microscopic composition variations. For example, a Co-(84 at %) Cr (16 at %) alloy for perpendicular media was shown to have an as-deposited Curie temperature of about 700°C compared with 355°C for the bulk alloy (Ishizuka et al., 1983); $\theta_c = 700°C$ for a bulk alloy would correspond to 9% Cr. The temperature dependence was reversible up to 400°C, but, upon annealing at 650°C, the film Curie temperature moved to even higher values. This suggests that composition may vary from grain-to-grain boundaries, and these variations may become more pronounced with heat treatment owing to diffusion processes. In the same study, perpendicular anisotropy energy was also shown to vary reversibly below a given annealing temperature, but the specific reversible temperature dependence changed character as the annealing temperature was increased. The reversible anisotropy energy temperature dependence was of the order of several hundred $J/(m^3 \cdot °C)$. Similar studies have been reported for plated Co-Ni-P longitudinal media (Ouchi and Iwasaki, 1972) and cobalt- and copper-doped γ-Fe_2O_3-sputtered films (Nishimoto and Aoyama, 1980).

The reversible temperature variations in magnetic properties do not themselves constitute a serious stability issue. However, if a recording system must operate reliably within a broad temperature range, then even reversible changes will result in recording performance which depends on ambient conditions. If the system operates within a rather narrow range of temperatures, as in a high-performance rigid-disk computer file, such changes in performance may be small. Nevertheless it is necessary to design the media with the expected operating temperature in mind, since the magnetic properties may be significantly different than at room temperature.

In practice, media designs which have strong temperature dependencies are undesirable. Elemental magnetic materials like cobalt or iron have Curie temperatures which are so high that thermal magnetization variations are minor under practical operating conditions. In alloys, however, the addition of nonmagnetic constituents (like Cr in Co) does not simply dilute the magnetic moment, but actually lowers θ_c. If the Curie temperature is thus lowered to near the operating temperature, the saturation magnetization can change rapidly, as seen by Eq. (3.36). Bulk Co-Cr alloys with 20% Cr have Curie temperatures of only about 150°C (Bozorth, 1951). Similar considerations apply to anisotropy constants which may cause strong coercivity temperature variations.

Irreversible thermal effects pose even more hazardous stability problems. For the most part, these phenomena result from atomic diffusion, as suggested for the Co-Cr study above. Such atomic rearrangement may occur in films at temperatures above that of deposition because, during deposition, the films grow too fast for atomic arrangement to reach thermal equilibrium. Consequently, when the film is later heated, there is a driving force toward the equilibrium state. In some materials, the phase equilibrium itself may have a strong temperature dependence.

Films of Co-P and Co-Ni-P can be deposited in amorphous or crystalline form. In amorphous Co-P, magnetic properties undergo changes at low temperatures because of the relaxation of built-in stresses (Riveiro and de Frutos, 1983). As the temperature is raised, the films go through phosphorus diffusion, compositional relaxation, and finally recrystallization. Even in the crystalline state, a phosphorus-rich grain boundary continues to segregate and grow by diffusion. Similar effects have been observed in Co-Ni-P films (Maeda, 1982), Co-Pt films (Aboaf et al., 1983), and iron oxide films (Borrelli et al., 1972). In the latter case it was vacancy defects, rather than impurity atoms, which diffused to preferred sites, causing the coercivity to increase with time.

Such relaxation phenomena often show thermally activated Arrhenius behavior, where the time constant for relaxation is inversely proportional to the relevant diffusion constant. Depending on the diffusion activation energy, the relaxation may have a strong or weak temperature dependence. For Co-P, stress relaxation has been shown to have an activation energy of about 0.2 eV, and phosphorus diffusion about 0.75 eV (Riveiro and de Frutos, 1983). Vacancy diffusion in iron oxide showed an activation energy of about 1.0 eV (Borrelli et al., 1972).

When relaxation phenomena exhibit thermally activated behavior, it is possible to make reasonable estimates of stability at recording system operating temperatures by measuring the relaxation rate at elevated temperatures. However, not all thermal behaviors possess a well-defined activation energy. Nor are the relaxation phenomena always dependent on temperature alone. Magnetic fields and chemical ambient may also play a role.

3.3.3.2 Stress effects.   When films are deposited they may possess built-in stresses, ranging from large tensile to large compressive stresses, depending on the process and the substrate used. These stresses can affect the magnetic properties, mechanical properties (adhesion and wear), or corrosion properties. As mentioned above, such built-in stresses may show long- or short-term relaxation, with accompanying performance instabilities. Even when the built-in stress shows little tendency to relax thermally, mechanical interaction of the head slider and the medium can induce abrupt relaxation via cracking or delaminating. Stress can also exacerbate corrosion.

In addition to stress relaxation effects, magnetic films which are sufficiently magnetostrictive may be subject to data loss caused by mechanical interaction of the head and the medium. For a material with isotropic saturation magnetostriction constant (λ_s), and saturation magnetization M_s, the magnetic field equivalent of a stress σ is

$$H_{\text{eq}} = \frac{3\lambda_s}{\mu_0 M_s} \sigma \tag{3.37}$$

To generate a field equivalent of 80 kA/m (1000 Oe) on a film with $M_s = 600$ kA/m (emu/cm^3) and $\lambda_s = 20 \times 10^{-6}$, the required stress is 10^9 N/m^2(0.1 g/μm^2). Although this is a very large stress, it is well within the realm of possibility for a corner of a head slider impacting a rigid disk. If the medium coercivity is comparable to the stress-equivalent field, written data could be partially or totally erased. Thus, even though film media are protected by an overcoat against data loss by physically wearing away the magnetic layer, erasure may still occur in highly magnetostrictive materials.

Magnetostriction is a very composition-dependent parameter. Figure 3.58 shows magnetostriction and magnetization values for chromium alloys of iron, cobalt, and nickel, as well as cobalt-platinum alloys (Aboaf and Klokholm, 1981; Aboaf et al., 1983). By using ternary alloys, magnetostriction can be made near zero over a range of Cr content, thereby allowing greater flexibility in designing for magnetization, corrosion resistance, and magnetostriction.

3.3.3.3 Corrosion. Corrosion is probably the overriding stability concern for metal-film media. While gross changes in magnetic properties can occur, the more significant problems stem from microscopic corrosion sites. Such corrosion can produce magnetic defects resulting in noise or other signal aberrations, as well as mechanical defects associated with corrosion products. Conventional techniques for measuring corrosion are inadequate for evaluating recording system reliability, because most of these methods rely on macroscopic detection, such as weight gain, to quantify corrosion. Although these techniques provide useful relative information, they can be misleading. Corrosion is usually initiated at local compositional inhomogeneities, stress concentrations, or pinholes, which are statistically sparse. Such sites may become fully corroded, forming hazardous mechanical defects or signal dropouts, and yet be undetectable by macroscopic corrosion measurements. Moreover, the corrosion kinetics at these sites may be very different from those indicated by the average macroscopic rates.

Unlike with purely thermal instabilities, it is difficult to define mean-

(a)

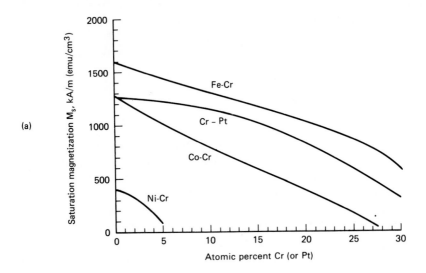

(b)

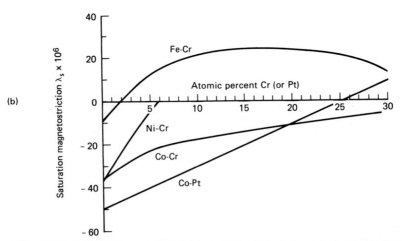

Figure 3.58 (*a*) Saturation magnetization and (*b*) saturation magnetostriction for various transition metals alloyed with chromium or platinum, as a function of nonmagnetic additive. The films are sputtered (*after Aboaf and Klokholm, 1981; Aboaf et al., 1983*).

ingful accelerated corrosion tests. In general, corrosion depends on temperature, relative humidity, and the species of corrosive gases in the atmosphere. It is best to correlate any accelerated corrosion test with actual data under real operating conditions. Unfortunately this is often impractical because the correlation can take as long to complete as the market life of the proposed product. Just as it is difficult to define valid conditions for accelerated corrosion testing, it is also difficult to define the conditions of failure. Corrosion failure could be defined as the point at which the error rate exceeds some value, or the point at which a catastrophic head-medium crash occurs. In either case, the most suitable detector of failure is a recording head; yet to put an entire recording system in an accelerated corrosion environmental chamber could induce failure unrelated to actual corrosion. For example, high concentrations of pollutant gases, or high relative humidity, can cause adsorption or condensation on the medium, inducing mechanical failure without necessarily producing corrosion.

Corrosion studies have been made on a number of film media, including Co-P (Judge et al., 1965b); Co-Ni-P with Rh overcoat (Garrison, 1983); Co-Pt and Co-Ni-Pt (Yanagisawa et al., 1983); a variety of Co-based alloys with different undercoats and substrates (Dubin et al., 1982); Fe-Co-Cr (Phipps et al., 1984); and perpendicular Co-Cr on different substrates, with and without Ni-Fe underlayers (Dubin and Winn, 1983). The reader is referred to this and other literature for details, because the results must be evaluated within the context of the individual tests, and to specify the detailed test conditions here would be too lengthy. However, several aspects of metal-film corrosion do stand out in these studies and deserve mention: (1) Sulfur- and chlorine-containing atmospheres are the most corrosive for metal-film media. (2) Relative humidity is the single most critical factor; corrosion is relatively low even in sulfur- and chlorine-containing gases when humidity is low but proceeds rapidly when it is high. (3) Corrosion accelerates with increased temperature and constant relative humidity; however, since in most situations relative humidity decreases with increased temperature, somewhat elevated temperature operation favors decreased corrosion. (4) Pinholes in multilayered, dissimilar metal media may show enhanced corrosion due to galvanic action. (5) Applied lubricants or adsorbed organic contaminants from the atmosphere can retard corrosion.

3.3.3.4 Mechanical wear. Mechanical wear has been discussed in Sec. 3.3.2.4 on overcoats and is further discussed in Chap. 7 on head-medium interface. Here only the instability aspect of wear is addressed. Without protective overcoats, all film media suffer signal degradation with time due to wear of the magnetic layer by the recording head, or they suffer catastrophic mechanical failure. With an overcoat, wear of the magnetic layer is not an issue but overcoat wear is of major concern.

Overcoat films which wear by delamination are undesirable, because they leave the magnetic layer exposed to damage at best, and risk a catastrophic head-medium crash at worst. Overcoats which wear by gradual loss of thickness are tolerable, provided the debris generated is not hazardous to mechanical operation and provided the overcoat is thick enough to endure the design life of the recording system. Wear of the overcoat can be virtually eliminated by the presence of a surface lubricant; however, such a lubricant itself can carry with it instabilities. For a lubricant to be useful it should be self-replenishing. For example, liquid lubricant in sufficient quantity can flow by capillary action to replenish an area depleted by rubbing contact. However, nonporous film media cannot retain more than a one- or two-molecule thickness of lubricant (at least on a spinning disk), and, even if more lubricant could be retained, a serious stiction problem would result. Lubricant which is only one or two molecules thick behaves more like a solid lubricant than a liquid, and as a result may not provide adequate replenishing action. Thus, with repeated head-medium rubbing, wear of the overcoat may accelerate.

Overcoats which are reasonably wear-resistant without lubricant are generally sought. However, the wear properties of most overcoats, as gauged by friction and stiction measurements, are often extremely dependent on the ambient and prior history. Adsorbed organic contaminants may act as a lubricant, but, with repeated contact, such adsorbates are removed, causing friction and wear to accelerate. In other cases adsorbed water may act in just the opposite way, producing high friction and stiction values until worn away. Either of these situations represents time- and ambient-dependent wear properties and therefore a potential reliability problem.

3.4   Media Applications

3.4.1   Particulate media

3.4.1.1   Audio tapes. The requirements for analog audio tapes are strongly influenced by the anhysteretic recording process which is employed for linearization and noise reduction. The long-wavelength sensitivity of audio tapes is determined by the anhysteretic susceptibility. The key properties of audio tapes are the sensitivity, the frequency response, the signal-to-noise ratio (dynamic range), and the print-through. Equally important are modulation noise, dropouts, and wear resistance, properties associated with the mechanical homogeneity of the medium.

In order to ensure tape and recorder compatibility, a careful standardization has been established with internationally accepted reference tapes that set the ac bias and the signal current as well as the equalization. Thus, the range of permissible coercivity, of remanence, and of thickness is limited within about 10 percent. A fine-tuning has to be made with respect to the degree of particle alignment which controls the remanence squareness, and with respect to the switching-field distribution (Köster et al., 1981). Coating thicknesses much larger than twice the recording head gap length are not favorable for the balance between long- and short-wavelength output. Typical values are 10 μm for reel-to-reel and 5 μm for compact cassette tape. Once the magnetic tape parameters are properly set at long wavelengths, the surface smoothness and the switching-field distribution primarily determine the frequency response. The response should be somewhat positive rather than negative with respect to the reference tape. The main focus of product improvement is on the reduction of the particulate noise through a reduction in particle size. Smaller particles mean a larger number of particles per unit volume, to which the signal-to-noise power ratio at long wavelengths is directly proportional. This route is limited by the print-through signal, which must be kept below a certain level, but progress can be made by changes in the preparation of the magnetic particles with respect to their morphology and their size distribution.

Major improvements in signal-to-noise ratio and frequency response through higher values of coercivity and remanence, or through a significantly smoother coating surface, require a new standardization. Therefore, there are now four different standards in existence for compact cassette recorders, apart from the standards for reel-to-reel machines. Historically, the latter have changed with the reductions made in tape speed. The cassette standards differ in coercivity, which has been increased successively in order to improve short-wavelength output. The increases in coercivity also involved increases in specific saturation moment of the particles from gamma ferric oxide, through chromium dioxide, to iron. Typical values of saturation remanence $M_r(\infty)$, coercivity H_c, switching-field distribution Δh_r, and number of particles per unit volume N, are shown in Table 3.9 for home reel-to-reel, as well as IEC I, II, and IV standard compact cassette audio tapes.

Only the first outer 1 or 2 μm of the magnetic coating contribute to the short-wavelength signal in compact cassette recording, which employs a tape speed of 47.5 mm/s. Therefore, it is of value to decouple the long- and short-wavelength recording by using a double-layer tape. Such tapes allow separate optimization of the outer layer for short, and the inner layer for long, wavelengths. Typically, the coercivity of the outer layer is 1.5 times that of the inner layer. This technique was first introduced for the IEC III standard, but is now successfully used in tape products meeting all four compact cassette standards. The next logical steps were to use smaller particles in the outer layer and larger ones in the inner layer for a better compromise between noise and print-through, with the coercivities of the layers being the same or having the ratio mentioned above.

The tapes for digitally encoded audio recording very much resemble those which are used for video recording application, and will be discussed in the following section.

3.4.1.2 Instrumentation and video tapes. Instrumentation and video tapes have in common that they are much more demanding in surface smoothness, and in low error rates, than audio tapes. They differ in their way of application. Instrumentation tape is moved over a multiple-track head, as in professional audio or computer tape applications. Video tape has to withstand narrow-track heads, which are mounted on a drum that runs at a circumferential speed of the order of 4 m/s. Thus, the mechanical wear situation, and consequently the formulation of the magnetic coatings, are quite different.

TABLE 3.9   Magnetic Material, Saturation Remanence $M_r(\infty)$, Coercivity H_c, Switching-Field Distribution Δh_r and Number of Particles per Unit Volume, N, of Various Particulate Magnetic Recording Media

Application	Material	$M_r(\infty)$, kA/m (emu/cm^3)	H_c, kA/m (4π Oe)	Δh_r	N, $10^3/\mu m^3$
Reel-to-reel audio tape	γ-Fe$_2$O$_3$	100–120	23–28	0.30–0.35	0.3
Audio tape IEC I	γ-Fe$_2$O$_3$	120–140	27–32	0.25–0.35	0.6
Audio tape IEC II	CrO$_2$	120–140	38–42	0.25–0.35	1.4
	γ-Fe$_2$O$_3$ + Co	120–140	45–52	0.25–0.35	0.6
Audio tape IEC IV	Fe	230–260	80–95	0.30–0.37	3
Professional video tape	γ-Fe$_2$O$_3$	75	24	0.4	0.1
	CrO$_2$	110	42	0.3	1.5
	γ-Fe$_2$O$_3$ + Co	90	52	0.35	1
Home video tape	CrO$_2$	110	45–50	0.35	2
	γ-Fe$_2$O$_3$ + Co	105	52–57	0.35	1
	Fe	220	110–120	0.38	4
Instrumentation tape	γ-Fe$_2$O$_3$	90	27	0.35	0.6
	γ-Fe$_2$O$_3$ + Co	105	56	0.50	0.8
Computer tape	γ-Fe$_2$O$_3$	87	23	0.30	0.16
	CrO$_2$	120	40	0.29	1.4
Flexible disk	γ-Fe$_2$O$_3$	56	27	0.34	0.3
	γ-Fe$_2$O$_3$ + Co	60	50	0.34	0.5
Computer disk	γ-Fe$_2$O$_3$	56	26–30	0.30	0.3
	γ-Fe$_2$O$_3$ + Co	60	44–55	0.30	0.5

Instrumentation and 2-in-wide video tapes are designed for professional tape recorders based on standards which were established some 30 years ago. Consequently, these tapes historically relied on rather conservative gamma ferric oxide particles, except that instrumentation tapes have now been supplemented by tapes with a higher coercivity (Table 3.9). Also, the 1-in professional video recorders, introduced 7 years ago, use tapes with modern chromium dioxide or cobalt-modified gamma iron oxide particles. These tapes have intermediate remanence and coercivity values. Surface homogeneity and few dropouts are the key properties, in addition to the usual requirements with respect to wear and friction.

The half-inch home video tapes must meet a delicate balance between providing the surface smoothness needed for a good signal-to-noise ratio, without causing blocking of the tape on its path through the cassette housing and the recorder. The video color information is recorded separately in a frequency band below the luminance channel. Recording takes place anhysteretically with the luminance signal providing the bias field. Consequently, all parameters mentioned for audio tapes are applicable and must be properly adjusted in order to comply with the standard set by the recorder system. Further, the same long-wavelength surface inhomogeneities which are mentioned in Sec. 3.2.3 may contribute to an undue modulation of the color signal.

Next-generation camera-type or improved-quality home video, digitally encoded audio, and professional video recording systems are expected to go to wavelengths as short as 0.5 μm. This asks for the high remanence and coercivity values which are found with metal particle tapes. Barium ferrite platelets, oriented for perpendicular recording (Fujiwara, 1983), lend themselves to an extremely flat frequency response which may result in similar output levels as those of metal-particle tapes at the target wavelengths below 0.5 μm. However, until these particles are commercially available, it appears that metal-particle tapes will play an important role in this area of application. The main rival is metal-film evaporated tape. Typical data of different types of video tapes are listed in Table 3.9.

3.4.1.3 Computer tapes and flexible disks.

The standards of the reel-to-reel computer tape date back to the early sixties. Progress took place in the recording densities of the tape drives, which imposed increasingly stringent demands on surface smoothness and low dropout rates, but with little else changing. It was only recently that a cartridge-type tape drive was developed, using a chromium dioxide tape for much higher track and longitudinal recording densities. The corresponding magnetic tape data are shown in Table 3.9. A number of smaller, low-end cassette and cartridge drives are being used, but these are not demanding with respect to the recording tape involved and will not be included here. Great importance is attached to the archival behavior of computer tapes and tape reels

(Bertram and Eshel, 1979), which are still the most important carriers of archival data.

For more than a decade, flexible disks have constituted an important means of peripheral data storage for small computer applications. Apart from rigid disks, the flexible disk is the most prominent example of a recording process which, ideally, employs full saturation of the magnetic coating throughout its thickness at all wavelengths. This is different from video recording, where the magnetic coating has a thickness of approximately 4 μm, and the record current is adjusted for maximum output for the shortest wavelength used. This implies that the magnetic coating is only magnetized through a depth of about 0.5 μm, half the recorded wavelength. In most types of tape recording, previously written information can be erased with an additional erase head before writing new data. This is rather difficult to implement in disk configurations. Therefore, the thickness of the magnetic coating of flexible disks must be less than half the shortest recorded wavelength. This leads to a coating thickness δ of the most advanced, smaller-size flexible disks of about 1 μm. This must be accurately controlled within 10 percent in order to assure the specified frequency response, usually called *resolution* in digital data recording. With gamma ferric oxide particles, and a volumetric packing fraction of about 0.4, the saturation remanence $M_r(\infty)$ of flexible disks is limited to values of about 70 kA/m (emu/cm^3). Thus the remanent flux per unit track width, $\mu_0 M_r(\infty)\delta$, has been reduced with increasing linear recording density through reductions of δ. The flux has decreased from 280 nWb/m (mMx/cm) of the first flexible disks, to 120 nWb/m (12 Mx/m) of the 3.5-inch-diameter flexible-disk cartridge, and may soon be as low as 70 nWb/m (mMx/cm). Rigid-disk media run as low as 55 nWb/m (mMx/cm). Consequently, the linear recording density limit of particulate media disks is determined, not only by the thinnest coating thickness that can be made, but also by the minimum signal level needed. It is of interest to note that similar flux levels are encountered in video recording. The magnetic particles used are predominantly made of gamma ferric oxide which, more recently, has a cobalt surface modification for higher coercivities. Metal particles may eventually be used for higher signal levels at thin coating thicknesses, as is the case in magnetic still-picture photography where a small 4.7-mm-diameter disk is used for the recording of wavelengths of the order of 0.5 μm. Barium ferrite may also be used for such applications

In contrast to magnetic tape media, care must be taken in the coating process of flexible-disk web material that the particle axes are randomly distributed in the plain of the coating; otherwise a circumferential signal modulation occurs. In addition, any other surface structures along the direction of coating must be eliminated, since the head regularly travels transverse to them, a situation that may lead to dropouts and extra signals. Abrasion and wear play an important role because the head may run on one track for many repeated times. Finally, a certain light absorption is specified which is not that easily met, since the drive manufacturers tend to use infrared light diodes in the index hole detection circuit. Typical magnetic properties are listed in Table 3.9.

3.4.1.4 Rigid disks. The properties of rigid disks are centered around the conditions needed for the low-flying recording head whose height above the disk may soon be 0.2 μm or less. On the one hand, it is necessary to provide a superfine surface finish and lubrication for interference-free flying, as well as good short-wavelength recording characteristics. On the other hand, it is essential to avoid stiction between disk and head that may prevent the drive spindle from turning after a standstill. The magnetic coating of typically 0.7-μm thickness uses elongated gamma ferric iron particles, which are oriented circumferentially, and whose coercivity has been pushed to the limit of the available ferrite-head fields by surface-cobalt modification. The magnetic data are given in Table 3.9. These disks, which may be limited to linear recording densities of about 600 fr/mm (15 kfr/in), are produced in a variety of diameters that range from 3.5 to 14 in. The smaller-diameter disks are used in small fixed-disk drives that are directly exchangeable with flexible-disk drives in small computer applications.

3.4.1.5 Other applications for particulate media. There is a wealth of special applications of particulate recording media, in the form of cards and stripes or small disks, that covers applications such as suburban traffic tickets, credit and access cards, accounting sheets, point-of-sale registration labels, word processing, and many more. Most of them constitute low-density recording modes with moderate requirements with respect to the magnetic recording media.

3.4.2 Film media

Although film media have been studied in the laboratory for over two decades, product applications have begun to appear at a significant level only in the 1980s. Most of these applications are for computer data storage on rigid disks. The one exception to this is evaporated metal tape for video recording. The early-entry products often use the higher storage density capability of film media to achieve miniaturization of devices, or storage compactness, rather than higher total storage capacities. Nearly all commercially available film disks are 130 mm in diameter or smaller, compared with the large-system disk drives which use 210 mm to 14-in-diameter disks. One reason is that small-format disks are more amenable to developing a new technology. More significantly, the personal computer market has been a major driving force in the development of film recording media as a means of maximizing storage capacity in a limited available space. Similarly, the video recording market has led to the development of metal-film tape to achieve compact portable video cameras and playback systems. The use of miniature magnetic disks to record still images, in competition with conventional still photography, is another potential application. Initial video and still-photography recorders employ particulate magnetic disks, but the drive toward higher storage capacity without sacrificing compactness points to the possible future use of film media.

3.4.2.1 Film disks. Early film-disk products were not designed to exceed the potential performance using particulate technology. This conservative approach reflects the need to prove the reliability of this new technology in products before pushing the performance levels to higher limits. It is reasonable to expect that the introduction of film media will not change the general trend of storage density increase, because achieving higher areal density depends on many factors in addition to the medium itself, such as head technology, mechanical head-disk interface control, and substrate quality. Rather, the use of film media is merely one advancement which will allow the trend to continue further. In fact, it is reasonable to expect that improvements in particulate media, as well as in other recording components, will result in the coexistence of film and particulate media, at least until the reliability of film media becomes thoroughly established.

All the film-disk products introduced to date have employed longitudinal recording media, mostly plated Co-P or Co-Ni-P magnetic films, although sputtered magnetic films are beginning to be applied. The magnetic films are typically 50 to 100 nm thick, with coercivities in the range of 50 to 100 kA/m (625 to 1250 Oe). Substrates of Al-Mg, usually coated with a plated Ni-P undercoat, are universally used. Most commercial film disks employ an overcoat, often 20 to 80 nm of sputtered carbon. Proprietary surface lubrication is frequently used as well. Track densities tend to be about 40 tracks per millimeter or lower, this being limited largely by head and servo technologies. Thus, much emphasis has been put on linear density gains for compact, low-cost system use. Linear densities on the order of 400 to 600 fr/mm are achieved and the use of run-length limited codes enables the bit densities to be 600 to 900 bits per millimeter.

3.4.2.2 Film video tape. Compact video camera recorders have been marketed using 8-mm tapes having either metal particulate coatings or metal-evaporated films. The use of two approaches again signifies the uncertainty about the long-term product reliability associated with metal films. The metal-film tapes promise potentially higher performance, but the reliability of metal-particle coatings is relatively better known, having been used in audio applications for a number of years.

The metal-film tape is fabricated by evaporation of a Co-Ni alloy onto a polyester web, such that the incident vapor stream impinging on the substrate forms a large angle (approximately 80°) relative to the substrate normal. The method of deposition was described earlier in this chapter. The oblique evaporation process is capable of high throughputs on the order of 100 m of web per minute. Magnetic layer thicknesses of 100 to 300 nm are used for this medium, with coercivities of 60 to 80 kA/m (750 to 1000 Oe). Pure Co-Ni films are corrosion-susceptible and show inadequate wear durability. Therefore an overcoat is required. However, another approach (Kunieda et al., 1984) has been outlined, in which oxygen is leaked into the vacuum chamber at a controlled rate during the Co-Ni evaporation. This has the effect of forming an oxide top layer of 10 to 30 nm, and probably introduces oxidized grain boundaries. This oxygenation improves durability and corrosion resistance and decreases noise. It also increases coercivity at the expense of saturation magnetization and hysteresis loop squareness. Such metal-film tapes are capable of recording the submicrometer wavelengths needed for video signals.

3.4.2.3 Perpendicular film media.

Commercial products which utilize the perpendicular mode of recording have been slow in emerging, but its use may not be far away. In fact, perpendicular media may see their earliest commercial use in video applications, where sinusoidal recording with gapped heads does not encounter the pulse shape problems encountered in digital pulse recording with ring heads on perpendicular media. If such perpendicular recording were introduced via particulate technology, through use of barium ferrite platelets, the use of perpendicular film media would be a logical follow-up.

Application of perpendicular media (film or particulate) to digital pulse recording on computer disks seems to be more dependent on the read-write head design than on solving media problems. One-sided recording on disks, using a probe main pole on the medium side of the disk and a large auxiliary pole on the other side, is best suited to flexible media. Recording on perpendicular media with a conventional ring head is also most suited to flexible media, because in this configuration the perpendicular media show superiority to longitudinal media only for small head-to-medium spacings. Such small spacings are most easily realized on flexible substrates. The ring-head approach has the advantage of practical two-sided recording, but calls for new approaches to signal processing.

The commercial use of perpendicular film media may also be spurred as much by signal-to-noise ratio differences as by raw linear density (Belk et al., 1985). Similarly, the technology and economics of head fabrication could motivate the use of perpendicular media, such as a combined ferrite and film head (Tsui et al., 1985) which performs much the same as a true probe head but is efficient and simple to fabricate.

References

Aboaf, J. A., and E. Klokholm, "Magnetic Properties of Thin Films of 3d Transition Metals Alloyed with Cr," *IEEE Trans. Magn.*, **MAG-17**, 3160 (1981).

Aboaf, J. A., S. R. Herd, and E. Klokholm, "Magnetic Properties and Structure of Cobalt-Platinum Thin Films," *IEEE Trans. Magn.*, **MAG-19**, 1514 (1983).

Alstad, J. K., and M. K. Haynes, "Asperity Heights on Magnetic Tape Derived from Measured Signal Drop Out Lengths," *IEEE Trans. Magn.*, **MAG-14**, 749 (1978).

Andriamandroso, D., G. Demazean, M. Pouchard, and P. Hagenmuller, "New Ferromagnetic Materials for Magnetic Recording: The Iron Carbonnitrides," *J. Solid State Chem.*, **54**, 54 (1984).

Arai, Y., and A. Nahara, "Magnetic Recording Media and Process of Producing Them, Fluoropolymer Overcoating," U.S. Patent, 4,419,404, 1983.

Arnoldussen, T. C., E. M. Rossi, A. Ting, A. Brunsch, J. Schneider, and G. Trippel, "Obliquely Evaporated Iron-Cobalt and Iron-Cobalt-Chromium Thin Film Recording Media," *IEEE Trans. Magn.*, **MAG-20**, 821 (1984).

Bate, G., "Angular Variation of the Magnetic Properties of Partially Aligned γ-Fe$_2$O$_3$ Particles," *J. Appl. Phys.*, **32S**, 2395 (1961).

Bate, G., and L. P. Dunn, "On the Design of Magnets for the Orientation of Particles in Tapes," *IEEE Trans. Magn.*, **MAG-16**, 1123 (1980).

Bate, G., and J. A. Williams, "The Cylindrical Symmetry of the Angular Distribution of Particles in Magnetic Tape," *IEEE Trans. Magn.*, **MAG-14**, 869 (1978).

Bate, G., *Ferromagnetic Materials*, ed., E.P. Wohlfarth, N. Holland Publ., Amsterdam, 1980, pp 381–507.

Baugh, R. A., E. S. Murdock, and B. R. Natarajan, "Measurement of Noise in Magnetic Media," *IEEE Trans. Magn.*, **MAG-19**, 1722 (1983).

Belk, N. R., P. K. George, and G. S. Mowry, "Noise in High-Performance Thin-Film Longitudinal Magnetic Recording Media," *Trans. Magn.*, **MAG-21**, 1350 (1985).

Berkowitz, A. E., W. J. Schuele, and P. J. Flanders, "Interference of Crystallite Size on the Magnetic Properties of Acicular γ-Fe$_2$O$_3$ Particles," *J. Appl. Phys.*, **39**, 1261 (1968).

Bertram, H. N., Private Communication (1976).

Bertram, H. N., "Anisotropy of Well Oriented γ-Fe$_2$O$_3$ Magnetic Tape," *AIP Conf. Proc.*, **18**, 1113 (1978).

Bertram, H. N., and A. Bhatia, "The Effect of Interactions on the Saturation Remanence of Particulate Assemblies," *IEEE Trans. Magn.*, **MAG-9**, 127 (1973).

Bertram, H. N., and E. F. Chuddihy, "Kinetics of the Humid Ageing of Magnetic Recording Tape," *IEEE Trans. Magn.*, **MAG-18**, 993 (1982).

Bertram, H. N., and A. Eshel, "Recording Media Archival Attributes (Magnetic)," Final Rep. Contract, AFSC Rome Air Dev. Center, No. F30602:78:C-0181, 1979.

Bertram, N., M. Stafford, and D. Mills, "The Print Through Phenomenon (Magnetic Recording), *J. Audio Eng. Soc.*, **28**, 690 (1980).

Bickford, L. R., J. M. Brownlow, and R. F. Penoyer, "Magnetocrystalline Anisotropy in Cobalt Substituted Magnetic Single Crystals," *Proc. Inst. Elec. Eng. (London)*, **104B**, 238 (1956).

Birss, R. R., "The Saturation Magnetization of Ferromagnetics," *Adv. Phys.*, **8**, 252 (1959).

Birss, R. R., *Symmetry and Magnetism*, North-Holland, Amsterdam, 1965.

Bonn, T. H., and D. C. Wendell, Jr., "Electrodeposition of a Magnetic Coating," U.S. Patent, 2,644,787, 1953.

Borrelli, N. F., S. L. Chen, and J. A. Murphy, "Magnetic and Optical Properties of Thin Films in the System $(1 - x)$Fe$_3$O$_4 \cdot x$Fe$_{8/3}$O$_4$," *IEEE Trans. Magn.*, **MAG-8**, 648 (1972).

Bozorth, R. M., "Determination of Ferromagnetic Anisotropy in Single Crystals and in Polycrystalline Sheets," *Phys. Rev.*, **50**, 1076 (1936).

Bozorth, R. M., *Ferromagnetism*, Van Nostrand, New York, 1951, p. 289.

Brenner, A., and G. E. Riddell, "Nickel Plating on Steel by Chemical Reduction," *J. Res. Natl. Bur. Stand.*, **37**, 31 (1946).

Brown, W. H., R. E. Ansel, L. Laskin, and S. R. Schmid, "Electron Beam Curing of Magnetic Media Coatings," *Proc. Symp. Magn. Media Mfg. Meth.*, MMIS, Chicago, 1983.

Burgess, K. A., "The Role of Carbon Black in Magnetic Media," *Proc. Symp. Magn. Media Mfg. Meth.*, MMIS, Chicago, 1983.

Chen, H. Y., D. M. Hiller, J. E. Hudson, and C. J. A. Westenbroek, "Advances in Properties and Manufacturing of Chromium Dioxide," *IEEE Trans. Magn.*, **MAG-20**, 24 (1984).

Chen, T., "The Micromagnetic Properties of High-Coercivity Metallic Thin Films and Their Effects on the Limit of Packing Density in Digital Recording," *IEEE Trans. Magn.*, **MAG-17**, 1181 (1981).

Chen, T., and G. B. Charlan, "A Comparison of the Uniaxial Anisotropy in Sputtered Co-Re and CoCr Perpendicular Recording Media," *J. Appl. Phys.*, **54**, 5103 (1983).

Chikazumi, S., *Physics of Magnetism*, Wiley, New York, 1974.

Chin, C., and P. B. Mee, "Stiction at the Winchester Head-Disk Interface," *IEEE Trans. Magn.*, **MAG-19**, 1659 (1983).

Chopra, K. L., *Thin Film Phenomena*, McGraw-Hill, New York, 1969.

Cloud, W. H., D. S. Schreiber, and K. R. Babcock, "X-Ray and Magnetic Structure of CrO_2 Single Crystals," *J. Appl. Phys.*, **33**, 1193 (1962).

Craik, D. J., and R. S. Tebble, *Ferromagnetism and Ferromagnetic Domains*, North-Holland, Amsterdam, 1965.

Dasgupta, S., "Characterization of Magnetic Dispersions: Rheological, Mechanical and Magnetic Properties," *IEEE Trans. Magn.*, **MAG-20**, 7 (1984).

Daval, J., and D. Randet, "Electron Microscopy on High Coercive Force CoCr Composite Films," *IEEE Trans. Magn.*, **MAG-6**, 768 (1970).

Doss, S., Private Communication (1985).

Dressler, D. D., and J. H. Judy, "A Study of Digitally Recorded Transitions in Thin Magnetic Films," *IEEE Trans. Magn.*, **MAG-10**, 674 (1974).

Dubin, R. R., and K. D. Winn, "Behavior of Perpendicular Recording Materials in Cl and S Containing Atmospheres," *IEEE Trans. Magn.*, **MAG-19**, 1665 (1983).

Dubin, R. R., K. D. Winn, L. P. Davis, and R. A. Cutler, "Degradation of Co-based Thin-Film Recording Materials in Selected Corrosive Environments," *J. Appl. Phys.*, **53**, 2579 (1982).

Eagle, D. F., and J. C. Mallinson, "On the Coercivity of Fe_2O_3 Particles," *J. Appl. Phys.*, **38**, 995 (1967).

Feng, J. S. Y., C. H. Bajorek, and M. A. Nicolet, "Magnetite Thin Films," *IEEE Trans. Magn.*, **MAG-8**, 277 (1972).

Feuerstein, A., and M. Mayr, "High Vacuum Evaporation of Ferromagnetic Materials— A New Production Technology For Magnetic Tapes," *IEEE Trans. Magn.*, **MAG-20**, 51 (1984).

Fisher, R. D., and W. H. Chilton, "Preparation and Magnetic Characteristics of Chemically Deposited Cobalt for High Density Storage," *J. Electrochem. Soc.*, **109**, 485 (1962).

Fisher, R. D., L. Herte, and A. Lang, "Recording Performance and Magnetic Characteristics of Sputtered Cobalt-Nickel-Tungsten Films," *IEEE Trans. Magn.*, **MAG-17**, 3190 (1981).

Fisher, R. D., V. S. Au-Yeung, and B. B. Sabo, "Perpendicular Anisotropy Constants and Anisotropy Energy of Oriented Cobalt-Chromium Alloys," *IEEE Trans. Magn.*, **MAG-20**, 806 (1984).

Flanders, P. J., "Stress-Induced Playback Loss and Switching Field Reduction in Recording Materials," *J. Appl. Phys.*, **50**, 2390 (1979).

Flanders, P. J., "The Temperature Dependence of the Reduced Anisotropy in Magnetic Recording Particles," *IEEE Trans. Magn.*, **MAG-19**, 2683 (1983).

Flanders, P. J., G. Kaganowicz, and Y. Takei, "Magnetostriction and Stress-Induced Playback Loss in Magnetic Tapes," *IEEE Trans. Magn.*, **MAG-15**, 1065 (1979).

Freiser, M. J., "On the Zigzag Form of Charged Domain Walls," *IBM J. Res. Dev.*, **23**, 330 (1979).

Fujiwara, T., "Barium Ferrite Particulate Tapes for Perpendicular Magnetic Recording," *Symp. Magn. Media Mfg. Meth.*, Honolulu, Hawaii, 1983.

Gans, R., "Über das Verhalten isotroper Ferromagnetika," *Ann. Phys.*, **15**, 28 (1932).

Garrison, M. C., "Effects of Absorbed Films on Galvanic Corrosion in Metallic Thin Film Media," *IEEE Trans. Magn.*, **MAG-19**, 1683 (1983).

Greenberg, H. J., R. L. Stephens, and F. E. Talke, "Dimensional Stability of 'Floppy' Disks," *IEEE Trans. Magn.*, **MAG-13**, 1397 (1977).

Grundy, P. J., M. Ali, and C. A. Faunce, "Electron Microscopy of Co-Cr Films," *IEEE Trans. Magn.*, **MAG-20**, 794 (1984).

Gustard, B., and H. Vriend, "A Study of the Orientation of Magnetic Particles in γFe_2O_3 and CrO_2 Recording Tapes Using an X-Ray Pole Figure Technique," *IEEE Trans. Magn.*, **MAG-5**, 326 (1969).

Gustard, B., and M. R. Wright, "A New γ-Fe_2O_3 Particle Exhibiting Improved Orientation," *IEEE Trans. Magn.*, **MAG-8**, 426 (1972).

Hattori, S., Y. Ishii, M. Shinohara, and T. Nakagawa, "Magnetic Recording Characteristics of Sputtered γ-Fe_2O_3 Thin Film Disks," *IEEE Trans. Magn.*, **MAG-15**, 1549 (1979).

Heller, J., "Deposition of Magnetite Films by Reactive Sputtering of Iron Oxide," *IEEE Trans. Magn.*, **MAG-12**, 396 (1976).

Hibst, H., "Hexagonal Ferrites from Melts and Ageous Solutions, Magnetic Recording Media," *Angew. Chem. Int. Ed. Engl.*, **21**, 270 (1982).

Hintermann, H. E., "Thin Solid Films to Combat Friction, Wear, and Corrosion," *J. Vac. Sci. Technol.*, **B2**, 816 (1984).

Hokkyo, J., K. Hayakawa, I. Saito, and K. Shirane, "A New W-Shaped Single-Pole Head and a High Density Flexible Disk Perpendicular Magnetic Recording System," *IEEE Trans. Magn.*, **MAG-20**, 72 (1984).

Holloway, A. J., "Polyester Film, Surface Definition and Control," *Symp. Magn. Media Mfg. Meth.* Honolulu, Hawaii, 1983.

Hsieh, E. J., R. F. Soohoo, and M. F. Kelly, "A Lorentz Microscopic Study of Head-On Domain Walls," *IEEE Trans. Magn.*, **MAG-10**, 304 (1974).

Hughes, G. F., "Magnetization Reversal in Cobalt-Phosphorus Films," *J. Appl. Phys.*, **54**, 5306 (1983).

Huisman, H. F., "Particle Interactions and H_c: An Experimental Approach," *IEEE Trans. Magn.*, **MAG-18**, 1095 (1982).

Huisman, H. F., and C. J. F. M. Rasenberg, "Mercury Porosimetry Analysis of Magnetic Coatings," *IEEE Trans. Magn.*, **MAG-20**, 13 (1984).

Imaoka, Y., "Ageing Effects of Ferromagnetic Iron Oxides," *J. Electrochem. Soc., Japan*, **33**, 1 (1965).

Ishii, O., and I. Hatakeyama, "Os Doped γ-Fe_2O_3 Thin Films Having High Coercivity and Coercive Squareness," *J. Appl. Phys.*, **55**, 2269 (1984).

Ishii, Y., A. Terada, O. Ishii, S. Ohta, S. Hattori, and K. Makino, "New Preparation Process for Sputtered γ-Fe_2O_3 Thin Film Disks," *IEEE Trans. Magn.*, **MAG-16**, 1114 (1980).

Ishizuka, M., T. Komoda, T. Tsuchimoto, M. Yoshikawa, S. Ishio, and M. Takahashi, "Perpendicular Anisotropy in Co-Cr Films," *J. Magn. Magn. Mater.*, **35**, 286 (1983).

Itoh, F., M. Satou, and Y. Yamazuki, "Anomalous Increase of Coercivity in Iron Oxide Powder Coated with Sodium Polyphosphate," *IEEE Trans. Magn.*, **MAG-13**, 1385 (1977).

Itoh, F., T. Namikawa, and M. Satou, "On the Coercivity and Chemical Structure in Nonstochiometric Iron Oxide Particle," *Proc. Int. Conf. Ferrites*, 537 (1980).

Iwasaki, S. I., and Y. Nakamura, "An Analysis for the Magnetization Mode for High Density Magnetic Recording," *IEEE Trans. Magn.*, **MAG-13**, 1272 (1977).

Izawa, F., "Theoretical Study on Stress-Induced Demagnetization in Magnetic Recording Media," *IEEE Trans. Magn.*, **MAG-20**, 523 (1984).

Jacobs, I. S., and C. P. Bean, "An Approach to Elongated Fine Particle Magnets," *Phys. Rev.*, **100**, 1060 (1955).

Jacobs, I. S., and C. P. Bean, *Magnetism*, Academic Press, New York, 1963, vol. III, p. 271.

Jakusch, H., Private Communication (1984).

Jaycock, M. J., *Dispersion of Powders in Liquids*, 3d ed., Applied Science, London, 1981, p. 51.

Jeschke, J., "Verfahren zur Herstellung hartmagnetischer γ-Ferrite für Magnettonbänder," *Bild Ton*, **7**, 318 (1954).

Johnson, C. E., and W. F. Brown, Jr., "Stoner Wohlfarth Calculation on Particles with Both Magnetocrystalline and Shape Anisotropy," *J. Appl. Phys.*, **30S**, 320 (1959).

Johnson, C. E., and W. F. Brown, Jr., "Theoretical Magnetization Curves for Particles with Cubic Anisotropy," *J. Appl. Phys.*, **32S**, 243 (1961).

Judge, J. S., J. R. Morrison, and D. E. Speliotis, "Very High Coercivity Chemically Deposited Co-Ni Films," *J. Appl. Phys.*, **36**, 948 (1965a).

Judge, J. S., J. R. Morrison, D. E. Speliotis, and G. Bate, "Magnetic Properties and Corrosion Behavior of Thin Electroless Co-P Deposits," *J. Electrochem. Soc.*, **112**, 681 (1965b).

Judge, J. S., J. R. Morrison, and D. E. Speliotis, "Flexible Recording Surfaces of Electrodeposited Cobalt-Nickel-Phosphorus," *Plating*, **53**, 441 (1966).

Kamiya, I., Y. Makino, and Y. Sugiura, "Induced Magnetic Anisotropy in Cobalt-Ferrite Fine Particle," *Proc. Int. Conf. Ferrites*, 483 (1980).

Kaneko, M., "Change in Coercivity of γ-Fe_2O_3-Fe_3O_4 Solid Solutions with Magnetostrictive Anisotropy," *IEEE Trans. Magn.*, **MAG-16**, 1319 (1980).

Kay, E., R. A. Sigsbee, G. L. Bona, M. Taborelli, and H. C. Siegmann, "Magnetic Depth Profiling and Characterization of Fe Oxide Films by Kerr Rotation and Spin Polarized Photoemission," *Intl. Conf. Magn.*, *1985 Digests*, 148 (1985).

Khalafalla, D., and A. H. Morrish, "Investigation of Ferrimagnetic Cobalt-Doped Gamma-Ferric Oxide Micropowders," *J. Appl. Phys.*, **43**, 624 (1972).

King, F. K., "Datapoint Thin Film Media," *IEEE Trans. Magn.*, **MAG-17**, 1376 (1981).

Kishimoto, M., "Effect of Fe^{2+} Content on the Instability of Coercivity of Cobalt-Substituted Acicular Iron Oxides," *IEEE Trans. Magn.*, **MAG-15**, 906 (1979).

Kishimoto, M., S. Kitaoha, H. Andoh, M. Amemiya, and F. Hayama, "On the Coercivity of Cobalt-Ferrite Epitaxial Iron Oxide," *IEEE Trans. Magn.*, **MAG-17**, 3029 (1981).

Kneller, E., *Ferromagnetismus*, Springer-Verlag, Berlin, 1962.

Kneller, E., *Handbuch der Physik*, Springer-Verlag, Berlin, 1966, band XVIII/2, p. 438.

Kneller, E., "Magnetic Interaction Effects in Fine Particle Assemblies and Thin Films," *J. Appl. Phys.*, **39**, 945 (1968).

Kneller, E., *Magnetism and Metallurgy*, Academic Press, New York, 1969, vol. I, p. 365.

Kneller, E., "Static and Anhysteretic Magnetic Properties of Tapes," *IEEE Trans. Magn.*, **MAG-16**, 36 (1980).

Kneller, E., and F. E. Luborsky, "Particle Size Dependence of Coercivity and Remanence of Single Domain Particles," *J. Appl. Phys.*, **34**, 656 (1963).

Knowles, J. E., "The Measurement of The Anisotropy Field of Single Tape Particles," *IEEE Trans. Magn.*, **MAG-20**, 84 (1984).

Knudsen, J. K., "Stretched Surface Recording Disk for Use with a Flying Head," *IEEE Trans. Magn.*, **MAG-21**, 2588 (1985).

Kobayashi, K., and G. Ishida, "Magnetic and Structural Properties of Rh Substituted Co-Cr Alloy Films with Perpendicular Magnetic Anisotropy," *J. Appl. Phys.*, **52**, 2453 (1981).

Kobayashi, K., J. Toda, and T. Yamamoto, "High Density Perpendicular Magnetic Recording on Rigid Disks," *Fujitsu Sci. Tech. J.*, **19**, 99 (1983).

Kojima, H., and K. Hanada, "Origin of Coercivity Changes during the Oxidation of Fe_3O_4 to γ-Fe_2O_3," *IEEE Trans. Magn.*, **MAG-16**, 11 (1980).

Kolecki, J. C., "Microhardness Studies on Thin Carbon Films Grown on P-Type 100 Silicon," NASA Technical Memorandum 82980, 1982.

Köster, E., "Reversible Susceptibility of an Assembly of Single Domain Particles and Their Magnetic Anisotropy," *J. Appl. Phys.*, **41**, 3332 (1970).

Köster, E., "Magnetic Anisotropy of Cobalt-Doped Gamma Ferric Oxide," *IEEE Trans. Magn.*, **MAG-8**, 428 (1972).

Köster, E., "Temperature Dependence of Magnetic Properties of Chromium-Dioxide and Cobalt-Doped Gamma-Ferric-Oxide Particles," *IERE Conf. Proc.*, **26**, 213 (1973).

Köster, E., "A Contribution to Anhysteretic Remanence and AC Bias Recording," *IEEE Trans. Magn.*, **MAG-11**, 1185 (1975).

Köster, E., "Recommendation of a Simple and Universally Applicable Method for Measuring the Switching Field Distribution of Magnetic Recording Media," *IEEE Trans. Magn.*, **MAG-20**, 81 (1984).

Köster, E., and W. Steck, "Anhysteretic Growth of Exchange Fixed Remanent States," *IEEE Trans. Magn.*, **MAG-12,** 755 (1976).

Köster, E., H. Jakusch, and U. Kullmann, "Switching Field Distribution and AC Bias Recording," *IEEE Trans. Magn.*, **MAG-17,** 2550 (1981).

Köster, E., P. Deigner, P. Schäfer, K. Uhl, and R. Falk, "Device for Magnetically Orienting and Magnetizable Particles of Magnetic Recording Media in a Preferred Direction," U.S. Patent 0043,822,44, 1983.

Krones, F., *Technik der Magnetspeicher*, Springer, Berlin, 1960, p. 474.

Kubo, O., T. Ido, and H. Yokoyama, "Properties of Ba-Ferrite Particles for Perpendicular Magnetic Recording Media," *IEEE Trans. Magn.*, **MAG-18,** 1122 (1982).

Kullmann, U., E. Köster, and C. Dorsch, "Amorphous CoSm Thin Films: A New Material for High Density Longitudinal Recording," *IEEE Trans. Magn.*, **MAG-20,** 420 (1984a).

Kullmann, U., E. Köster, and B. Meyer, "Magnetic Anisotropy of Ir-Doped CrO_2," *IEEE Trans. Magn.*, **MAG-20,** 742 (1984b).

Kunieda, T., K. Shinohara, and A. Tomago, "Metal Evaporated Videotape," *IERE Proc.*, **37** (1984).

Landskroener, P. A., "Conventional Coating Methods," *Symp. Magn. Media Mfg. Meth.*, Honolulu, Hawaii, 1983.

Lee, E. W., "Magnetostriction and Magnetochemical Effects," *Rep. Prog. Phys.*, **18,** 184 (1955).

Lee, T. D., and P. A. Papin, "Analysis of Dropouts in Video Tapes," *Trans. Magn.*, **MAG-18,** 1092 (1982).

Lemke, J. U., "An Isotropic Particulate Medium with Additive Hilbert and Fourier Field Components," *J. Appl. Phys.*, **53,** 2561 (1982).

Lindner, R. E., and P. B. Mee, "ESCA Determination of Fluorocarbon Lubricant Film Thickness on Magnetic Disk Media," *IEEE Trans. Magn.*, **MAG-18,** 1073 (1982).

Lloyd, J. C., and R. S. Smith, "Structural and Magnetic Properties of Permalloy Films," *J. Appl. Phys.*, **30S,** 274 (1959).

Luborsky, F. E., "Development of Elongated Particle Magnets," *J. Appl. Phys.*, **32S,** 171 (1961).

Luborsky, F. E., "High Coercive Force Films of Cobalt-Nickel with Additions of Group VA and VIB Elements," *IEEE Trans. Magn.*, **MAG-6,** 502 (1970).

Maeda, H., "High Coercivity Co and Co-Ni Alloy Films," *J. Appl. Phys.*, **53,** 3735 (1982).

Maestro, P., D. Andriamandroso, G. Demazean, M. Pouchard, and P. Hagenmuller, "New Improvements of CrO_2-Related Magnetic Recording Materials," *IEEE Trans. Magn.*, **MAG-18,** 1000 (1982).

Maissel, L. I., and R. Glang, *Handbook of Thin Film Technology*, McGraw-Hill, New York, 1970.

Maller, V. A. J., and B. K. Middleton, "A Simplified Model of the Writing Process in Saturation Magnetic Recording," *Radio Electron. Eng.*, **44,** 281 (1974).

Manly, W. A., Jr., "Erasure of Signals on Magnetic Recording Media," *IEEE Trans. Magn.*, **MAG-12,** 758 (1976).

Matsumoto, S., T. Koga, F. Fukai, and S. Nakatani, "Production of Acicular Ferric Oxide," U.S. Patent 4,202,871, 1980.

Matsuoka, M., M. Naoe, and Y. Hoshi, "Ba-Ferrite-Thin-Film Disk for Perpendicular Magnetic Recording," *J. Appl. Phys.*, **57,** 4040 (1985).

Mayer, D. H., "On the Abrasivity of γ-Fe_2O_3 and CrO_2 Magnetic Tapes," *IEEE Trans. Magn.*, **MAG-10,** 657 (1974).

Mee, C. D., *The Physics of Magnetic Recording*, North-Holland, Amsterdam, 1964.

Meijers, F. B., "Rigid Disk Coating Parameters," *Symp. Magn. Media Mfg. Meth.*, Honolulu, Hawaii, 1983.

Meiklejohn, W. H., "New Magnetic Anisotropy," *Phys. Rev.*, **105,** 904 (1957).

Meiklejohn, W. H., and C. P. Bean, "New Magnetic Anisotropy," *Phys. Rev.*, **102,** 1413 (1956).

Middelhoek, S., "Domain Walls in Thin Ni-Fe Films," *J. Appl. Phys.*, **34,** 1054 (1963).

Mihalik, R. S., "The Relationship between Polyurethane Binders and Various Binder Additives and Modifiers," *Symp. Magn. Media Mfg. Meth.*, Honolulu, Hawaii, 1983.

Mills, D., H. Kristensen, and V. Santos, "Control of Modulation Noise in Magnetic Recording Tape," *52d AES Convention*, Reprint No. 1084, 1975.

Missenbach, F. S., "Premixing, Milling, Filtering and Related Procedures," *Symp. Magn. Media Mfg. Meth.*, Honolulu, Hawaii, 1983.

Miyoshi, K., and D. H. Buckley, *Tribology and Mechanics of Magnetic Storage Systems*, American Society of Lubrication Engineers, Park Ridge, Illinois, 1984, p. 13.

Monteil, J. B., and P. Dougier, "A New Preparation Process of Cobalt Modified Iron Oxide," *Proc. Int. Conf. Ferrites*, 532 (1980).

Moradzadeh, Y., "Chemically Deposited Cobalt Phosphorus Films for Magnetic Recording," *J. Electrochem. Soc.*, **112**, 891 (1965).

Morrish, A. H., and K. Haneda, "Surface Magnetic Properties of Fine Particles," *J. Magn. Magn. Mater.*, **35**, 105 (1983).

Morrish, A. H., and S. P. Yu, "Dependence of the Coercive Force on the Density of Some Iron Oxide Powers," *J. Appl. Phys.*, **26**, 1049 (1955).

Nakamura, K., Y. Ohta, A. Itoh, and C. Hayashi, "Magnetic Properties of Thin Films Prepared by Continuous Vapor Deposition," *IEEE Trans. Magn.*, **MAG-18**, 1077 (1982).

Néel, L., "Propriétés d'un Ferromagnétic Cubique en Grain Fines," *C. R. Acad. Sci. Paris*, **224**, 1488 (1947).

Néel, L., "Anisotropie Magnetique Superficielle et Structurelle d'Orientation," *J. Phys. Radium*, **15**, 225 (1954).

Néel, L., "Energy of Block Walls in Thin Films" (French), *C. R. Acad. Sci. Paris*, **241**, 533 (1955).

Néel, L., "Remarks on the Theory of the Magnetic Properties of Thin Films and Fine Particles" (French), *J. Phys. Radium*, **17**, 250 (1956).

Newman, J. J., "Orientation of Magnetic Particles Assemblies," *IEEE Trans. Magn.*, **MAG-14**, 866 (1978).

Nishimoto, K., and M. Aoyama, "Preparation of γ-Fe_2O_3 Thin Film Disks by Reactive Evaporation and Their Read/Write Characteristics," *Proc. Int. Conf. Ferrites*, 588 (1980).

Ohkoshi, M., H. Toba, S. Honda, and T. Kusuda, "Electron Microscopy of Co-Cr Sputtered Films," *J. Magn. Magn. Mater.*, **35**, 266 (1983).

Ohtsubo, A., Y. Satoh, T. Masuko, M. Hirama, and E. Abe, "A Perpendicular Orientation of a Coated Particulate Medium by an AC-Demagnetization Process," *IEEE Trans. Magn.*, **MAG-20**, 751 (1984).

Oppegard, A. L., F. J. Darnell, and H. C. Miller, "Magnetic Properties of Single Domain Iron and Iron-Cobalt Particles Prepared by Boronhydride Reduction," *J. Appl. Phys.*, **32S**, 184 (1961).

Osborn, J. A., "Demagnetizing Factors of the General Ellipsoid," *Phys. Rev.*, **67**, 351 (1945).

Ouchi, K., and S. Iwasaki, "Thermomagnetic Behavior of Hard Magnetic Thin Films," *IEEE Trans. Magn.*, **MAG-8**, 473 (1972).

Parfitt, G. D., *Dispersion of Powder in Liquids*, 3d ed., Applied Science, London, 1981, p. 1.

Patton, T. C., *Paint Flow and Pigment Dispersion*, 2d ed., Wiley, New York, 1979.

Phipps, P. B., S. J. Lewis, and D. W. Rice, "The Magnetic and Corrosion Properties of Iron Cobalt Chromium Films," *J. Appl. Phys.*, **55**, 2257 (1984).

Pourbaix, M., *Atlas of Electrochemical Equilibria in Aqueous Solutions*, National Association Corrosion Engineers, Houston, 1974.

Rasenberg, C. J. F. M., and H. F. Huisman, "Measurement of the Critical Pigment Volume Concentration of Particulate Magnetic Tape," *IEEE Trans. Magn.*, **MAG-20**, 748 (1984).

Reimer, V., "Ein Modell aus Permanentmagneten zum Verstandnis der Ummagnetisierung in dunnen Schichten," *Z. Angew. Phys.*, **17**, 196 (1964).

Riveiro, J. M., and J. M. de Frutos, "Magnetic Annealing in Electrodeposited Co-P Amorphous Alloys," *J. Magn. Magn. Mater.*, **37**, 155 (1983).

Rodbell, D. S., "Magnetocrystalline Anisotropy of Single Crystal CrO_2," *J. Phys. Soc. Jpn.*, **21**, 1224 (1966).

Rossi, E. M., G. McDonnough, A. Tietze, T. Arnoldussen, A. Brunsch, S. Doss, M. Henneberg, F. Lin, R. Lyn, A. Ting, and G. Trippel, "Vacuum-Deposited Thin-Metal Film Disk," *J. Appl. Phys.*, **15**, 2254 (1984).

Salmon, O. N., R. E. Fayling, G. E. Gurr, and V. H. Halling, "Thermodynamics of Post-Erasure Signal Effects in γ-Fe_2O_3 Magnetic Recording Tapes," *IEEE Trans. Magn.*, **MAG-15**, 1315 (1979).

Schönfinger, E., M. Schwarzmann, and E. Köster, "Verfahren zur Herstellung von kobaltdotiertem γ-Eisen(III)-Oxiden," West German Patent 002243231, 1972.

Schuele, W. J., "Coercive Force of Angle of Incidence Films," *J. Appl. Phys.*, **35**, 2558 (1964).

Sharpe, J., T. Sutherst, and E. Cooper, "Machine Calendering," Technical Division of the British Paper and Board Industry Federation, London, 1978.

Sharrock, M. P., "Particle-Size Effects on the Switching Behavior of Uniaxial and Multiaxial Magnetic Recording Materials," *IEEE Trans. Magn.*, **MAG-20**, 754 (1984).

Shimizu, S., N. Umeki, T. Ubeori, N. Horiishi, Y. Okuda, Y. Yuhara, H. Kosaka, A. Takedoi, and K. Yaguchi, Japanese Patents 753,7667 and 753,7668, 1975.

Shtrikman, S., and D. Treves, "The Coercive Force and Rotational Hysteresis of Elongated Ferromagnetic Particles," *J. Phys. Radium*, **20**, 286 (1959).

Smaller, P., and J. S. Newman, "The Effect of Interactions of the Magnetic Properties of a Particulate Medium," *IEEE Trans. Magn.*, **MAG-6**, 804 (1970).

Smit, J., and H. P. J. Wijn, *Ferrites*, Wiley, New York, 1959.

Smith, D. O., M. S. Cohen, and G. P. Weiss, "Oblique Incidence Anisotropy in Evaporated Permalloy Films," *J. Appl. Phys.*, **31**, 1755 (1960).

Soohoo, R. F., "Influence of Particle Interaction on Coercivity and Squareness of Thin Film Recording Media," *J. Appl. Phys.*, **52**, 2459 (1981).

Speliotis, D. E., J. R. Morrison, and G. Bate, "Thermomagnetic Properties of Fine Particles of Cobalt-Substituted γ-Fe_2O_3," *Proc. 64th Conf. Magn.*, Nottingham, 1964, p. 623.

Speliotis, D. E., G. Bate, J. K. Alstad, and J. R. Morrison, "Hard Magnetic Films of Iron, Cobalt, and Nickel," *J. Appl. Phys.*, **36**, 972 (1965).

Stavn, M. J., and A. H. Morrish, "Magnetization of a Two-Component Stoner-Wohlfarth Particle," *IEEE Trans. Magn.*, **MAG-15**, 1235 (1979).

Steck, W., "Preparation and Properties of Particles for Magnetic Recording," *J. Phys. (Paris) Collogne C6 Suppl.*, **46**, C6–33 (1985).

Stoner, E. C., and E. P. Wohlfarth, "A Mechanism of Magnetic Hysteresis in Heterogeneous Alloys," *Philos. Trans. R. Soc. London*, **A240**, 599 (1948).

Suganuma, Y., H. Tanaka, M. Yanagisawa, F. Goto, and S. Hatano, "Production Process and High Density Recording Characteristics of Plated Disks," *IEEE Trans. Magn.*, **MAG-18**, 1215 (1982).

Sugihara, H., Y. Taketoni, T. Uehori, and Y. Imaoka, "The Behavior of Surface Hydroxyl Group of Magnetic Iron Oxide Particles," *Proc. Int. Conf. Ferrites*, 545 (1980).

Sumiya, K., S. Watatani, F. Hayama, K. Hakamae, and T. Matsumoto, "The Orientation of Magnetic Particles for High Density Recording," *J. Magn. Magn. Mater.*, **31–34**, 937 (1983).

Sumiya, K., N. Hirayama, F. Hayama, and T. Matsumoto, "Determination of Dispersability and Stability of Magnetic Paint by Rotation-Vibration Method," *IEEE Trans. Magn.*, **MAG-20**, 745 (1984).

Takada, T., Y. Ikeda, H. Yoshinaga, and Y. Bando, "A New Preparation Method of the Oriented Ferrite Magnets," *Proc. Int. Conf. Ferrites*, 275 (1970).

Tasaki, A., M. Ota, S. Kashu, and C. Hayashi, "Metal Tapes Using Ultra Fine Powder Prepared by Gas Evaporation Method," *IEEE Trans. Magn.*, **MAG-15**, 1540 (1979).

Tasaki, A., K. Tagawa, E. Kita, S. Harada, and T. Kusunose, "Recording Tapes Using Iron Nitride Fine Powder," *IEEE Trans. Magn.*, **MAG-17**, 3026 (1981).

Tasaki, A., N. Saegusa, and M. Ota, "Ultra Fine Magnetic Metal Particles—Research in Japan," *IEEE Trans. Magn.*, **MAG-19**, 1731 (1983).

Tebble, R. S., and D. J. Craik, *Magnetic Materials*, Wiley, New York, 1969.

Terada, A., O. Ishii, and K. Kobayashi, "Pressure-Induced Signal Loss in Fe_3O_4 and Gamma Fe_2O_3 Thin Film Disks," *IEEE Trans. Magn.*, **MAG-19**, 12 (1983).

Thomas, J. R., "Preparation and Magnetic Properties of Colloidal Cobalt Particles," *J. Appl. Phys.*, **37**, 2914 (1966).

Thornton, J. A., "High Rate Thick Film Growth," *Annu. Rev. Mater. Sci.*, **7**, 239 (1977).

Tochihara, S., Y. Imaoka, and M. Namikawa, "Accidental Printing Effect of Magnetic Recording Tapes Using Ultrafine Particles of Acicular γ-Fe_2O_3," *IEEE Trans. Magn.*, **MAG-6**, 808 (1970).

Tokuoka, Y., S. Umeki, and Y. Imaoka, "Anisotropy of Cobalt-Adsorbed γ-Fe_2O_3 Particles," *J. Phys. Suppl.*, **38**, 337 (1977).

Tong, H. C., R. Ferrier, P. Chang, J. Tzeng, and K. L. Parker, "The Micromagnetics of Thin-Film Disk Recording Tracks," *IEEE Trans. Magn.*, **MAG-20**, 1831 (1984).

Tonge, D. G., and E. P. Wohlfarth, "The Remanent Magnetization of Single Domain Ferromagnetic Particles: II. Mixed Uniaxial and Cubic Anisotropies," *Philos. Mag.*, **3**, 536 (1958).

Tsui, R., H. Hamilton, R. Anderson, C. Baldwin, and P. Simon, "Perpendicular Recording Performance of Thin Film Probe Heads and Double-Layer Co-Cr Media," *Digest of Intermag. Conf.*, **GA4** (1985).

Uesaka, Y., M. Koizumi, N. Tsumita, O. Kitakami, and H. Fujiwara, "Noise from Underlayer of Perpendicular Magnetic Recording Medium," *J. Appl. Phys.*, **57**, 3925 (1984).

Umeki, S., S. Saito, and Y. Imaoka, "A New High Coercive Magnetic Particle for Recording Tape," *IEEE Trans. Magn.*, **MAG-10**, 655 (1974).

Vossen, J. L., and W. Kern, *Thin Film Processes*, Academic Press, New York, 1978.

Wheeler, D. A., *Dispersion of Powders and Liquids*, 3d ed., Applied Science, London, 1981, p. 327.

Wielinga, T., "Investigations on Perpendicular Magnetic Recording," Thesis, Twente University of Technology, the Netherlands, 1983.

Williams, J. L., and P. H. Markusch, "The Chemistry of Isocyanates and Polyurethanes," *Symp. Magn. Media Mfg. Meth.*, Honolulu, Hawaii, 1983.

Williams, M. L., and R. L. Comstock, "An Analytic Model of the Write Process in Digital Magnetic Recording," *AIP Conf. Proc. Magn. Magn. Mater.*, **5**, 738 (1971).

Witherell, F. E., "Surface and Near-Surface Chemical Analysis of Cobalt-Treated Iron Oxide," *IEEE Trans. Magn.*, **MAG-20**, 739 (1984).

Wohlfarth, E. P., "The Effect of Particle Interaction on Coercive Force of Ferromagnetic Micropowders," *Proc. R. Soc. London*, **A-232**, 208 (1955).

Wohlfarth, E. P., "Angular Variation of the Coercivity of Partially Aligned Elongated Ferromagnetic Particles," *J. Appl. Phys.*, **30S**, 117 (1959).

Wohlfarth, E. P., *Magnetism*, Academic Press, New York, 1963, vol. III, p. 351.

Wohlfarth, E. P., and D. G. Tonge, "The Remanent Magnetization of Single Domain Ferromagnetic Particles," *Philos. Mag.*, **2**, 1333 (1957).

Woodward, J. G., "Stress Demagnetization in Videotapes," *IEEE Trans. Magn.*, **MAG-18**, 1812 (1982).

Wuori, E. R., and J. H. Judy, "Initial Layer Effects in CoCr Films," *IEEE Trans. Magn.*, **MAG-20**, 774 (1984).

Wuori, E. R., and J. H. Judy, "Particle-like Magnetic Behavior of RF-sputtered CoCr Films," *J. Appl. Phys.*, **57**, 4010 (1985).

Yanagisawa, M., N. Shiota, H. Yamaguchi, and Y. Suganuma, "Corrosion-Resisting Co-Pt Thin Film Medium for High Density Recording," *IEEE Trans. Magn.*, **MAG-19**, 1638 (1983).

Chapter

4

Recording Heads

Robert E. Jones, Jr. and C. Denis Mee
IBM Corporation, San Jose, California

4.1 Types of Recording Elements

Despite the many years of development of transducers for magnetic recording, almost all the head designs for video, audio, instrumentation, and data recording products are based on the familiar inductive coil and magnetic core head design. The evolution of materials for head cores from alloy laminations to ferrites and thin films has had a significant effect on the fabrication processes for head structures. The general trend has been to require smaller dimensions in gap length, track width, and core volume as recording densities and bandwidths have increased. The impact of major applications for magnetic recording heads on their design is discussed in Sec. 4.4 and in the relevant chapters of Volume II.

Inductive recording heads have been designed to perform the three major functions in recording: recording (writing), reproducing (reading), and erasing. For reasons of cost, and also sometimes to control the location of reading, writing, and erasing, all three functions are often performed by the same head. This leads to a compromise in the optimization of the head design for those functions. For stationary-head applications, such as tape drives, separate heads are usually used for reading and writing, and often for erasing. However, in moving-head applications such as disk files and helical-scan video recorders, the requirements for low head mass and simple connection technology have emphasized designs with single inductive elements performing all functions. Normally, in disk files, the function of erasing is achieved by overwriting the previous information.

Recording densities have increased dramatically in all applications of magnetic recording, and the reading signal has been correspondingly reduced. High-magnetization film media have offset this reduction in some applications, but eventually head designs limit recording performance because of the combined requirements of wider bandwidths and higher recording densities. Consequently, there is a trend toward using separate transducers for reading and writing even in some disk applications. The same trend to separate transducers is emerging in perpendicular recording. Once this change is accepted, then heads can be designed either for optimum reading or for optimum writing. Improved performance over inductive designs in reading is possible by using magnetoresistive or Hall effect transducers.

Surface magnetic flux on recorded media can also be detected by magnetooptical effects. Magnetooptical readout is receiving more developmental attention, as storage densities increase, and is even being applied to the conventional storage media by the use of a magnetically soft, but magnetooptically active, material in close proximity to the recorded track (Nomura, 1985). This material (e.g., magnetic garnet) is readily magnetized by the surface flux of the recorded medium and is interrogated by a polarized laser beam. Recording can also be performed with a laser beam, for instance, by heating to the Curie temperature in the presence of a localized magnetic field. The reader is referred to Volume II, Chap. 9, for a detailed discussion of magnetooptical recording and transducers.

4.1.1 Inductive recording heads

The gap in the pole pieces of a recording head is designed to produce a field amplitude capable of recording the storage medium to a sufficient depth, normally considered to be equal to or greater than the reading depth corresponding to the recorded wavelength. The pole geometry and materials are designed to provide adequate field strength at the signal frequency, and a rapid decrement of the writing field along the direction of the medium motion in order to maximize the short-wavelength recording efficiency. Normally, the optimum writing-head gap is longer than the optimum reading-head gap. A limiting factor, as writing track widths become narrower, is the relative increase in side writing that occurs for long-gap heads.

Magnetic flux is delivered to the writing poles by a magnetic yoke that normally has a greater cross-sectional area than the poles in order to avoid saturation of the yoke region. The writing coil around the yoke is usually designed to be close to the poles to improve the efficiency of the inductive coupling between the coil and the poles.

Some designs of inductive writing heads for perpendicular recording have portions of the yoke and coil on the opposite side of the recording medium from the writing pole. The relatively open-circuit flux path dictates different trade-offs in design compared with the ring head. Furthermore, the introduction of soft magnetic underlayers in the recording medium changes the recording efficiency and the wavelength response of the head, and thus the soft magnetic underlayers become part of the head circuit.

4.1.2 Inductive reproducing heads

Although basically identical in design to inductive recording heads, inductive reproducing heads are inherently low-flux-density devices and, therefore, can use lower-flux-density and higher-permeability core and pole materials. At low flux densities, uncontrolled domain switching in a reproducing head can add detectable noise, and this effect becomes more noticeable as the magnetic volume of the head core is reduced. The design trend in inductive reproducing heads is to maximize the reproduced voltage by using as many turns as possible. Restriction on turns occurs due to lowering of the head resonance frequency and less efficient coupling of the larger volume coils. In very high-track-density applications the uncertainty of head positioning can cause reading of the adjacent track. Head designs with separate recording and reproducing heads can reduce this sensitivity through the use of a reduced reading-head track width and, in the case of sequential track recording, through use of heads with different gap azimuth alignment for adjacent tracks.

4.1.3 Erase heads

Erase heads have the least critical design requirements since they require high field amplitude but not high spatial resolution. For very low-cost erasing, static-field heads have been designed using a permanent magnet or an inductive head with direct current. Usually, alternating fields are applied and produce lower-noise erased media than dc-field erasing. High-magnetization materials and very long gaps are used in inductive ring-head designs to produce large erasing fields. However, even when the recorded medium experiences many field reversals during its passage close to a ring-head erasing field, the noise level achievable is not as low as can be achieved with bulk erasure of the recorded medium by a separate large ac-field electromagnet. This incomplete erasing achieved by a single-gap erase head has been explained as a re-recording effect due to the magnetic fields from the recorded medium combining with the erasing field to produce a low-level recording as the medium leaves the erasing field. This effect has been reduced by providing multiple-gap erase heads which spread the erasing field along the recording medium (Sawada and Yoneda, 1985).

4.1.4   Flux-sensitive heads

Flux-sensitive reproducing heads depend on flux rather than the rate of change of flux as is the case with inductive heads. The most highly developed flux-sensitive reproducing head uses the change in resistance of a magnetic material that accompanies a change in magnetization direction. To satisfy the requirement for maximum resistance change while using narrow tracks and closely packed bits, the magnetoresistive sensing element is designed as a narrow thin-film stripe perpendicular to the plane of the recorded medium with its length equal to the track width. When this element is located at the head surface, the wavelength response can be enhanced by placing magnetic shields close and parallel to the element. Since these shields short-circuit the long-wavelength recorded flux away from the sensing element, this design does not have a wide-wavelength response. Other designs, of somewhat reduced sensitivity, use conventional soft magnetic poles and a ring core with the element placed in series with the core. The core-flux response to the recorded wavelength is now similar to that of the ring head. If linear amplitude response is required, it is then necessary to apply a bias field transverse to the resistance measurement direction, and techniques for doing this are discussed in Sec. 4.4.3. Magnetoresistive reproducing heads using nickel-iron film elements are inherently wideband, and their low impedance matches well to low-noise preamplifiers.

Nickel-iron and other alloy films may also be used in flux-sensitive reading heads by measuring the Hall voltage along a current-carrying stripe of similar dimensions to the magnetoresistive stripe. For example, a "one-sided" design is feasible in which current enters the middle of the stripe and flows in opposite directions to each end. The output voltage is lower than that obtained in the magnetoresistive head, but it is linear with amplitude (Fluitman and Groenland, 1981). Nonmagnetic semiconductors such as indium-antimonide, with high electron mobility, have been designed as Hall effect field detectors. Films of about 1-μm thickness are sandwiched between two ferrite poles which guide the flux from the recording medium to the sensing element. In this design, current, field, and Hall voltage are orthogonal. Two-channel audio heads have been designed with the Hall elements in the rear gap of a ring-head structure (Kotera et al., 1979).

Finally, flux-sensitive heads have been designed using the flux gate principle of modulating the reluctance of a ring core in a balanced magnetic circuit that is part of the ring-core circuit. With this approach, it is possible to increase the voltage induced in the head coil by using a modulation frequency higher than the signal frequency. A common approach to applying the modulation field is to pass the modulation signal conductor through a hole in the ring core (Daniel, 1955). The core will then be magnetized circularly around the current-carrying conductor. Thus the core can be driven into saturation locally without causing the modulation flux to appear at the recording gap. This type of head has been applied to long-wavelength, low-frequency recording where considerable increase of output voltage is obtainable. Thin-film versions of this design have been considered with very high modulation frequencies with application to higher-frequency recording (Hamilton, 1983).

4.2   Head Performance

The subject of recording head performance starts with a description of the physics of the head recording and reproducing mechanisms and the models which describe head performance. Only the recording fields arising from the head and the closely related subject of head sensitivity functions are described: the recording and playback performance, involving media and electronics, are treated in the relevant application chapters in Volume II. Within each of the following subsections the text proceeds from the more conventional metal and ferrite inductive-head technology, to the more innovative and speculative inductive perpendicular recording heads sensing heads.

4.2.1   Head fields and sensitivity functions

4.2.1.1   Infinite-pole-length inductive heads.   A conventional inductive recording head roughly consists of a slit toroid of high-permeability magnetic material wound by several conductor turns as shown schematically in Fig. 4.1. Any treatment of recording head performance necessarily begins with a discussion of the fields in the vicinity of the recording gap. These fields determine the nature of written transitions, although a full

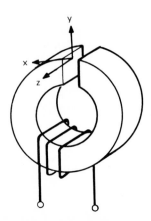

Figure 4.1 Approximate configuration of a conventional inductive recording head.

treatment of the writing process must involve many other factors, such as the media and the recording electronics, as well as spacing between the head and media. The relationship of the head fields to output voltage is less obvious; however, the powerful reciprocity theorem states that the output voltage e is related to the field produced by a current i passing through the head windings, $\mathbf{H}(x, y, z)$, by the integral over all space:

$$
\begin{aligned}
e &= -\frac{\mu_0}{i} \int_v \mathbf{H}(x, y, z) \cdot \frac{\partial \mathbf{M}(x, y, z)}{\partial t} \, dv \\
&= \frac{\mu_0 V}{i} \int_v \mathbf{H}(x, y, z) \cdot \frac{\partial \mathbf{M}(x, y, z)}{\partial x} \, dv
\end{aligned}
\tag{4.1}
$$

where $\mathbf{M}(x, y, z)$ is the magnetization of the medium, μ_0 is the permeability of space, and V is the medium velocity in the x direction. A more complete discussion of the principle of reciprocity may be found in Chap. 2.

If it is assumed that head and media are of infinite width, techniques for solving two-dimensional potential problems can be used. The two-dimensional problem for the case of an infinite-permeability head with pole tips that extend to infinity can be solved by using a conformal mapping Schwarz-Christoffel transformation (Westmijze, 1953). These results are shown in Fig. 4.2. The fields at pole-tip corners approach infinity according to this model, a result which cannot hold for real heads whose materials have neither infinite permeability nor infinite saturation magnetization. However, fields higher than the average field in the deep-gap region ($x = 0$, $y = -\infty$) can be found at distances less than about one-tenth of the gap dimension away from the corners. This concentration of high fields is of importance only in the case of wide gaps reading thin media at a very close spacing.

The results of the Schwarz-Christoffel transformation do not permit a closed-form evaluation of the integral in Eq. (4.1) since neither the magnetic potential nor the field can be written as an explicit function of the coordinates (x, y). Thus fields at a given position can be found only by interpolation, and the integral for the output can be evaluated only by numerical methods.

Expresssions for the magnetic potential U and fields in the upper half plane, $y \geq 0$, can be found if the potentials along the line $y = 0$ and at $x^2 + y^2 = \infty$ are known (and if proper mathematical constraints on that boundary potential are obeyed):

$$
U(x, y) = \frac{1}{\pi} \int_{-\infty}^{\infty} \frac{U(x', 0)y}{(x - x')^2 + y^2} \, dx'
\tag{4.2}
$$

Useful expressions (Karlqvist, 1954) for the magnetic fields can be obtained by assuming a constant potential (implied by an infinite permeability) on each pole, and a constant field H_g between them at $y = 0$. The

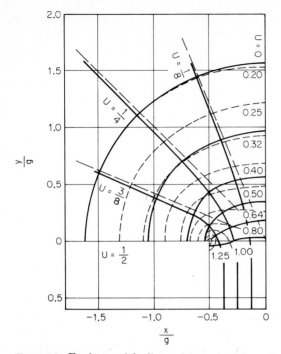

Figure 4.2 Equipotentials, lines of force, and lines of constant field strength (dashed curves). The center of the gap is at $x/g = 0$ and the potentials on the poles are $\pm \frac{1}{2}$ *(Westmijze, 1953).*

results for the longitudinal and perpendicular field components of a ring head are

$$H_x = \frac{H_g}{\pi} \left(\arctan \frac{g/2 + x}{y} + \arctan \frac{g/2 - x}{y} \right) \tag{4.3}$$

and

$$H_y = -\frac{H_g}{2\pi} \ln \frac{(x + g/2)^2 + y^2}{(x - g/2)^2 + y^2} \tag{4.4}$$

The limiting equations for the Karlqvist expressions and the conformal mapping result are the same at great distances from the pole corners (Westmijze, 1953). The maximum longitudinal recording field, at distances y greater than about $y = 0.2g$, is then for both models given approximately by the result

$$H_x(0, y) = \frac{2H_g}{\pi} \arctan \frac{g}{2y} \tag{4.5}$$

This field dependence, which is important in characterizing the head's ability to saturate longitudinally oriented media, is shown in Fig. 4.3.

An exact two-dimensional description of the fields around an infinite-permeability head has been developed using a Fourier method (Fan, 1961). The results are in the form of a summation of an infinite series of integral terms. These results have been compared with the Karlqvist expressions, and the conditions in which the two results are in good agreement have been established (Baird, 1980). Figure 4.4 shows a comparison of computed perpendicular field components for $y = 0.4$ μm and $g = 2.0$ μm. The agreement between the two expressions improves as the ratio y/g increases.

Semiempirical analytical expressions for the fields outside an infinite-pole-tip head have been derived (Szczech et al., 1983). The longitudinal field in the gap of a large-scale model was measured and the results fitted to a function of the form

$$H_x(x, 0) = H_s \left[A + \frac{Bg^2}{(Cg)^2 - x^2} \right] \tag{4.6}$$

with $H_s = H(0, 0)$ and the coefficients given by $A = 0.835$, $B = 0.0433$, and $C = 0.512$. The fields above the pole tips were then found from integrals

$$H_x(x, y) = \frac{1}{\pi} \int_{-\infty}^{\infty} H_x(x', 0) \frac{y}{(x - x')^2 + y^2} \, dx' \tag{4.7}$$

and $$H_y(x, y) = -\frac{1}{\pi} \int_{-\infty}^{\infty} H_x(x', 0) \frac{x - x'}{(x - x')^2 + y^2} \, dx' \tag{4.8}$$

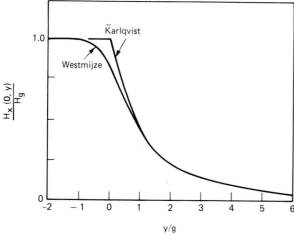

Figure 4.3 Calculated field strength H_x (0, y) for the Karlqvist expression, Eq. (4.5), compared with Westmijze's result *(Westmijze, 1953)*.

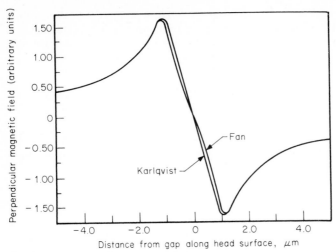

Figure 4.4 Calculated magnetic field H_y near the gap for a gap length of 2 μm and $y = 0.4$ μm *(Baird, 1980)*.

The resulting expressions, although more complex than Karlqvist's equations, reproduce many of the features of the conformal mapping solution fields, as shown in Fig. 4.5 (Szczech et al., 1983).

4.2.1.2 Finite-width inductive heads.   The limits to obtaining high track densities with narrow recording heads are determined by, among other things, the effects of side-fringing fields. These fields, which affect the written track width, can partially erase adjacent tracks, and, as can be shown by the reciprocity theorem, increase the reading width of the head and the pickup of signals from adjacent tracks.

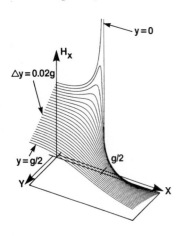

Figure 4.5 Three-dimensional plot of H_x for various y values using the field expression of Szczech *(Szczech et al., 1983)*.

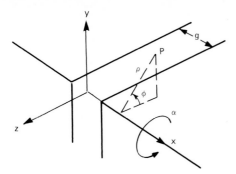

Figure 4.6 Polar and rectangular coordinate systems at the side of a semi-infinite head.

The potential of an infinite-permeability head with a wedge-shaped side geometry shown in Fig. 4.6 has been derived by several workers (van Herk, 1977; Lindholm, 1977; Hughes and Bloomberg, 1977; Ichiyama, 1977). In the initial instance the head is assumed to be one-sided, extending to infinity in the $-z$ direction. The calculation of the potential outside the head represents a three-dimensional boundary-value problem, the solution of which can be written as

$$U(\mathbf{R}) = \int \int_s U(\mathbf{R}') \, \nabla' G(\mathbf{R}, \mathbf{R}') \cdot \mathbf{n} \, dS \tag{4.9}$$

where $G(\mathbf{R}, \mathbf{R}')$ = Green's function for configuration used
$\qquad$ $\mathbf{R}, \mathbf{R}'$ = position vectors
$\qquad$ S = surface on which potential $U(\mathbf{R}')$ is given
$\qquad$ $\mathbf{n}$ = unit vector normal to surface pointing outside head

For a zero gap length and the two poles at magnetostatic potentials $\pm U_0/2$, the potential $\overline{U}$ at any point outside the wedge-shaped head is given in the cylindrical coordinates shown in Fig. 4.6 by

$$\overline{U}(\rho, \phi, x) = -\frac{U_0}{\pi} \arctan\left[\csc\frac{\pi\phi}{\alpha} \sinh\left(\frac{\pi}{\alpha} \sinh^{-1}\frac{x}{\rho}\right)\right] \tag{4.10}$$

This solution for zero gap length can be extended to a finite gap length g if the potential in the gap at the head surfaces has a linear dependence on x, and only on x, as with the two-dimensional Karlqvist approximation. The finite gap potential U is then related to the potential given in Eq. (4.10) by an integral

$$U(\rho, \phi, x) = \frac{1}{g} \int_{-g/2}^{g/2} \overline{U}(\rho, \phi, x - x') \, dx' \tag{4.11}$$

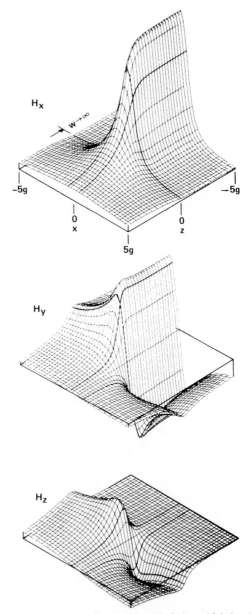

Figure 4.7 Magnetic fields at $y = g/2$ for a semi-infinite-width head. Components are longitudinal (H_x), vertical (H_y), and transverse (H_z). The side of the head is at $z = 0$. Projected onto each field surface are outlines of the top of the head *(Lindholm, 1977)*.

The magnetic field is obtained from this equation by taking the gradient of the potential

$$\mathbf{H} = -\nabla U \tag{4.12}$$

Calculated fields in the vicinity of the head edge are shown in Fig. 4.7 (Lindholm, 1977) for the case $\alpha = 3\pi/2$. The importance of the three field components cannot be stated without knowing the magnetomotive potential difference between the pole tips, the hysteresis loops of the media in all directions, the positions of adjacent tracks, and so on. However, several general features can be noted in comparing these results with those of the two-dimensional head. In addition to the usual longitudinal and perpendicular fields H_x and H_y, there is a cross-track component H_z whose direction is in the same sense as that of the perpendicular field. That is, H_z is directed outward from the head surface on the same surface where H_y is directed outward. This field is concentrated at the edges of the head. The longitudinal field H_x begins a modest drop off inside the physical width of the poles. Outside the poles the half width of the H_x peak increases approximately in proportion to distance z to the side of the head, indicating a lower resolution in reading off-track recorded information. The peak amplitude of the longitudinal field decreases approximately in inverse proportion to the same distance. The results for a head with a one-sided, semi-finite width can be used to construct good approximations to the field for heads of finite width for which no analytical solution is available. The side effects are localized near the head edges. If the width w of a perpendicular-sided head ($\alpha = 3\pi/2$) is greater than six times the spac-

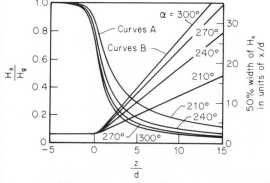

Figure 4.8 Maximum value of H_x (curves A) and half width of H_x (curves B) with the head side angle as a parameter, versus z/d, where $y = d$ is the vertical spacing from the head (*van Herk, 1977*). The side of the head is at $z/d = 0$ and H_g is the deep gap field.

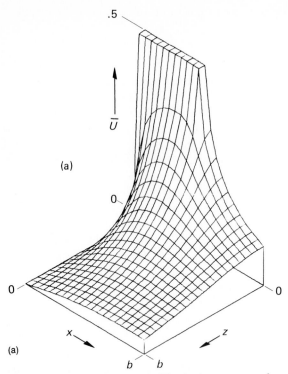

.5

$\overline{U}$

(a)

0

(a)

0   0

x   z

b   b

Figure 4.9 Potential at $y = 0$, $U(x, 0, z)$, versus x and z over a quarter of the geometry of a rectangular head, $d/b = 1$, $w/b = 0.1$. (a) Unshielded; (b) with side shields: $b_s/b = 1.5$, $c_s/b = 0.1$, $w_s/b = 0.05$ (shaded region is portion where $U = 0$ on top of the shield) *(Lindholm, 1980)*.

ing y to the media, the fields on both edges are within 1 percent of the results calculated for the semi-infinite-width case. For heads with sloping sides, $\pi < \alpha < 3\pi/2$, which are of considerable technological importance, H_x drops off less rapidly with z (Fig. 4.8), causing these heads to write and read wider than perpendicular-sided heads of the same width (van Herk, 1977). For high track densities, this generally leads to a preference for perpendicular-sided head configurations.

The effects of shields on potentials at the side of a head are illustrated in Fig. 4.9 for the rectangular head geometry shown in Fig. 4.10 (Lindholm, 1980). The potentials were computed numerically using a summation equivalent to the integration of Eq. (4.9) over the surface of the head. The shield is assumed to be at zero potential, halfway between the potentials of the two poles. In general, the effects of the side shields are to limit the lateral extent of the x and z field components, as shown

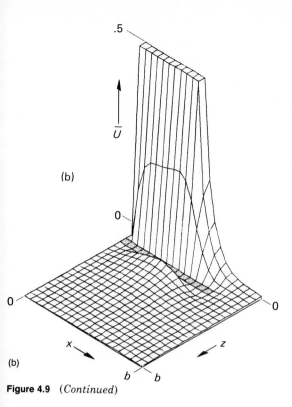

(b)

(b)

Figure 4.9 (*Continued*)

in Fig. 4.9. At the same time the z component of fields in the region between the head and the shields becomes more intense. The rapid descent of the potential toward the shield intensifies the longitudinal field component near the leading and trailing edges relative to the fields present when the head is unshielded. The net effect of such shields depends on the mode of recording and the properties of the media.

4.2.1.3 Finite-pole-length heads.   The models considered to this point have been for pole tips of infinite extent in the track direction. In practice, these models will not hold, and local fields near the leading and trailing edges must be taken into account. A first approximation to these edge fields for a two-dimensional infinite-permeability head, with a zero gap dimension, was obtained from a Schwarz-Christoffel conformal mapping (Westmijze, 1953). The potential along the line $y = 0$ is shown schematically in Fig. 4.11. At the corners the gradient of this potential, the longitudinal field H_x, is directed in the opposite sense from the gap field and goes to infinity at the corner approximately as $(L/2 - x)^{-1/3}$, where $L/2$ is the distance from the center of the gap to the leading or trailing edge.

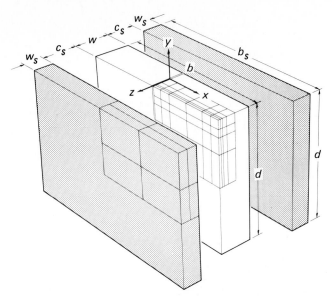

Figure 4.10 Rectangular head with optional side shields (shaded). Nomenclature: head length, b; head width, w; shield length, b_s; shield width, w_s; shield clearance, c_s; and depth, d (same for both head and shields) *(Lindholm, 1980)*.

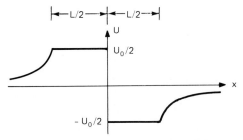

Figure 4.11 Distribution of the potential along the x axis for a head with $g = 0$ and a total length L *(Westmijze, 1953)*.

At a fixed distance, $x - L/2$, from the pole edge the amplitude of the field is approximately inversely proportional to $L^{2/3}$, indicating that these fields diminish for large poles. The fields near the corner are shown schematically in Fig. 4.12 (van Herk, 1980b). When this type of sensitivity function is correlated with a sinusoidally recorded magnetization using Eq. (4.1) the negative portions of the sensitivity function at the edges can either add or subtract from the output associated with the gap, depending on whether the magnetization at the edge is in phase or out of phase with the magnetization at the gap. These wavelength-dependent oscillations in head output can be decreased by guiding flexible media away from the edges of the head (Westmijze, 1953) or by beveling the head edges (Lindholm, 1979). In these ways the fields associated with the edges are made less intense either by increasing the distance to the media or by reducing the high concentration of flux associated with a corner.

The fields around three-dimensional heads have been computed using a surface-integral technique, and it has been shown (Fig. 4.13) that the x field components associated with the leading and trailing edges of heads

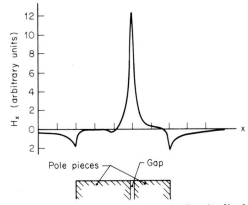

Figure 4.12 Schematic plot of the longitudinal field component, H_x, versus x for a head with finite pole tips *(Van Herk, 1980b)*.

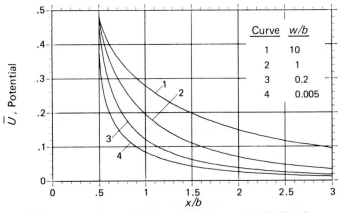

Figure 4.13 Magnetic potential at $y = 0$ for an unshielded head versus x/b, where b represents the length of the head, for various widths w. The head geometry is shown in Fig. 4.10 *(Lindholm, 1980)*.

are accentuated as the width of the head is diminished (Lindholm, 1980). Apparently the fields associated with a corner in three dimensions are even stronger than those associated with a corner in two dimensions.

With the film head, the total length of the head may be only a few times larger than the gap dimension, as shown in Fig. 4.14. In this case, the reverse fields associated with the leading and trailing edges cut into the extent of the fields associated with the gap. In this way the gap field is narrowed and gradient of the field made sharper, significantly affecting resolution in both reading and writing. Using the Schwarz-Christoffel transformation solution for fields around infinite-permeability, infinite-width heads with a finite gap dimension g and a finite pole length p (Elabd, 1963), the x and y fields have been computed, and the results for several p/g ratios are shown in Figs. 4.14 and 4.15 (Potter et al., 1971).

As with most conformal mapping solutions, the field results are not closed-form analytic expressions which can be used to evaluate the integral expression for output given by Eq. (4.1). Such expressions for infinite-permeability heads have been obtained by approximating the potential at $y = 0$ outside the pole tips, $|x| \geq g/2 + p$, by

$$U(x, 0) = \pm \frac{1}{4}\left(1 + \frac{g + 2p}{2x}\right) \tag{4.13}$$

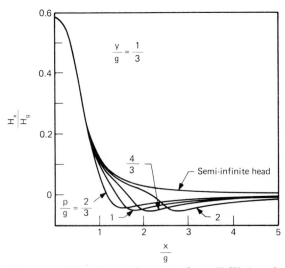

Figure 4.14 The effect of the ratio p/g on H_x/H_g for $y/g = \frac{1}{3}$; p is the pole-tip dimension from the edge of the gap to the corner of the pole tip, and H_g is the average longitudinal field in the gap at $y = 0$ *(Potter et al., 1971)*.

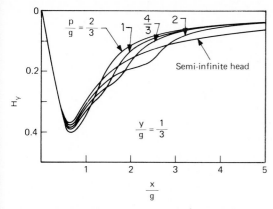

Figure 4.15 The effect of the ratio p/g on H_y/H_g for $y/g = \frac{1}{3}$; H_g is the average longitudinal field in the gap at $y = 0$ *(Potter et al., 1971).*

and assuming a constant field between the pole tips at $y = 0$ (Potter, 1975). A constant potential will exist at the pole surface, and the potential at the pole tip is taken to be $-\frac{1}{2}$ on the positive x axis ($x > 0$, $y = 0$) and $+\frac{1}{2}$ on the negative x axis. The potential along the line $y = 0$ is then completely described as shown in Fig. 4.16, and Eq. (4.2) can be used to find closed-form expressions for the potential and its associated fields.

An improved closed-form analytical approximation to the Schwarz-Christoffel solution potential is obtained by using expressions of the form

$$U(x, 0) = \pm \frac{1}{2(1 + \gamma)} \left[\exp \frac{-\alpha(x - g/2 - p)}{g/2 - p} \right. \\ \left. + \gamma \exp \frac{-\beta(x - g/2 - p)}{g/2 + p} \right] \tag{4.14}$$

for $|x| \geq g/2 + p$. The parameters α, β, and γ are chosen to optimize the fit to the conformal mapping fields. Asymmetric structures with different pole-tip thicknesses are accommodated by using different p parameters for the two sides of the head (Ichiyama, 1975).

An empirical potential has been deduced based on agreement with measured values of H_x outside the pole tips of a large-scale model head (Szczech, 1979). The potential is found by integration and fitting the form

$$U(x, 0) = \pm \frac{\Delta U_0}{2} \left(\frac{C_1}{x - C_2} + C_0 \right) \quad \text{for } |x| > \frac{g}{2} + p \tag{4.15}$$

where $C_1 = p/2$, $C_2 = 1.1g/2$, and $C_0 = 0.41$. Unlike the previous potential curves, the potential immediately beyond the pole tip does not equal

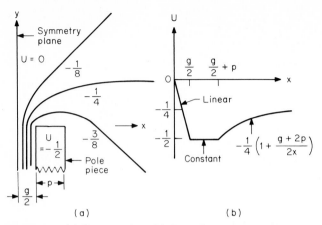

Figure 4.16 (*a*) Cross section of finite-pole-tip head with several equipotential lines schematically shown; (*b*) assumed magnetic potential at $y = 0$ *(Potter, 1975).*

$-U_0/2$, providing a discontinuity in potential at the pole-tip surface. As might be expected, the resulting closed-form expressions for H_y and H_x are somewhat lengthier than the Karlqvist equations, but still require only a modest computational power. The agreement between calculated and measured values of the horizontal field component H_x is shown in Fig. 4.17 for an asymmetrical head with one pole tip three times the thickness of the other.

With a properly chosen recording medium, recording mode, and head-to-medium spacing, the finite-pole-head waveform can give rise to small peak shifts in digital recording. Assuming that peak shift is due entirely to the superposition of adjacent pulses, the dibit peak shift τ can be approximated by

$$V_\tau \approx \frac{de/dx|_{x=-S}}{d^2e/dx^2|_{x=0}} \tag{4.16}$$

where V is the velocity of the medium, e is the output voltage, $x = 0$ corresponds to the position of a peak, and the two symmetrical dibits are spaced a nominal distance S apart. When the peaks are widely spread out, with the adjacent peak lying outside the negative minimum in the pulse forms, the peak shift will be negative and the peaks are shifted closer together. If the peak in one pulse coincides exactly with the minimum in the adjacent pulse, there is no peak shift. Finally, if the peaks approach more closely, peak shift rises rapidly, exceeding the bit shift for an infinite-pole-tip head because of the greater derivative of pulse amplitude.

4.2.1.4   Perpendicular recording inductive heads.   Numerous studies have been conducted using conventional ring heads and finite-pole-tip heads with perpendicular recording media (for example, Iwasaki et al., 1979; Iwasaki, 1980; Potter and Beardsley, 1980). However, the more extensive theoretical and experimental efforts have been directed toward the use of a single-pole head (Iwasaki and Nakamura, 1977) to accentuate the perpendicular field component. Much of this work has been done using a relatively large auxiliary pole, or a deposited high-permeability layer, beneath the recording layer. This auxiliary pole or deposited underlayer, in a sense, is part of the head structure. It modifies the head-field configurations and intensifies the perpendicular field by providing a return path for flux linking the pickup coils.

The exact solution for the fields from a single-pole head of arbitrary dimensions in proximity to an infinitely permeable layer has been derived using a Schwarz-Christoffel conformal mapping (Steinback et al., 1981). The calculated results with the geometry shown in Fig. 4.18 are displayed in Fig. 4.19 for the case $y/s = 0.001$ and $T/s = 0.28$. The presence of the infinite-permeability layer is taken into account by introducing an image pole at a spacing $2s$ from the single reference pole. As with previous Schwarz-Christoffel mappings, the vertical fields rise sharply to infinity at the corners of the probe. Such a result is physically unrealistic, arising from the assumptions of infinite permeability and infinite saturation magnetization; however, there is ample experimental evidence from scale models of field maxima near those corners (Steinback et al., 1981).

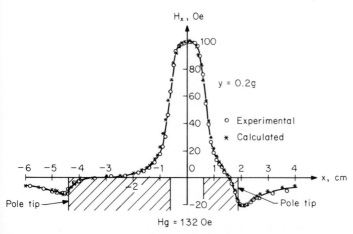

Figure 4.17   Comparison of calculated and experimental horizontal field components at $y = 0.2g$ for an asymmetrical scale model with one pole tip three times larger than the other *(Szczech, 1979)*.

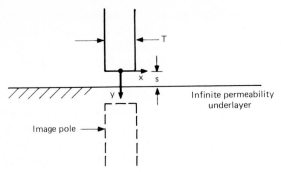

Figure 4.18 Geometry for a single-pole head over an infinite-permeability underlayer.

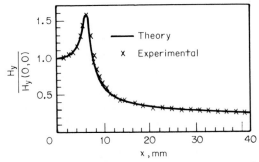

Figure 4.19 Comparison of measured and calculated H_y field components for a scale model; $T = 7.15$ mm, $T/s = 0.28$, $y/s = 0.001$ *(Steinback et al., 1981)*.

Approximate expressions for $H_y(x, y)$ and $H_x(x, y)$ in the region $0 \leq y \leq s$ can be found from the mapping solution by fitting the results for field along the line $y = 0$ to appropriate simple forms with adjustable coefficients. The approximations are then made by evaluating definite integrals involving these forms from $x = -\infty$ to $x = +\infty$ over this line. The results, although complex and tedious to write out, are closed-form expressions which are convenient for computations. A comparison of the conformal mapping and calculated y components of field is shown in Fig. 4.20 (Szczech et al., 1982). These results are compared with field measurements near a large-scale model as shown in Fig. 4.21.

In regions in which there are no magnetic poles, the potential U is a solution of Laplace's equation, $\nabla^2 U = 0$. Numerical solutions can be found for the potential in the vicinity of a single-pole head by iterating finite-difference expressions for $\nabla^2 U = 0$, subject to the appropriate fixed boundary conditions (Szczech and Palmquist, 1984). Results are obtained both with and without an infinitely permeable magnetic sublayer. A potential of 1 is assumed for the single pole, and zero potential is assigned either to the surface of the infinitely permeable sublayer or at great distances from the pole, taken in this instance to be a distance of $5T$ from the pole. After iteration to a convergent solution, the resulting potential is numerically differentiated to obtain H_x and H_y. The results for H_y in the region $0 \le y \le \frac{1}{2}T$, with and without an infinitely permeable magnetic sublayer, are shown in Figs. 4.22 and 4.23. The infinitely permeable layer in this instance was assumed to be at $y = \frac{1}{2}T$.

Comparing these results shows that H_y fields are significantly stronger for the single-pole head with an underlayer. The singularity at the corner $(x, y) = (T/2, 0)$ is again present, but the field maxima associated with this corner appear to vanish at about $y = 0.15T$.

These results can be compared with the H_x and H_y fields computed by the same methods for an infinite-pole-tip ring head and a finite-pole-tip film head. The perpendicular field component, with and without an infinitely permeable magnetic sublayer, for a finite-pole-tip head is shown in Figs. 4.24 and 4.25 (Szczech and Palmquist, 1984).

A complementary relationship between the fields from an infinite-pole-tip ring head with a gap dimension $g = T$ and a single-pole head of thickness T was recognized with the initial revival of interest in perpendicular recording (Iwasaki and Nakamura, 1977). It was then proposed that the

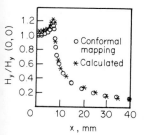

Figure 4.20 Comparison of conformal mapping and calculated H_y field components for a scale model; $T = 7.15$ mm, $T/s = 2.5$, $y/s = 0.35$ *(Szczech et al., 1982)*.

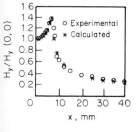

Figure 4.21 Comparison of experimental and calculated H_y field components for a scale model; $T = 7.15$ mm. $T/s = 0.28$, $y/s = 0.04$ *(Szczech et al., 1982)*.

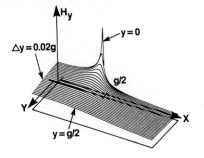

Figure 4.22 y component of field versus x and y for a single-pole head *(Szczech and Palmquist, 1984).*

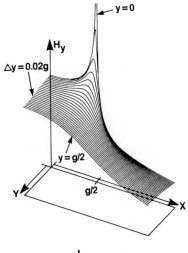

Figure 4.23 y component of field versus x and y for a single-pole head with a high-permeability underlayer at $y = T/2$ *(Sczcech and Palmquist, 1984).*

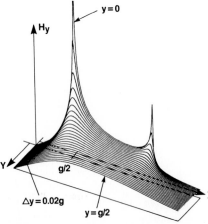

Figure 4.24 y component of field versus x and y for a thin-film head with $p = g$ *(Szczech and Palmquist, 1984).*

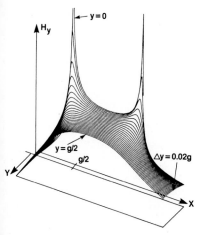

Figure 4.25 y component of field versus x and y for a thin-film head with $p = g$ and a high-permeability underlayer at $y = g/2$ *(Sczcech and Palmquist, 1984).*

H_y fields from the single-pole head would have the form of the H_x fields for the Karlqvist head model, at least as a first approximation. Later (Iwasaki et al., 1981) it was assumed the H_y field of a single probe would have the same form as the H_x field calculated by Fan (Fan, 1961).

It can be shown that a uniformly perpendicularly magnetized single-pole head gives rise to fields which can be simply related to the fields of a ring-head model in which the gap region is uniformly longitudinally magnetized. The fields H_{px} and H_{py} from the uniformly perpendicularly magnetized single-pole head without an infinite sublayer are given by

$$H_{px}(x, y) = -H_{ry}(x, y) \tag{4.17}$$

and $\qquad H_{py}(x, y) = H_{rx}(x, y) \tag{4.18}$

where it is to be understood that the pole thickness T is substituted for the gap dimension g in H_{rx} and H_{ry}, which are the longitudinal and perpendicular field components given by Eqs. (4.3) and (4.4) (Mallinson and Bertram, 1984).

These expressions, which probably represent good approximations for y values greater than a few tenths of the pole thickness, can be equally well viewed as resulting from a uniform magnetic pole density at the top of the pole, uniform magnetization in the pole, or uniform magnetic electric current sheets on the sides of the pole (Mallinson, 1981; Mallinson and Bertram, 1984). The magnetic poles on the opposite end of the probe can be assumed to be at a great distance L', and hence give rise to broad field distributions H'_{px} and H'_{py} whose features are not important in the short range:

$$H'_{px}(x, y) = H_{ry}(x, y + L') \tag{4.19}$$

and $H'_{py}(x, y) = -H_{rx}(x, y + L') \tag{4.20}$

Similarly, if the pole is positioned opposite an infinitely permeable layer at a spacing s, the total field in the region $0 \le y \le s$ will be a result of the fields given above and two other expressions representing the pole's image on the other side of the infinite-permeability layer.

The degree to which an assumption of uniform magnetization is valid will naturally depend on how uniformly fields from current sources are applied to the probe. For example, for a very thin infinite-permeability probe, only partially covered by coil turns, the field distribution will depend not only on the thickness of the pole tip but on the extent of the vertical region not covered by turns. In this case magnetic poles are distributed over the uncovered region of the probe (Minuhin, 1984).

Computations using a magnetic field surface-integral equation method (van Herk, 1980a, 1980b; Luitjens and van Herk, 1982) have been carried out for three-dimensional models of the single-pole head with and without the infinitely permeable underlayer. The single pole in this case consists of a thin slab of infinite-permeability magnetic material with $T = 0.1$ and height and width equal to 1.0. The head is excited by a homogenous magnetic field in the y direction which is not included in the calculated results. Figure 4.26 shows the resultant H_y fields. As with two-dimensional models, there are pronounced singularities at the leading and trailing

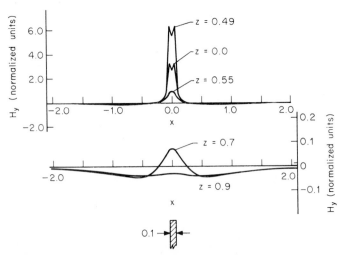

Figure 4.26 Perpendicular component of a magnetic field for a single-pole head. Center of the head is at $z = 0$, the geometrical edge is at $z = 0.5$, and the field is shown for $y = 0.01$ *(Luitjens and van Herk, 1982).*

edges of the pole which become less pronounced with increasing y, and finally vanish. Even more pronounced field maxima exist near the side edges of the head ($z = 0.49$). However, H_y drops off rapidly to the side of the track ($z = 0.55$), suggesting favorable side writing and reading characteristics.

Figure 4.26 also shows a feature reminiscent of the finite-pole-tip head: a region surrounding the sides and edges of the head in which the H_y fields are of opposite sign. These areas represent the return paths of flux lines from the vicinity of the pole face to the remote end of the head. This feature, which contributes to the steep field gradients near the probe edges, may, however, be significantly modified by other return paths for flux provided by permeable underlayers or second poles.

As with two-dimensional models, the y component of field is strongly enhanced by the presence of an infinite-permeability layer near the pole, as shown in Fig. 4.27. This is true both at the center of the head, $z = 0$, and at the side, $z = 0.7$. Computations also show that the y component of field varies less rapidly with y in the presence of an infinitely permeable underlayer.

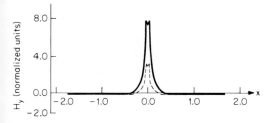

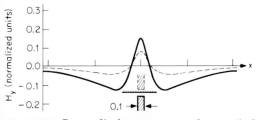

Figure 4.27 Perpendicular component of magnetic field for a single-pole head with an infinite-permeability underlayer at $y = 0.05$ (solid line) and without such an underlayer (dashed line). Center of the head is at $z = 0$, the geometrical edge is at $z = 0.5$, and the field is shown for $y = 0.01$ *(Luitjens and van Herk, 1982)*.

4.2.1.5 Magnetoresistive reproducing heads.

In 1971 a new class of reading heads was introduced based on the magnetoresistive effect in films of materials such as Ni-Fe, Ni-Co, and Co-Fe (Hunt, 1971). These devices have characteristics which are fundamentally different from those of inductive heads. In its simplest form the head consists of a narrow stripe of height h and length w, mounted in a plane perpendicular to the recording media, and connected to leads at each end carrying a sense current I_s, as shown in Fig. 4.28. As a result of the magnetoresistive effect, the resistivity of each portion of this stripe will depend on the angle θ between the direction of magnetization and the current-density vector:

$$\rho = \rho_0 + \Delta\rho \cos^2 \theta \tag{4.21}$$

For most materials of interest the increment of resistivity, $\Delta\rho$, is in the order of 2 to 6 percent of the base resistivity ρ_0.

In most embodiments, $\theta = 0$ in the quiescent state owing to magnetic anisotropy characterized by an anisotropy field H_k. Under the influence of a local field H_y, the magnetization rotates through an angle θ such that

$$\sin \theta = \frac{H_y}{H_k + H_d} \tag{4.22}$$

where H_d is the local demagnetization field, which depends on the proximity to the edges of the stripe and the stripe's magnetization. In general, H_y will equal the sum of two terms, a field h_y from the medium and a bias field H_b which is applied to linearize the head response. Both fields will depend on y and z. By assuming that the bias and demagnetizing fields are constant, and that only the field from the medium h_y depends on y and z, the expression for small signals is obtained by averaging over the stripe:

$$e = 2I_s R_0 \frac{\Delta\rho}{\rho_0} \frac{H_b}{(H_k + H_d)^2} \int \int h_y(y, z) \frac{dy}{h} \frac{dz}{w} \tag{4.23}$$

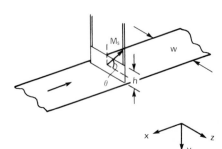

Figure   4.28 Magnetoresistive sensor stripe geometry.

where R_0 is the base resistance of the stripe and where quadratic terms in H and constant terms have been neglected. Under these conditions, the device essentially responds to the average applied field, and because of this, the resolution of the unshielded stripe is limited. The width at half-peak amplitude of the signal from an isolated transition will be in the order of the stripe height h (Thompson, 1974). For current practical film deposition and head fabrication procedures, this dimension cannot be less than about 3 to 5 μm.

Signal levels for the magnetoresistive head are generally much higher than for conventional inductive heads. Furthermore, the output of the magnetoresistive head depends only on the instantaneous fields from the media, and is independent of the media velocity or the time rate of change of the fields. This offers a significant advantage for reading low-velocity media. The response of a magnetoresistive head stripe has been analyzed more rigorously by taking into account local variations in demagnetizing and exchange fields (Anderson et al., 1972). Computations of the fractional changes of resistance for an unbiased stripe in a uniform applied field are shown in Fig. 4.29 along with a comparison with experimental data. The results show a slow approach to complete saturation at fields significantly above the value of field, $H_y = H_k + H_d$, which would result in saturation according to Eq. (4.22). The optimum bias field, the field at which $d(\Delta R)/dH$ is maximum, is also significantly changed by nonuniform demagnetizing fields. Analysis shows that, near the stripe edges, the demagnetizing fields are large, implying through Eq. (4.22) a small rotation angle θ and little sensitivity to external fields. This results in an insensitive "dead zone" at the head's surface and an apparent increase in head-media separation in fitting experimental results (Jeffers and Karsh, 1984). This apparent increase in spacing also is associated with an effective decrease in resolution. Despite these intrinsic problems, the output from a magnetoresistive stripe can be equalized with a properly designed linear amplifier to allow reproduction of 3150-fr/mm (80-kfr/in) recorded

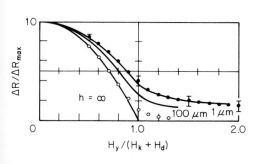

Figure 4.29 Numerical prediction (solid curves) of $\Delta R / \Delta R_{max}$ versus normalized y field with h as a parameter. ΔR_{max} is the change in resistance corresponding to saturation. The demagnetizing field H_d is approximately equal to $tB_s/h\mu_0$, where t is the sensor thickness and B_s is the film saturation magnetic induction. Open circles are for measurements of a 20-nm-thick, 3-cm-diameter film, and the solid circles are for 9- to 19.7-nm-thick stripes with h = 3 to 100 μm (Anderson et al., 1972).

signals. Another technique for adapting magnetoresistive stripes for high-resolution playback is to apply a bias field strong enough to saturate the center of the stripe (Uchida et al., 1982). In this way only the narrow edge of the stripe near the media is sensitive, effectively reducing the stripe height and thereby increasing resolution. Unfortunately the stripe output is also reduced.

The desire to provide an improved resolution led to the shielded magnetoresistive head shown in Fig. 4.30 (Potter, 1974). The purpose of the shields is to prevent the magnetoresistive stripe from experiencing the fields from the media until the recorded transitions are close, in the order of one gap dimension away. A sensitivity function of particular simplicity can be derived for this geometry, subject to the usual assumptions associated with the Karlqvist expressions. If it is assumed that the head materials have an infinite permeability, and that an energized coil is wound about the magnetoresistive element as shown in Fig. 4.30, a current in these turns will generate fields that trace the sensitivity to recorded transitions. Following Karlqvist, the fields between the magnetoresistive element and the shields are taken to be constant, implying the potential at the head surface shown in Fig. 4.30. The shields are assumed to extend to infinity and the ampere-turns associated with the imaginary coils are taken to be I_0, implying that the field strength between the shields and the magnetoresistive stripe is equal to $2I_0/(b - t)$, where b is the space

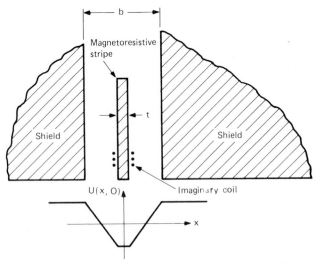

Figure 4.30 Cross section of a shielded magnetoresistive head, showing the coil that is imagined to exist for the purposes of computing the sensitivity function. Scalar potential U at $y = 0$ corresponding to Potter's model is shown below (*Potter, 1974*).

between the shields and t is the thickness of the magnetoresistive stripe. Since the potential is completely prescribed along the line $y = 0$, Eq. (4.2) can be used to derive the potential for $y > 0$. The potential gradient then provides the sensitivity function fields, H_{mrx} and H_{mry}, which, not surprisingly, are related to Karlqvist's expressions, written here as H_{rx} and H_{ry}, for the fields of the ring head:

$$H_{mrx}(x, y) = H_{rx}\left(x + \frac{b + t}{4}, y\right) - H_{rx}\left(x - \frac{b + t}{4}, y\right) \tag{4.24}$$

and

$$H_{mry}(x, y) = H_{ry}\left(x + \frac{b + t}{4}, y\right) - H_{ry}\left(x - \frac{b + t}{4}, y\right) \tag{4.25}$$

Since the magnetoresistive element senses flux rather than the time derivative of flux, the reciprocity integral for magnetoresistive sensors analogous to Eq. (4.1) for inductive heads is

$$e \propto \phi = \frac{\mu_0}{I_0} \int_v \mathbf{H}(x, y, z) \cdot \mathbf{M}(x, y, z) \, dv \tag{4.26}$$

where ϕ is the flux entering the element at $y = 0$. An equivalent expression which makes clearer the analogy between flux and time derivative of flux sensing can be derived from Gauss' divergence theorem, noting that $\nabla \cdot U\mathbf{M} = \nabla U \cdot \mathbf{M} + U\nabla \cdot \mathbf{M}$:

$$e \propto \phi = \frac{\mu_0}{I_0} \int_v U(x, y, z)\nabla \cdot \mathbf{M}(x, y, z) \, dv \tag{4.27}$$

The relationship between ϕ and the signal e is complex, involving the local sensitivity of the stripe and the flow of flux to the surrounding shields.

Comparing Eq. (4.27) with Eq. (4.1) for the case of longitudinally recorded media, it can be seen that the potential U plays a role for the magnetoresistive head somewhat analogous to that of fields $\mathbf{H}$ for the inductive head. It is appropriate in this case to speak of a "potential sensitivity function." The shape of the potential sensitivity function will give a good approximation to the shape of the output for a very narrow transition of magnetization in a thin media.

Where the edge of the magnetoresistive element is not planar with the face of the sensor, but recessed, the potential sensitivity function broadens slightly and drops off approximately exponentially with the amount of recession (Davies et al., 1975; Middleton et al., 1979; Middleton, 1980). Displacement of the magnetoresistive element toward one of the shields will cause an asymmetry in the potential shape at small y spacings. It has been shown that, at large spacings, this asymmetry diminishes and the peak value of potential is only slightly increased by the displacement.

The effects of finite shield lengths have been investigated through computations (Kelley and Ketcham, 1978) and with a large-scale model (Middleton et al., 1979). The computations were performed using numerical solutions of the integral form of Maxwell's equations. Relative breadths of output voltage pulse shapes are shown for several shield lengths in Fig. 4.31. Both the half width and the low-amplitude tails of isolated transitions read by shielded magnetoresistive stripes are broader for small shield dimensions, particularly for shield dimensions below about 2 μm. Very thin shields tend to give the relatively low-resolution characteristic of unshielded stripes.

The exact conformal mapping solution for the potential around the symmetrical shielded magnetoresistive head has been worked out for the case of an infinitely thin sensor (Heim, 1983). In addition to confirming earlier results, this analysis shows that the potential has undershoots near the edges of the shield of the type characteristic of the finite-pole-tip inductive head. Although generally smaller in the case of the shielded magnetoresistive element, these undershoots can still influence recording performance as pulse crowding begins. The potential between the sensor and the shields also has been computed exactly to show that the dependence of potential on x is not precisely linear (Lindholm, 1981). The effects of this small distortion may be expected to be negligible for many considerations, but the distortion can be evoked to explain the absence of a gap null with shielded magnetoresistive sensors (Schwartz and Decker, 1979).

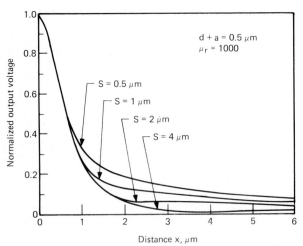

Figure 4.31 Isolated pulses computed for various values of shield length S. Permeability of the shield assumed to be 1000; $g = 1.02$ μm, $t = 0.02$ μm, d is the head-to-media spacing, and a is the transition length parameter for an assumed arctangent media transition *(Kelley and Ketcham, 1978)*.

4.3 Head Materials

4.3.1 Magnetic, electrical, and mechanical properties

The importance of many properties of head materials, for example, permeability and saturation magnetization, is clear from discussions of head function. Other properties can be appreciated only with a knowledge of fabrication techniques and the recording environment. A number of head core material properties which must be taken in account are as follows:

1. Permeability $\mu\mu_0$, the ratio of magnetic induction B to field H, is a key parameter for read and write performance. The relative permeability μ is sensitive to purity, thermal history, mechanical cold working, and the wear environment.

2. Saturation magnetic induction B_s dictates the maximum flux density which can be obtained in the head poles and is important for the writing function.

3. Coercivity H_c is the field necessary to reverse the magnetization and decrease the magnetic induction to zero. It is a measure of the ease of switching the magnetization and is sensitive to the material-handling history. It is also a measure of the openness of the B-H loop and therefore of the heat generation during magnetization cycles.

4. Remanent magnetic induction B_r is the magnetic induction remaining with zero applied field. Its value is closely related to coercivity and permeability and controls the residual field in and around the head, after magnetization during writing has ceased.

5. Electrical resistivity ρ influences the high-frequency performance of a head, which is reduced by eddy currents (Cullity, 1972). Eddy currents shield the interior of the head from the penetration of applied fields, reduce the effective permeability, and modify local fields around the gap. For a semi-infinite magnetic material, the maximum field amplitude follows a decaying exponential dependence, $\exp(-x/\delta)$, where δ is the "penetration depth" and x is the distance from the surface. The penetration depends on material properties through the relationship $\delta \propto \sqrt{\rho/f\mu\mu_0}$.

6. An intrinsic damping phenomenon associated with spin resonance, characterized by a frequency f_c, dictates the ability of ferrite magnetic materials to follow applied fields at high frequencies. This frequency, approximately 1 to 10 MHz for common head materials, increases as the ratio $B_s/\mu\mu_0$ increases.

7. The magnetoresistance coefficient $\Delta\rho/\rho$ is the fractional change of resistance associated with changing magnetization from parallel to perpendicular to current flow. This coefficient is one measure of a material's suitability for a magnetoresistive sensor.

8. The thermal coefficients of all the properties listed above are important because of head heating. It is generally desirable to have materials with a high Curie temperature, since properties at relatively low temperatures tend to be less sensitive to temperature excursions.

9. When magnetic head materials are in the form of films or single crystals, or have a pronounced crystalline orientation texture, for example, as a result of a rolling orientation, properties such as permeability, coercivity, and residual magnetic induction need to be specified as a function of orientation. For films with a uniaxial magnetic anisotropy, the parameter most frequently associated with this anisotropy is the anisotropy field H_k, which is the field (neglecting demagnetizing fields) at which an anisotropic film saturates in the hard-axis direction. This field is related to the saturation magnetic induction and permeability in the hard-axis direction by the relationship $H_k = B_s/\mu\mu_0$.

10. Head materials are frequently subject to strain, either during their original formation, as a result of head machining and lapping processes, or through wear, abrasion, and stress during use. This leads to a consideration of magnetostriction and the parameters which characterize changes of magnetic properties as a function of strain. The magnetostriction coefficient λ relates the magnitude and sign of strain, along a given crystalline direction or magnetic anisotropy axis, to a change in magnetic properties (see, for example, Bozorth, 1951). Since stresses are inevitable during head fabrication or in use, zero-magnetostriction materials are generally preferred.

11. Purely mechanical properties are important in forming and using the head. Hardness, measured, for example, in units of Vickers hardness, is sometimes an indication of durability. Wear is typically measured as a removal rate for head material in a particular environment with a particular medium. It depends not only on material properties, but also on wear interface characteristics such as friction, hardness of the media, and corrosive constituents of the environment. Even with the head material alone, many other aspects of material science are involved, such as the yield strength, porosity, and susceptibility to chipping. With relatively soft materials, there is the tendency for the material to distort or yield during the head manufacturing process, while for hard materials porosity and the tendency to chip are crucial.

4.3.2 Magnetic alloys

The earliest recording heads were made with laminated magnetic alloys with perhaps the three most prominent examples being molybdenum Permalloy (4 wt % Mo, 17 wt % Fe-Ni), Alfenol (16 wt % Al-Fe), and Sendust (5.4 wt % Al, 9.6 wt % Si-Fe) (see, for example, Chin and Wernick, 1980). As shown in Table 4.1, the initial permeabilities of these materials are high at low frequencies and their coercivities are low, although a wide

TABLE 4.1   Magnetic Metallic Materials

Material	μ†	H_c, A/m (Oe)	B_s, T (kG)	ρ, $\mu \Omega \cdot$ cm	Vickers hardness
4% Mo Permalloy	11,000	2.0 (0.025)	0.8 (8)	100	120
Alfenol	4,000	3.0 (0.038)	0.8 (8)	150	290
Sendust	8,000	2.0 (0.025)	1.0 (10)	85	480

†Measured at 1 K Hz with 0.2-mm-thick material.
SOURCE: Chin and Wernick (1980, p. 164).

range can be found, depending on thermal history and work hardening. Both molybdenum Permalloy and Sendust can be made with near zero magnetostriction, rendering them relatively insensitive to strain and with high saturation magnetization and good writing performance. The resistivity of molybdenum Permalloy is low, leading to losses of permeability due to eddy currents even at moderate frequencies unless the head is made in laminate form (Cullity, 1972). The principal advantage of Sendust is its hardness (Table 4.1). This hardness unfortunately also renders it more difficult to work, although special methods for forming ribbons of material with excellent magnetic properties have been reported (Ohmori et al., 1980; Tsuya et al., 1981).

Alfenol is an alloy with properties intermediate between those of molybdenum Permalloy and Sendust. It is somewhat easier to form than Sendust and has increased hardness and resistivity in comparison to molybdenum Permalloy. The permeability of Alfenol is somewhat lower than either of the other materials, but its magnetic performance is still adequate for many applications.

Many other alloys, such as Alperm (17 wt % Al-Fe) and Mumetal (4 wt % Mo, 5 wt % Cu, 77 wt % Ni-Fe), have been used successfully in heads, and efforts continue to develop better alloys. For example, alloys near the 4 wt % Mo, 79 wt % Ni-Fe composition with niobium and titanium additions have been shown to have an enhanced hardness (Vickers hardness, 240 to 300) while still retaining the high permeability and low magnetostriction of Permalloy (Miyazaki et al., 1972).

4.3.3   Amorphous magnetic alloys

Ferromagnetic amorphous alloys are a relatively new class of materials typically formed of 75 to 85 at % transition element (Fe, Co, Ni), with the balance being a glass-forming metalloid (B, C, Si, P, or Al). A thorough review of this class of materials is available (Luborsky, 1980). Like glasses, these materials have no structural order that can be detected, except that associated with near-neighbor interaction, and consequently are free of some of the characteristics of crystalline materials, such as crystalline anisotropy, porosity associated with crystal grains, and stresses associated with grain growth. Amorphous magnetic materials generally have smaller permeability losses due to eddy currents than crystalline materials, since the resistivities of these alloys are typically two to four times larger than the corresponding transition metal crystalline alloys without the metalloid glass-forming elements. Amorphous magnetic alloys have shown desirable permeabilities and saturation magnetic inductions for head applications, along with low magnetostriction, suitable hardness, and stability against recrystallization and corrosion. However, all these desirable properties may not be associated with a single composition (Mukasa, 1985).

Table 4.2 shows Vickers hardnesses and wear rates for a number of amorphous alloys in comparison with Fe-Si-Al (Sendust). For Co-Nb-B alloys it is desirable to keep the boron content below about 10% to achieve a wear rate lower than that of Sendust.

Co-Fe-Si-B amorphous alloys are potentially good materials for video heads, with a higher saturation magnetic induction than Mn-Zn ferrite heads, and a higher permeability at 5 MHz than Sendust heads (Shiiki et al., 1981; Matsuura et al., 1983). Amorphous Co-Zr alloys (10 wt % Zr) produced by magnetron sputtering have been used to fabricate film heads (Yamada et al., 1984). In sputtered form the saturated magnetic induction of these materials is higher than that of Ni-Fe Permalloy (1.4 versus 1.0 T), as is its relative permeability at 10 MHz (3500 versus 2000), and hardness (Vickers hardness of 650 versus approximately 300).

A serious concern regarding amorphous alloy heads is their stability in the presence of heat generated by tape friction. Amorphous Fe-Co-Si-B materials wear faster than either ferrite or Sendust, despite having a greater hardness.

In fact, as can be noted in Table 4.2, the wear and hardness of amorphous materials are not correlated in any simple fashion. This may be because the amorphous alloy surface is crystallized, or possibly oxidized,

TABLE 4.2   Vickers Hardness and Wear Rates for Various Amorphous Alloys

Composition	Vickers hardness	Relative wear rate
Fe-Si-Al	560	1
$(Co_{85.5}Nb_{14.5})_{98}B_2$	850	0.3
$(Co_{85.5}Nb_{14.5})_{95}B_5$	800	0.4
$(Co_{85.5}Nb_{14.5})_{90}B_{10}$	900	0.6
$Fe_{2.5}Co_{71.5}Mn_3Si_8B_{15}$	900	2
$Fe_{80}P_{13}C_7$	760	3
$Fe_{40}Ni_{40}P_{14}B_6$	750	4

SOURCE: Sakakima et al. (1981).

TABLE 4.3   Ferrite Properties

Property	Hot-pressed Ni-Zn ferrite	Hot-pressed Mn-Zn ferrite	Single-crystal Mn-Zn ferrite
μ	300–1500	3000–10,000	400–1000
H_c, A/m (Oe)	11.8–27.6 (0.15–0.35)	11.8–15.8 (0.15–0.20)	3.95 (0.05)
B_s, T (kG)	0.4–0.46 (4–4.6)	0.4–0.6 (4–6)	0.4 (4)
ρ, $\Omega \cdot$ cm	$\sim 10^5$	~ 5	> 0.5
θ_c, °C	150–200	90–300	100–265
Vickers hardness	900	700	

during wear experiments to either increase or decrease the wear rate (Ozawa et al., 1984).

4.3.4   Ferrite materials

The resistivity of ferrite materials (Table 4.3) is at least three orders of magnitude greater than that of most metallic magnetic materials, so that eddy currents and associated permeability losses are relatively small. As a consequence, these materials have dominated the field of high-frequency head applications for the last 20 years. Ferrites are also hard, with Vickers hardnesses in the range of 550 to 900, and behave well when head-media contact occurs. If properly made, ferrites can be precision-machined to close tolerances with no tendency to distort; and being oxides rather than metals, they are immune from attack by atmospheric gases, including water. Chipping in the region of the write-read gap is the most severe deterioration encountered in ferrite-head cores. This becomes relatively more serious as track widths are reduced. A description of ferrite properties and preparation techniques can be found in a recent review (Slick, 1980). A review of ferrite video heads can be found in Chap. 5 of Volume II.

There are two categories of high-permeability ferrite of commercial interest: nickel-zinc (Ni-Zn) ferrites, whose stoichiometric composition is $(\text{NiO})_x(\text{ZnO})_{1-x}(\text{Fe}_2\text{O}_3)$, and manganese-zinc (Mn-Zn) ferrites, $(\text{MnO})_x(\text{ZnO})_{1-x}(\text{Fe}_2\text{O}_3)$, both with the spinel crystal structure. Compositions of Mn-Zn ferrite, optimized for high permeability, may also contain additional ferrous (Fe^{2+}) ions replacing divalent manganese, or zinc. These ferrous ions further increase electrical resistivity and decrease eddy-current losses relative to the stoichiometric material. The properties of both materials are influenced by the nickel-to-zinc and manganese-to-zinc ratios. For small additions of zinc, permeabilities and saturation magnetic induction are increased and coercivities and Curie temperatures decreased until optimum magnetic properties are obtained in the range x = 0.3 to 0.7.

Although there can be advantages to porosity, for example, to provide voids in the ferrite body that inhibit domain wall motion and decrease losses, every other consideration favors the lowest porosity that can be provided practically. Wear and chipping are diminished with low porosity, and it is easier to hold precise head geometries. Toward this end, dense hot-pressed ferrites were developed (Sugaya, 1968). These materials are sintered under uniaxial compression and, as a result, can have grain sizes in the order of 70 μm and porosities of 0.1 percent or less. Hot isostatic pressing is another technique recently developed to further decrease porosity (Takama and Ito, 1979).

Ni-Zn ferrite is to be preferred over Mn-Zn ferrite for very high-frequency operation because its relatively high resistivity suppresses permeability losses due to eddy currents (Table 4.3). On the other hand, Mn-Zn ferrite is generally preferred at frequencies below a few megahertz because it has a lower coercivity, higher permeability, and higher saturation induction. Despite the fact that Ni-Zn ferrite is a harder material, it tends to have a thicker "dead layer" (or nonmagnetic layer) due to lapping or wear damage. The dead layer increases the effective flying height in the order of tens of nanometers.

Single-crystal Mn-Zn ferrites were developed to improve the porosity still further, and in several ways they represent an ideal magnetic and mechanical material. The microcracking problem of polycrystalline ferrite is largely avoided, and residual stresses and dead layers from machining can be removed by annealing. If specified crystal orientations are used, single-crystal heads also exhibit excellent wear characteristics. However, magnetostriction-induced noise due to media contact can be a problem with single-crystal heads. With certain polycrystalline ferrites, the existence of many grains and grain boundaries suppresses magnetostriction noise. Compositions can be chosen to minimize the magnetostriction coefficient, but it is difficult to achieve a zero value along with a high saturation magnetic induction. The choice of single-crystal orientation is a compromise between permeability, wear rate, and magnetostriction noise (Hirota et al., 1980). Permeabilities may vary as much as 2 to 1 and wear rates 3 to 1 with different orientations.

With regard to magnetic properties, the most serious problem with ferrites is their inherently low values of saturation magnetic induction. There is a theoretical upper limit for ferrite of 0.6 T (6 kG), which is significantly lower than that for metallic magnetic materials. Sendust, for example, has a saturation magnetic induction of 1.0 T (10 kG), approximately twice that of any currently available ferrite. Another limitation, which is not nearly so obvious, is associated with the manufacturing techniques that must be used in making ferrite heads. The precision grinding and lapping processes used tend to impose a scale and associated geometry which restrict design options. It remains to be seen whether new techniques such as ion milling can effectively offset these limitations (Toshima et al., 1979).

4.3.5 Film-head materials

Film heads contain a variety of materials formed by techniques that are extensions of the thin-film technology of earlier magnetic microelectronic devices (Maissel and Glang, 1970; Cullidy, 1972). The properties of these materials are frequently sensitive to their method of deposition, and the reader should consult the original papers cited below for descriptions of those methods and properties.

4.3.5.1 Magnetic films. With a few exceptions, film heads have been made of approximately 80 wt % Ni with 20 wt % Fe, the material used for magnetic film devices over the past 30 years. The nature of film Ni-Fe is significantly different from bulk Ni-Fe because of its thickness, the techniques used to form it, and limitations on annealing and heat treatment. It is deposited by evaporation, sputtering, or plating (see Table 4.4) with impurities due to residual gases, the sputtering gas, or the plating bath. Because of constraints of the film substrate and other film constituents, high-temperature heat treatments and outgassing that might be used with bulk metals are not possible. As a result, very high permeabilities are rarely possible. Furthermore, because of these constraints it has not been possible to employ ferrites in thin-film form. Anisotropy fields H_k, for Ni-Fe films deposited under optimal conditions, fall in the range 200 to 400 A/m (2.5 to 5.0 Oe) and saturation inductions are near 1.0 T (10 kG), implying low-frequency relative permeabilities of 2000 to 4000. Eddy-current losses, which limit the useful frequency range of bulk Perm-

TABLE 4.4 Film-Head Magnetic Materials

Material	References
Inductive Heads	
Ni-Fe, evaporated	Kaske et al., 1971; Lazzari, 1978
Ni-Fe, sputtered	Hanazono et al., 1982; Potzlberger, 1984
Ni-Fe, plated	Romankiw and Simon, 1975; Jones, 1980
Ni-Fe-Cr/SiO$_2$, evaporated	Lazzari and Melnick, 1971
Co-Zr amorphous, sputtered	Yamada et al., 1984
Cu-Mo Permalloy, sputtered	Nakamura and Iwasaki, 1982
Mu metal, sputtered	Berghof and Gatzen, 1980
Fe-Si-Ru/Ni-Fe, sputtered	Shiiki et al., 1984
Magnetoresistive Heads	
Ni-Fe, evaporated	Hunt, 1971; Bajorek et al., 1974a
Ni-Co, evaporated	Bajorek et al., 1974a
Fe$_3$O$_4$, permanent magnet bias, sputtered	Bajorek and Thompson, 1975
Mn-Fe, unidirectional bias, evaporated	Hempstead et al., 1978
α-Fe$_2$O$_3$, unidirectional bias, sputtered	Bajorek and Thompson, 1975

alloy, are diminished in the film form so film heads can still operate at a frequency of 10 MHz and beyond (Calcagno and Thompson, 1975).

A tabulation of other magnetic film-head materials is given in Table 4.4, along with examples of references describing their use. Laminated Ni-Fe-Cr/SiO$_2$ offers the advantage of suppressing edge domains which can give rise to irreproducible head response pulse shapes (Jones, 1979). Sputtered amorphous Co-Zr, annealed in a rotating field to provide a low anisotropy, has a higher saturation magnetic induction, 1.4 T, and relative permeability (3500, at 10 MHz) than Ni-Fe. Other alternatives to Ni-Fe offer, in some combination, an improved high-frequency permeability associated with a higher resistivity, greater hardness, or a higher saturation magnetic induction.

Magnetoresistive heads also have been made with a variety of materials. In addition to Ni-Fe films, Ni-Co has also been used as a sensor film because of its higher saturation magnetization and broader linear range of response. Sputtered Fe$_3$O$_4$ has been used for permanent magnetic bias, and sputtered α-Fe$_2$O$_3$ and evaporated Mn-Fe films have been used for the unidirectional biasing of magnetoresistive heads.

4.3.5.2 Conductors. With the exception of silver, most of the high-conductivity metals have been used in fabricating film heads (Table 4.5). High conductivity and ease of fabrication are the primary requirements for conductors. At very high current densities, electromigration can be important. This is particularly true for aluminum, leading to the introduction of small amounts of copper to suppress electromigration along grain boundaries. Titanium has been used in intimate contact with Ni-Fe in shunt-biased magnetoresistive heads because of its reduced tendency to interdiffuse with Ni-Fe (Chow et al., 1979).

4.3.5.3 Insulation and gap materials. The most frequently used insulation and gap materials in film heads are SiO$_2$, SiO, and Al$_2$O$_3$ (see Table 4.6), which are reasonably durable and hard, and particularly suitable for environments in which wear or corrosion is a problem. Silicon dioxide is

TABLE 4.5   Film-Head Conductors

Material	References
Copper, evaporated	Lazzari and Melnick, 1971
Copper, sputtered	Berghof and Gatzen, 1980
Copper, plated	Romankiw and Simon, 1975; Jones, 1980
Gold, plated	Jones, 1980
Mo-Au-Mo, sputtered	Hanazono et al., 1979
Aluminum, 4% copper	Miura et al., 1980
Titanium	Brock and Shelledy, 1974; Shelledy and Brock, 1975

TABLE 4.6   Film-Head Insulation and Gap Materials

Material	References
SiO_2, evaporated	Lazzari and Melnick, 1971; Lazzari, 1978
SiO_2, sputtered	van Lier et al., 1976; Miura, 1980
SiO, evaporated	Kaminaka et al., 1980
Al_2O_3, sputtered	Berghof and Gatzen, 1980
Cured photoresist	Hanazono et al., 1982
Polyimide	Hanazono et al., 1982

the principal insulation for silicon-integrated circuits and has an extensive technology base associated with it; SiO can be used if evaporation is the chosen deposition technique; and Al_2O_3 may be preferred because it has the highest thermal expansion coefficient of the three, hence will have smaller thermally induced stresses.

Under some circumstances, organic insulators such as cured photoresist and polyimide can be used, although they are much softer and less thermally stable materials. Cured photoresist is particularly convenient to deposit; however, properly cured polyimide is probably a more stable material at high temperatures. Both these materials are more suitable as insulation layers than as gap materials.

4.3.5.4   Substrate materials.   Unlike conventional heads, film heads must be deposited on a substrate which must exhibit good wear characteristics, high thermal conductivity, and, in some instances, good soft magnetic characteristics. The substrate must also be capable of being precision-machined into an air-bearing or media-interface form. Several designs have been built which have a hybrid thin-film bulk magnetic material construction. In these cases, ferrite is chosen to serve as a pole or shield, and the balance of the structure is in film form. Sometimes Ni-Zn ferrite is preferred over Mn-Zn ferrite because its high resistivity does not lead to a shorting out of deposited-film conductors. A relatively new material, Al_2O_3-TiC, has been employed for heads where extraordinary wear durability is required (Table 4.7). The hardness of these substrates (Vickers hardness of ~ 2100) is significantly greater than that of ferrites.

TABLE 4.7   Film-Head Substrate Materials

Material	References
Ni-Zn ferrite	Druyvesteyn et al., 1983; Kanai et al., 1979
Al_2O_3-TiC	Jones, 1980; Yamada et al., 1984
Glass	Lazzari and Melnick, 1971
Oxidized silicon	Lazzari and Melnick, 1971

4.4   Materials Constraints on Head Performance

4.4.1   Effect of finite permeability on performance

For an inductive head core with infinite permeability the magnetomotive force potential drop $\int \mathbf{H} \cdot d\mathbf{l}$ between the pole tips will be equal to the product of the number of turns in the head times the current in the turns, ni. For finite permeability, however, the ratio of the potential difference to ni will be a quantity η, known as the *head efficiency:*

$$\eta = \frac{1}{ni} \int_{\text{gap}} \mathbf{H} \cdot d\mathbf{l} \tag{4.28}$$

The efficiency is a measure of the write-field strength for a given current and, by the reciprocity theorem, a measure of signal amplitude.

In principle, the efficiency can be determined by computations of the fields inside and outside the head caused by a current in the turns. However, as shown earlier, computations of this sort are generally complex and difficult to extend from one geometry to another, and so it is often expedient to resort to simpler models in estimating the effects of head structure and permeabilities. Simple approximate expressions for the efficiency can be obtained by considering the head as a magnetic circuit and applying Ampère's law, giving

$$gH_g + \int_c \mathbf{H} \cdot d\mathbf{l} = ni \tag{4.29}$$

where the second term corresponds to the line integral within the magnetic material and the product gH_g corresponds to the line integral in the gap region. Through use of Eqs. (4.28) and (4.29), the efficiency in this case is given by

$$\eta = \frac{gH_g}{gH_g + \displaystyle\int_c \mathbf{H} \cdot d\mathbf{l}} \tag{4.30}$$

This expression shows that the efficiency is reduced by the line integral segment in the magnetic portion of the head. If a uniform cross section in the gap region, A_g, and in the magnetic portion of the circuit, A_c, is assumed, the fields are given by

$$\mu\mu_0 HA = BA = \phi \tag{4.31}$$

where ϕ is the flux which is continuous around the magnetic circuit. Introducing this expression into Eq. (4.30) gives

$$\eta = \frac{g/\mu_0 A_g}{g/\mu_0 A_g + l_c/\mu\mu_0 A_c} \tag{4.32}$$

where l_c is the length of the line integral in the magnetic material whose permeability is $\mu\mu_0$, and μ_0 is the permeability of space. The individual terms $g/\mu_0 A_g$ and $l_c/\mu\mu_0 A_c$ represent the magnetic reluctances of the gap region and the magnetic regions of the head, respectively. A finite reluctance in the magnetic portion of the head effectively divides the magnetomotive force created by the current in the turns, so only a fraction η of the magnetomotive force drop occurs between the pole tips of the head. This fraction can approach unity only as the product of permeability and magnetic cross section approaches infinity, or as the magnetic path length approaches zero. Since the gap dimension g and the gap field H_g are often fixed by other constraints, such as recording resolution and media coercivity, considerations of how to obtain suitable permeabilities and magnetic geometries are basic to head design.

It is evident from Eq. (4.32) that any feature which increases the reluctance associated with the magnetic portions of the head, for example, air gaps or boundary regions with low permeability (referred to as *dead layers*), will decrease head efficiency. Similarly, saturation of portions of the magnetic material will effectively increase the reluctance of the magnetic circuit and result in a low efficiency for current increments above the point of saturation.

Flux paths parallel to the gap will also decrease the efficiency. Such fringing or leakage paths can be taken to correspond to a parallel reluctance which decreases the effective reluctances of the gap region, causing an increase in the relative magnitude of the magnetomotive force drop in the magnetic portions of the head. This in turn decreases the efficiency of the head.

Several approximate expressions have been derived for the reluctances associated with flux emanating from pole tips into the media and for flux leakage to the sides and behind the gap region. These can be incorporated in more elaborate reluctance networks to predict efficiency as a function of more complex head geometrical parameters (Unger and Fritzsch, 1970; Walker, 1972; Sansom, 1976; Jorgensen, 1980).

The head inductance, which is important in characterizing its frequency response, also can be derived from magnetic circuit models. For example, by combining Eqs. (4.29) and (4.31) for the case of uniform cross sections in both the gap and magnetic regions, one finds

$$\frac{ni}{g/\mu_0 A_g + l_c/\mu\mu_0 A_c} = \phi \tag{4.33}$$

Since inductance L is defined by the relationship

$$-L\frac{di}{dt} = e = -n\frac{d\phi}{dt} \tag{4.34}$$

this leads to a simple expression

$$L = \frac{n^2}{g/\mu_0 A_g + l_c/\mu\mu_0 A_c} \tag{4.35}$$

which shows that a low inductance requires a high reluctance in the magnetic portion of the head and/or a high gap reluctance.

4.4.2 Transmission-line analysis of magnetic circuits

A special variant of the magnetic circuit analysis has been used with film-head structures. In this case, thin planar magnetic films, with a relatively high reluctance per unit length, are in close proximity to one another, corresponding to a relatively low reluctance per unit length for flux leakage between films. This geometry suggests a treatment which has been called a *transmission-line analysis* because it results in a differential equation analogous to those governing current and voltage in transmission lines. Although generally only relatively simple geometries are treated, a complete description of the flux flow, local fields, and the efficiency of the head results (Thompson, 1974). The original analysis of a single-turn head was derived from the integral forms of Maxwell's equation with time variations of flux taken into account (Paton, 1971). However, a simpler, more frequently used, static analysis starts from Ampère's law and results in expressions valid at low frequencies. Considering the line integral of mag-

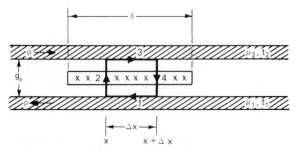

Figure 4.32 Cross section of a thin-film head showing a line integral enclosing a current i' *(Jones, 1978).*

netic field around a portion of a current-carrying stripe, as shown in Fig. 4.32, the portions of the line integral corresponding to legs 1 and 3 are

$$H_1 \, \Delta l_1 = \frac{\phi \, \Delta x}{w \mu_0 \mu_1 t_1} \tag{4.36}$$

and

$$H_3 \, \Delta l_3 = \frac{\phi \, \Delta x}{w \mu_0 \mu_2 t_2} \tag{4.37}$$

where ϕ = flux
w = head width
μ_1, μ_2 = relative permeabilities of the two films
t_1, t_2 = thicknesses of the two films

The portions of the integral corresponding to leakage in legs 2 and 4 are

$$H_2 \, \Delta l_2 = \frac{g_a}{\mu_0 w} \frac{d\phi}{dx} (x) \tag{4.38}$$

and

$$H_4 \, \Delta l_4 = -\frac{g_a}{\mu_0 w} \frac{d\phi}{dx} (x + \Delta x) \tag{4.39}$$

where g_a is the separation between magnetic layers. For a total current i, the current enclosed by the line integral is

$$i' = \frac{\Delta x i}{\delta} \tag{4.40}$$

where δ is the conductor stripe width.

Combining Eqs. (4.36) to (4.40) and taking the limit $\Delta x \to 0$ yields the governing differential equation

$$\frac{d^2\phi}{dx^2} - \frac{\phi}{\lambda_a^2} = \frac{\alpha}{\lambda_a^2} i \tag{4.41}$$

where

$$\frac{1}{\lambda_a^2} = \frac{1/\mu_1 t_1 + 1/\mu_2 t_2}{g_a} \tag{4.42}$$

and

$$\frac{\alpha}{\lambda_a^2} = \frac{\mu_0 w}{\delta g_a} \tag{4.43}$$

In general, solutions of Eq. (4.41) are linear combinations of terms of the form $\exp(\pm x/\lambda_a)$ with coefficients given by the boundary conditions selected (Jones, 1978).

An analysis of the flux flow in a single-turn head has been developed in which a copper conductor completely fills the area within the yoke. Eddy currents in the conductor were also taken into account in the resultant expression for head efficiency (Paton, 1971). The predicted low-frequency efficiency is shown to be in reasonable agreement with experimental results (Valstyn and Shew, 1973). This model has been extended by considering a single-turn head with four regions: a coil region, a gap region (which has no conductor), and two permeable pole regions. Eddy currents are assumed to be in the magnetic poles as well as in the conductor. The head impedance and efficiency are modified considerably, particularly at high frequencies. The efficiency is shown first to rise with increasing frequency due to conductor eddy currents resisting flux leakage from pole to pole across the conductor. At still higher frequencies, when the pole skin depth approaches the pole thickness, the efficiency diminishes owing to a lower effective permeability caused by eddy currents in the poles (Hughes, 1983). The zero-frequency limit of this model was described earlier (Miura et al., 1978).

Equations governing the flux in multiturn film heads can be derived by extending the analysis of the single-turn head to the case where there are two regions: a thin short-gap region, where dimensions are dictated by recording performance, and a thick back region created by the combined thicknesses of multiple conductors and intervening insulation layers (Jones, 1978). The efficiency of an individual turn can be shown to depend on its proximity to the pole-tip region. It can also be shown that head performance is critically dependent on the depth ("throat height") of the pole-tip region.

The transmission-line equations have also been solved using a finite-element analysis where the differential equations are converted to difference equations and solved numerically by direct integration (Katz, 1978). Within this framework, it is possible to simulate the effects of continuous changes of pole thickness and pole width by assigning a different cross section to each serial element in the film structure. Increased widths and thicknesses in the back region of the head were shown to be effective in increasing head efficiency.

Experiments have generally shown a surprising agreement with the predictions for multiturn heads, if reasonable values of permeability are used (van Lier et al., 1976; Anderson and Jones, 1981). More elaborate analyses involving numerical solutions of the integral form of Maxwell's equation have also generally supported the simpler transmission-line model (Kelley and Valstyn, 1980).

The transmission-line analysis has been applied to an inductive tape head, in which one pole is a deeply grooved ferrite substrate and the other, a Ni-Fe film. This permits a selection of planar dimensions, groove depth, film thicknesses, and gap depth for efficient reading and writing with such a head (Kanai et al., 1979). The transmission-line model also has been generalized by formulating two-dimensional partial differential equations which are solved by finite-element methods. Proper design of the width of the rear flux closure of film heads may greatly increase the writing current for the onset of head saturation. The case of an anisotropic permeability was also analyzed, and shown to have little effect on performance (Yeh, 1982*b*).

Transmission-line analysis of ferrite core heads has demonstrated the importance of leakage fluxes in reducing ring-head design efficiency. If the head pole surface is chamfered, so that only the gap region is close to the recording medium, the recording efficiency is improved (Corcoran and Pope, 1985).

4.4.3 Magnetoresistive head output and bias techniques

The output of a magnetoresistive stripe is a complicated function of the local fields in the stripe emanating from the recording media, the biasing used, and the proximity of shields or yokes which define the head resolution. In principle, the biasing is applied to provide the maximum possible linear output by rotating the magnetization approximately 45° with respect to current flow. However, the local bias depends on the method of application and also on the flux leakage paths away from the sensor. With the original magnetoresistive sensor (Hunt, 1971), a bias was applied by an external field, and, under some circumstances, this may be feasible. In most cases, particularly those where the sensor is shielded, other means are necessary. The shielding which prevents the sensing of transitions outside the shields will also render an external field ineffective. Furthermore, an external field cannot be introduced without a complex head structure and a risk of disturbing recorded data.

Many biasing schemes have been proposed; the abundance of these ideas is not only evidence of the ingenuity of designers in this field but also testimony to the fact that no single biasing scheme simultaneously offers all the advantages of simplicity, output, amplitude, linearity, and reproducibility.

Biased magnetoresistive sensors have also been frequently used in a center-tapped structure, with the two ends of the sensor biased in an opposite sense. Under the influence of a field from the medium, the magnetization and current vectors will in one case be moved closer together and in the other case farther apart. The signals from the two ends of the sensor are then sensed differentially. Such an arrangement, in effect, suppresses quadratic responses to the field from the recorded medium and extends the linear response range of the sensor, particularly in cases where it is systematically underbiased (O'Day and Shelledy, 1974; Shelledy and Brock, 1975; Gorter, 1977; Jeffers, 1979; and Metzdorf et al., 1982).

The largest and most complex class of biasing schemes involve applying a biasing field through an auxiliary microstructure in close proximity to the magnetoresistive stripe.

With shunt biasing, a current-carrying conductor is placed adjacent to the magnetoresistive (MR) stripe as shown in Fig. 4.33. In the earliest use of this configuration (Brock and Shelledy, 1974; O'Day and Shelledy, 1974; Shelledy and Brock, 1975) the bias conductor and magnetoresistive element are in electrical contact with a common current supply, providing both the sense current in the MR stripe and the current in the bias conductor. The relative current levels in the two stripes are thus controlled by their respective thicknesses and conductivities. Since there is a sizable portion of the total current flow in the conductor, the response per unit input current is reduced and the Joule heat associated with a given sensor response is increased. Both metal layers can, however, be deposited and configured in a common step, a basic fabrication simplicity which has made this an attractive choice for tape-head applications.

Through use of a transmission-line analysis, it has been shown (Shelledy and Brock, 1975) that the field produced by the bias conductor is high in the center of the stripe (see Fig. 4.34) and diminishes approximately to zero at the edges of the stripe. Since it is the edge nearest the surface of the head that is most sensitive to fields from the media, the current may be increased until the center portion of the stripe is saturated. Figure 4.34 shows the results of a transmission-line analysis for the saturated and unsaturated cases (O'Connor et al., 1985). The field entering the stripe from the medium leaks out of the stripe to the adjacent shields, following a differential equation whose coefficients depend on film thicknesses, spacings, and permeability. The signal field penetrates either to the end of the stripe, for the unsaturated case, or to the boundary of the saturated region. The change of resistance which corresponds to the change in voltage is an integral over the height of the stripe involving local electrical resistivity changes, which are a function of the local field.

Other current-bias schemes have been proposed using conductor stripes electrically insulated from, but adjacent to, the magnetoresistive sensor (Nepela and Potter, 1975; Druyvesteyn et al., 1981b). These structures,

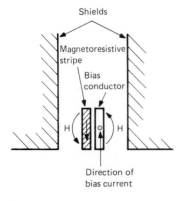

Figure 4.33 Configuration of a shunt-biased, shielded magnetoresistive head.

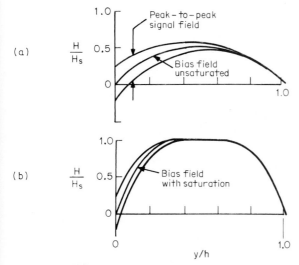

(a)

(b)

Figure 4.34 H/H_s in a shunt-biased magnetoresistive stripe versus y/h (O'Connor et al., 1985). (a) Unsaturated stripe; (b) stripe saturated at center.

although more responsive in terms of voltage output per unit sense current, still generate excessive Joule heating near the sensor, and have the additional disadvantage of requiring a much more complex film structure and at least two more leads to provide the biasing current.

A variant on the idea of shunt bias has been provided by placing the magnetoresistive sensor asymmetrically between the shields (Thompson, 1974; Shelledy and Brock, 1975; Brock et al., 1976). As shown in Fig. 4.35, the resulting bias can be considered to be the result of sensor current being imaged in the high-permeability shields. This type of bias provides an asymmetrical output pulse and a smaller bias field than a pure shunt bias, but does not suffer a loss of signal from the shunting effect of the bias conductor, nor is Joule heating as great. This type of bias can also be combined easily with the shunt bias described previously.

Soft-film biasing, as first described, calls for the placement of a thin, electrically insulated, soft magnetic film adjacent to the magnetoresistive sensor, as shown in Fig. 4.36 (Beaulieu and Nepela, 1975). When a sense current flows in the magnetoresistive element, the soft bias film is magnetized and thereby sets up a magnetizing field which biases the magnetoresistive element. If the magnetoresistive sense current is large enough, and if the soft-bias-film thickness, saturation magnetization, and saturation field are properly chosen, the bias film will be driven into saturation.

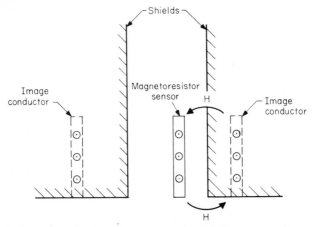

Figure 4.35 Image conductors associated with asymmetrical placement of a MR sensor between shields.

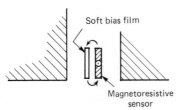

Figure 4.36 Configuration of a soft-film-biased, shielded magnetoresistive head.

Once the bias film is saturated, the bias field acting on the magnetoresistive element will be largely independent of the sense current and the amplitude of fields from the medium. The bias field should be enough to rotate the average magnetization of the sensor by about 45°. The soft bias film offsets, at least partially, the effects of demagnetizing fields associated with the magnetoresistive stripe. Results indicate that the soft-film bias provides a more uniform bias over the sensor height than does an external applied field (Jeffers et al., 1985).

In a number of instances, a soft-film bias has effectively been combined with a shunt-bias conductor. The structure shown in Fig. 4.37 was initially proposed, in which two identical magnetoresistive sensors are positioned side by side with currents flowing in the same direction in both sensors (Voegeli, 1975). In this case, both the current and the magnetization of each sensor contribute to the biasing field magnetizing its neighbor. An output from the coupled sensors is obtained from a differential amplifier which provides common mode rejection of noise, for example, thermal noise arising from head-media contact (Gorter et al., 1974; Hempstead, 1974, 1975). Experimental results with coupled film heads show that, because of relatively large signal amplitudes and low off-track sensitivity, track densities greater than 1000 tracks per centimeter are achievable (Kelley et al., 1981).

Soft-film bias and shunt bias have also been combined by placing an insulated conductor between the two essentially identical sensors (O'Day, 1973; Jeffers, 1979). In another instance the biases are combined without intervening insulation layers (Bajorek et al., 1977).

Proposals for biasing magnetoresistive sensors have been made which either use a hard magnetic layer adjacent to the sensor or provide an exchange coupling with an antiferromagnetic layer in intimate contact with the sensor (Bajorek et al., 1974c). With hard-magnet biasing, the distribution of magnetization produced is more uniform across the width of the sensor than that produced by either shunt or soft-film bias, as shown in Fig. 4.38 (Thompson, 1974). This is advantageous, since it is desirable to have a well-biased region near the edge of the sensor where the media signal field is greatest. The bias flux can be thought of as being injected at the upper and lower edges of the sensor from the edges of the permanent-magnet structure. Ideally, if the permanent-magnet material is uniformly magnetized, the coupling is perfect, and if the saturation magnetization–thickness product is properly chosen, it should be possible to bias the sensor well throughout its width (Bajorek et al., 1974b).

With exchange-coupling bias, the sensor film is in intimate contact and exchange-coupled at the atomic level to an antiferromagnetic film. The most frequently studied example of this bias is the Ni-Fe/Fe-Mn couple. In the ideal cases, the behavior of the Ni-Fe layer experiencing exchange coupling can be described as if there were a uniform field applied to the otherwise normal Ni-Fe film (Bajorek et al., 1974c; Hempstead et al., 1978). In principle, this technique can provide an effective bias field on

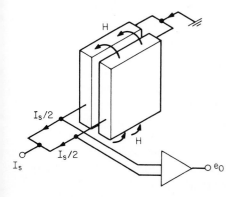

Figure 4.37 Current flow and signal detection in a pair of coupled film-biased MR sensors *(Voegeli, 1975)*.

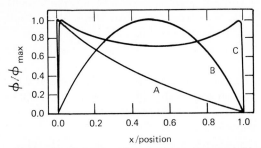

Figure 4.38 Bias flux for permanent-magnet bias (curve A), shunt soft film or external field bias (curve B), and signal flux entering near the media. The signal flux (curve C) is assumed to leak off to adjacent shields following the transmission-line theory *(Thompson, 1974)*.

the order of 50 Oe more uniformly than that provided by hard-magnet bias. Exchange-coupling bias has chiefly been used to provide longitudinal bias; that is, an effective field parallel to the track width direction, along with some other means for transverse bias. The effect of this longitudinal field is to give the sensor stripe a unidirectional, rather than uniaxial, anisotropy, which assists in diminishing detrimental domain wall switching (Hempstead et al., 1978; Tsang and Fontana, 1982).

Both the hard bias and exchange-coupling bias approaches, although nearly ideal from a design point of view, pose challenging material science problems. The magnetic properties of hard bias films (Bajorek and Thompson, 1975), and exchange-coupled surfaces and films (Hempstead et al., 1978; Tsang et al., 1981; Tsang and Lee, 1982), must be reproducible, stable, and free of local inhomogeneities, which induce irregular domain movement in magnetoresistive sensors, and cause Barkhausen noise.

With canted current or "barber-pole" bias, linearization of the signal response is obtained by forcing the sense current to flow at an angle (approximately 45°) with respect to the magnetization direction in the sensor, as shown in Fig. 4.39. This is accomplished by overlaying the sensor with slanted strips of a conductive material so that, over most of the sensor area, the current flows perpendicular to the strips in the desired direction (Kuijk et al., 1975; Gorter, 1977; Kuijk, 1977). Only in relatively small areas near the top and bottom of the stripe (an example of which is shown as hatched in Fig. 4.39) will the current direction remain relatively close to the magnetization direction, so the magnetoresistive material is in an underbiased state, tending to give a quadratic, rather than linear, response. This underbiased condition can be at least partially offset by increasing the inclination of the conductor stripes to greater than 45° (Tsang and Fontana, 1982). A clear advantage of this type of bias is the relative simplicity of the film structure that must be deposited. The bias is largely determined by the planar geometry of the conductive material. No additional composite film structures or materials other than the sensor itself are required, nor is it necessary to generate a field transverse to the magnetization direction. Analysis of the response of the barber-pole device shows that the range of linearity of signal response is increased compared to the shunt-bias sensor. Still another advantage of this bias technique is that, over portions of the stripe, an auxiliary field in the direction of magnetization is generated by current in the conductive strips. As in the case of exchange coupling, this introduces a unidirectional, rather than uniaxial, anisotropy, which tends to stabilize a single reproducible domain structure in the sensor (Feng et al., 1977). A field component transverse to the original magnetization vector is also produced by the current in the conductive strips. The conductor stripes in the barber-pole structure should generally be narrow to maximize the signal amplitude associated with a given track width, and the uncovered areas of the sensor should also be narrow to limit the underbiased sensor area near the media. Both considerations favor extending the photolithographic art to as high a resolution as possible in making these devices. The responses of the two ends of the two-lead sensor shown in Fig. 4.39 to side reading will be significantly different (Yeh, 1982a; Tsang, 1984). For optimum high-density applications, it probably will be necessary to offset the geometric center of the sensor relative to the written track center to minimize side reading.

4.4.4 Domains in inductive film heads

In the previous analysis it was assumed that magnetic materials could be characterized by a relative permeability μ, where $\mu\mu_0 = B/H$. In fact, magnetic materials are composed of individual domains with local magnetiza-

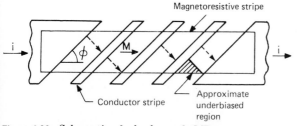

Figure 4.39 Schematic of a barber-pole MR sensor.

tions equal to the saturation magnetization of the material. When a head is used for writing or reading, the rotation of magnetization within these domains, or the shift of domain walls, constitutes the head response to magnetic fields. The assumption of a constant ratio of induction to field is at best a first approximation to this response.

An example of a domain structure for an inductive film head-yoke geometry is shown in Fig. 4.40. With an easy axis in the indicated direction, most of the pattern will constitute domains with magnetizations parallel to the easy axis, but in opposite directions, separated by 180° Bloch walls. Because of the high energy associated with free poles, closure domains form at the periphery of the yoke with magnetizations parallel to the pattern edges. All domain walls are ideally positioned so the component of magnetization perpendicular to the wall is the same in adjacent domains, implying there are no magnetic poles at these walls.

In the absence of applied fields, the lowest-energy head domain pattern corresponds to the smallest sum of wall energy, anisotropy energy, and magnetostriction energy terms. Energetically favorable domain configurations representing this minimum sum have been determined for rectangular (Druyvesteyn et al., 1979; Soohoo, 1982) and trapezoidal (Narishigi et al., 1985) head patterns, and found to be in reasonably good agreement with observations. The analysis of rectangular patterns has been extended to include magnetostatic energy contributions that occur in the presence of applied fields (Druyvesteyn et al., 1981a).

With a larger anisotropy, or higher values of H_k, the energy per unit area of closure domains is larger and these domains tend to be smaller. This is illustrated by Fig. 4.41 (Narishigi et al., 1984) for materials with

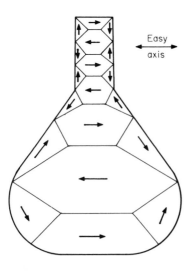

Easy
axis

Figure 4.40 Domain configuration in an inductive film head.

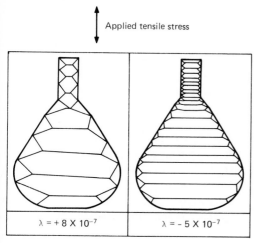

↑ Applied tensile stress

λ = + 8 X 10⁻⁷   λ = − 5 X 10⁻⁷

Figure 4.41 Domain configurations in positive and negative magnetostriction inductive film heads (*Narishigi et al., 1984*).

positive and negative values of magnetostriction. With tension applied in the hard-axis direction, the effective anisotropy energy of the positive magnetostriction material is decreased and large closure domains become energetically favorable. With a negative magnetostriction, and the same stress applied, the effective anisotropy energy increases and closure domains are small. [For a description of the relationships between tensile and compressive stresses, magnetostriction, and anisotropy energy, see Maissel and Glang (1970), Chap. 17.] In real heads, domains will not always occur in the ideal patterns shown in Figs. 4.40 and 4.41. Domain structures are often fixed by accidental features such as nicks at pattern edges and nonmagnetic inclusions and voids in the films. The magnetization pattern will also be influenced by the presence of other magnetic material, such as a second magnetic yoke layer. Domain images obtained using a scanning electron microscope (type II magnetic constrast mode) have shown a number of irregular domains originating at edge defects in head patterns, in addition to normal closure domains (Mee, 1980).

Domain walls are slow in responding to applied fields in comparison to magnetization rotation and, in many instances, they limit the dynamic response of film heads. It has been argued that time constants associated with closure domain wall motion are directly proportional to domain size (Jones, 1979; Soohoo, 1982), implying that large domains are particularly slow in responding. Magnetization processes associated with large closure domains also have been shown to be relatively unstable or "noisy," perhaps because of the longer distances the walls travel, or because of the larger fraction of the response controlled by domain wall motion (Narishigi et al., 1984). This has generally led to a preference for materials with small closure domains. For example, negative magnetostriction materials are better for heads with tensile stresses parallel to the hard-axis direction, since this gives smaller dimensions and consequently smoother head responses (Narishigi et al., 1984). Since domains can be easily reconfigured during writing, head responses in subsequent reading cycles can also be irreproducible as a result of changes in domain patterns (Jones, 1979).

Although the central portion of the yoke pattern may be free for closure domains, it is likely that the magnetization rotation process in this region will still be constrained by domain wall motion. The nature of this constraint has been analyzed (Ohashi, 1985), and it has been shown that the magnetization within the 180° domains curls near the closure domain wall to prevent the occurrence of magnetic poles. The length of this constrained region was found to be comparable to the domain spacing. Since this represents a considerable portion of the track width for most film heads, much of the head response should take place with a response time associated with the closure domain wall motion. According to this model, the switching dynamics of heads with closure domains consist of two steps: a rotation of magnetization in the central 180° domains, with magnetization curling and resultant magnetic pole distribution near the closure domain walls, followed by a closure domain wall motion driven by these distributed poles.

Experiments with single-pole film heads have shown that an optimum anisotropy field, H_k, exists to provide the maximum effective permeability for a given track width (Nakamura et al., 1985). This implies that there exists an optimum size for closure domains. If H_k is too small, edge domains are large and effectively block the response of the outer portions of the head pole. On the other hand, large values of H_k imply low values of hard-axis relative permeability ($\mu = B_s/H_k\mu_0$) even for the central portions of the pole. An optimum value of H_k appears to be about 480 A/m (6 Oe) for a 50-μm-wide pole corresponding to a 30 percent extension of closure domains across the pole width. Apparent permeabilities also depend on track width, a dependence evidently related to domain structure (Druyvesteyn et al., 1981a).

The local amplitude of domain wall motion in response to coil currents has been observed with a SEM lock-in technique in both the upper-pole piece of partially completed heads and the pole-tip surfaces of completed heads (Wells and Savoy, 1981). As might be expected, the amplitude of motion varied with applied current, showing the largest excursions in regions where models would indicate the greatest concentration of flux. Random changes of pole-tip magnetization were seen which were consistent with irregular domain motions in the head. Similar instabilities, particularly with positive magnetostriction (large domain) pole tips, have been observed using a Kerr magnetooptic technique (Narishigi et al. 1984).

A Kerr magnetooptic apparatus also has been used to measure local variations of magnetization as a function of frequency (Re and Kryder, 1984; Narishigi et al., 1984). Different responses are interpreted in terms of the domain structure of the yoke, and it was concluded that wall motion limits the head magnetization dynamics. At the center of the top surface of the head, the magnetization was 90° out of phase with the drive current at 8.4 MHz, while near the yoke edge this phase difference occurred at 4.7 MHz (Re and Kryder, 1984). These response frequencies, which are much lower than those associated with magnetization rotation, are at least in qualitative agreement with domain dynamics models (Jones, 1979; Soo-hoo, 1982; Ohashi, 1985).

It has been appreciated for some time that closure domains in head structures can be largely eliminated by using magnetic-nonmagnetic layer laminated films (Lazzari and Melnick, 1971). Flux closure paths in these films will be provided by adjacent layers magnetized in an antiparallel arrangement, and the closure domain walls will disappear. Domains in multiple-layer pole structures have been observed to largely consist of widely spaced 180° walls (Feng and Thompson, 1977; Mee, 1980). In these films, rotation rather than wall motion should control magnetization dynamics. As an added benefit, eddy-current losses in such films can be significantly improved in the 50- to 100-MHz range if the nonmagnetic layers are an insulating material (Feng and Thompson, 1977).

4.4.5 Domains in magnetoresistive heads

Many aspects of domain formation are the same for both inductive and magnetoresistive film heads. The magnetization is forced to lie in the plane of the film because the large demagnetization factor perpendicular to the film and head is broken up into more or less uniformly magnetized regions separated by narrow walls in which the magnetization changes orientation continuously.

Nevertheless, there are significant differences between domains in the devices. Magnetoresistive sensors are thin, in the order of a few tens of nanometers, and give rise to Néel and cross-tie walls rather than the Bloch walls in inductive heads. Film thicknesses in inductive heads are on the order of micrometers. The magnetization in Bloch walls rotates around an axis perpendicular to the wall, whereas in both Néel and cross-tie walls this rotation is around an axis perpendicular to the plane of the film. [A review of domain wall structures can be found in Middelhoek (1961).]

The planar shape of the magnetoresistive sensor also is different. Magnetoresistive stripes are narrow (y direction), in the order of 3 to 10 μm, and for narrow-track applications, the stripe length (z direction) may be less than a factor of 2 greater. As a result, demagnetizing fields can be expected to affect greatly the domain structures involved despite the thinness of the sensor films.

Extensive investigations have been carried out to correlate magnetoresistive sensor response with domain structure changes (Decker and Tsang, 1980; Tsang and Decker, 1981, 1982; Tsang, 1984). Figure 4.42, curve A, shows schematically the output for a 20 by 60 μm rectangular unbiased sensor as the transverse field is decreased from saturation (positive) values. The easy axis of the Ni-Fe material is parallel to the length (60 μm) of the film, which is also the direction of current flow. A relatively smooth response curve, $D(H)$, is traced as the field is initially decreased, with only occasional abrupt changes in resistance due to Barkhausen noise. This region ends near zero field in a noisy region, $T(H)$, characterized by a large number of abrupt decreases. As the field is increased to saturation in the negative direction, the response again follows a curve, $S(H)$, which is smooth. For comparison, an ideal response (Anderson et al., 1972) without domain effects is shown as curve B in Fig. 4.42.

Simultaneous measurements of both the magnetoresistive response and domain patterns have enabled a correlation between the two to be made (Tsang and Decker, 1981). Figure 4.43a to d shows the domain patterns of a 40-nm-thick rectangular stripe as the field is reduced from saturation. At a positive field of about 1.2 kA/m (15 Oe), ripple patterns start to appear, indicating the influence of demagnetizing fields near the pattern edges. The orientations of magnetization within the Néel walls in this case vary over a relatively small range of angles, θ_D. As the field approaches zero (Fig. 4.43), the ripple patterns become a well-defined buckling

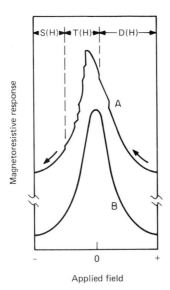

Figure 4.42 Hard-axis response of a 20 by 60 μm rectangle *(Decker and Tsang, 1980).*

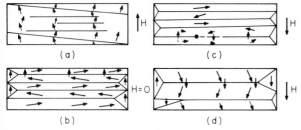

Figure 4.43   Typical domain patterns in a hard-axis cycle *(Tsang and Decker, 1981)*. (*a*) Ripples at high field; (*b*) onset of wall-state transition; (*b*) well-defined buckling domain structure at low fields; *(d)* closure pattern after N^+ N^- transitions.

domain structure with a regular array of closure domains at the ends to lower the magnetostatic energy. The range of angles, θ_D, in the longitudinal domain walls become larger, approaching 180°, with an associated increase in wall energy. With a higher wall energy, the original high density of walls is no longer favored and discontinuous changes in the domain structure are observed. Pairs of walls tend to merge abruptly and disappear. These rearrangements, plus motion of the closure domain walls, give rise to discontinuous increases ("noise") in the magnetoresistance characteristic in the otherwise quiet region, $D(H)$ (Fig. 4.42). Such noise is particularly detrimental to performance since it can occur in the portion of the characteristic used with a biased sensor. As indicated in Fig. 4.43b, the total magnetization at zero field remains at an angle with respect to the sensor's easy axis.

As the field is reversed, the domain magnetizations continue to rotate until in some cases the range of angles, θ_D, in the walls, may exceed 180°. At this point it becomes energetically favorable for the Néel wall to make a transition to a state in which the range of angles does not exceed 180°. This is called the N^+ to N^- transition (Decker and Tsang, 1980). The Néel walls break up to form long N^+ segments separated by N^- segments, which are indicated as dashed lines in Fig. 4.43. The propagation of N^- walls and the associated changes in magnetization give rise to abrupt changes in resistance which occur throughout the region $T(H)$ (Fig. 4.42) and, in some instances, at small positive values of field. After completion of the N^+ and N^- transitions, reverse saturation is finally achieved by further rotation of the magnetization and smooth expansion of the favored closure domains. This process is associated with a relatively quiet segment, $S(H)$, of the response curve shown in Fig. 4.42.

With this outline of changes with applied field it becomes clear why fields in the easy-axis direction induce a quiet magnetoresistive response. Easy-axis fields give rise to a unidirectional, rather than a uniaxial, anisotropy, inhibiting the formation of buckling domain patterns. Several means of applying these fields are described in Sec. 4.4.3. Also it is generally desirable to provide extensive electrical shorting contacts at the ends of the sensor, rendering the sensor relatively insensitive to the motion of closure domains. Finally, it is important to bias the magnetoresistive head properly so that excursions to near zero field do not induce noisy Néel wall-state transitions.

4.5 Head Applications

The wide scope of applications for magnetic recording systems has generated the need for a range of head designs covering signal bandwidths from the audio-frequency band up to very high data recording bandwidths. Head designs have also been influenced by the correspondingly large range of head-medium velocities, and by the different requirements for materials compatibility between head cores and the recording medium. The evolution of head designs is considered from the point of view of advances in head core materials. Initially, magnetic alloy laminations were used for head cores. Ferrite-head cores, with high-frequency capability, were introduced in the 1960s and have coexisted with alloy laminations. Alloy film cores have been introduced during the last decade, allowing heads to be fabricated by using processes similar to those developed in the semiconductor industry. New advances in head designs have emerged which combine high-magnetization alloy film pole tips with bulk ferrite cores. Films have also enabled the introduction of galvomagnetic reproducing head designs, using magnetoresistance or Hall effect detection, with higher sensitivities than inductive reproducing heads.

4.5.1 Laminated alloy core heads

High-magnetization magnetic alloys have been preferred for head core materials since large recording fields can be generated without limitations due to magnetic saturation effects. However, since these alloys are good conductors, it is sometimes necessary to laminate the head core structure to minimize losses due to eddy currents.

Design considerations involve not only electrical and mechanical performance but simplicity for manufacturing. Laminated cores are usually designed as ringlike structures, in which the rear gap is made as short as possible (Fig. 4.44). The front gap length is determined by the functional requirements (write, read, erase, or combination) and by the mechanical and electrical constraints on the gap depth, which should be sufficient for adequate wear life but small enough to provide the required gap reluctance. Laminations with a thickness as low as 25 μm have been used to reduce high-frequency eddy-current losses. The laminations are glued together and coils are wound on the bonded stacks. The stacks are then assembled with a nonmagnetic spacer in the front gap, and subsequently the top surface is lapped and polished to the desired contour.

Lamination heads have been used for many years for audio recording applications. Nickel-iron (Ni-Fe) alloy laminations (with additions of Mo, Cu, Cr) and nonmagnetic spacer shims (e.g., Cu-Be) satisfied the mechanical and electrical requirements. As shorter gaps were required for short-wavelength recording and reproducing, other gap-spacer materials were used which could be deposited by evaporation or sputtering. Typical examples of deposited gap materials are glass and silica. The shorter gap lengths (< 1 μm), required to resolve very short wavelengths, are inefficient in recording head designs when wideband recording is used. As a consequence, high flux densities in the pole-tip region are required to produce sufficient recording field throughout the coating thickness, leading to the possibility of pole-tip saturation. Under such conditions, alloy heads are preferred to ferrite heads because of their higher saturation magnetization, and are more suitable to record the higher coercivities of modern oxide and metal particle tapes and disks.

4.5.1.1 Application of laminated heads in digital recording. Digital tape recording heads are designed to handle high-frequency signals and operate at relatively high tape speeds. The head core efficiency is determined

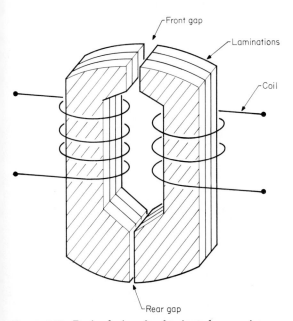

Figure 4.44 Basic design for laminated core ring head.

by the ratio of the reluctance of the flux path across the gap to the reluctance around the coil-carrying core at the signal frequency. For frequencies around 10 MHz, a lamination thickness of 8 μm or less is required to achieve permeabilities of 1000 in nickel-iron alloys. The head contour for the tape path is designed to minimize head-to-medium spacing and head wear. The airfoil between the head and tape is reduced by providing slots in the magnetic pole surface parallel to the direction of tape motion.

Multitrack laminated heads are used in high-density digital recording on tape. Reduction of cross-talk between tracks and collinearity of recording gaps are important design requirements. The design approach for multitrack laminated heads is to build the head stacks in two halves separately (Fig. 4.45). These stacks include laminations, coils, and magnetic shields between adjacent tracks. The shields are required to attenuate a range of field frequencies and amplitudes, and normally are built up with laminated structures of soft ferromagnetic sheets and good electrical conductors. Cross-talk can be reduced further by adjusting the head coil locations and the winding sense of adjacent coils. Sometimes this approach has been extended to providing bucking coils on adjacent core halves. There is a further possible cross-talk mechanism during recording which is not addressed by these changes. This effect amounts to an anhysteretic magnetization of the tape by the cross-talk signal from adjacent tracks, when the signal on the measured track is of higher frequency (e.g., all 1s)

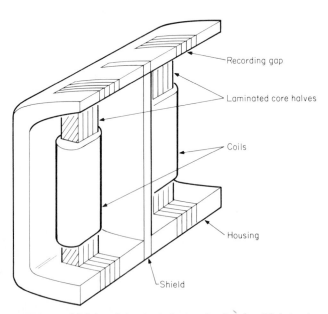

Figure 4.45 Multitrack laminated core ring heads with interelement shield.

and thereby acts as a bias. This effect can cause the recording cross-talk to be substantially higher than the playback cross-talk. Most of this cross-talk can be removed by a suitable electronic canceling circuit in the record or playback circuit. The lowest-power approach is to place the cancellation circuit in the playback channel (Tanaka, 1984).

Combined dual-element recording and reproducing heads have also found application in high-performance tape recorders. A special requirement in the case of digital tape recording (see Chap. 4 of Volume II) is the function of read checking immediately after writing. Preferably this would be done on a bit-by-bit basis, but this would require the read and write gaps to be less than one bit length apart, and still have sufficient feedthrough reduction to allow the reading of a low-level bit field immediately after writing a high-field transition. In practice, laminated read and write heads have been designed using a single C-shape lamination stack with a coil wound on it. The flux closure paths for the heads are provided by laminated stacks without coils. In this design, the read and write gaps have been placed in line along the tape path and are separated by a few millimeters. The center head section also acts as a shield to minimize feedthrough, and further reduction is sometimes obtained by using a high-permeability shield on the remote side of the tape, opposite the read and write gaps.

4.5.1.2 Applications of laminated heads in analog recording.

Probably the most challenging application for multitrack laminated head designs is the need to record more tracks in parallel, and at shorter recording wavelengths, as required for instrumentation recording (see Chap. 7 of Volume II). Since relatively high tape velocities are used, the wearing of Permalloy laminations by the tape places additional limitations on the efficiency of laminated read heads. To achieve reading efficiency at short wavelengths requires submicrometer gap lengths and reduced gap depth. Magnetic ferrite is a suitable core material which provides improved wearing capability, along with dimensional tolerance control and high-frequency performance. Ferrite-head designs have dominated the high-frequency applications, but laminated core heads have maintained large-volume usage in lower-frequency (and lower-tape-speed) applications such as audio recording.

The performance of laminated heads is limited by their frequency response and the read-head gap lengths achievable with adequate tolerance control. Furthermore, for very high track densities with multitrack heads, the tolerances on positioning stacks of laminations in mechanical housings become tighter, along with the tolerances on track widths for both the writing and the reading tracks. As an example of the state of the art, techniques have been developed for assembling multitrack heads with more than one track per millimeter (Cullum and Dorreboom, 1982).

   Numerous attempts have been made to simplify the construction of laminated recording heads, and, for very narrow tracks, the single-lamination design is eventually feasible. Such a simplified design approach is appropriate for future high-track-density video recording. On the other hand, digital data tape recording still uses a relatively wide track, and simplification of head design has taken a different course. Laminated head designs for wide-track data recording have evolved into ferrite core designs and, more recently, into dual-element thin-film designs.

   Despite the earlier trend to displace alloy laminated heads with higher-frequency ferrites, the trend to short-wavelength recording has turned attention back to high-magnetization alloys to provide sufficiently high recording fields from narrow-gap head designs (Mukasa, 1985). This is due to the increase in recording media coercivities required to support very short wavelengths. Such is the case in 8-mm video recording applications which have spurred the development of high-magnetization soft magnetic materials such as amorphous alloys. For instance, ribbons of amorphous metallic glasses [$M_s \approx 700$ to 1000 kA/m (emu/cm^3)] have been produced, by rapid quenching, with thicknesses intermediate between deposited thin films and the rolled alloy laminations described earlier. High permeabilities have been obtained in ribbons 10 to 24 μm thick, and this has led to single-lamination designs for narrow-track applications where the track width is determined by the lamination thickness (Matsuura et al., 1983). Figure 4.46 shows the design of a video recording head using a single lamination of amorphous $Fe_{4.7}Co_{70.3}Si_{15}B_{10}$. The per-

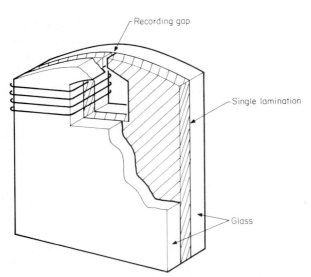

Figure 4.46   Single-lamination video recording head.

formance of this head is superior to that of a ferrite video head in its ability to record on high-coercivity media. Although superior to nickel-iron alloys, amorphous alloys are inferior to crystalline alloys such as Sendust in their resistance to media wear. This drawback has been reduced by additions of Nb and Cr which have doubled the wear resistance of amorphous alloys and improved the corrosion resistance. The main objective of the additional components in the alloys is to improve the durability and environmental stability without compromising the high magnetization; in this regard, an amorphous alloy of Co-Ni-B has been developed (Sakakima et al., 1981). Zero magnetostriction has been obtained in these alloys by further small additions of iron or manganese.

Crystalline alloys, in thin sheet form, have also been developed for use as single-lamination video heads. With somewhat lower saturation magnetizations than the amorphous alloys [that is, $M_s \approx 580$ to 750 kA/m (emu/cm^3)], the alloys of Fe-Si-Al (Sendust) are also capable of recording beyond the capability of ferrite heads on high-coercivity media. Techniques for producing Sendust by squeeze casting have been developed which allow head cores to be fabricated with little final mechanical working. Through use of this fabrication method, head wear has been substantially reduced and approaches that of ferrite heads. Since the fabrication method is inexpensive, it has been applied to audio compact-cassette heads as well as video heads. It is also observed that the noise level of Sendust heads is similar to that of Permalloy heads and lower than that of ferrite heads (Senno et al., 1977).

4.5.2 Ferrite heads

4.5.2.1 Applications. As magnetic ferrites were developed in the 1960s, the combined requirements of wide electrical bandwidth and improved head-wearing characteristics promoted a change from earlier laminated alloy cores to high-resistivity ferrite cores. In particular, the growth of instrumentation, video, and data recording applications stressed the need for durable, inexpensive heads capable of handling signals in the 10- to 100-MHz range. For the past 20 years or so, developments in ferrites and ferrite materials processing have ensured a dominant position for ferrite heads in this still growing area of applications. On the other hand, improvements in heads and media have led to substantial reductions in recording and reproducing losses, and the opportunity to record even shorter wavelengths. This trend has stressed a limitation in magnetic ferrites, that is, the inherently low saturation magnetization. This limitation has been alleviated through the use of magnetic alloy pole tips, either on the head-tape surface or on the pole-gap surface. With some attendant processing complications, such pole tips have extended the ability of ferrite core heads to record on higher-coercivity media.

The high-frequency responses required by most applications are satisfied by head designs using manganese-zinc and nickel-zinc ferrite cores. Other applications, such as high-density 8-mm video tapes, high-performance rigid-disk files, high-performance audio recorders, and some instrumentation recorders, require a combination of high fields and high frequencies which pose more difficult design challenges. Not only have these requirements prompted the development of ferrite heads with alloy pole tips, but also designs which use metal alloy films for both pole tips and cores. The evolution to thin-film head designs will be described in subsequent sections of this chapter.

4.5.2.2 Rigid-disk file applications. The application of ferrite-head designs to rigid-disk files began in the mid-1960s as data rates rose to above 2 Mb/s. Initially, ferrite was used to replace Permalloy lamination heads in a self-acting, stainless-steel slider (see Chap. 7), and subsequently was mounted in a mechanically similar ceramic material, alumina, to provide an improved head-medium interface. Next, in the early 1970s nickel-zinc ferrite sliders were developed and provided a low-cost ferrite-head design which has been used to the present day. This new slider design, Fig. 4.47a (Warner, 1974), is machined from a single block of ferrite and comprises a flat air-bearing surface with a taper on the leading edge. The magnetic element, with its coil, is mounted on the trailing edge of a center rail in the slider block. Two wider outer rails provide the bearing surface, which enables the slider to rest on a stationary disk surface and then adopt a stable flying position when the disk is spinning. Also, during the 1970s, hot-pressed manganese-zinc ferrite material was developed for head applications; it offers higher saturation magnetization, which improves the capability of the head to overwrite the previously recorded information (Kanai et al., 1973). This ferrite has been developed in a small-grain structure to allow narrow-track heads (less than 20 μm) to be produced without grain chipping occurring in the gap region. At the same time, low coercivity, high permeability, and high saturation magnetization have been maintained (Hirota et al., 1980). More recent designs of ferrite heads for rigid-disk applications have used a nonmagnetic slider with a small ferrite head inserted into the trailing edge of a two-rail slider design, Fig 4.47b. This approach has reduced the sensitivity of the head to external fields.

The taper-flat ferrite slider and recording head combination can be designed for multitrack disk file application as is required for fast-access fixed-head designs. A multitrack head (Solyst, 1971), using the slider as a common half core, was designed using batch fabrication techniques as indicated in Fig. 4.48.

As track width is reduced, the mechanical cutting and grinding techniques become too coarse and damaging to maintain sufficient control on

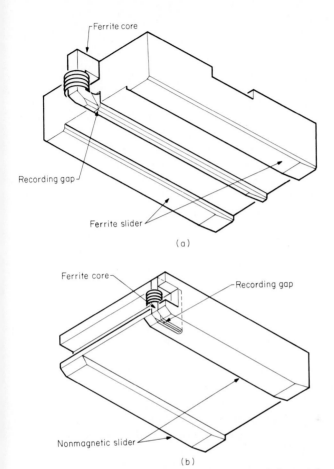

Figure 4.47 Ferrite slider with inductive write-read element for rigid-disk recording. (*a*) Ferrite slider design; (*b*) composite design with nonmagnetic slider.

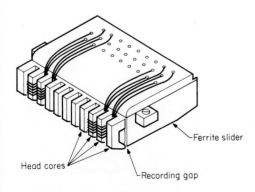

Figure 4.48 Multitrack ferrite head for rigid-disk application (*Solyst, 1971*).

tolerances. Ion-beam milling is used with some success to provide shallow but accurately defined slider rails. With multitrack batch fabrication techniques (Fig. 4.48), the slider surface in the gap region can be milled away, leaving tracks as narrow as 5 μm. Since no mechanical stresses are involved, the remaining track has unspoiled magnetic properties (Nakanishi et al., 1979). Auxiliary slots are required in the slider surface, away from the gap region, to adjust the flying height. These are produced mechanically or by laser-beam machining. Furthermore, in order to define the track width for long-wavelength recording, the head magnetic material at the side of the track is removed to a greater depth than can be achieved with ion-beam etching. In this case, further material may also be removed with the laser-beam machining tool. With the use of ion-beam etching technology, it is now possible to etch complex air-bearing surfaces, defined by photolithography. For instance, slider cavities with negative air-bearing force may be designed; these lead to improved takeoff performance when the start-stop, in-contact mode of operation is used (Susuki et al., 1981). Finally, the ion-beam etching technique produces a bevel of about 70° at the edge of the ferrite poles. This is superior to mechanical grinding of the track edges, where 60° in a 30-μm track width is achieved with difficulty. The advances in ferrite materials and in slider fabrication technology have enabled ferrite heads to remain a viable contender for use in all types of rigid-disk files.

Future advances will occur with improvements in the ferrite core material to extend the frequency range of ferrite heads. To improve the high-frequency permeability of ferrite materials having high saturation magnetization will require increases in the material resistivity. As the gap width is reduced in disk file applications, there will be a need to protect the magnetic poles with compatible nonmagnetic material in the pole-face region. Design advances in ferrite heads for rigid-disk file application generally are in the direction of miniaturized taper-flat sliders having improved frequency response and lower flying. Additional modifications are also included, such as the provision of a crowned surface to reduce the area of the slider surface in contact with the recording surface when at rest. As these surfaces become smoother, for low-flying head designs, the tendency for sticking between head and disk increases, and this is relieved by the crowned-surface design for the slider (Kaneko and Koshimoto, 1982).

4.5.2.3 Flexible-disk file applications.

Ferrite heads have been designed for application in flexible-disk drives since the early development of such files in the 1970s. As flexible-disk drive technology has evolved from 200-mm-diameter disks to those of diameter less than 100 mm, the head-disk velocity and the head output have correspondingly been reduced. Linear recording densities have also advanced, requiring gap lengths to be reduced to around 1 μm. The result is that polycrystalline ferrite core materials have sufficed to date, but these will give way to higher magnetization pole-tip materials as shorter wavelengths are required. In parallel with these advances, the development of heads for flexible-disk applications is following video recorder head evolution, and manganese-zinc ferrite single-crystal heads are being applied.

Flexible-media head designs are heavily influenced by the cost of the design. Multigap structures have evolved due to the requirement to erase previously recorded information in adjacent track regions. Although relatively low track densities are used, it is advantageous to erase previously recorded information, since reducing the interference from adjacent tracks is more important than suffering a noise increase due to reducing the track width by the erase heads. Two dc-erase heads are provided to erase the zones to the side of the read-write tracks. A typical design for a read-write ferrite head with separate trailing "tunnel erase" heads is shown in Fig. 4.49. This head element is mounted in a slider button having a spherical contour, and is held in contact with the flexible recording medium by means of a pressure pad on the opposite side of the recording medium. Later designs of flexible-disk files record on both sides of the flexible disk, and the head design combines two read-write elements and two pressure pads. The overall design has to provide sufficient separation between the head units, and between the erase and read-write elements, to avoid track-to-track cross-talk and erase-write head cross-talk. Additional design details for flexible-disk heads are given in Chap. 3 of Volume II.

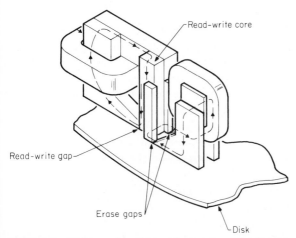

Figure 4.49 Write-read and tunnel erase head for flexible disk *(Baasch and Luecke, 1981)*.

4.5.2.4 Ferrite video heads. Over the last 20 years or so, the dramatic increases in recording density for video recording have been due, in part, to improvements in ferrite-head technology for producing narrow gaps and narrow tracks. Gap lengths have been reduced by nearly an order of magnitude from the 2-μm gap heads fabricated in the mid-1960s. Track widths have been reduced even more, from 180 to 10 μm, during the same period. The need for high-saturation magnetization ferrites to record very short wavelengths on high-coercivity media has spurred the evolution to manganese-zinc ferrite cores in either sintered polycrystalline or single-crystal form. To date, the mechanical and magnetic requirements for ferrite video recording heads have been met best by single-crystal cores, although fine-grain polycrystalline cores have also been used. Polycrystalline cores suffered from gap crumbling caused by microcracks in the cores. Hot-pressed ferrites were developed to improve the mechanical properties, but the most durable head poles have been produced in single-crystal cores. Single-crystal ferrite-head cores are lapped with abrasives that produce a surface-modified layer with deteriorated magnetic properties. This layer of reduced permeability is usually less than 1 μm after lapping and 0.05 μm after subsequent polishing. In video heads, noise is generated on playback by the rubbing of the tape against the head, and this is caused by magnetization fluctuations due to magnetostrictive effects, and by local heating due to contact friction. The noise in single-crystal ferrite heads due to tape rubbing has been minimized by suitable choice of crystal orientation. Using sputtered-film gaps and a specified core crystal orientation, single-crystal heads exhibit long-wearing characteristics and provide at least 1000 h playing time in home video cassette recorders. Under clean conditions, with unworn tapes, this lifetime can be substantially extended. Improved consistency of head performance has been obtained by using polycrystalline cores with single-crystal pole tips.

A typical design for a home video recorder head is shown in Fig. 4.50a. The single-crystal ferrite core is narrowed in the region of the read-write gap to provide the narrow track (Fig. 4.50b). An important feature of the high track density of a home video head is the tilting of the read-write gap with respect to the transverse direction of the track. As shown in Fig. 4.50b, the gap is aligned at an angle of about 7° to the transverse direction. In most helical-scan recorders, two heads are provided on the rotating drum, and they record alternate tracks. Since the gap alignment of the second head is $-7°$, there is a 14° difference in the magnetization directions between adjacent tracks; this leads to a reduction in the signal picked up by the heads from adjacent tracks, and allows the guard band between tracks to be eliminated.

The evolution of video heads for miniaturized 8-mm video tape recorders, along with the trend to improve picture resolution, has maintained the need for reductions in the track width and gap length and therefore in the core size of video heads. It has also focused attention on higher-coercivity media. At this point, the saturation magnetization limitations of ferrite pole tips have been extended through the use of high-magne-tization hard alloy pole pieces. The first approach to a solution was to use pole tips of iron-aluminum-silicon which were glued to the surface of the ferrite core material (Fig. 4.51a). For a number of years in the early development of video recording, this type of recording head was in use. Nevertheless, the pole tips exhibited shorter life than the ferrite they replaced and required replacement after a few hundred hours of operation. More recently, sputtered alloy films have been deposited on ferrite gap faces to

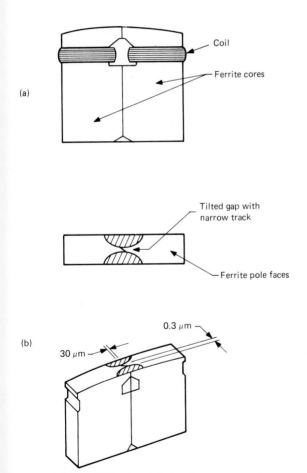

Figure 4.50 (a) Video head design for consumer tape recorder; (b) tilted recording gap design.

produce a high-magnetization recording gap region as shown in Fig. 4.51*b* (Jeffers et al., 1982). With this approach, both the pole tips and the nonmagnetic gap material are sputtered onto the gap faces. This advance has allowed the reduction of the gap length without a corresponding reduction in recording field, leading to a significant increase in the short-wavelength recording amplitude. While the deposited-film head design has the advantages of low cost and mechanical stability, it is necessary to ensure that good magnetic contact occurs between the alloy pole tips and the ferrite core. Otherwise, undesirable secondary pole effects occur if the additional gaps are of comparable magnitude to the main reproducing gap. Some of these problems can be designed out by making the alloy poles of different thicknesses, but the effect is best reduced by ensuring good magnetic contact (Ruigrok, 1984). Finally, designs using a single-alloy pole tip on one of the ferrite poles can make use of the resulting asymmetrical field pattern. If the trailing pole is alloy, the advantages are a higher recording field and a reduction of the reading interference (Ruigrok, 1984). On the other hand, if the leading-edge pole is alloy, a high field with reduced trailing-edge gradient results, and such designs have been proposed for erase heads for high-coercivity recording media (Yamashita et al., 1984).

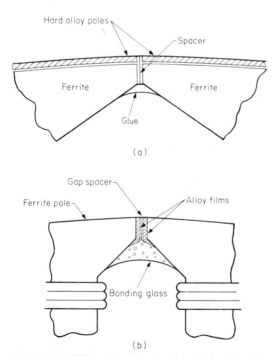

Figure 4.51   Video recording head pole designs. (*a*) Alloy pole-tipped ferrite head; (*b*) alloy film pole tips on gap faces.

4.5.2.5   Ferrite heads for audio recording.   Ferrite-head cores offer a low-cost design for audio tape recording. Unfortunately, the noise level of early polycrystalline manganese-zinc ferrite heads was substantially higher than in heads made with Permalloy laminations. As mentioned earlier, some ferrites are sensitive to the small mechanical shocks generated when a tape is run in contact with the head pole surface. Magnetostrictive coupling of these mechanical disturbances to the magnetization structure in the core is probably the source of noise in some ferrite heads (Watanabe, 1974). The noise has been reduced by optimizing both the ferrite composition and the grain size with the result that acceptable noise levels are obtained. Hot-pressed ferrite heads have been applied extensively to multitrack audio cassette designs (Nomura et al., 1973). In these designs for stereo recording and reproducing, four tracks are required with very low cross-talk between the pairs of stereo heads. Achievement of cross-talk below 45 dB was possible due to reduction of the face-to-face area of adjacent cores and by the provision of high-permeability shields between cores. Probably the greatest drawback with ferrite-head cores has been the development of reduced permeability in the surface layer after repeated playback of tapes. Furthermore, high-coercivity tapes have been introduced into audio recording applications; metal particle tapes, for instance, produce increased playback signal but require larger recording fields. All-metal heads or the combination of high-magnetization alloy pole tips and ferrite cores can produce larger recording fields and are applicable for audio recording heads.

4.5.3   Film heads

The trend to film-head designs has been driven by the desire to capitalize on semiconductor-like processing technology with the aim of reducing the customized fabrication steps for individual heads. Batch fabrication techniques have also been developed for bulk core materials with the same objective in mind. Film heads have the combined advantages of close control of pole-piece dimensions and extended frequency response.

Applications for film heads have occurred where high-precision multitrack designs are required, such as in digital audio recording and in digital data recording. As the trend continues toward high bit density and high track density, the desire for high-magnetization head cores increases, and thin-film designs become more attractive. In some cases, optimum simplicity and performance are achieved in designs which combine ferrites and thin films in the core; in particular, such combination designs have occurred for multitrack audio heads and for perpendicular recording head applications.

4.5.3.1   Film heads for rigid-disk applications.   In advancing the design of rigid-disk recording components, the requirements on the magnetic head have stressed combined improvements toward narrower track widths, reduced flying heights, wider frequency response, increased writing fields, narrower gaps for improved reading resolution, and improved side-reading response. These combined requirements have emphasized the development of film-head designs for disk file applications. Film heads can take advantage of the high-magnetization alloy poles, high-frequency response of thin films, the reduced inductance due to a small volume of magnetic core material, and the improved wavelength response due to the use of finite pole lengths.

Early film-head designs used a single-turn coil with Permalloy film cores. Two major design approaches were developed, one using films in the same plane as the recording medium, and the other using films perpendicular to it, as shown in Fig. 4.52. In parallel with these early designs, the development of air-bearing sliders, on which the head elements were deposited, was evolving toward hydrodynamic sliders which were in contact with the disk at rest, and lifted to a small spacing as the disk velocity increased to its operating level. Thus, the sliders were in rubbing contact with the disk during starting and stopping. Vulnerability to wearing of Permalloy films is greater for the horizontal film-head design (Fig. 4.52a),

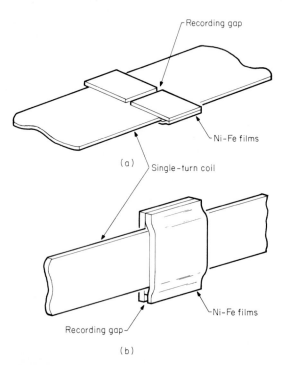

Figure 4.52 Single-turn film head. (a) Horizontal design; (b) vertical design.

and development directions have concentrated on the perpendicular film designs of Fig. 4.52b.

Because of the small coil geometry of the single-turn design, it is natural that applications to fixed-head multitrack disk files were produced first. Multitrack single-turn read-write head designs, with high positional accuracy of the track locations, were obtained by depositing the film-head structures on the trailing edge of silicon sliders (Chynoweth et al., 1973). The track density achieved at that time was 4 tracks per millimeter and a recording density of 173 fr/mm (4400 fr/in) at a flying height of 2 μm. Four heads were mounted on each disk surface. Later designs (see e.g., Taranto et al., 1978) increased the number of head elements per slider to 16 and used ceramic slider materials. Some of the early designs of single-turn read-write heads required a matching transformer for each head element.

The primary application area for recording heads for rigid-disk files has occurred in designs using a single element per slider. Here, space for the head element is less constrained than in multitrack head applications, and designs using multiple turns have predominated. Although the fabrication process is more complicated than with the single-turn design, insulated single-layer multiturn coils have been developed with many turns. An example of a multiturn head design is shown in Fig. 4.53a, and a scanning electron micrograph of a 18-turn coil is shown in Fig. 4.53b. Multilayer, multiturn coils have also evolved with increased turns (Miura et al., 1980). Some design options for multilayer coils are shown in Fig. 4.53c. The advantage of the multilayer coil lies in the reduced losses in the core due to a shorter core length, and a greater separation of the film core legs. Subsequently, as confidence was gained in the processing of film-head structures, multitrack versions were developed for fixed-head file use.

Future advances in film-head design will require further miniaturizing of the head dimensions. The frequency response of heads of the design type shown in Fig. 4.53a shows that performance degradation is small up to 50 MHz. Thus, for immediate future applications, frequency response is not a limiting factor, but film thickness could be reduced by lamination where extended responses are required. On the other hand, smaller gap lengths, as required for short-wavelength recording, will limit the writing-field magnitude. In the future, the gap will need to be increased to write the higher-coercivity media required for very short-wavelength recording. Fortunately, materials with a higher magnetization than Permalloy have been developed in film form and may be used to increase the writing field, and also the ability of the head to write over previously recorded information. For instance, sputtered amorphous films of cobalt-zirconium (10 wt % Zr) have been applied to a film head for rigid-disk application, resulting in an improvement in overwrite capability of about 14 dB (Yamada et al., 1984).

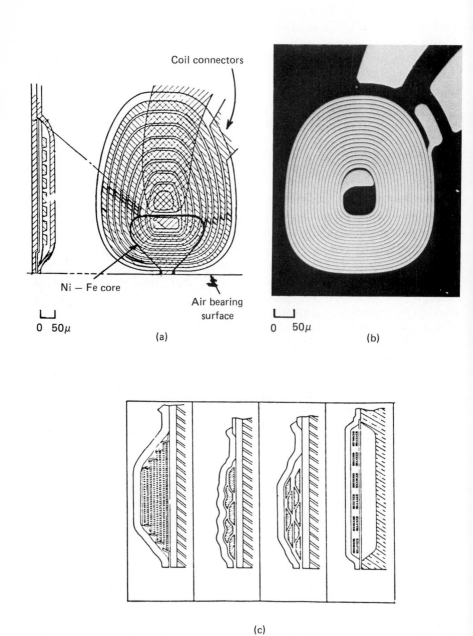

Figure 4.53 Multiturn film head. (*a*) Eight-turn design; (*b*) 18-turn coil *(Courtesy of J. Lee);* (*c*) multilayer coil designs *(after Miura , 1983)*.

Design optimization of the pole lengths for future high-density recording will favor relatively longer pole lengths in order to increase the recording field for high-coercivity media. As the pole length is increased relative to the bit length, both resolution and peak shift are reduced, and overwrite capability is improved. A compromise pole length is usually chosen which favors peak-shift reduction (Kakehi et al., 1982).

4.5.3.2 Film heads for tape applications. Many directions have been pursued for designing high-track-density heads with accurate positional control of the tracks, adequate reading and writing sensitivity, and, in some cases, separate read and write capability. A leading application for such high-track-density heads occurs in digital audio recording where track densities of about 12 tracks per millimeter have been developed in 20-track heads (Yohda et al., 1985). Designs of film heads for high-density tape recording seek to minimize recording currents to levels handled by large-scale integration write drivers, by providing multiple turns. Offsetting this is the need to place the tracks as close together as possible, and this limits the spread of the head coil in the plane of the films. Consequently, multilayer, multiturn coil structures have been developed for digital audio heads.

Another possibility for film technology in tape heads is to combine the batch fabrication of deposited coils with ferrite yokes. Separate read- and write-head assemblies can be obtained by using two coil structures on their ferrite substrates, with a common ferrite yoke to form the closure paths as already described for earlier ferrite-head designs (Brock and Shelledy, 1975). Relatively low track densities are achieved, using a single spiral coil, and multilayer coils have been designed to increase track density while maintaining sufficient turns in the read-write coil.

The use of magnetic ferrite, rather than nonmagnetic substrate materials (such as Al_2O_3-TiC), allows the substrate to become a part of the head core and thus simplifies the design of the film-head element. Ferrites have well-proven wear characteristics and are available in large-area substrates. A significant improvement in the recording efficiency of a combined film-ferrite head core can be obtained by removing the ferrite core material in the vicinity of the head turns. As shown in Fig. 4.54a and b, a semicircular groove is etched in the ferrite core and subsequently filled with glass. The deposited coil structure is close to the overlying Permalloy film, but is separated by about 50 μm from the underlying ferrite core. The improvement in head writing efficiency, resulting from the avoidance of saturation of the head core, is about a factor of 2 for the grooved ferrite design (Kanai et al., 1979).

In a further development of the grooved ferrite film-head design for digital audio recording, a high-saturation magnetization sputtered Sendust film is used to complete the head circuit. In addition, the signal-writ-

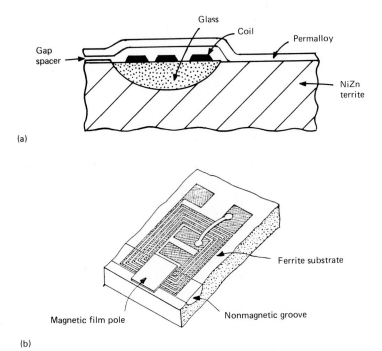

(a)

(b)

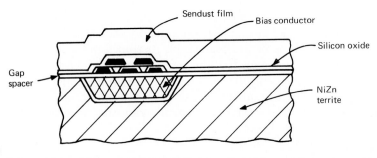

(c)

Figure 4.54 Head design with grooved ferrite pole piece. (*a*) Single-layer coil grooved substrate design; (*b*) head element design example; (*c*) two-layer coil grooved substrate design.

ing circuit is reduced by providing a bias field from a separate bias conductor in the groove (Fig. 4.54c) (Wakabayashi et al., 1982). The temperature sensitivity of the Sendust film is lower than that of Permalloy and, with bias recording, it was possible to design a 37-track head of this design without excessive heating.

4.5.3.3 Flux-sensing film-head applications. The main attribute of film magnetoresistive elements for reading recording fields is that the reading voltage is larger than that obtained in a film head with a few turns, especially for low-frequency signals. Compared with inductive film heads, the flux-sensing elements are compact, since no turns are required. Therefore these elements are becoming preferred as track widths are reduced, and are especially suited for multitrack designs. Even in single-track heads, the combination of relatively large signals and a low-impedance source (10 to 100 Ω) is an advantageous combination. Furthermore, the very thin layers used in magnetoresistive elements allow the design of a combination of an inductive writing and magnetoresistive reading head, in which the write and read elements can be physically close together. It is even possible to design a combined head in which the magnetoresistive reading element is included in the gap of the writing-head poles.

In order to achieve high resolution, the head is designed with Permalloy shields on either side of the current-carrying magnetoresistive element (Fig. 4.55a). Alternatively, the element is placed in series with two Permalloy or ferrite pole pieces (Fig. 4.55b and c). In the latter case, the high-permeability poles guide the surface flux from the recorded medium to magnetize the magnetoresistive element in a part of the head yoke remote from the medium. This design approach has the advantage of removing the thin-film sensing element from wear due to head-medium contact.

Hall element films (e.g., indium antimonide) have been applied as the sensing element instead of a Permalloy magnetoresistive element, and audio heads have been designed based on this approach (Kotera et al., 1979). For relatively low-frequency applications, such as audio playback, magnetoresistive head designs (Fig. 4.55c) have been improved through the use of magnetic feedback to the element. This is provided by a conductor, placed near the element, carrying the readout signal current, which applies a field in the opposite direction to the reading field. In this way, the noise of the magnetoresistive reading head is reduced substantially, since the magnetization of the element is held near zero and Barkhausen noise pulses are substantially avoided.

The use of magnetoresistive elements has been applied to digital audio systems. For a typical 2-Mb/s data rate, multitrack heads have been designed with a range of track and bit densities. For a 32-track recording and a bit rate of 62.5 kb/s per track, the shortest recorded wavelength is 6 and 1.5 μm for tape speeds of 190 and 47.5 mm/s, respectively. Adequate

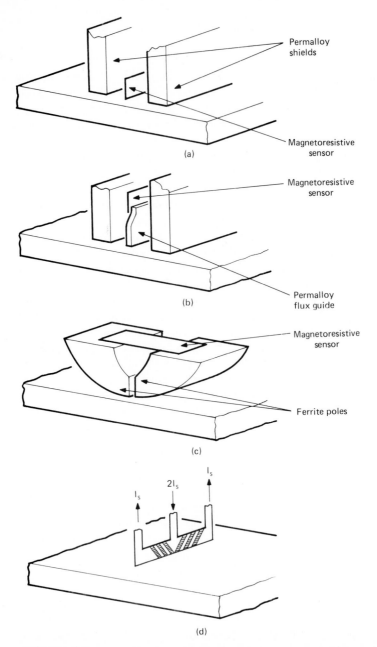

Figure 4.55 Magnetoresistive read-head designs. (*a*) Sensor with Permalloy shields; (*b*) sensor with soft-film flux guide; (*c*) sensor with ferrite poles; (*d*) barber-pole sensor.

response has been obtained from a magnetoresistive reading element with a reading gap length of 0.25 μm and a tape speed of 47.5 mm/s (Druyvesteyn et al., 1981*b*).

Multitrack magnetoresistive heads have also been applied to low-speed audio recorders and data cassettes. In one design, a symmetrical barberpole structure is used with the read current flowing in opposite directions for adjacent halves of the track, as shown in Fig. 4.55*d* (Metzdorf et al., 1982). This technique produces a linear reading characteristic across the whole track. Multitrack configurations of the element can be produced with high track density (10 tracks per millimeter) by allowing neighboring sensors to use one conductor in common. For increased reading resolution, soft magnetic film shields are deposited on each side of the sensor element. Another approach to increasing the reading resolution, without using shields, is to select the biasing level of the element to be optimum at the edge adjacent to the recording medium; spacing loss is thereby minimized (Uchida et al., 1982).

4.5.3.4 Dual-element read-write film heads. The perpendicular design of film heads (Fig. 4.53*a*) for longitudinal recording is geometrically amenable to a dual-element system in which, for instance, separate write and read heads are deposited on top of each other with an intervening spacer of a deposited insulating film. Potential functional advantages of dual-element heads are the ability to read immediately after writing, and to write with a wider track than the reading track to allow for tracking misregistrations. Furthermore, the two elements may be optimized for their specific function of writing, reading, erasing, or combinations thereof. Thus, added function is gained at the expense of head design complication and cost.

For dual-element inductive film heads, a limiting design factor is the difficulty of aligning the poles of the superimposed elements. Mask alignment accuracy is improved by decreasing the vertical distance between mask levels, at the expense of increasing the feedthrough between the two elements. Difficulties with feedthrough reduction (Chen et al., 1981) have deterred applications for dual-element inductive film heads.

On the other hand, combinations of inductive writing heads and magnetoresistive reading elements have been designed which have adequate feedthrough, and these have been applied to both data and audio recording. An experimental data head using a single-turn inductive writing element and a shielded magnetoresistive reading element has been designed in which a single pole serves the combined function of read-element shield and write-element pole as shown in Fig. 4.56 (Bajorek et al., 1974*b*). This design avoids the intervening separation layer between the read and write element, but many layers are required for this combination head design. The biased magnetoresistive element requires three layers (Fig. 4.56) that are, however, defined by a single photolithographic step. Separate gold layers are used for the write-head gap space and for the single-turn head current conductor. A simplifying alternative is to replace the substrate and first Permalloy shield with a magnetic ferrite substrate which serves both functions.

Ferrite substrates have been used as the starting point for combined multitrack dual-element heads for digital audio recording. In these designs, the inductive writing head is first deposited on the ferrite substrate. As described earlier (Fig. 4.54), a grooved ferrite substrate produces an efficient first pole on which to deposit the write-head windings and the second Permalloy pole. The magnetoresistive reading head can then be deposited on top of the writing head, with a suitable nonconducting smoothing layer interposed.

Another approach to combining a film magnetoresistive reading head and a film inductive writing head is to place the reading element in the gap of the writing-head yoke. This approach is feasible with the barberpole configuration (Fig. 4.39) for biasing since no external bias field is required. Experimental heads of this design use the perpendicular core arrangement with single or multiple write-element turns and the magnetoresistive element in the gap. Heads of this type have been applied to both tape and rigid-disk recording.

Yet another approach to a combined inductive writing head and magnetoresistive reading head places the writing turn or turns in the gap region of the head core, and the magnetoresistive element in the core circuit remote from the gap, as shown in Fig. 4.57 (Koel et al., 1981). In this

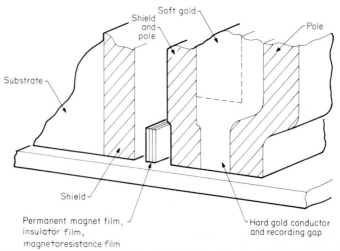

Figure 4.56 Dual-element film head: inductive write head and magnetoresistive read head *(Bajorek et al., 1974b).*

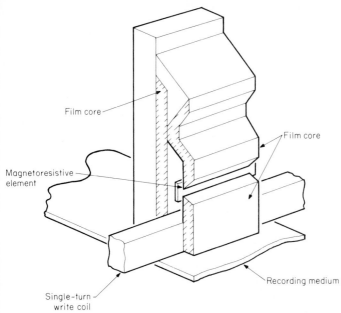

Film core

Film core

Magnetoresistive element

Recording medium

Single-turn write coil

Figure 4.57 Inductive write and magnetoresistive read dual-element head *(Koel et al., 1981)*.

design, the role of the head poles during reading is to guide the read-head flux to the magnetoresistive element. Thus, in contrast to the previous magnetoresistive reading element, the long-wavelength recording of an audio signal is read with high efficiency.

4.5.3.5 Perpendicular recording heads. Applications of perpendicular recording heads are under development for both rigid and flexible recording media. Two major design directions have emerged for single-pole, perpendicular read-write heads. In the first design (e.g., Nakamura and Iwasaki, 1982), the single main pole is driven by an auxiliary pole placed opposite to it (Fig. 4.58). In this design, the head output is enhanced by the addition of a soft magnetic nickel-iron layer in the recording medium under the storage film of cobalt-chromium. The main pole thickness determines the resolution of the head, and the amplitude of the replayed signal depends on the pole geometry and its initial permeability. Amorphous films of cobalt-zirconium or cobalt-zirconium-niobium have been developed for the main pole design, offering about 50 percent improvement both in initial permeability and in saturation magnetization compared with molybdenum Permalloy. Narrow-track recordings have been demonstrated along with very high linear densities (Iwasaki et al., 1983). As the recording pole is narrowed, the design of the pole material has to be modified to avoid the complications of relatively large closure domains.

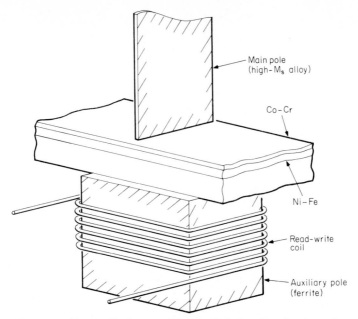

Figure 4.58 Perpendicular recording head design. Read-write pole-head design and double-layer recording medium.

Lamination of the pole and increase in the material anisotropy field both help to alleviate this problem.

An alternative design for a read-write perpendicular recording head uses a single pole without the auxiliary pole on the opposite side of the recording medium. Instead, the single main pole is attached to a W-shaped ferrite core as shown in Fig. 4.59, and the energizing coil is wound on the center leg of this core. Thus, the main pole is magnetized from its end remote to the head-medium interface, as distinct from the previous design where the pole is magnetized from the end in contact with the recording medium. In this W-shaped head design, the function of the soft Permalloy underlayer is to complete the magnetic circuit from the tip of the main pole, through the recording medium, along the soft magnetic backing layer to the outer legs of the W-shaped core, via the recording medium ahead and behind the main pole. The large area of the outer legs of the head ensures that the flux density through the recording medium is low enough not to disturb any recorded signal.

The sensitivity of the W-shaped head depends on the choice of head dimensions and film permeabilities to achieve low reluctance in the film sections of the magnetic circuit. Thus the length of the main pole is made as short as possible, consistent with avoiding recording from the surrounding ferrite poles in the center leg of the W-shaped core. Also, the reluctance of the Permalloy return path through the soft film in the recording medium has to be controlled to be as low as possible. In some designs, a soft ferrite shunt is also placed on the remote side of the recording medium. This has some beneficial effect magnetically, but destroys the ease-of-operation advantage of the single-sided head. A correctly designed W-shaped head gives a similar recording sensitivity and a 3- to 6-dB improvement in reading sensitivity compared with a typical ring-type video head (Hokkyo et al., 1984).

The W-shaped pole head has been applied to digital recording on a flexible disk. For equivalent peak shift, the linear density recorded on a perpendicular medium was increased approximately an order of magnitude over the capability of a ring head with the same medium. At the same time, the greater penetration of the recording field into the recording medium improves the overwrite capability; for example, the overwrite of a 2.6-kb/mm signal on a 1.3-kb/mm original signal is better than 30 dB (Hokkyo et al., 1984).

The idea of a single-sided head may be further simplified by using a single ferrite return path for the reading-head flux, rather than the double return path of the W-shaped core. An example of this approach is shown for a magnetoresistive element reading head for perpendicular recording in Fig. 4.60. In this design, a flux guide main pole is used, as described earlier for longitudinal recording, in order to isolate the magnetoresistive element and the recording medium. Again, the soft magnetic backing film to the perpendicularly oriented cobalt-chromium recording medium is an

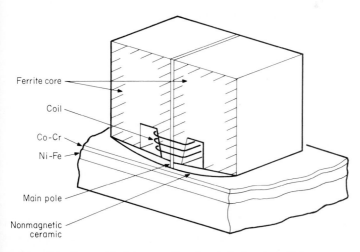

Figure 4.59 Perpendicular recording head design for single-sided operation.

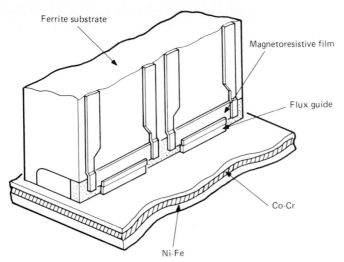

Figure 4.60 Composite pole head for perpendicular recording with magnetoresistive reproducing element. *(after Yohda et al., 1985).*

integral part of the head magnetic circuit, and therefore uniform magnetic properties in this soft film are important for low-noise head operation. Very high recording densities up to 8 kfr/mm have been achieved in this type of head design (Takahashi et al., 1983).

Finally, inductive hybrid pole heads have also been designed for perpendicular recording (Shinagawa et al., 1982). The general design is similar to the hybrid longitudinal recording head described in Fig. 4.54, but with a much larger recording gap of the order of 10 to 30 μm. This design, in combination with a dual-layer perpendicular recording medium, has been applied to rigid-disk recording systems (Tsui et al., 1985), and to digital audio tape recording as shown in Fig. 4.61 (Yohda et al., 1985). Compared with the more open magnetic path, when auxiliary pole recording without a soft magnetic backing layer is used, these designs show high efficiency recording (~0.3 to 0.4 ampere-turns) and low cross-talk between adjacent elements.

For perpendicular recording on flexible disks, low-cost components are favored, and single-layer Co-Cr recording media have been developed. Adequate writing on a single-layer medium of coercivity about 820 Oe has been achieved with a simple film ring-head design using high-saturation magnetization Co-Zr-Nb films for the head poles. The efficiency of the head is improved by depositing the film structure onto ferrite substrates that form part of the ultimate head structure, as shown in Fig. 4.62 (Okuwaki et al., 1985).

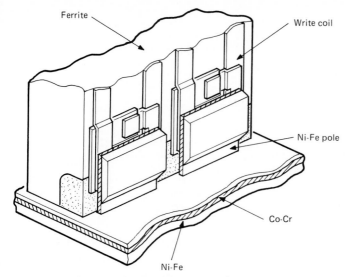

Figure 4.61 Composite pole inductive writing head for perpendicular recording *(after Yohda et al., 1985)*.

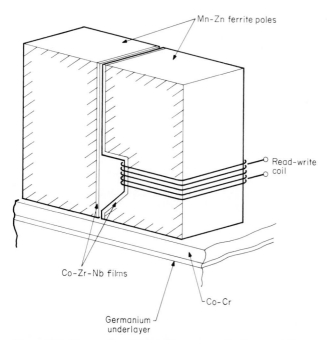

Figure 4.62 Composite ring head for perpendicular recording on flexible disks.

References

Anderson, N. C., and R. E. Jones, Jr., "Substrate Testing of Film Heads," *IEEE Trans. Magn.*, **MAG-17**, 2896 (1981).

Anderson, R. L., C. H. Bajorek, and D. A. Thompson, "Numerical Analysis of a Magnetoresistive Transducer for Magnetic Recording Applications," *AIP Conf. Proc., Magn. Magn. Mater.*, **10**, 1445 (1972).

Baasch, H. J., and F. S. Luecke, "Read-Write and Tunnel Erase Magnetic Head Assembly," U.S. Patent 4,276,574, 1981.

Baird, A. W., "An Evaluation and Approximation of the Fan Equation Describing Magnetic Fields Near Recording Heads," *IEEE Trans. Magn.*, **MAG-16**, 1350 (1980).

Bajorek, C. H., and D. A. Thompson, "Permanent Magnetic Films for Biasing of Magnetoresistive Transducers," *IEEE Trans. Magn.*, **MAG-11**, 1209 (1975).

Bajorek, C. H., C. Coker, L. T. Romankiw, and D. A. Thompson, "Hand-Held Magnetoresistive Transducer," *IBM J. Res. Dev.*, **18**, 541 (1974a).

Bajorek, C. H., S. Krongelb, L. T. Romankiw, and D. A. Thompson, "An Integrated Magnetoresistive Read, Inductive Write High Density Recording Head," *AIP Conf. Proc.*, **24**, 548 (1974b).

Bajorek, C. H., L. T. Romankiw, and D. A. Thompson, "Self-Biased Magnetoresistive Sensor," U.S. Patent 3,840,898 (1974c).

Bajorek, C. H., R. D. Hempstead, S. Krongelb, and A. F. Mayadas, "Magnetoresistive Sandwich Including Sensor Electrically Parallel with Electrical Shunt and Magnetic Biasing Layers," U.S. Patent 4,024,489, 1977.

Beaulieu, T. J., and D. A. Nepela, "Induced Bias Magnetoresistive Read Transducer," U.S. Patent 3,864,751, 1975.

Berghof, W., and H. H. Gatzen, "Sputter Deposited Thin-Film Multilayer Head," *IEEE Trans. Magn.*, **MAG-16**, 782 (1980).

Bozorth, R. M. *Ferromagnetism*, Van Nostrand, New York, 1951.

Brock, G. W., and F. B. Shelledy, "Internally Biased Magnetoresistive Magnetic Transducer," U.S. Patent 3,813,692, 1974.

Brock, G. W., and F. B. Shelledy, "Batch-Fabricated Heads from an Operational Standpoint," *IEEE Trans. Magn.*, **MAG-11**, 1218 (1975).

Brock, G. W., F. B. Shelledy, S. H. Smith, and R. F. M. Thornley, "Shielded Magnetoresistive Magnetic Transducer," U.S. Patent 3,940,797, 1976.

Calcagno, P. A., and D. A. Thompson, "Semiautomatic Permeance Tester for Thick Magnetic-Films," *Rev. Sci. Instrum.*, **46**, 904 (1975).

Chen, T. H., R. E. Jones, Jr., P. T. Chang, R. W. Cole, and K. L. Deckert, "The Coupling Effect Between the Dual Elements of a Superimposed Head," *IEEE Trans. Magn.*, **MAG-17**, 2905 (1981).

Chin, G. Y., and J. H. Wernick, "Soft Magnet Metallic Materials," *Ferromagnetic Materials*, North-Holland, Amsterdam, 1980, vol. 2.

Chow, L. G., S. K. Decker, D. J. Pocker, G. C. Pendley, and J. Papadopoulos, "Interdiffusion of Titanium Permalloy Thin Films," *IEEE Trans. Magn.*, **MAG-15**, 1833 (1979).

Chynoweth, W., J. Jordan, and W. Kayser, "A Transducer-Per-Track Recording System with Batch-Fabricated Magnetic Film Read/Write Transducers," *Honeywell Comp. J.*, **2**, 103 (1973).

Corcoran, J., and N. Pope, "Transmission Line Model for Magnetic Heads Including Complex Permeability," Pap. 4D8, *Int. Conf. Magn.*, San Francisco, 1985.

Cullity, C. D., *Introduction to Magnetic Materials*, Addison-Wesley, Reading, Mass., 1972.

Cullum, D. F., and J. Dorreboom, "High-Density Multitrack Magnetic Head," U.S. Patent 4,346,418, 1982.

Daniel, E. D., "A Flux-Sensitive Reproducing Head for Magnetic Recording Systems," *Proc. Inst. Electr. Eng. (London)*, **102**, 442 (1955).

Davies, A. V., and B. K. Middleton, "The Resolution of Vertical Magneto-Resistive Readout Heads," *IEEE Trans. Magn.*, **MAG-11**, 1689 (1975).

Decker, S., and C. Tsang, "Magnetoresistive Response of Small Permalloy Features," *IEEE Trans. Magn.*, **MAG-16**, 643 (1980).

Druyvesteyn, W. F., L. Postma, and G. Somers, "Wafer Testing of Thin Film Record and Reproduce Heads," *IEEE Trans. Magn.*, **MAG-15**, 1613 (1979).

Druyvesteyn, W. F., E. L. M. Raemaekers, R. D. J. Verhaar, J. de Wilde, J. H. J. Fluitman, and J. P. J. Groenland, "Magnetic Behavior of Narrow Track Thin-Film Heads," *J. Appl. Phys.*, **52**, 2462 (1981a).

Druyvesteyn, W. F., J. A. C. Van Ooyen, E. L. M. Raemaekers, L. Postma, J. J. M. Ruigrok, and J. de Wilde, "Magnetoresistive Heads," *IEEE Trans. Magn.*, **MAG-17**, 2884 (1981b).

Druyvesteyn, W. F., L. Postma, G. H. J. Somers, and J. de Wilde, "Thin-Film Read Head for Analog Audio Application," *IEEE Trans. Magn.*, **MAG-19**, 1748 (1983).

Elabd, L., "A Study of the Field Around Magnetic Heads of Finite Length," *IEEE Trans. Audio*, **AU-11**, 21 (1963).

Fan, G. J., "A Study of the Playback Process of a Magnetic Ring Head," *IBM J. Res. Dev.*, **5**, 321 (1961).

Feng, J. S. Y., and D. A. Thompson, "Permeability of Narrow Permalloy Stripes," *IEEE Trans. Magn.*, **MAG-13**, 1521 (1977).

Feng, J. S. Y., L. T. Romankiw, and D. A. Thompson, "Magnetic Self-Bias in the Barber Pole MR Structure," *IEEE Trans. Magn.*, **MAG-13**, 1466 (1977).

Fluitman, J. H. J., and J. P. J. Groenland, "Comparison of a Shielded 'One-Sided' Planar Hall-Transducer with an MR Head," *IEEE Trans. Magn.*, **MAG-17**, 2893 (1981).

Gorter, F. W., "Magnetoresistive Magnetic Head," U.S. Patent 4,040,113, 1977.

Gorter, F. W., J. S. L. Potgiesser, and D. L. A. Tjaden, "Magnetoresistive Reading of Information," *IEEE Trans. Magn.*, **MAG-10**, 899 (1974).

Hamilton, H. J., "Magnetic Head and Multitrack Transducer for Perpendicular Recording and Method for Fabricating," U.S. Patent 4,423,450, 1983.

Hanazono, M., K. Kawakami, S. Narishige, O. Asai, E. Kaneko, K. Okuda, K. Ono, H. Tsuchiya, and W. Hayakawa, "Fabrication of 8 Turn Multitrack Thin Film Heads," *IEEE Trans. Magn.*, **MAG-15**, 1616 (1979).

Hanazono, M., S. Narishige, K. Kawakami, N. Saito, and M. Takagi, "Fabrication of a Thin Film Head Using Polyimide Resin and Sputtered NiFe Film," *J. Appl. Phys.*, **53**, 2608 (1982).

Heim, D. E., "The Sensitivity Function for Shielded Magnetoresistive Heads by Conformal Mapping," *IEEE Trans. Magn.*, **MAG-19**, 1620 (1983).

Hempstead, R. D., "Thermally Induced Pulses in Magnetoresistive Heads," *IBM J. Res. Dev.*, **18**, 547 (1974).

Hempstead, R. D., "Analysis of Thermal Spike Cancellation," *IEEE Trans. Magn.*, **MAG-11**, 1224 (1975).

Hempstead, R. D., S. Krongelb, and D. A. Thompson, "Unidirectional Anisotropy in Nickel-Iron Films by Exchange Coupling with Antiferromagnetic Films," *IEEE Trans. Magn.*, **MAG-14**, 521 (1978).

Hirota, E., K. Hirota, and K. Kugimiya, "Recent Development of Ferrite Heads and Their Materials," *Proc. Int. Conf. Ferrites*, 667 (1980).

Hokkyo, J., K. Hayakawa, I. Saito, and K. Shirane, "A New W-Shaped Single-Pole Head and a High Density Flexible Disk Perpendicular Magnetic Recording System," *IEEE Trans. Magn.*, **MAG-20**, 72 (1984).

Hughes, G. F., "Thin-Film Recording Head Efficiency and Noise," *J. Appl. Phys.*, **54**, 4168 (1983).

Hughes, G. F., and D. S. Bloomberg, "Recording Head Side Read/Write Effects," *IEEE Trans. Mag.*, **MAG-13**, 1457 (1977).

Hunt, R., "A Magnetoresistive Readout Transducer," *IEEE Trans. Magn.*, **MAG-7**, 150 (1971).

Ichiyama, Y., "Reproducing Characteristics of Thin Film Heads," *IEEE Trans. Magn.*, **MAG-11**, 1203 (1975).

Ichiyama, Y., "Analytic Expressions for the Side Fringe Field of Narrow Track Heads," *IEEE Trans. Magn.*, **MAG-13**, 1688 (1977).

Iwasaki, S., "Perpendicular Magnetic Recording," *IEEE Trans. Magn.*, **MAG-16**, 71 (1980).

Iwasaki, S., and Y. Nakamura, "An Analysis for the Magnetization Mode for High Density Magnetic Recording," *IEEE Trans. Magn.*, **MAG-13**, 1272 (1977).

Iwasaki, S., Y. Nakamura, and K. Ouchi, "Perpendicular Magnetic Recording with a Composite Anisotropy Film," *IEEE Trans. Magn.*, **MAG-15,** 1456 (1979).

Iwasaki, S., Y. Nakamura, and H. Muraoka, "Wavelength Response of Perpendicular Magnetic Recording," *IEEE Trans. Magn.*, **MAG-17,** 2535 (1981).

Iwasaki, S., Y. Nakamura, S. Yamamoto, and K. Yamakawa, "Perpendicular Recording by a Narrow Track Single Pole Head," *IEEE Trans. Magn.*, **MAG-19,** 1713 (1983).

Jeffers, F. W., "Magnetoresistive Transducers with Canted Easy Axis," *IEEE Trans. Magn.*, **MAG-15,** 1628 (1979).

Jeffers, F., and H. Karsh, "Unshielded Magnetoresistive Heads in Very High-Density Recording," *IEEE Trans. Magn.*, **MAG-20,** 703 (1984).

Jeffers, F. J., R. J. McClure, W. W. French, and N. J. Griffith, "Metal-In-Gap Record Head," *IEEE Trans. Magn.*, **MAG-18,** 1146 (1982).

Jeffers, F., J. Freeman, R. Toussaint, N. Smith, D. Wachenschwanz, S. Shtrikman, and W. Doyle, "Soft-Adjacent-Layer Self-Biased Magnetoresistive Heads in High Density Recording," *IEEE Trans. Magn.*, **MAG-21,** 1563 (1985).

Jones, R. E., Jr., "Analysis of the Efficiency and Inductance of Multiturn Thin Film Magnetic Recording Heads," *IEEE Trans. Magn.*, **MAG-14,** 509 (1978).

Jones, R. E., Jr., "Domain Effects in the Thin Film Head," *IEEE Trans. Magn.*, **MAG-15,** 1619 (1979).

Jones, R. E., Jr., "IBM 3370 Film Head Design and Fabrication," *IBM Disk Storage Technol.*, **GA 26-1665-0,** 6 (1980).

Jorgensen, F., *The Complete Handbook of Magnetic Recording,* TAB Books, Blue Ridge Summit, Penn., 1980.

Kakehi, A., M. Oshiki, T. Aikawa, M. Sasaki, and T. Kozai, "A Thin Film Head for High Density Recording," *IEEE Trans. Magn.*, **MAG-18,** 1131 (1982).

Kaminaka, N., N. Nomura, K. Kanai, and E. Hirota, "Thin Film Disk Heads" (Japanese), *IECE Tech. Group Mtg. Magn. Rec. Jpn.*, MR80-24 (1980).

Kanai, K., R. Sasaki, and H. Sugaya, "Type 3330 Flying Head Using HPF," *Natl. Tech. Rep.*, **19,** 578 (1973).

Kanai, K., N. Naminaka, N. Nouchi, N. Nomura, and E. Hirota, "High Track Density Thin-Film Tape Heads," *IEEE Trans. Magn.*, **MAG-15,** 1130 (1979).

Kaneko, R., and Y. Koshimoto, "Technology in Compact and High Recording Density Disk Storage," *IEEE Trans. Magn.*, **MAG-18,** 1221 (1982).

Karlqvist, O., "Calculation of the Magnetic Field in the Ferromagnetic Layer of a Magnetic Drum," *Trans. R. Inst. Technol. (Stockholm),* 1 (1954).

Kaske, A. D., P. E. Oberg, M. C. Paul, and G. F. Sauter, "Vapor-Deposited Thin-Film Recording Heads," *IEEE Trans. Magn.*, **MAG-7,** 675 (1971).

Katz, E. R., "Finite Element Analysis of the Vertical Multi-turn Thin-Film Head," *IEEE Trans. Magn.*, **MAG-14,** 506 (1978).

Kelley, G. V., and R. A. Ketcham, "An Analysis of the Effect of Shield Length on the Performance of Magnetoresistive Heads," *IEEE Trans. Magn.*, **MAG-14,** 515 (1978).

Kelley, G. V., and E. P. Valstyn, "Numerical Analysis of Writing and Reading with Multiturn Film Heads," *IEEE Trans. Magn.*, **MAG-16,** 788 (1980).

Kelley, G. V., J. Freeman, H. Copenhaver, R. A. Ketcham, and E. P. Valstyn, "High-Track-Density, Coupled-Film Magnetoresistive Head," *IEEE Trans. Magn.*, **MAG-17,** 2890 (1981).

Koel, G. J., F. W. Gorter, and J. T. Gerkema, "Thin Film Magnetic Head for Reading and Writing Information," U.S. Patent 4,300,177, 1981.

Kotera, N., J. Shigeta, K. Naria, T. Oi, K. Hayashi, and K. Sato, "A Low-Noise InSb Thin Film Hall Element: Fabrication, Device Modeling, and Audio Application," *IEEE Trans Magn.*, **MAG-15,** 1946 (1979).

Kuijk, K. E., "Magnetoresistive Magnetic Head," U.S. Patent 4,052,748, 1977.

Kuijk, K. E., W. J. van Gestel, and F. W. Gorter, "The Barber Pole: A Linear Magnetoresistive Head," *IEEE Trans. Magn.*, **MAG-11,** 1215 (1975).

Lazzari, J. P., "Integrated Head Concepts," *IEEE Trans. Magn.*, **MAG-14,** 503 (1978).

Lazzari, J. P., and I. Melnick, "Integrated Magnetic Recording Heads," *IEEE Trans. Magn.*, **MAG-7,** 146 (1971).

Lindholm, D. A., "Magnetic Fields of Finite Track Width Heads," *IEEE Trans. Magn.*, **MAG-13,** 1460 (1977).

Lindholm, D. A., "Long-Wavelength Response of Magnetic Heads with Beveled Outer Edges," *J. Audio Eng. Soc.*, **27**, 542 (1979).

Lindholm, D. A. "Effect of Track Width and Side Shields on the Long Wavelength Response of Rectangular Magnetic Heads," *IEEE Trans. Magn.*, **MAG-16**, 430 (1980).

Lindholm, D. A., "Application of Higher Order Boundary Integral Equations to Two-Dimensional Magnetic Head Problems," *IEEE Trans. Magn.*, **MAG-17**, 2445 (1981).

Dimensional Magnetic Head Problems," *IEEE Trans. Magn.*, **MAG-17**, 2445 (1981).

Luborksy, F. E., "Amorphous Ferromagnets," *Ferromagnetic Materials*, North-Holland, Amsterdam, 1980, vol. 1.

Luitjens, S. B., and A. van Herk, "A Discussion on the Crosstalk in Longitudinal and Perpendicular Recording,"*IEEE Trans. Magn.*, **MAG-18**, 1804 (1982).

Maissel, L. I., and R. Glang, *Handbook of Thin Film Technology*, McGraw-Hill, New York, 1970.

Mallinson, J. C., "On the Properties of Two-Dimensional Dipoles and Magnetized Bodies," *IEEE Trans. Magn.*, **MAG-17**, 2453 (1981).

Mallinson, J. C., and N. H. Bertram, "A Theoretical and Experimental Comparison of the Longitudinal and Vertical Modes of Magnetic Recording," *IEEE Trans. Magn.*, **MAG-20**, 461 (1984).

Matsuura, K., K. Oyamada, and T. Yazaki, "Amorphous Video Head for High Coercive Tape," *IEEE Trans. Magn.*, **MAG-19**, 1623 (1983).

Mee, P. B., "SEM Observations of Domain Configurations in Thin Film Head Pole Structures," *J. Appl. Phys.*, **51**, 861 (1980).

Metzdorf, W., M. Boehner, and H. Haudek, "The Design of Magnetoresistive Multitrack Read Heads for Magnetic Tapes," *IEEE Trans. Magn.*, **MAG-18**, 763 (1982).

Middelhoek, S., "Ferromagnetic Domains in Thin NiFe Films," Thesis, University of Amsterdam, 1961.

Middleton, B. K. "Modelling the Digital Magnetic Recording Behavior of Shielded Magneto-Resistive Replay Heads with Displaced Elements," *Radio Electron. Eng.*, **50**, 419 (1980).

Middleton, B. K., A. K. Davies, and D. J. Sansom, "The Modelling of Shielded Magneto-Resistive Replay Head Performance," *Conf. Video Data Recording, IERE Conf. Proc.*, **43**, 353 (1979).

Minuhin, V. B., "Comparison of Sensitivity Functions for Ideal Probe and Ring-Type Heads," *IEEE Trans. Magn.*, **MAG-20**, 488 (1984).

Miura, Y., A. Kawakami, and S. Sakai, "An Analysis of the Write Performance on Thin Film Head," *IEEE Trans. Magn.*, **MAG-14**, 512 (1978).

Miura, Y., Y. Takahaski, F. Kume, J. Toda, S. Tsutsumi, and S. Kawakami, "Fabrication of Multi-Turn Thin Film Head," *IEEE Trans. Magn.*, **MAG-16**, 779 (1980).

Miura, Y., "Thin Film Heads," *JARECT, Recent Magn., Electr.*, **MAG-10**, 77 (1983).

Miyazaki, T., R. Sawada, and Y. Ishijima, "New Magnetic Alloys for Magnetic Recording Heads," *IEEE Trans. Magn.*, **MAG-8**, 501 (1972).

Mukasa, K., "Recent Magnetic Recording Technology II—Magnetic Heads," *J. Inst. Telev. Eng-Japan*, **39**, 295 (1985).

Nakamura, Y., and S. Iwasaki, "Reproducing Characteristics of Perpendicular Magnetic Head," *IEEE Trans. Magn.*, **MAG-18**, 1167 (1982).

Nakamura Y., K. Yamakawa, and S. Iwasaki, "Analysis of Domain Structure of Single Pole Perpendicular Head," *IEEE Trans. Magn.*, **MAG-21**, 1578 (1985).

Nakanishi, T., T. Toshima, K. Yanagisawa, and N. Tsuzuki, "Narrow Track Magnetic Head Fabricated by Ion-Etching Method," *IEEE Trans. Magn.*, **MAG-15**, 1060 (1979).

Narishigi, S., M. Hanazono, M. Tokagi, and S. Kuwatsuka, "Measurements of the Magnetic Characteristics of Thin Film Heads Using Magneto-Optic Method," *IEEE Trans. Magn.*, **MAG-20**, 848 (1984).

Narishigi, S., T. Imagawa, and M. Hanazono, "Thin Film Heads for Hard Disk Drive—Magnetic Domain Structure of Magnetic Core" (Japanese), *39th Conf. Magn. Soc. Jpn.*, **39-5**, 1 (1985).

Nepela, D. A., and R. I. Potter, "Head Assembly for Recording and Reading, Employing Inductive and Magnetoresistive Elements," U.S. Patent 3,877,945, 1975.

Nomura, T., "A New Magneto-Optic Readout Head Using Bi-Substituted Magnetic Garnet Film," *IEEE Trans. Magn.,* **MAG-21,** 1545 (1985).

Nomura, Y., T. Tanaka, H. Chiba, and E. Hirota, "Development of 4-Track 4-Channel Cassette Head Made of Hot-Pressed Ferrite," *Electron. Commun. Jpn.,* **56-C,** 86 (1973).

O'Connor, D. J., F. B. Shelledy, and D. E. Heim, "Mathematical Model of a Magneto-resistive Read Head for a Magnetic Tape Drive," *IEEE Trans. Magn.,* **MAG-21,** 1560 (1985).

O'Day, R. L., "Balanced Magnetic Head," *IBM Tech. Discl. Bull.,* **15**(9), 2680 (1973).

O'Day, R. L., and F. B. Shelledy, "Internally Biased Magnetoresistive Magnetic Transducer," U.S. Patent 3,814,863, 1974.

Ohashi, K., "Mechanism of 90° Wall Motion in Thin Film Heads," *IEEE Trans. Magn.,* **MAG-21,** 1581 (1985).

Ohmori, K., K. Arai, and N. Tsuya, "Ribbon-Form Sendust Alloy Made by Rapid Quenching," *Appl. Phys.,* **21,** 335 (1980).

Okuwaki, T., F. Kugiya, N. Kumasaka, K. Yoshida, N. Tsumita, and T. Tamura, "5.25 Inch Floppy Disk Drive Using Perpendicular Magnetic Recording," *IEEE Trans. Magn.,* **MAG-21,** 1365 (1985).

Ozawa, K., H. Wakasugi, and K. Tanaka, "Friction and Wear of Magnetic Heads and Amorphous Metal Sliding Against Magnetic Tapes," *IEEE Trans. Magn.,* **MAG-20,** 425 (1984).

Paton, A., "Analysis of the Efficiency of Thin-Film Magnetic Recording Heads," *J. Appl. Phys.,* **42,** 5868 (1971).

Potter, R. I., "Digital Magnetic Recording Theory," *IEEE Trans. Magn.,* **MAG-10,** 502 (1974).

Potter, R. I., "Analytic Expressions for the Fringe Field of Finite Pole-Tip Length Recording Heads," *IEEE Trans. Magn.,* **MAG-11,** 80 (1975).

Potter, R. I., and I. A. Beardsley, "Self-Consistent Computer Calculations for Perpendicular Magnetic Recording," *IEEE Trans. Magn.,* **MAG-16,** 967 (1980).

Potter, R. I., R. J. Schmulian, and K. Hartman, "Self Consistently Computed Magnetization Patterns in Thin Magnetic Recording Media," *IEEE Trans. Magn.,* **MAG-7,** 689 (1971).

Potzlberger, H. W., "Magnetron Sputtering of Permalloy for Thin Film Heads," *IEEE Trans. Magn.,* **MAG-20,** 851 (1984).

Re, M. E., and M. H. Kryder, "Magneto-Optic Investigation of Thin-Film Recording Heads," *J. Appl. Phys.,* **55,** 2245 (1984).

Romankiw, L. T., and P. Simon, "Batch Fabrication of Thin Film Magnetic Recording Heads: A Literature Review and Process Description for Vertical Single Turn Heads," *IEEE Trans. Magn.,* **MAG-11,** 50 (1975).

Ruigrok, J. M., "Analysis of Metal-In-Gap Heads," *IEEE Trans. Magn.,* **MAG-29,** 872 (1984).

Sakakima, H., Y. Yanagiuchi, M. Satomi, H. Senno, and E. Hirota, "Improvement in Amorphous Magnetic Alloys for Magnetic Head Core," *Proc. Conf. Rapidly Quenched Metals,* Sendai, Japan, 1981, p. 941.

Sansom, D. J., "Recording Head Design Calculations," *IEEE Trans. Magn.,* **MAG-12,** 230 (1976).

Sawada, T., and K. Yoneda, "AC Erase Head for Cassette Recorder," *IEEE Trans. Magn.,* **MAG-21,** 2104 (1985).

Schwartz, T. A., and S. K. Decker, "Comparison of Calculated and Actual Density Responses of a Magnetoresistive Head," *IEEE Trans. Magn.,* **MAG-16,** 1622 (1979).

Senno, H., Y. Yanagiuchi, M. Satomi, E. Hirota, and S. Hayakawa, "Newly Developed Fe-Si-Al Alloy Heads by Squeeze Casting," *IEEE Trans. Magn.,* **MAG-13,** 1475 (1977).

Shelledy, F. B., and G. W. Brock, "A Linear Self-Biased Magnetoresistive Head," *IEEE Trans. Magn.,* **MAG-11,** 1206 (1975).

Shiiki, K., S. Otomo, and M. Kudo, "Magnetic Properties, Aging Effects and Application Potential for Magnetic-Heads of Co-Fe-Si-B Amorphous Alloys," *J. Appl. Phys.,* **52,** 2483 (1981).

Shiiki, K., Y. Shinoshi, K. Shinagawa, N. Numasaka, and M. Kudo, "Probe Type Thin Film Head for Perpendicular Magnetic Recording," *IEEE Trans. Magn.*, **MAG-20,** 839 (1984).

Shinagawa, K., H. Fujiwara, F. Kugiya, T. Okuwaki, and M. Kudo, "Simulation of Perpendicular Recording on Co-Cr Media with a Thin Permalloy Film–Ferrite Composite Head," *J. Appl. Phys.*, **53,** 2585 (1982).

Slick, P. I., "Ferrites for Non-Microwave Applications," *Ferromagnetic Materials,* North-Holland, Amsterdam, 1980, vol. 2.

Solyst, E., "Multichannel Magnetic Head with Common Leg," U.S. Patent 3,579,214, 1971.

Soohoo, R. F., "Switching Dynamics in a Thin-Film Recording Head," *IEEE Trans. Magn.*, **MAG-18,** 1128 (1982).

Steinback, M., J. A. Gerber, and T. J. Szczech, "Exact Solution for the Field of a Perpendicular Head," *IEEE Trans. Magn.*, **MAG-17,** 3117 (1981).

Sugaya, H., "Newly Developed Hot-Pressed Ferrite Head," *IEEE Trans. Magn.*, **MAG-4,** 295 (1968).

Suzuki, S., J. Toriu, C. Fukao, and H. Oda, "High Density Magnetic Recording Heads for Disk," *IEEE Trans. Magn.*, **MAG-17,** 2899 (1981).

Szczech, T. J. "Analytic Expressions for Field Components of Nonsymmetrical Finite Pole Tip Length Magnetic Head Based on Measurements on Large-Scale Models," *IEEE Trans. Magn.*, **MAG-15,** 1319 (1979).

Szczech, T. J., and K. E. Palmquist, "A 3-D Comparison of the Fields from Six Basic Head Configurations," *Video and Data Recording Conference,* Southhampton, England, 1984.

Szczech, T. J., M. Steinback, and M. Jodeit, "Equations for the Field Components of a Perpendicular Magnetic Head," *IEEE Trans. Magn.*, **MAG-18,** 229 (1982).

Szczech, T. J., D. M. Perry, and K. E. Palmquist, "Improved Field Equations for Ring Heads," *IEEE Trans. Magn.*, **MAG-19,** 1740 (1983).

Takahashi, K., S. Sasaki, K. Kanai, F. Kobayashi, "A Method of Reproduction in Perpendicular Magnetic Recording," *Inst. Electr. Commun. Eng., Jpn.*, 199 (1983).

Takama, E., and M. Ito, "New Mn-Zn Ferrite Fabricated by Hot Isostatic Pressing," *IEEE Trans. Magn.*, **MAG-15,** 1858 (1979).

Tanaka, Y., "Multitrack Magnetic Head for a Tape Player," U.S. Patent 4,322,764, 1982.

Taranto, J., R. Stromsta, and R. Weir, "Application of Thin Film Head Technology to a High Performance Head/Track Disk File," *IEEE Trans. Magn.*, **MAG-14,** 188 (1978).

Thompson, D. A., "Magnetoresistive Transducers in High-Density Magnetic Recording," *AIP Conf. Proc., Magn. Magn. Mater.*, **24,** 528 (1974).

Toshima, T., T. Nakanishi, and K. Yanagisawa, "Magnetic Head Fabricated by Improved Ion Etching Method," *IEEE Trans. Magn.*, **MAG-15,** 1637 (1979).

Tsang, C., "Magnetics of Small Magnetoresistive Sensors," *J. Appl. Phys.*, **55,** 2226 (1984).

Tsang, C., and S. Decker, "The Origins of Barkhausen Noise in Small Permalloy Magnetoresistive Sensors," *J. Appl. Phys.*, **52,** 2465 (1981).

Tsang, C., and S. Decker, "Study of Domain Formation in Small Permalloy Magnetoresistive Elements," *J. Appl. Phys.*, **53,** 2602 (1982).

Tsang, C., and R. E. Fontana, "Fabrication and Wafer Testing of Barber-Pole and Exchange-Biased Narrow-Track MR Sensors," *IEEE Trans. Magn.*, **MAG-18,** 1149 (1982).

Tsang, C., and K. Lee, "Temperature Dependence of Unidirectional Anisotropy Effects in the Permalloy-FeMn Systems," *J. Appl. Phys.*, **53,** 2605 (1982).

Tsang, C., N. Heiman, and K. Lee, "Exchange Induced Unidirectional Anisotropy at FeMn-$Ni_{80}F_{20}$ Interfaces," *J. Appl. Phys.*, **52,** 2471 (1981).

Tsui, R., H. Hamilton, R. Anderson, C. Baldwin, and P. Simon, "Perpendicular Recording Performance of Thin Film Probe Heads and Double-Layer Co-Cr Media," *Dig. Intermag. Conf.*, GA4 (1985).

Tsuya, N., T. Tsukagoshi, K. Arai, K. Ogasawara, K. Ohmori, and S. Yosuda, "Magnetic Recording Head Using Ribbon-Sendust," *IEEE Trans. Magn.*, **MAG-17,** 3111 (1981).

Uchida, H., S. Imakoshi, Y. Soda, T. Sekiya, and H. Takino, "A Non-Shielded MR Head with Improved Resolution," *IEEE Trans. Magn.*, **MAG-18,** 1152 (1982).

Unger, E., and K. Fritzsch, "Calculation of the Stray Reluctance of Gaps in Magnetic Circuits," *J. Audio Eng. Soc.*, **18,** 641 (1970).

Valstyn, E. P., and L. F. Shew, "Performance of Single-Turn Film Heads," *IEEE Trans. Magn.*, **MAG-9,** 317 (1973).

van Herk, A., "Side Fringing Fields and Write and Read Crosstalk of Narrow Magnetic Recording Heads," *IEEE Trans. Magn.*, **MAG-13,** 1021 (1977).

van Herk, A., "Three-Dimensional Analysis of Magnetic Fields in Recording-Head Configurations," Thesis, Delft University of Technology, the Netherlands, 1980a.

van Herk, A., "Three-dimensional Computation of the Field of Magnetic Recording Heads," *IEEE Trans. Magn.*, **MAG-16,** 890 (1980b).

van Lier, J. C., G. J. Koel, W. J van Gestel, L. Postma, J. T. Gerkema, F. W. Gorter, and W. F. Druyvesteyn, "Combined Thin-Film Magnetoresistive Read, Inductive Write Heads," *IEEE Trans. Magn.*, **MAG-12,** 716 (1976).

Voegeli, O., "Magnetoresistive Read Head Assembly Having Matched Elements for Common Mode Rejection," U.S. Patent 3,860,965, 1975.

Wakabayashi, N., I. Abe, and H. Miyairi, "A Thin Film Multi-Track Recording Head," *IEEE Trans. Magn.*, **MAG-18,** 1140 (1982).

Walker, P. A., "A Systems Overview of Transducers," *Ann. N.Y. Acad. Sci.*, **189,** 144 (1972).

Warner, M. W., "Flying Magnetic Transducer Assembly Having Three Rails," U.S. Patent 3,823,416, 1974.

Watanabe, H., "Noise Analysis of Ferrite Head in Audio Tape Recording," *IEEE Trans. Magn.*, **MAG-10,** 903 (1974).

Wells, O. C., and R. J. Savoy, "Magnetic Domains in Thin-Film Recording Heads as Observed in the SEM by a Lock-In Technique," *IEEE Trans. Magn.*, **MAG-17,** 1253 (1981).

Westmijze, W. K., "Studies on Magnetic Recording," *Philips Res. Rep.*, **8,** 161 (1953).

Yamada, K., T. Maruyama, H. Tanaka, H. Kaneko, I. Kagaya, and S. Ito, "A Thin Film Head for High Density Magnetic Recording Using CoZr Amorphous Films," *J. Appl. Phys.*, **55,** 2235 (1984).

Yamashita, K., G. Takeuchi, K. Iwabuchi, Y. Kubota, and H. Miyairi, "Sendust Sputtered Flying Erase Head," *IEEE Trans. Magn.*, **MAG-20,** 869 (1984).

Yeh, N. H. "Analysis of Thin Film Heads with a Generalized Transmission Line Model," *IEEE Trans. Magn.*, **MAG-18,** 233 (1982a).

Yeh, N. H., "Asymmetric Crosstalk of Magnetoresistive Head," *IEEE Trans. Magn.*, **MAG-18,** 1155 (1982b).

Yohda, H., K. Takahashi, S. Sasaki, and N. Kaminaka, "Multi-Channel Thin Film Head for Perpendicular Magnetic Recording," *Natl. Tech. Rep.*, **31,** 136 (1985).

Recording Limitations

John C. Mallinson

Center for Magnetic Recording Research
University of California, San Diego
La Jolla, California

5.1 Distinction between Noise, Interference, and Distortion

The output signal of a magnetic recorder is not perfect but is accompanied to some extent by three faults: noise, interference, and distortion. To understand the limitations of a recording channel, it is essential to distinguish clearly between these phenomena.

Noise is due to an uncertainty in some phenomenon and is, therefore, treated statistically. Thus magnetic medium noise arises from a lack of knowledge of, or randomness in, the positions or directions of magnetization of the individual particles. Head and electronics noises arise from fluctuations of magnetic domain walls or electric charge carriers. Medium noise is fixed spatially, and, consequently, given a precision transport, each replay of a recorded medium produces exactly the same noise waveform. The head-medium motion transforms the spatial randomness into a pseudo-temporal effect. Head and electronics noises are, on the other hand, purely temporal; these noises never repeat exactly.

Interference is frequently confused with noise even though it is of completely different origin. Interference is due to the reception or reproduction of signals other than those intended. Several examples are discussed in this chapter: crosstalk, incomplete erasure, feedthrough, misregistration, sidefringing, and print-through. Interference is completely deterministic and may, in principle, be calculated exactly.

Distortion is also frequently confused with noise although it, too, is deterministic. Recording channels and systems show distortion because they are not ideal linear channels. There are two ways that the inevitable distortion is minimized in recording systems. For a linear, ac-biased recording system, such as that shown in Fig. 5.1, ac bias makes the magnetic recording channel itself almost linear and distortionless for small signals. In a nonlinear, unbiased recording system, such as that shown in Fig. 5.2, modulation schemes, for example, frequency or pulse-code (digital) modulation, are used to make the system almost distortionless for particular signals. In unbiased recording the channel is highly nonlinear, but is required to preserve exactly only a specific feature of the recorded waveform, zero crossings, for example.

Noise, interference, and distortion all combine to render the output signal an imperfect replica of the input signal. It is for this reason that the phenomena are often confused. The task of the recording system designer is to reduce these faults to tolerable levels so that some predetermined criteria of performance are met. Common criteria are signal-to-noise ratio (SNR), third harmonic distortion (THD), and bit error rate (BER). In this context, it is important to realize that there are no truly digital or binary devices; at the device level, the performance is always governed by a number of analog considerations. These considerations are virtually indistinguishable in, for example, a frequency-modulated analog video cassette recorder and a run-length coded digital disk recorder.

In linear recorders, the input signal current and ac bias are applied directly to the write head. The magnitudes of the noise, interference, and

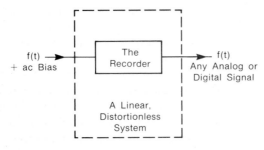

Figure 5.1 A linear, distortionless, ac-biased recording system.

distortion are controlled solely by the parameters of the recording channel. In nonlinear recorders using modulation schemes, the effects of the noise, interference, and distortion in the recording channel are additionally controlled by the parameters of the modulation scheme used. In all modulation schemes, a trade-off is made between bandwidth and immunity to channel faults. By recording a greater bandwidth than that of the input signal, the signal-to-noise ratio may be increased. In professional frequency-modulated video recorders, for example, almost three times the video bandwidth is recorded on tape. To reduce the quantizing errors inherent in analog-to-binary digital conversion, digital recorders employ the greatest increase in bandwidth. Thus the 16-bit resolution of a digital audio recorder requires approximately a 16-fold increase in bandwidth.

In analog-modulation schemes, the system signal-to-noise ratio bears a direct relationship to the medium noise. With digital modulation, however, the cumulative effect of noise, interference, and distortion leads only to digit errors. In a binary digital system, corruption of the output signal can only confuse 1s with 0s. The limited lexicon (two states only) of binary systems greatly facilitates the implementation of error detection and correction (EDAC) coding. Since error detection and correction is a purely logical or mathematical procedure, it is possible, in principle, to achieve perfect correction of bit errors. Thus, apart from quantizing errors, digital recording systems can be made to operate virtually perfectly, notwithstanding the recording channel faults.

In Sec. 5.5, a forecast is made for the future of magnetic recording. It seems that binary digital recording will be used increasingly because of its ability to achieve near-perfect operation. The increasing ease with which error correction can be implemented will lead systems designers to considerably higher areal densities with concomitantly higher bit error rates. Whereas today the recording medium noise is usually greater than other

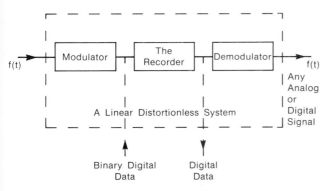

Figure 5.2   A nonlinear, unbiased recorder in a linear, distortionless system.

noises in the system, it is expected that, in the much higher areal-density systems of the future, the reproduce head noise will become dominant. At least a 10-fold increase in areal density is forecast in the next decade.

5.2 Noise

There are three principal contributors to the noise power of a recording channel: the electronic noise, the reproduce-head noise, and the recording medium noise. The noise power spectra of these noises for a 2-MHz-bandwidth instrumentation recorder are sketched in Fig. 5.3. The electronic noise is generally negligible at medium and high frequencies, but increases as $1/f$ at very low frequencies. The head noise behaves conversely; at low frequencies, it is usually negligible but increases to be significant at higher frequencies. In most analog recorders (audio, instrumentation, video), the recording medium noise is, by design, dominant over most of the band. Thus, all analog recorders are medium-noise-limited. Since in reality similar analog considerations govern the operation of digital recorders, it follows that most of them (digital audio, digital video, computer flexible and rigid disks) are, or should be, presently medium-noise-dominated. With increasing areal densities achieved by narrower track widths and with higher data rates, it is believed, however, that head noise will eventually become the dominant form of noise in future digital recorders.

5.2.1 Electronic noise

Electronic noise is caused by random fluctuations in time of the electric charge carriers. Essentially all the electronics noise is generated by the first-stage amplifier in the reproduce circuitry. Textbooks and semicon-

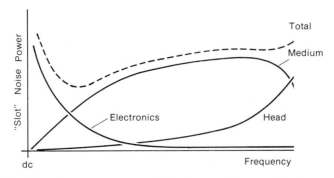

Figure 5.3 The relative contributions of medium, head, and electronic noise in a 2-MHz instrumentation recorder.

ductor manufacturers' handbooks cover the basic principles and give equivalent circuit diagrams. Since these equivalent diagrams have both voltage and current noise power generators, it follows that the best performance (lowest noise figure) is obtained when the amplifier is driven by a source of optimum impedance. The noise figure is the decibel ratio of the integrated wideband output noise power to that of an ideal, noiseless amplifier and, in well-designed sytems, is 1 to 2 dB only. For amplifiers designed to operate at frequencies of around 1 MHz, a typical optimum source impedance is 100 Ω, and, further, it matters little whether the source impedance is real (dissipative) or imaginary (inductive). Electronic noise is well understood, is not unique to magnetic recording, and can be made almost negligible; therefore, the subject is not pursued further here.

5.2.2   Head noise

The magnetic flux in both write and read heads is subject to thermally induced fluctuations in time. Because the flux densities are high (> 100 mWb/m^2, or 1000 G) in write heads, the resultant noise is negligible. In read heads, however, the flux densities are low (< 1 mWb/m^2, or 10 G), and read-head noise is often appreciable. The key to understanding read-head noise is the Nyquist noise theorem, which states that "any device, which dissipates energy when connected to a power source, will generate noise power as a passive device." The magnitude of the noise power is

$$e_N^2 = 4kT \, \mathrm{Re} \, Z \, \Delta f \qquad \text{W into a hypothetical 1-}\Omega \text{ load} \qquad (5.1)$$

where k = Boltzmann's constant
T = absolute temperature
$\mathrm{Re} \, Z$ = real part of the impedance, Ω
Δf = frequency, Hz

A familiar example is the Johnson noise power of a resistor, given by

$$e_N^2 = 4kTR \, \Delta f \qquad (5.2)$$

where R is the resistance (ohms).

The crucial factor about the Nyquist theorem is that only the dissipative part of the impedance generates noise. Accordingly, only the dissipative part of the read-head impedance generates noise. In any system where the response is not precisely in phase with the driving function, there is dissipation or heat generation. This includes all systems which display hysteresis, for example, magnetic materials.

Consider the initial (i.e., low-level) hysteresis loops shown in Fig. 5.4. The loss, or heat generated per unit volume per cycle, is equal to the area enclosed by the loop. In the interests of mathematical simplicity, it is

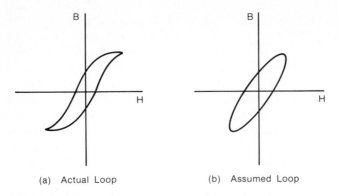

(a) Actual Loop   (b) Assumed Loop

Figure 5.4 (*a*) The actual low-level hysteresis loop and (*b*) the assumed elliptical Lissajous loop of the head core.

usual to assume that these initial loops are elliptical in shape so that the complex permeability

$$\mu^* = \mu' - j\mu'' \tag{5.3}$$

has two parts: μ', the real part, which is exactly in phase, and μ'' , the imaginary part, which is exactly 90° out of phase. The complex notation serves solely to keep track of the phase in subsequent algebraic manipulations.

Because physical systems are causal, that is, they cannot respond before being driven, certain relationships linking μ' and μ'' exist; they are known as Kramers-Kronig, Bode, and Hilbert transforms, respectively. Only the Hilbert transform formalism is given here:

$$\mu'(\omega) = \frac{1}{\pi} \int_{-\infty}^{\infty} \frac{\mu''(\omega') \, d\omega'}{\omega - \omega'} \tag{5.4a}$$

and

$$\mu''(\omega) = \frac{1}{\pi} \int_{-\infty}^{\infty} \frac{\mu'(\omega') \, d\omega'}{\omega - \omega'} \tag{5.4b}$$

where $\mu'(\omega)$ and $\mu''(\omega)$ are the real and imaginary parts of the permeability spectrum. Typical μ' and μ'' spectra are shown in Fig. 5.5. Note that if either part is known at every frequency, then the other part is completely determined. This fact holds whatever the origin (domain wall damping, eddy currents, etc.) of the hysteresis or losses.

Consider now a circular toroid made of a lossy material with cross-sectional area A and circumference l, and wound with N turns. This toroid will serve as a simple model for the read head. The complex inductance is

$$L^* = C \frac{n^2 A \mu^*}{l} \tag{5.5}$$

where C is a proportionality constant, and the impedance $Z^* = j\omega L^*$ is, dropping the proportionality constant,

$$Z^* = \frac{j\omega n^2 A\mu'}{l} + \frac{\omega n^2 A\mu''}{l} \tag{5.6}$$

The first term represents a pure inductor. Inductors merely store energy when excited; they are dissipationless and, therefore, noiseless. The second term, the real part of the impedance, is a dissipator and generates noise according to the Nyquist theorem. At frequencies well below that where $\mu''(\omega)$ peaks, the real part of the impedance, and therefore the noise power spectrum, increases with increasing frequency. This explains why head noise becomes of increasing importance at high frequencies.

In general, the dissipative impedance of a head has two parts: the hysteretic loss term discussed above and the resistance of the coil. In manganese-zinc ferrite video heads, the coil resistance is almost negligible ($<$ 1 Ω), and the noise is generated principally by the hysteresis losses. With a complex inductance of several microhenries, the dissipative impedance at 10 MHz is usually of the order of 10 Ω. The efficiency of a head is independent of scale, but from Eq. (5.6), the impedance is proportional to the scale; therefore, improvements in heads attend reductions in size. Thus a half-scale head would have one-half the impedance and noise power. To retain the optimum impedance match to the preamplifier, the designer might then choose to increase the number of turns by $\sqrt{2}$, which returns the impedance and noise power to their full-scale values. Either choice yields a twofold increase in signal-to-head-noise power ratio, with the latter choice giving the best signal-to-electronic-noise power ratio. Factors which limit such size reductions include the reduced bandwidth due to long-wavelength head "bumps" caused by finite pole lengths, the increased mechanical fragility, and the difficulties of fabrication of smaller heads.

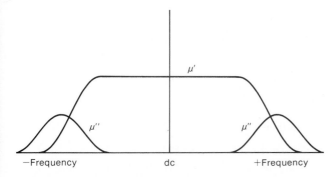

Figure 5.5 The magnitude of the real (μ') and imaginary (μ'') parts of the permeability.

The converse situation exists in film heads. Here the inductance is negligible ($< 10^{-8}$ H), but the coil resistance is in the range of 10 to 20 Ω. Thus, at frequencies of 10 to 20 MHz, the noise power of current film and ferrite video heads is comparable. Since the scaling rule for resistors is the opposite of inductors, namely, doubling the size halves the resistance, it is clear that improved film heads will need larger coils.

In the future, as areal densities and data rates increase, it is anticipated that head noise will become the dominant form of noise in many recording systems. The most effective way to increase areal density is, as discussed below, to increase the track density. Narrower tracks, which generate less medium noise, and higher frequencies or data rates both make the relative contribution of head noise greater. It may be argued that narrower-track recording should use narrower heads, which also have reduced noise. For a wide variety of mechanical and magnetic reasons, however, the region close to the gap, which defines the track width, is usually narrower than the body of the head. In ferrite video heads, the cores are notched, and in film heads, the pole tips are often an order of magnitude narrower than the coil area. As a practical matter then, as track width varies, the head width stays almost unchanged.

Two other forms of head noise occur in some circumstances. In ferrite heads, rubbing noise is generated if the material has nonzero magnetostriction. The acoustic waves generated by the head-tape rubbing contact cause fluctuations in the magnetic flux threading the head coil. While the excitation of the acoustic waves is thought to be random, thus injecting white noise, the properties of the head (size, elastic constants) often cause the rubbing noise to peak at certain mechanical frequencies. Rubbing noise is, however, reduced when polycrystalline materials are used, because the random orientations of the crystallites permits some cancellation of the magnetostrictive effect. In film heads, Barkhausen noise can be observed. This is caused by large changes in the domain wall structure in the thin films. The distinction between Barkhausen and the normal hysteretic noise discussed above is not well defined and is simply one of magnitude: many small domain wall jumps give normal hysteresis. Barkhausen noise is reduced by careful micromagnetic design; the use of multilayer film structures and the application of easy-axis magnetic fields have been explored (Decker and Tsang, 1980).

5.2.3   Additive medium noise

Recording medium noise is due to the uncertainty or randomness of some property of the medium. Any variation or fluctuation from point to point will produce medium noise. The most basic random process in recording media is the uncertainty in the positions of the magnetic particles or grains. If these particles are packed according to Poisson statistics, that is, equal probabilities in equal volumes, then the noise is additive and not multiplicative. Additive noise is present whether a signal is recorded or not. Multiplicative noise, also called modulation noise, varies with the recorded signal level. Modulation noise arises wherever the statistics of the causative phenomenon are not Poisson statistics. Examples range from local clustering or ordering of the particle magnetizations to variations in head-to-medium velocity.

In this section, the theory of additive noise will be developed and applied to the problems of calculating the signal-to-noise ratios of audio, instrumentation, and video recorders. These ratios will be compared with the signal-to-quantizing-noise ratio of binary pulse-code-modulated (digital) recorders. Because the statistics of the non-Poisson phenomena giving modulation noise cannot be predicted, no general theory exists.

Consider the algebraic manipulation

$$\overline{(e_1 + e_2)^2} = \overline{e_1^2} + \overline{e_2^2} + \overline{2e_1e_2} \tag{5.7}$$

where e_1 and e_2 are voltage waveforms in time and the bar indicates time averaging. If e_1 and e_2 are not correlated, that is, they bear no fixed relationship to each other, then the term $\overline{2e_1e_2}$ is equal to zero, and the square of the sum is equal to the sum of the squares. With correlated or coherent, signals, the power is obtained by adding the voltages and then squaring. With uncorrelated sources, the total noise power is obtained by adding the individual noise powers. This rule finds application, for example, in assessing the total noise power in a recording channel: the medium-noise power plus the head-noise power plus the electronic-noise power equals the total-noise power.

Figure 5.6 shows a single magnetic particle passing a reproduce head. A fraction of its flux threads the coil producing the voltage and power waveforms shown. Each and every particle in the recording medium produces a similar power pulse. If the particles are packed as a Poisson process, the individual power pulses are uncorrelated, and the total noise power P is simply given by addition or integration. Assuming that the read head has 100 percent efficiency, one turn, and no gap loss, the result of performing the integration is

$$P = 4\pi m^2 N w V^2 \frac{\delta(d + \delta/2)}{d^2(d + \delta)^2} \tag{5.8}$$

where m = the dipole moment of a magnetic particle
$\quad\ N$ = number of particles per unit volume
$\quad\ w$ = track width
$\quad\ V$ = head-medium relative velocity
$\quad\ \delta$ = coating thickness
$\quad\ d$ = head-to-medium spacing

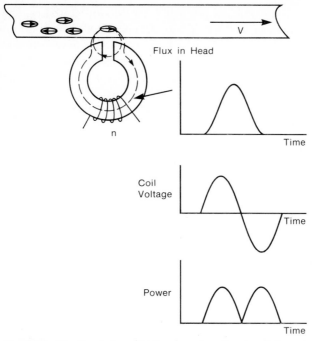

Figure 5.6 The flux induced in the reproduce head, and the voltage and power in the coil of the head, as a single magnetic particle passes the gap.

The power P is given in watts into a hypothetical 1-Ω load (Mee, 1964). Equation (5.8) shows that the noise power increases with the dipole moment and the number of particles; thus the more magnetic the medium, the greater the noise. The power also increases as w and V^2, which illustrates the difference between a coherent and an incoherent process; the particles across the width are uncorrelated, whereas all the particles are subject to the same velocity. The noise power decreases with increasing head-to-medium spacing, as it must. Finally, the power is almost independent of the coating thickness because only the near layers contribute significantly to the head flux.

Unfortunately, knowing the total noise power does not reveal the noise power spectral density $e^2(k)$, where

$$P = \int_{-\infty}^{\infty} e^2(k)\, dk \tag{5.9}$$

To deduce the noise power spectral density, the most direct analysis uses the Wiener autocorrelation theorem. This theorem states that the noise power spectral density is the Fourier cosine transform of the autocorrelation function $F(x)$; thus

$$e^2(k) = \int_{-\infty}^{\infty} F(x) \cos kx \, dx \qquad (5.10)$$

where
$$F(x) = \frac{1}{L} \int_{-L/2}^{L/2} f(x')f(x' - x) \, dx' \qquad (5.11)$$

Completion of these operations results in

$$e^2(k) = 4\pi m^2 N w V^2 |k| (1 - e^{-2|k|\delta}) e^{-2|k|d} \, \Delta k \qquad (5.12)$$

where Δk is the wave number interval (or slot) over which the power spectrum is measured (Mann, 1957; Mallinson, 1969).

It is instructive, at this point, to compare the noise power spectrum with the maximum possible signal power spectrum $E^2(k)$. The signal power spectrum is, for a 100 percent efficient, one-turn, zero-gap approximation head, simply one-quarter the square value of the spectrum corresponding to uniform, sinusoidal magnetization (Wallace, 1951). Thus

$$E^2(k) = \tfrac{1}{4}[4\pi m N w V(1 - e^{-|k|\delta}) e^{|k|d}]^2 \qquad (5.13)$$

with units, again, watts into a hypothetical 1-Ω load. The signal and noise power spectra are plotted versus frequency ($f = Vk/2\pi$) in Fig. 5.7. Both power spectra increase at 6 dB, or a factor of 4, per octave at the longer wavelengths; but at wavelengths comparable with the coating thickness,

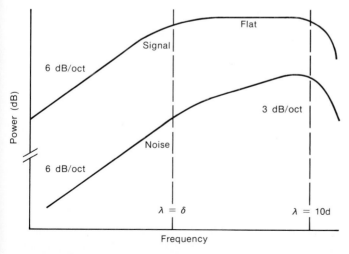

Figure 5.7 The signal and medium-noise power spectra for an ideal medium.

differences in slope occur. The signal power spectrum becomes flat due to the onset of the thickness loss term. The noise power spectrum, on the other hand, now slopes to 3 dB, or a factor of 2, per octave. This difference has profound implications upon the optimal design of recorders. At small wavelengths at which appreciable spacing loss is suffered, say $\lambda < 10d$, both spectra roll over according to the same exponential term.

At this point, we can obtain, by division, the narrow-band, or slot, signal-to-noise ratio (SNR). This is the signal power divided by the noise power, in a slot Δk, at some wave number k:

$$(\text{SNR})_{\text{slot}} = \frac{\pi N w (1 - e^{-|k|\delta})^2}{|k|(1 - e^{-2|k|\delta})\,\Delta k} \tag{5.14}$$

Several interesting facts now appear. The slot signal-to-noise ratio is independent of the particle moment, because all signal-to-medium-noise ratios essentially involve particle statistics and are counting exercises. The ratio is independent of head-to-medium velocity and head-to-medium spacing for the same reason; while these changes may alter the absolute values of the signal and the noise powers, they do not affect the underlying statistics.

In order to have a nearly distortionless output signal, all recorders must be equalized; that is, electrical filter or compensation networks are employed to correct the signal spectrum. For audio and instrumentation recorders, which are linear analog recorders using ac bias, it is necessary to make the overall channel amplitude transfer function flat, that is, constant, between some lower and upper frequency limits. The required equalization is the reciprocal of the signal spectrum, so that the overall transfer function is given by $E(k)[E(k)]^{-1} = 1$, a constant. The resulting expression for the wideband signal-to-noise ratio is

$$(\text{SNR})_{\text{wide}} = \pi N w \left(\int_{k_{\min}}^{k_{\max}} |k| \coth \frac{|k|\delta}{2}\, dk \right)^{-1} \tag{5.15}$$

which at short wavelengths, $kd > 1$, reduces to

$$(\text{SNR})_{\text{wide}} \approx 2\pi N w (k_{\max}^2 - k_{\min}^2) \tag{5.16}$$

and for large-bandwidth systems, $k_{\max} >> k_{\min}$, becomes

$$(\text{SNR})_{\text{wide}} \approx 2\pi N w k_{\max}^2 = \frac{N w \lambda_{\min}^2}{2\pi} \tag{5.17}$$

The underlying reason for the simplicity of this result is shown in Fig. 5.8. The volume of medium effectively sensed by the reproduce head is strictly limited; in the longitudinal direction, it is only that half wavelength which actually spans the reproduce-head gap which is sensed. In the perpendicular direction, the spacing loss limits contributions to a depth equal to a small fraction (say 0.3) of a wavelength. Thus the volume, shown shaded, of medium sensed at any instant is proportional to $(\lambda/2)(\lambda/3)w$, or $w\lambda^2/6$, and the number of particles sensed is proportional to $Nw\lambda^2/6$. The signal-to-medium-noise ratio involves a statistical calculation in which the mean signal power is analogous to the square of the mean value of the distribution, and the noise power is analogous to the mean-square deviation or variance. The wideband signal-to-noise ratio is therefore determined by the number of particles being sensed at any instant by the read head. If the ratio is, say, 40 dB, that means 10,000 particles are being sensed in the shaded volume; 30 dB corresponds to 1000, 20 dB to 100, and so on.

Another important conclusion to be drawn from Fig. 5.8 and Eq. (5.17) is that it is in principle better, when seeking higher areal densities, to use higher track densities than it is to use shorter wavelengths. Halving the track width costs 3 dB in wideband signal-to-noise ratio, whereas halving the wavelength costs at least 6 dB. In many recorders, halving the wavelength will cost more than 6 dB because it may no longer be possible to keep the channel signal-to-medium-noise limited. This conclusion is unfortunate in that decreasing head-medium velocities is always the easiest thing to do in the quest for higher areal densities. Reengineering the whole machine to utilize narrower track-width heads is more difficult.

Given a particular recording format, the track width and minimum wavelength are fixed. Thus, the only improvement possible is in the medium, and, according to the analysis above, the only parameter of consequence is the number of particles per unit volume. A change from γ-Fe_2O_3 to metallic iron will not change the signal-to-noise ratio if the particle density is unchanged. It follows that the correct way to exploit metallic iron, or other higher-energy particles, is to use smaller particles which are equally stable against superparamagnetic decay. Thus, for metallic iron, particles of one-sixteenth the volume, with 16 times the par-

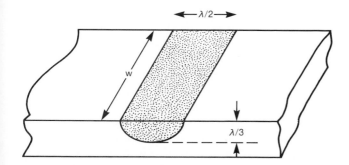

Figure 5.8 The volume sensed by the read head; it is $\lambda/2$ long, $\lambda/3$ deep, and w wide.

ticle density and 12-dB-higher wideband signal-to-noise ratio, should be achievable in medium-noise-limited recorders.

It is of interest to compare the calculated signal-to-noise ratios with measurements made on various recorders. In linear recorders using ac bias, a maximum signal level of only 20 percent of the maximum is allowed in order to limit the third harmonic distortion (THD) to 1 percent. This means that only one particle out of five in the coating is used to record the signal, which imposes a factor of 25, or a 14-dB penalty, on the attainable ratio because all the particles still contribute to the noise. A professional audio recorder has a bandwidth of 40 Hz to 15 kHz, a track width of approximately 2 mm, and a tape speed of 190 mm/s (7.5 in/s). Using standard γ-Fe_2O_3 tape with 6×10^{10} particles per cubic millimeter, the measured wideband ratio is 56 to 58 dB; Eq. (5.17) yields 55 dB. An instrumentation recorder has a 400-Hz to 2-MHz bandwidth, 1.25-mm track width, 3-m/s (120-in/s) speed, and a measured 34-dB ratio versus the computed value of 36 dB.

Frequency-modulated (FM) video recorders are nonlinear machines where distortion is controlled by using a modulation scheme. Detection of the FM waveform is performed on the zero crossings, and it is critical that every zero crossing be correctly placed. This is achieved by ensuring that the overall transfer function of the channel has "straight-line" amplitude and phase characteristics. Because the FM signal is analog, with an infinitude of permitted zero-crossings, equalizing an FM channel is a considerably more demanding task than equalizing a digital channel with a finite and usually small number of pulse positions. With the requisite straight-line equalization, the video baseband signal-to-noise ratio is

$$(SNR)_{video} \approx \frac{3 N w V^2 (\Delta f)^2}{8 \pi f_c f_s^{\,3}} \tag{5.18}$$

where f_c is the carrier frequency, Δf is the change in f_c for a 1-V video input, and f_s is the video bandwidth (Mallinson, 1975). For a type C professional video recorder, f_c is 9 MHz, Δf is 3 MHz, f_s is 4.5 MHz, w is 200 μm, and V is 25 m/s (1000 in/s), and approximately 4 dB of video preemphasis is used. The computed signal-to-noise ratio is 38 dB when normal γ-Fe_2O_3 tape is used. To this figure, 4 dB is to be added for video preemphasis and 9 dB (i.e., $20 \log 2\sqrt{2}$) to convert from the mean signal power used above to the peak-to-peak signal power customarily used in the video industry. The net figure calculated is then 51 dB, which concurs well with the 50 to 52-dB values measured on such recorders.

The agreement between the calculated and measured ratios is to some degree fortuitous. The magnitude of the signal power spectrum measured is always considerably different from that given by Eq. (5.13) (Smaller, 1969). Specifically, the short-wavelength signal power is smaller because of the existence of writing-process losses. Similarly, the noise power spectrum measured does not follow that given by Eq. (5.12), but, at short wavelengths, the actual noise power in particulate media is much smaller than expected. At long wavelengths, the noise spectrum also deviates from that expected. It appears that phenomena closely analogous to a $1/f$ noise process occur at extremely long wavelengths. At medium wavelengths, both signal and noise spectra are as expected; at short wavelengths, the lower-than-predicted signal and the lower-than-predicted noise fortuitously compensate nearly exactly.

In the interest of completeness, a brief discussion of quantizing noise in digital systems is given. Digital systems can never be absolutely accurate when an analog signal is to be recorded. An analog-to-digital converter tests the analog signal against a number of preset digital levels and chooses the digital level which is closest. Error, though inevitable, may, in principle, be made arbitrarily small by increasing the resolution, that is, the number of levels, of the analog-to-digital converter. Practical engineering factors limit the number of levels to approximately 10^5 corresponding to the number of combinations possible in a 16-bit word. The error is called *quantizing noise,* although it is not a noise of physical origin but rather is an instrumentation error inherent in the digitization of the analog signal. When the source data are already digital, then, of course, no apparent quantizing error occurs.

Suppose there are N bits corresponding to 2^N quantizing levels. The mean-square quantizing error for equally probable input signals is one-twelfth of the interval. It follows that the signal power to quantizing-noise power ratio is

$$(\text{SNR})_{\text{quant}} = (6N + 7.8) \quad \text{dB} \tag{5.19}$$

Thus, for $N = 8$, the proposed standard for digital video recorders, the ratio is 56 dB. For $N = 14$, as in digital audio tape recorders (and optical compact disks), the ratio is 92 dB, a figure that could be achieved with an analog recorder of the type discussed above only if, for example, the track width were to be increased by a factor of 10,000! For a discussion of the relative merits of the various modulation schemes, the reader is referred to the standard textbooks. The concepts are not unique to magnetic recording and are not pursued further here.

5.2.4 Multiplicative (or modulation) noise

In the case of Poisson statistics, the autocorrelation function of an erased tape ($M_r = 0$) is just the sum of contributions from all the individual particles correlating with themselves; all terms due to particles correlating with other particles sum to zero. The additive noise discussed above is given by Fourier transformation of the self terms. As the medium becomes coherently magnetized, the nonself terms no longer sum to zero, and the extra term which appears in the Fourier transform is the signal power spectrum.

When Poisson statistics are not obeyed, the situation is more complicated because a third contribution, arising from the nonself terms, appears in the autocorrelation function. The third part can have either or both negative or positive parts corresponding to reductions and additions to the medium-noise power spectrum. The magnitude of the third part varies with the signal power, most likely as M^2 or $1 - M^2$. Multiplication in the spatial domain of the autocorrelation function becomes, upon Fourier transformation, convolution in the frequency domain. Accordingly, the output power spectrum of a recorder has the typical features shown in Fig. 5.9. Note how the convolution has caused the appearance of characteristic "modulation"-noise sidebands, or skirts, around the signal. Depending upon the statistics, modulation noise may either increase or decrease the total medium-noise power, and the magnitude of the effect may either increase or decrease with recorded signal power (Thurlings, 1985).

Very little useful theory exists because the particular physical phenomena are poorly understood and the appropriate statistics are not known. For particulate media, the two most fundamental modulation noises arise from deviations from Poisson statistics in the position or packing of the particles and in the patterns or ordering of the particle magnetization directions (Satake and Hokkyo, 1974; Thurlings, 1980, 1983). Consider the case where the particles are not packed with exactly equal probabilities in equal volumes. The fluctuations in particle packing fraction, often termed *clumping, chaining,* or *roping,* may be due in part to the magnetic attractions between particles during the dispersion and coating processes. It is to be expected that the magnitude of the modulation noise due to clump-

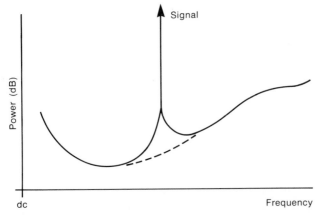

Figure 5.9 The total noise spectrum showing the modulation noise "skirts" around the signal.

ing will increase from zero in the erased state to a maximum when the tape magnetization is dc-saturated.

In the second effect, magnetic interaction fields may be expected to control the spatial patterns in which the particles are magnetized (Arratia and Bertram, 1984). Thus, the particle magnetizations might form closed rings or arrays. Although the effect of this form of modulation noise could add to or subtract from the total medium noise, the magnitude of the effect is anticipated to decrease as the tape magnetization increases and become zero when the tape is saturated.

Experimentally, it is found that the modulation-noise behavior differs in particulate tapes and thin metallic films. In particulate media, the total noise in the erased state is considerably lower (-10 dB) than that expected from additive noise theory, and it increases with increasing magnetization as expected for non-Poisson statistics for particle position. In metallic thin films, the total noise in the erased state is close to that expected from additive noise theory, but it decreases with increasing dc magnetization. In metallic thin films, it is also found that a form of modulation noise is associated with written magnetic transitions (Tanaka et al., 1982; Baugh et al., 1983; Terada et al., 1983). Zigzag domain walls separating regions of opposite magnetization have been observed (Yoshida et al., 1983) and are believed to be a source of transition noise in longitudinally magnetized thin films (Belk et al., 1985).

There are many other causes of modulation noise. Among the most important are variations in the tape surface roughness, the tape coating thickness, the substrate roughness, the head-to-tape spacing, and the head-to-tape velocity. Some causes of modulation noise are coherent across the track width. Examples include packing fraction variations due to gravure coating, certain types of head-to-tape spacing, and velocity variations. In these cases, the signal-to-modulation-noise ratio will be independent of track width.

5.3 Interference

Interference is the reception or reproduction of signals other than those intended. It is deterministic and thus may be measured or, in principle, calculated exactly. It may be reduced to arbitrarily small magnitude by proper design. In this section, the principal interferences encountered in recorders are discussed.

5.3.1 Crosstalk

Crosstalk is due to the unintentional leakage of flux from one channel to another in a multichannel read or write head stack. The leakage flux may have either polarity and is generally of the order of 1 percent (-40 dB) in immediately adjacent heads. Although the leakage flux may be calculated by using, for example, the finite-element method, considerable sophistication is required in the analysis to achieve the necessary accuracy.

Consider a multichannel head stack with only the center channel excited with write current. Because the leakage fluxes fall off approximately geometrically ($-$ 40, -80, -120 dB, etc.), crosstalk is essentially limited to the immediately adjacent tracks. When the adjacent heads carry write current, the flux in these gaps adds linearly to the leakage fluxes. Consequently, the writing field and the written magnetization pattern are corrupted by a small fraction of the signal from the adjacent track. Suppose that the leakage flux is -40 dB. With ac bias, the writing process is virtually linear, and, accordingly, the written magnetization will contain a -40-dB component of the adjacent track signal. When ac bias is not used, somewhat surprisingly, the written crosstalk magnetization is even higher; the crosstalk has been measured to be approximately a factor of 4 ($+12$ dB) higher (Tanaka, 1984). The origin of the apparent gain is not understood. Presumably a very similar effect occurs in consumer video recorders, where (in the color-under system) the color chrominance signals are recorded linearly at a low level relative to the frequency-modulated luminance signal.

By reciprocity, the writing and reading crosstalk performances of a given multichannel head are identical. If the writing current in a coil engenders -40-dB leakage flux in an adjacent head gap, then, during reproducing, flux will appear in the same head coil which is -40 dB below that entering the channel. If ac bias is used, the total (write plus read) system crosstalk is doubled to -34 dB. Without ac bias, the total system crosstalk becomes -22 dB. It should be noted that the phenomenon of concern in crosstalk is not the coupling from coil to coil in the multichannel head stack. The important aspects of crosstalk arise from coil to adjacent gap in writing and gap to adjacent coil in reading. Coil-to-coil phenomena involve entirely different flux paths and are more properly categorized as an example of feedthrough.

The measures taken to minimize crosstalk include adding magnetic shields, increasing the distance between the heads, and reducing the side areas of the heads. In this regard, film heads, with their extremely small side areas, have very low crosstalk without resorting to magnetic shields. Because crosstalk is a diminished version of the adjacent track signal with the same spectral properties as the signal, signal processing techniques cannot be applied to reduce crosstalk. In this sense, crosstalk is more troublesome than noise.

5.3.2 Incomplete erasure

There are several methods of erasing previously recorded information. The most effective is bulk erasure in which the complete tape reel is subjected to an ac field of decreasing magnitude. If the initial field magnitude is great enough to switch even the hardest (highest-nucleation-field) particles, very nearly perfect erasure can be achieved. Typically, the initial field magnitudes needed are about five times the coercive force. Instrumentation and professional video recorders use bulk erasure.

A less effective erasure method is the use of erase heads. These are generally similar in construction to read-write heads, but have a gap (or double gaps) of greater gap length. The longer gap lengths permit higher-magnitude erase fields at the far side of thick media. Wherever erase heads are used, only a very small length of tape is subjected to the erasing field at any instant. The erasure is, therefore, less effective than that achieved with bulk erasure because the regions just downstream of the erase-head gap can be re-recorded by the magnetic field arising from the unerased tape upstream of the gap (McKnight, 1963). The re-recording effect is expected to be greatest at relatively long wavelengths because then the upstream magnetic field is highest.

The erasure technique used in digital tape and disk recorders is much less effective. In these machines, there is no specific erase process; the new data are simply written over the old data. In longitudinal recording, it is found that overwriting a low-density recording with a high-density recording results in the least complete erasure; accordingly, most overwrite tests use two square waves, $f_2 > f_1$, with f_2 erasing the f_1 recording. Typically, f_1 is incompletely erased to about -30 dB only. It is not properly understood why the erasure is so poor. When square waves of writing-head flux are used, the field history of any point in the medium is a succession of fields having opposite polarities but the same magnitude. The magnitude differs, of course, at different depths in the medium. According to the Preisach formalism, the application of alternating-polarity fields of fixed magnitude causes the remanence to alternate between two levels, implying that perfect erasure of digital data should be possible. It seems likely, then, that a re-recording phenomenon similar to that occurring with erase heads is happening (Fayling et al., 1984; Wachenschwantz and Jeffers, 1985). This explains why it is more difficult to overwrite the low-density, long-wavelength data. Re-recording does not seem to explain, however, another puzzling phenomenon. In order to record optimally short-wavelength, high-bit-density data, it is necessary to use lower record currents and writing-field magnitudes. In turn it is thus found that satisfactory overwrite performance can be obtained only with thinner media. This had led to the widespread adoption of a scaling rule that the medium thickness and the bit interval should remain in the same ratio; higher linear densities necessitate thinner coatings. It may well be that certain mechanical effects are contributing to the incomplete erasure; thus misregistration, treated below in Sec. 5.3.4, and variations in flying height could act to limit the erasure attainable.

5.3.3 Feedthrough

Feedthrough refers to any unintentional coupling of signals which does not involve the recording medium. Thus there may be feedthrough from the write to the read electronics within a channel. Signals from one channel can leak in the adjacent channels, as in the misnamed transformer crosstalk between head coils. Feedthrough can occur via electromagnetic, inductive, capacitive, and resistive coupling. Feedthrough is mentioned here only in the interest of completeness. The methods used to reduce or eliminate feedthrough, such as shielding and grounding, are well known in electrical engineering and are not unique to magnetic recorders.

5.3.4 Misregistration

Misregistration interference is due to the inability of a recording system to maintain the relative positions of the heads and media exactly. In a fixed-head tape recorder, misregistration is principally associated with interchange, that is, playing a recorded tape upon other machines. In disk recorders, the imperfect reproducibility of the moving-head positioning system and differential thermal expansion are the principal causes of misregistration.

In longitudinal-track, fixed-head tape machines (audio, instrumentation, and digital), maximum peak-to-peak tracking errors of approximately 50 μm are typical; this strictly limits the minimum track widths usable. With scanning-head machines (professional and consumer video recorders), the relative position of the heads to the tracks is continuously servoed, and maximum tracking errors of perhaps 5 μm occur. In rigid-disk recorders, the maximum head positioning error is also about 5 μm.

When an erase head and guard bands are employed, the initial effect of misregistration is simply to reduce the true signal-to-noise ratio. If, however, the mistracking is severe enough to permit the reproduce head to cover the adjacent track, interference occurs because the signal is contaminated by another signal. At short wavelengths, the magnitudes of interfering signals are proportional to track widths covered by the reproduce head. At long wavelengths, an additional effect occurs which is included in the discussion on side fringing below.

When an erase head is not used, as in digital disk recorders, even seemingly small misregistrations can cause serious interference. Suppose that a new data track is overwritten with a mere 5-μm tracking error with a 25-μm-wide track width. Suppose that subsequently, perhaps after large-amplitude excursions of the head, the reproduce head is repositioned with a 5-μm error of the opposite sense. The reproduce head is now covering only 20 μm of the new overwritten data and 5 μm of old data. Thus the ratio of new data to old data is only 4 (12 dB), which is sufficiently low to have a serious impact on the channel performance.

In the interests of achieving extremely high areal recording densities, consumer video recorders do not use guard bands. In order to combat the effects of the inevitable misregistration, use is made of slant azimuth recording. In this technique, two record-reproduce heads are used alternately. The gap azimuth angles of the heads are set at positive and negative angles. Typically, angles of ± 5 to $10°$ are employed. This results in recordings where the slant angle of the phase fronts of the recorded signal alternates from track to track. When reproducing such a recording, the head with matching azimuth angle reproduces without incurring any loss in performance. If, however, this reproduce head is misregistered and partially covers an adjacent track, the azimuth angle of the head and the recorded phase fronts are misaligned by twice the slant angle. If the maximum phase error thus incurred is a multiple of 2π rad, complete cancellations of the adjacent track reproduce signal occurs. In consumer video recorders, the design is such that a maximum phase error of 2π rad occurs at the most probable tracking error with the FM luminance signal of maximum energy. Thus, the objectionable effects of misregistration are greatly reduced.

5.3.5 Side fringing

Side-fringing interference occurs increasingly at longer wavelengths because a reproduce head can sense the fringing field which exists to the side, that is, not directly above, a recorded track. Consequently, a side-fringing signal is reproduced even when the head is not physically positioned above the track.

The side-fringing effect can best be understood with the aid of the reciprocity theorem. When the head is excited by a current in the coil, a fringing field exists, not only above the gap but also from the sides of the head. Essentially perfect but very complicated mathematical analyses of the three-dimensional fields around the gap sides are available (Lindholm, 1978). For our purposes, however, it is sufficient to regard the gap when viewed from the side as being virtually equivalent to a half head. The relevant half lies underneath the plane of the head surface; the missing half lies above that plane. To a good approximation, the side-fringing field is, therefore, one-half that in the usually considered xy plane. Upon Fourier transformation, the side-reading differential sensitivity is, by reciprocity, equal to $\frac{1}{2}e^{-kz}$, or $(55z/\lambda - 6)$dB, where z is the off-track distance. Upon integration, the side-fringing reproduce voltage spectrum from an adjacent track of width w with a guard bandwidth of b is

$$e(k) \approx \tfrac{1}{2}(1 - e^{-|k|w})e^{|k|b} \tag{5.20}$$

The similarity to the well-known thickness and spacing loss spectrum is obvious.

An interesting consequence of side fringing is that, when the reproduce spectrum of a head with a width much less than the track width of the recorded track (write wide, read narrow) is measured, the long-wavelength spectral slope turns out to be 3 dB per octave and not the usual 6 dB per octave. While the side-fringing interference is potentially worse than true noise, it is generally of lesser consequence than other interferences because the spectrum is weighted; negligible side-fringing effects occur at short wavelengths.

5.3.6   Print-through

Print-through occurs when the magnetic field from a layer of recorded tape magnetizes the layers of tape above and below it in a reel of tape. The basic physical process is closely related to superparamagnetism where thermal energy is sufficient to cause individual particles to switch (Tochihara et al., 1970; Bertram et al., 1980). Only a very small fraction (0.3 percent or -50 dB) of the tape particles contributes to print-through in most tapes. Print-through is principally a long-wavelength phenomenon; the maximum effect occurs when the wavelength is 2π times the distance between the layers of tape (i.e., the total tape thickness) (Daniel and Axon, 1950). Accordingly, negligible print-through occurs in FM video recorders where short wavelengths (< 12 μm) are used; in disk recorders the effect is absent because the disks are never closely packed.

Superparamagnetism occurs when the thermal energy kT is of the order of the energy required to switch a particle, $\Delta E = \frac{1}{2}M_sH_cv$. The first-order effect of the fringing field, H, from adjacent tape layers is to reduce the particle effective switching energy to $\frac{1}{2}M_s(H_c - H)v$. The relaxation time of the process is given by

$$\tau = \frac{1}{f_0} \exp \frac{\Delta E}{kT} \tag{5.21}$$

where f_0 is a frequency factor of about 10^9 Hz. When $\Delta E = 25kT$, $\tau = 100$ s; when $\Delta E = 40kT$ (corresponding to a 20 percent increase in particle linear dimensions), $\tau > 100$ years. Thus, a very sharp division separates the thermally stable from the thermally unstable particles.

In the measurement of print-through, a reel of tape is stored at elevated temperature; a standard industry test is 4 h at $+65°$C. The distribution function of particle sizes together with the effective superparamagnetism limiting volumes for both $+65°$C and room temperature is shown in Fig. 5.10. At $+65°$C, essentially all the particles below the $+65°$C limit will become magnetized by the fields from adjacent tape layers. Upon cooling back to room temperature, most of the particles between the two temperature limits will retain their print-through signal when the tape is

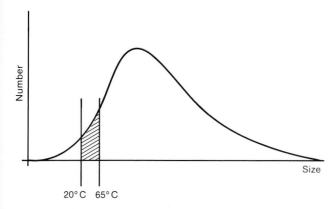

Figure 5.10 The particle-size distribution function, showing the superparamagnetic stability limits at 20 and 65°C.

unreeled and the printing fields vanish. Thus we see that the magnitude of print-through signal depends critically upon the distribution function of particle sizes, a property that is not easy to manipulate. This presents the tape designer with a difficult problem. If the particle size of a given magnetic material is reduced in order to increase the signal-to-noise ratio, then print-through will increase. Thus the wideband signal-to-noise ratio and the print-through levels of gamma ferric oxide professional audio tapes have remained essentially unchanged at 50 to 55 dB for many years.

An interesting phenomenon with regard to print-through is that, depending upon whether the tape is wound with magnetic coating inside or outside, differences in print-through level occur between the "pre-prints" and "postprints." The preprints occur outboard in the tape pack and vice versa for the postprints. The larger print-through always occurs in the layer of tape whose magnetic coating contacts the base film side of the printing layer. This has been explained in terms of the spatially rotating or Hilbert transform properties of the fringing-field outside tapes (Daniel, 1972; Mallinson, 1973).

5.4 Nonlinearity and Distortion

In linear recorders, ac bias makes the writing process linear for small signals. Since the reproduce process is linear, the complete recording channel is linear, and it has a well-defined, mathematically rigorous transfer function. When suitable equalization is added to linear recorders, the channel becomes distortionless, and the output signal is a noisy replica of the input signal. Analog audio and instrumentation recorders are linear recorders. In nonlinear recorders, no attempt is made to linearize the channel with ac bias; accordingly, the channel does not have a single-valued transfer function. In order to make the complete recording system distortionless for particular input signals, use is made of some form of modulation scheme; in analog video recorders this is frequency modulation, while in digital audio recorders pulse-code modulation (binary digital) is used. In all nonlinear machines, including digital data recorders, the output signal is highly distorted, but, given particular postequalization, certain features of the waveform can be recovered precisely. In FM video it is usual to recover the zero-crossing positions, whereas in digital recording the polarities of waveform peaks are recovered. The system's output signal is then reconstituted from these samples, and, subject to certain constraints, a noisy but distortion-free replica of the input signal is produced. Thus, in a nonlinear recorder, only certain features of the waveform after the equalizer are distortion-free, with the remainder of the waveform highly distorted.

5.4.1   A linear, distortionless system

A linear system has certain specific input-output characteristics. Mathematically, if

$$L[ax_1(t) + bx_2(t)] = aL[x_1(t)] + bL[x_2(t)] \tag{5.22}$$

the system is said to be linear. Here L is a time-domain operator such that the output for a signal input, $x(t)$, is $y(t) = L[x(t)]$, a and b are arbitrary scaling constants, and $x(t)$ and $y(t)$ are independent input signals. In a linear system, the output has no cross-modulation terms of the form $x(t)y(t)$; different input signals do not become mixed together. In reality, of course, systems approach linearity only for small signals. An example of this limitation is seen in the anhysteretic curve shown in Fig. 5.11. This

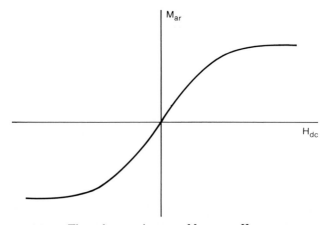

Figure 5.11 The anhysteretic curve, M_{ar} versus H_{dc}.

curve, frequently approximated by the error-function curve, introduces 1 and 3 percent third harmonic distortion (a measure of nonlinearity) at about 20 and 30 percent of maximum remanence levels, respectively (Fujiwara, 1979).

In order that the output of a linear system be distortionless, further conditions must be met. It is required that the linear operator L change in a specific way only for different frequency input signals. Suppose that the frequency-domain transfer function of the linear system is

$$L(\omega) = A(\omega)e^{-j\phi(\omega)} \qquad (5.23)$$

where $A(\omega)$ and $\phi(\omega)$ are the amplitude and phase transfer functions, respectively. For an input signal, band-limited between ω_1 and ω_2, distortion-free transmission through the linear system occurs if $A(\omega)$ is constant for $\omega_1 < \omega < \omega_2$, and $\phi(\omega)$ is a linear function of ω for $\omega_1 < \omega < \omega_2$, as shown in Fig. 5.12. The condition for $A(\omega)$ is usually called *flat amplitude response*, and that for $\phi(\omega)$ is called *constant group delay*, because all symbols are simply delayed by an equal time. The $\phi(\omega)$ condition is also, and most confusingly, called *linear phase* simply because it has a straight-line plot. The phase response, $\phi(\omega)$, must also have an intercept equal to an integer multiple of π when it is extrapolated to zero frequency.

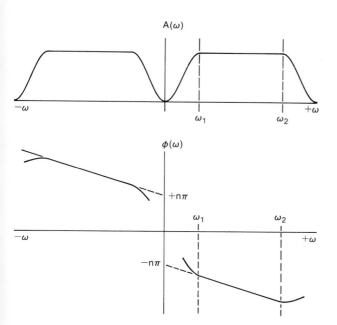

Figure 5.12 The amplitude $A(\omega)$ and phase $\phi(\omega)$ transfer functions required for the distortionless transmission of signals.

All inductive reproductive heads produce their output voltage by Faraday's law, that is, by differentiating the head flux. This introduces a $\pi/2$ phase shift; for sine-wave inputs, the output voltage is a cosine, that is, a sine wave lagging in phase angle by $\pi/2$. This phase distortion must be corrected in both linear and nonlinear recorders. In all linear recorders, the correction is made by integration. In nonlinear machines, integration is also used except in digital disk recorders where differentiation is employed. This introduces a further $\pi/2$ phase shift making π in total, which is equivalent to a mere polarity reversal.

In nonlinear recorders, mathematically rigorous transfer functions do not exist because the system is nonlinear in the sense discussed above. Nevertheless, it is found that, provided the input signal to the write head is limited to sequences of alternating-polarity step functions, the reproduce-head output voltage can be the linear superposition of the response to the individual input step functions. This is shown in Fig. 5.13. Requirements are that (1) the steps must alternate in polarity, (2) they must all be of the same amplitude, and (3) they must not come closer than some minimum bit interval. One of the reasons for the minimum bit-length limit is that the writing field adjacent to the write-head gap must rise to its full value within the minimum bit interval (Mallinson and Steele, 1969). At high frequencies or data rates, this implies that a specific prequalization operation must be performed, without which linear superposition would not hold. Of course, in true linear systems, the signal output waveform is, apart from noise, unchanged by varying the pre- and postequalizers as long as the product of their transfer function remains constant.

Given the correct conditions for linear superposition, a nonlinear recorder displays a pseudo-transfer function. This pseudo-transfer function fails the test of linearity because, for example, different input step-function amplitudes yield different output spectra; nevertheless, it is an extremely useful aid in understanding the system's behavior. It may be

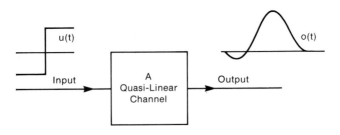

Figure 5.13 A quasi-linear channel displays linear superposition if the input and output are related by the following expressions. Input: $\Sigma_j \, (-1)^j \, u(t - j\Delta)$; output: $\Sigma_j \, (-1)^j \, o(t - j\Delta)$.

measured by (1) Fourier transformation of the isolated pulse response, (2) noting the amplitude of the fundamental of a varying-frequency square wave, and (3) noting the envelope of a pseudo-random square-wave sequence.

5.4.2 Amplitude response

The amplitude response spectrum is essentially given by calculations based upon the output expected from a medium magnetized sinusoidally along its length and uniformly through its depth (Wallace, 1951). In order to obtain a flat amplitude response in linear recorders, the required equalization is the inverse of this spectrum. Thus the output is boosted at both low and high frequencies.

In frequency-modulated video recorders, the overall pseudo-transfer function of the cascade, write process, reproduce process, and reproduce head must be postequalized to obtain the straight-line characteristic shown in Fig. 5.14 (Felix and Walsh, 1965). In this highly nonlinear system, the transfer functions of the preequalizer and the record head have already been accounted for as prerequisites for pseudolinearity. The straight-line equalization then ensures that the output waveform zero-crossing positions are undistorted.

In nonlinear digital recorders, which include digital audio, digital data tape, and digital data disk recorders, the task of the postequalizer is easier than in video recorders. In frequency modulation, the analog waveform has a continuum of zero-crossing positions to be preserved, whereas a binary digital channel has zero crossings allowed only at certain well-defined positions. These positions are determined by the channel code being used and are often specified in $[d, k]$ notation, where d is the minimum string or run of zeros allowed, and k is the maximum run allowed. It follows that many different postequalizers are acceptable in digital

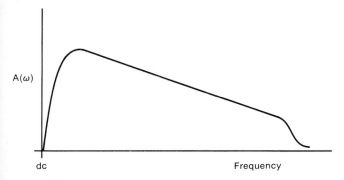

$A(\omega)$

dc   Frequency

Figure 5.14 The straight-line amplitude response required to preserve FM signal zero crossings.

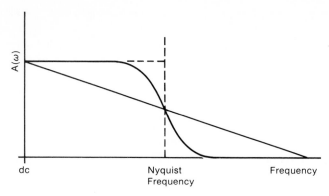

Figure 5.15 Examples of the infinite set of Nyquist amplitude responses which yield zero intersymbol interference.

recording. Generally, the transfer functions of the cascade, write process, reproduce process, reproduce head, and postequalizer must have one of the (infinite) set of Nyquist amplitude responses shown in Fig. 5.15 (Gibby and Smith, 1965). Signal-to-noise ratio and circuit complexity criteria govern the actual choices used in practical hardware.

5.4.3   Phase response

As discussed above, the standard output spectrum includes the phase shift due to Faraday's law. In complex notation, the voltage spectrum is preceded by the factor j ($= \sqrt{-1}$); in the signal power spectrum, of course, phase information is suppressed. In all recorders, it is necessary to correct for the phase distortion introduced by the j factor. As mentioned earlier, this is usually done by integration but, exceptionally, in digital disk recording, differentiation is employed. Since integrating pulses yields steps and differentiating pulses yields dipulses, as shown in Fig. 5.16, in both cases the subsequent (nonlinear) detector has to identify zero crossings.

There are other sources of phase distortion in magnetic recording. If the direction of magnetization in the recording medium is changed by an angle θ, the phase response changes by the same angle, as shown in Fig. 5.17 (Mallinson, 1981). Thus, when going from purely longitudinal recording to purely perpendicular recording, a phase shift of 90° occurs, and the corresponding isolated step-function output pulses are the Hilbert transform pair shown in Fig. 5.18. In distinction to the operations of integration and differentiation, the Hilbert transformation involves only a 90° phase change with no amplitude change. The transfer functions of differentiation, integration, and Hilbert transformation are $j\omega$, $(j\omega)^{-1}$, and $j\omega(|\omega|)^{-1}$, respectively.

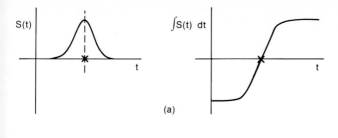

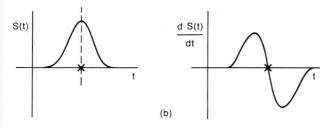

Figure 5.16 The identification of the peak of pulses, $S(t)$, by (a) integration and (b) differentiation involves, in both cases, finding the position of zero crossings.

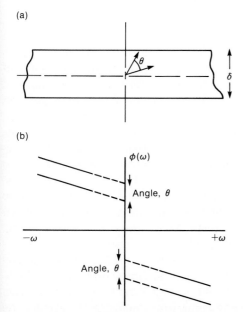

Figure 5.17 (a) The medium magnetization rotated by an angle θ and (b) the dc-extrapolated phase response, $\phi(\omega)$, changed by the same angle.

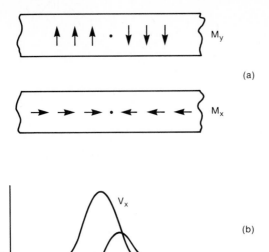

Figure 5.18 (*a*) Transitions of perpendicular and longitudinal magnetization. (*b*) Corresponding Hilbert pair of output voltage pulses.

Merely changing the phase response does not alter the energy in a pulse or signal. Phase-shifting equalizers can, therefore, be made and used for phase correction, and they have, in principle, no effect upon the signal-to-noise ratio. Accordingly, the odd-and-even symmetry Hilbert pair of pulses shown in Fig. 5.18 have identical information content, and preferences between the two must depend solely on considerations of circuit realization or complexity. When a nonlinear recorder is not properly equalized in both amplitude and phase, the pulse responses to step-function inputs overlap each other as shown in Fig. 5.19. The peak positions of the summed waveforms are displaced outward. The phenomenon is called *linear peak shift*. As long as the peak shift is caused solely by linear superposition, it is, in principle, correctable by proper postequalization, but with an inevitable reduction in signal-to-noise ratio.

At higher densities, when the minimum-bit-cell rule for linear superposition is violated, additional peak shifts of nonlinear origin arise which cannot be corrected completely by linear postequalizers. The origin of this nonlinear peak (or bit) shift is the demagnetizing field from the previously recorded bit (Mallinson and Steele, 1969). In order for linear superposition to hold, the magnetic field writing the nth bit must be solely the write-head field corresponding to the nth bit. At higher linear densities, the demagnetizing field from the $(n - 1)$th bit causes the total field writing the nth bit to be changed. In longitudinal recording, the total field is increased by the $(n - 1)$th bit demagnetizing field. In perpendicular recording, the total field is decreased below the value it would have in the absence of the $(n - 1)$th bit. It follows that, in longitudinal and perpendicular recording, nonlinear bit shifts of opposite sense are incurred. The actual recorded transition position, determined by the point where the total field equals the medium's coercivity, is recorded early in longitudinal and late in perpendicular recording. Thus, nonlinear effects pull longitudinal transitions together and push perpendicular transitions apart. It is found, therefore, that the total peak shift in longitudinal recording is, at high linear densities, less than that expected by linear superposition alone; the converse is expected for perpendicular recording (Bertram and Fielder, 1983).

In general, a linear or nonlinear peak shift of either polarity is undesirable because the peak positions become dependent on previous data. Many digital data disk recorders employ "prewrite compensation" in which the timing of each write-current step function is adjusted, early or late, depending upon the data pattern to be recorded. This is usually implemented with a read-only memory look-up table and is a rare example of the application of nonlinear equalization to a communications channel.

Finally, in audio tape duplication operations, considerable increases in production can be realized if it is possible to play the master tape both forward and backward and so eliminate the rewinding operation. Similarly, in computer tape drive operations, significant increases in data

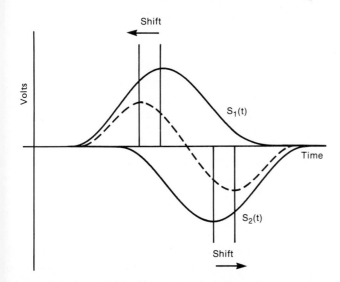

Figure 5.19 The peaks of the summed curve (dashed) are shifted outward: the peaks appear to repel each other.

throughput occur if it is possible to read the data equally well in both directions of tape movement. In order to have a perfect, but time-reversed, output signal in the reverse direction, it is necessary that the reproduce head and postequalizer cascade have a linear phase transfer function which intercepts zero frequency at an integer multiple of π; this makes the reproduce cascade impulse response an even function of time and thus indistinguishable from its reverse (Mallinson and Ferrier, 1974). It follows that the preequalizer used in recording the master must have a zero-frequency phase intercept which corrects any other phase factor, so that no phase distortion occurs for the overall system.

5.5   Future Areal Densities

In the future, increasing use will be made of binary digital recording. Apart from computer peripheral recorders, where the source data are already digital, most audio master recording and an increasingly great fraction of instrumentation telemetry are implemented digitally. It is anticipated that professional video recorders and, perhaps, consumer video recorders will eventually also become digital (see Chaps. 5 and 6 of Volume II).

The single most important reason for digital recording is the relative ease with which error detection and correction can be performed in binary digital systems. This not only permits indefinitely repeated error-free duplication of the recording, but it encourages the development of digital recording systems with considerably higher areal densities. Today, most digital recorders employ tape or disk formats which provide analog wide-band signal-to-noise ratios of approximately 30 dB. Concurrently, the interferences, overwrite, for example, are held at the -30-dB level. These figures permit the nonlinear digital detector a sufficient margin for the degradations due to tape or disk defects, so that raw (uncorrected) bit error rates in the range 10^{-6} to 10^{-9} are achieved. With advances in integrated-circuit technology (photolithography, large-scale integration, etc.), it seems inevitable that recording system designers will move to much higher digital areal densities and error rates; a target of an analog signal-to-noise ratio of 20 dB and a raw bit error rate of 10^{-3} to 10^{-4} appears to be a reasonable expectation for the next decade. Error detection and correction will then operate to yield the final error rate required for the particular application. For video, audio, and data recorders, the final error rates sought are of the order 10^{-6}, 10^{-9}, and 10^{-12}, respectively.

5.5.1   Head noise

In the quest for higher areal densities, both linear and track densities will be increased. Both actions reduce the signal level, and the latter the noise level, from the recording medium. Since in current high-density machines, the medium- and reproduce-head noise levels are of comparable magnitude, it seems likely that, in the future, the reproduce head will become the dominant source of noise.

In both ferrite video heads and film rigid-disk file heads, the dissipative impedance at frequencies close to 10 MHz is currently 15 to 20 Ω. By reducing the core size in ferrite heads and increasing the coil size in the film heads, the impedance can be reduced. In this discussion, it will be assumed that a modest decrease to 10 Ω is achieved.

5.5.2 Calibrated output spectra

In recording systems where the reproduce-head noise is dominant and of known magnitude, it is possible to estimate the wide band signal-to-noise ratio whenever absolutely calibrated data are available. In the data to be discussed, 0 dB corresponds to 1 nV rms per turn for each 1 μm of track width and 1 m/s of relative head-to-media velocity (Bertram and Fielder, 1983).

Figure 5.20 shows the absolute signal level needed to attain a 20-dB (a factor of 10 in voltage) signal-to-head noise ratio in a 10-MHz channel, which could support a 20×10^6 b/s data rate, when a 10-turn, 10-Ω, 100 percent efficient head is used. The five curves show the minimum signal levels needed for various track widths. The curves are 6 dB apart and rise at 6 dB per octave because, when the data rate is held constant, higher densities force lower speeds. These minimum signal levels are proportional also to the areal density (Mallinson, 1985).

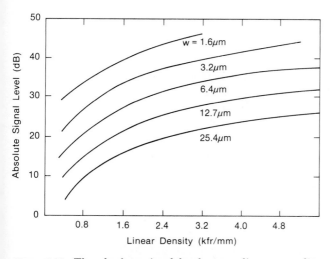

Figure 5.20 The absolute signal level versus linear recording density, with track width as a parameter.

The choice of a 20-Mb/s data rate has been made because it is representative of many different digital recording applications. Typical requirements are computer rigid disk, 5 to 25 Mb/s; digital audio, 1 to 2 Mb/s; digital instrumentation, 4 to 400 Mb/s; digital consumer video, 40 to 100 Mb/s; and digital professional video, 200 to 250 Mb/s. In the higher-rate applications, multiple parallel channels are used; the highest published rate is 117 Mb/s for a single channel (Coleman et al., 1984).

5.5.2.1 Flying height of 0.35 μm. Three rigid-disk media are shown in Fig. 5.21: a γ-Fe$_2$O$_3$ coating with a 24-kA/m (300-Oe) coercivity and 0.75-μm thickness, a Co-P metallic film with a 32-kA/m (400-Oe) coercivity and 0.1-μm thickness, and a special Co-P metallic film with a 72-kA/m (900-Oe) coercivity and a 0.027-μm thickness. The dotted line shows the criterion for the 25-μm track width and 20-dB signal-to-noise ratio. On 25-μm tracks, both the oxide and the regular metallic film disks support 1.1 kfr/mm (28 kfr/in); the special high-coercivity metallic film can operate at 1.8 kfr/mm (45 kfr/in).

5.5.2.2 Flying height of 0.1 μm. Data for the same three rigid disks used in Fig. 5.21 are shown in Fig. 5.22. Whereas a flying height of 0.35 μm may be measured by optical interference techniques, the value of 0.1 μm is somewhat speculative. No interference color or bands can be observed

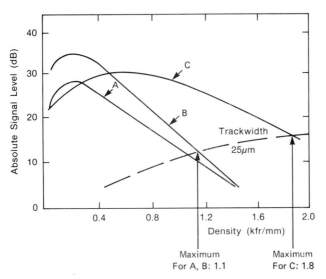

Figure 5.21 The absolute signal versus linear recording density for three recording media with a 0.35-μm head-to-medium spacing. Curve A: oxide medium, $H_c = 24$ kA/m, $\delta = 750$ nm; curve B: metal film, $H_c = 32$ kA/m, $\delta = 100$ nm, curve C: metal film, $H_c = 72$ kA/m, $\delta = 27$ nm.

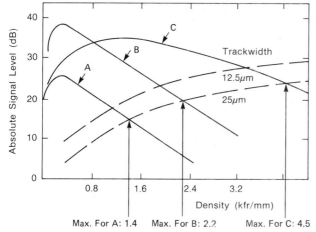

Figure 5.22 The same as Fig. 5.21, but with a head-to-medium spacing of 0.1μm.

below one-quarter of a wavelength of visible light; the value of 0.1 μm was obtained by extrapolation from higher speeds and heights. The reduction in flying height has little effect upon the performance of the relatively thick oxide. The metallic film disks, being thin and thus able to utilize the greatly increased writing-head magnetic field gradients associated with the smaller spacing, have greatly improved performance. The regular film disk now supports 2.2 kfr/mm (55 kfr/in), and the special high-coercivity disk supports 4.4 kfr/mm (112 kfr/in) on a 25-μm trackwidth.

5.5.2.3 Operating "in contact."   To avoid the mechanical complications associated with in-contact operation upon rigid disks, the data shown in Fig. 5.23 were all taken upon tapes. The four tapes were a cobalt surface-modified γ-Fe_2O_3 particle consumer video tape with a 52-kA/m (650-Oe) coercivity and 5-μm coating thickness, a cobalt bulk-diffused γ-Fe_2O_3 isotropic particle experimental tape with a 64-kA/m (800-Oe) coercivity and 5-μm coating thickness, a metallic Fe particle tape with a 116-kA/m (1450-Oe) coercivity and 5-μm coating thickness, and, finally, an evaporated cobalt-nickel metallic film tape with an 80-kA/m (1000-Oe) coercivity and 75-nm coating. The latter two tapes are of the kind developed for the 8-mm video cassette recorder. The regular consumer video tape has sufficient signal output to support operation on a 25-μm track width at 2 kfr/mm (50 kfr/in); this agrees with the specifications of a VHS video cassette recorder where the upper band-edge wavelength is 1 μm, corresponding to 0.5-μm bit cells or 2 kfr/mm. The experimental isotropic tape and the iron particle tape may be operated at 3 kfr/mm (75 kfr/in). This agrees with the 8-mm video cassette recorder specifications, which call for

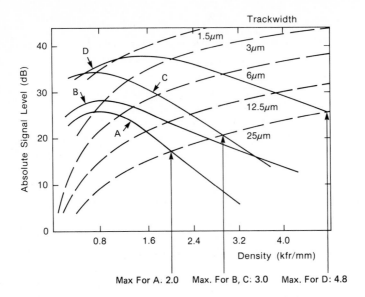

Max For A. 2.0   Max. For B, C: 3.0   Max. For D: 4.8

Figure 5.23   The absolute signal levels for four tapes operated with no deliberate head-to-medium spacing. Curve A: Co-modified Fe_2O_3, H_c = 52 kA/m, δ = 5 μm; curve B: isotropic cobalt-doped iron oxide, H_c = 64 kA/m, δ = 5 μm; curve C: iron particle tape, H_c = 116 kA/m, δ = 5 μm; curve D: evaporated metal tape, H_c = 80 kA/m, δ = 75 nm.

a minimum wavelength corresponding to 3.2 kfr/mm. The cobalt-nickel evaporated tape has exceptional performance, yielding over 4.8 kfr/mm potential linear density on a 25-μm track width.

It can be shown that 0 dB on this calibrated scale is equivalent to a sinusoidal peak magnetic flux density of 0.8 kA/m (10 Oe) entering the reproduce head; 40 dB thus corresponds to approximately 80 kA/m (1000 Oe). Since the fields emanating from a recording medium cannot exceed the coercivity, and the metal-film media all have coercivities in the range of 64 to 80 kA/m (800 to 1000 Oe), it seems that signal levels very close to the maximum possible are being observed. Higher signals would necessitate even higher coercivities.

It will be noted that the metallic films have extremely high output voltages at high bit densities. It has been found, both theoretically and experimentally, that this is true regardless of the orientation of the magnetization in the metallic film (Mallinson and Bertram, 1984). It is expected, moreover, that, in applications limited by reproduce-head noise, the system performance will be very nearly the same regardless of how a metallic film has been produced. Second-order differences due to differing deposition techniques are expected to affect only the medium noise, which will, eventually, be of little consequence.

5.5.3 The next decade

Areal densities are more difficult to forecast than linear densities because there is more uncertainty concerning the track widths and track densities which will be used. Current consumer video cassette recorders and computer rigid-disk recorders have approximately 15 μm minimum track widths. In the video recorders, no guard band is needed, and the maximum track densities are over 60 tracks per millimeter (1500 tracks per inch). It seems certain that substantial increases will be achieved through the use of embedded servo techniques, which are already employed. A track width of 6 μm with 160 tracks per millimeter (4000 tracks per inch) is a reasonable expectation in several years.

With computer rigid-disk systems, a substantial guard band, equal to half the track width, is the norm. Positioning and servoing systems are improving only incrementally, and track densities seem unlikely to substantially exceed 80 tracks per millimeter (2000 tracks per inch) in the short term.

Table 5.1 gives the areal densities expected given these assumptions about track densities. The principal parameter governing the performance of any magnetic recording system is, of course, the head-to-medium spacing. Accordingly, the anticipated densities are grouped by flying height. If the present flying heights of 250 to 400 nm prove to be irreducible for mechanical or reliability reasons, then 80×10^3 fr/mm^2 (50×10^6 fr/in^2) will be barely possible. If flying as low as 100 to 150 nm proves to be fea-

TABLE 5.1 Areal Bit Densities Predicted within the Next Decade for Various Flying Heights and Types of Media

Type of medium	Linear density, b/mm (25 b/in)	Track density, t/mm (25 t/in)	Areal density, b/mm$^2 \times 10^3$ (625 b/in^2)
350-nm Flying Height			
γ-Fe$_2$O$_3$	1000	40	40
Metal film, 32 kA/m (400 Oe)	1160	40	46
Metal film, 72 kA/m (900 Oe)	1800	40	72
100-nm Flying Height			
γ-Fe$_2$O$_3$	960	80	77
Metal film, 32 kA/m (400 Oe)	1900	80	150
Metal film, 72 kA/m (900 Oe)	3200	80	260
In-Contact Operation			
Co-γ-Fe$_2$O$_3$, VHS	1500	80	120
Co-γ-Fe$_2$O$_3$, isotropic	1900	80	150
Metal particle (MP)	1700	160	270
Metal evaporated (ME)	3000	160	480

sible, then densities exceeding 230×10^3 fr/mm^2 (150×10^6 fr/in^2) will become usable. For in-contact operations, approximately 160×10^3 fr/ mm^2 (100×10^6 fr/in^2) will be possible with the oxide particulate media. The elemental iron particle and evaporated cobalt-nickel metallic thin-film tapes have much higher signal levels which make feasible a higher track density of 160 tracks per millimeter (4000 tracks per inch) and will operate at areal densities of approximately 230×10^3 fr/mm^2 (150×10^6 fr/in^2) and 470×10^3 fr/mm (300×10^6 fr/in^2), respectively.

5.5.4 Ultimate areal densities

Normally, a decrease in track width results in an increase in areal density. Eventually, of course, this trend reverses, and it is of interest to determine what track width yields the maximum areal density. The minimum signal levels required to hold a 20-dB signal-to-head-noise ratio are proportional to areal density. Thus we seek the track width (dashed line) which intersects the peak of the medium signal curve. For the in-contact data for the evaporated cobalt-nickel thin metallic film, Fig. 5.23 shows that a track width of approximately 2.5 μm would be optimum. This would yield 400 tracks per millimeter (10,000 tracks per inch) at 1.6 kfr/mm (40 kfr/in) for an areal density of 620×10^3 fr/mm^2 (400×10^6 fr/in^2). For relatively "flat" signal spectra, the optimum conditions are not critical. The signal levels observed are close to the maximum possible for recording media with H_c = 80 kA/m (1000 Oe). We conclude that higher areal densities will be attainable, in the more distant future, in proportion to the coercivity. Thus if the coercivity were 160 kA/m (2000 Oe), an areal density of 1.24×10^6 fr/mm^2 (800×10^6 fr/in^2) would become potentially attainable.

References

Arratia, R. A., and H. N. Bertram, "Monte Carlo Simulation of Particulate Noise in Magnetic Recording," *IEEE Trans. Magn.*, **MAG-20**, 412 (1984).

Baugh, R. A., E. S. Murdock, and B. R. Najarajan, "Measurements of Noise in Magnetic Media," *IEEE Trans. Magn.*, **MAG-19**, 1722 (1983).

Belk, N. R., P. K. George, and G. S. Mowry, "Noise in High Performance Thin-Film Longitudinal Magnetic Recording Media," *IEEE Trans. Magn.*, **MAG-21**, 1350 (1985).

Bertram, H. N., and L. D. Fielder, "Amplitude and Bit Shift Comparisons in Thin Metallic Media," *IEEE Trans. Magn.*, **MAG-19**, 1605 (1983).

Bertram, H. N., M. Stafford, and D. Mills, "The Print-Through Phenomenon and Its Practical Consequences," *J. Audio Eng. Soc.*, **28**, 690 (1980).

Coleman, C., D. Lindholm, D. Petersen, and R. Wood, "High Data Rate Recording in a Single Channel," *Int. Conf. Video Data Recording, IERE Proc.*, **59**, 151 (1984).

Daniel, E. D., "Tape Noise in Audio Recording," *J. Audio Eng. Soc.*, **20**, 92 (1972).

Daniel, E. D., and P. E. Axon, "Accidental Printing in Magnetic Recording," *BBC Q.*, **4**, 241 (1950).

Decker, S. K., and C. Tsang, "Magneto-Resistive Response of Small Permalloy Features," *IEEE Trans. Magn.*, **MAG-16**, 643 (1980).

Fayling, R. F., T. J. Szczech, and E. F. Wollack, "A Model for Overwrite Modulation in Longitudinal Recording," *IEEE Trans. Magn.*, **MAG-20**, 718 (1984).

Felix, M. O., and H. Walsh, "F.M. Systems of Exceptional Bandwidth," *Proc. Inst. Electr. Eng.*, **112**, 1659 (1965).

Fujiwara, T., "Nonlinear Distortion in Long-Wavelength AC Bias Recording," *IEEE Trans. Magn.*, **MAG-15**, 894 (1979).

Gibby, R. A., and J. W. Smith, "Some Extensions of Nyquist's Telegraph Transmission Theory," *Bell Syst. Tech. J.*, **44**, 1487 (1965).

Lindholm, D. A., "Spacing Losses in Finite Track Width Reproducing Systems," *IEEE Trans. Magn.*, **MAG-14**, 55 (1978).

Mallinson, J. C., "Maximum Signal-to-Noise Ratio of a Tape Recorder," *IEEE Trans. Magn.*, **MAG-5**, 182 (1969).

Mallinson, J. C., "One-Sided Fluxes—A Magnetic Curiosity?" *IEEE Trans. Magn.*, **MAG-9**, 678 (1973).

Mallinson, J. C., "The Signal-to-Noise Ratio of a Frequency Modulated Video Recorder," *EBU Rev. Tech.*, **153**, 241 (1975).

Mallinson, J. C., "On the Properties of Two-Dimensional Dipoles and Magnetized Bodies," *IEEE Trans. Magn.*, **MAG-17**, 2453 (1981).

Mallinson, J. C., "The Next Decade in Magnetic Recording," *IEEE Trans. Magn.*, **MAG-21**, 1217 (1985).

Mallinson, J. C., and H. N. Bertram, "A Theoretical and Experimental Comparison of the Longitudinal and Vertical Modes of Magnetic Recording," *IEEE Trans. Magn.*, **MAG-20**, 461 (1984).

Mallinson, J. C., and H. Ferrier, "Motion Reversal Invariance in Tape Recorders," *IEEE Trans. Magn.*, **MAG-10**, 1084 (1974).

Mallinson, J. C., and C. W. Steele, "Theory of Linear Superposition in Tape Recording," *IEEE Trans. Magn.*, **MAG-5**, 886 (1969).

Mann, P. A., "Das Rauschen eines Magnettonbandes," *Arch. Electr. Ubertragung*, **11**, 97 (1957).

McKnight, J. G., "Erasure of Magnetic Tape," *J. Audio Eng. Soc.*, **11**, 223 (1963).

Mee, C. D., *The Physics of Magnetic Recording*, North-Holland, Amsterdam, 1964, p. 131.

Satake, S., and J. Hokkyo, "A Theoretical Analysis of the Erased Noise of Magnetic Tape," *IECE Tech. Group Meeting Magn. Rec., Jpn.*, **MR74-23**, 39 (1974).

Smaller, P., "Reproduce System Noise in Wide-Band Magnetic Recording Systems," *IEEE Trans. Magn.*, **MAG-1**, 357 (1969).

Tanaka, K., "Some Considerations on Crosstalk in Multihead Magnetic Digital Recording," *IEEE Trans. Magn.*, **MAG-20**, 160 (1984).

Tanaka, H., H. Gato, N. Shiota, and M. Yanagisawa, "Noise Characteristics in Plated Co-Ni-P Films for High Density Recording Medium," *J. Appl. Phys.*, **53**, 2576 (1982).

Terada, A., O. Ishii, S. Ohta, and T. Nakagawa, "Signal-to-Noise Ratio Studies on Gamma Fe_2O_3 Thin Film Recording Disks," *IEEE Trans. Magn.*, **MAG-19**, 7 (1983).

Thurlings, L. F. G., "Statistical Analysis of Signal and Noise in Magnetic Recording," *IEEE Trans. Magn.*, **MAG-16**, 507 (1980).

Thurlings, L. F. G., "On the Noise Power Spectral Density of Particulate Recording Media," *IEEE Trans. Magn.*, **MAG-19**, 84 (1983).

Thurlings, L. F. G., "Basic Properties of A.C. Noise," *IEEE Trans. Magn.*, **MAG-21**, 36 (1985).

Tochihara, S., Y. Imaoka, and M. Namikawa, "Accidental Printing Effect of Magnetic Recording Tapes Using Ultra-Fine Particles of Acicular γFe_2O_3," *IEEE Trans. Magn.*, **MAG-6**, 808 (1970).

Wachenschwanz, D., and F. Jeffers, "Overwrite as a Function of Record Gap Length," *IEEE Trans. Magn.*, **MAG-21**, 1380 (1985).

Wallace, R. L., "The Reproduction of Magnetically Recorded Signals," *Bell Syst. Tech. J.*, **30**, 1145 (1951).

Yoshida, K., T. Okuwaki, N. Osakabe, H. Tanabe, Y. Horiuchi, T. Matsuda, K. Shinagawa, A. Tonomura, and H. Fujiwara, "Observations of Recorded Magnetization Patterns by Electron Holography," *IEEE Trans. Magn.*, **MAG-19**, 1600 (1983).

6

Recording Measurements

James E. Monson

Harvey Mudd College
Claremont, California

This chapter treats various aspects of magnetic measurements and evaluation of system performance in relation to magnetic recording. It is assumed that the reader already has some familiarity with basic magnetic measurement techniques. Excellent texts are available which describe these techniques (Cullity, 1972; Zijlstra, 1967). Other useful sources include a brief review of measurements for digital recording media (Newman, 1978) and a text describing the principles and applications of scanning electron microscopy (Wells, 1974).

6.1 Media Properties

6.1.1 Hysteresis loop parameters

Chapter 3 on recording media presents definitions of the important hysteresis loop parameters and their relation to recording performance. These parameters are reviewed here in the context of how they are measured. Figure 6.1 shows a typical hysteresis loop measured quasi-statically with a maximum applied field of H_m. It is important to note the value of H_m because loop parameters are functions of the maximum applied field. $M(H_m)$ is the maximum magnetization observed on the loop. The saturation magnetization M_s is the limiting value $M(\infty)$ of $M(H_m)$ approached as the peak applied field is made higher and higher. The remanent magnetization $M_r(H_m)$ is the magnetization remaining when the applied field is reduced from H_m to zero. The retentivity $M_r(\infty)$ is the remanent magnetization which would be observed as the peak applied field is made infinitely large. The coercivity H_c is the field required to reduce the magnetization to zero; its measured value, $H_c(H_m)$, also depends upon H_m.

The remanence squareness $S(H_m)$ is the ratio of remanent magnetization $M_r(H_m)$ to maximum magnetization $M(H_m)$. Another characterization of squareness, the coercivity squareness S^*, is also illustrated in Fig. 6.1 (Williams and Comstock, 1971). The coercivity squareness is related to the slope of the M versus H curve at $H = H_c$ and is given by

$$S^* = 1 - \frac{M_r(H_m)/H_c(H_m)}{dM/dH} \tag{6.1}$$

The coercivity squareness approaches unity as dM/dH becomes infinitely large.

In addition to conventional hysteresis loops, remanence curves are measured to characterize recording media. The virgin remanence curve is measured by first erasing the medium in an ac field. In this condition of

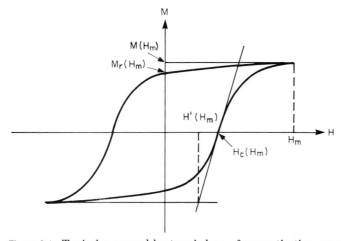

Figure 6.1 Typical measured hysteresis loop of magnetization versus applied field with peak value H_m. Graphical determination of remanence squareness, $S = M_r(H_m)/M(H_m)$, and coercivity squareness, $S^* = H'(H_m)/H_c(H_m)$, is shown.

zero magnetization, 50 percent of the particles are magnetized in the opposite sense to the remaining 50 percent. A small dc field is applied and then removed. The remanent magnetization $M_r(H)$ is measured and plotted versus H as the process is repeated and H increased. $H_{0.5}$ is the field at which 25 percent of the particles have switched.

$$\frac{M_r(H_{0.5})}{M_r(\infty)} = 0.50 \tag{6.2}$$

The reverse remanence curve is measured by saturating the sample and then applying a reverse field $-H$. The magnetization remaining after the reverse field is removed is plotted versus H. Figure 6.2 illustrates the mea-

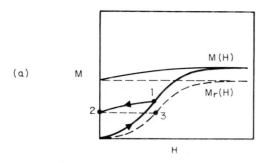

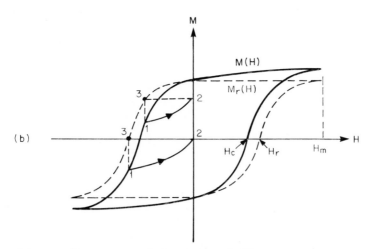

Figure 6.2 Measurement of (a) virgin and (b) reverse remanence curves (dashed). Magnetization paths from hysteresis loop to remanence curve are shown.

surement of both the virgin (Fig. 6.2a) and reverse remanence curves (Fig. 6.2b). The remanence coercivity H_r is the field which must be applied to a saturated sample to bring its remanent magnetization to zero.

Hysteresis measurements may be used to characterize the distribution of switching fields. A commonly used technique is to plot the derivative of the M-H loop versus the applied field H. The width ΔH between the half-amplitude points is normalized by H_c to give an estimate of the switching-field distribution suitable for comparing different media. This method is not sensitive to irreversible switching processes and may be very misleading when these are important. It is much better to define switching-field distribution in terms of the reverse remanence curve, M_r versus H, which measures irreversible magnetization changes. The parameter corresponding to the normalized half width of the dM/dH curve is

$$\Delta h_r = \frac{H_{0.75} - H_{0.25}}{H_r} \tag{6.3}$$

where $H_{0.25}$ and $H_{0.75}$ are the fields corresponding to 75 and 25 percent of the particles having been switched. These conditions are observed when the magnitudes of the reduced remanent magnetizations are equal to one-half.

$$\left| \frac{M_r(H_{0.25})}{M_r(\infty)} \right| = \left| \frac{M_r(H_{0.75})}{M_r(\infty)} \right| = 0.5 \tag{6.4}$$

The quantity $1 - S^*$, which can be obtained from the M-H loops, is a good estimate of Δh_r (Köster, 1984).

6.1.2 Anhysteretic parameters

Anhysteretic magnetization results when an alternating field is applied along with a dc field. There are two modes of applying the fields to a sample of recording medium. In the first, the alternating field is reduced slowly to zero while the dc field is held constant. This process is often called *ideal anhysteretic magnetization.* In the second, referred to as *modified anhysteretic magnetization,* both the alternating and dc fields are reduced slowly to zero at the same rate. The second case corresponds more closely to practical anhysteretic, or ac-biased, recording where both bias and signal currents are applied to the record head simultaneously. Figure 6.3a and b shows typical ideal and modified anhysteretic magnetization curves as a function of H_{dc}, the applied dc field. The anhysteretic susceptibility, χ_{ar}, is the slope of the curve $dM_{ar}(H_{\text{dc}})/dH_{\text{dc}}$.

The ac demagnetization curve may be measured by using the ideal anhysteretic mode (with $H_{\text{dc}} = 0$) initiated from the negative saturation

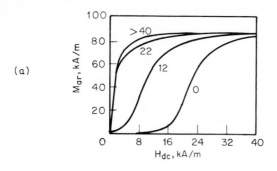

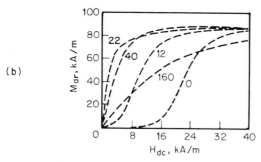

Figure 6.3 Plots of (*a*) ideal and (*b*) modified anhysteretic magnetization curves for various ac-bias levels in kiloamperes per meter *(Daniel and Levine, 1960).*

state instead of the ac-demagnetized state. The parameter H_{ac} is the peak applied ac field at which 25 percent of the particles have switched:

$$\frac{M_r(H_{ac})}{M_r(\infty)} = -0.5 \tag{6.5}$$

6.1.3 Anisotropy parameters

The most commonly measured anisotropy parameter is the orientation ratio, which is the ratio between the remanent magnetizations measured in two orthogonal directions. For conventional longitudinal media, the directions are chosen to be longitudinal, that is, along the direction of writing, and transverse to that direction but in the plane of the medium. For perpendicular media, the directions are taken to be in plane and perpendicular to the plane of the medium. The anisotropy constant K and anisotropy field H_k also characterize the anisotropic behavior of the material.

6.1.4 Instruments

6.1.4.1 Vibrating sample magnetometer. The vibrating sample magnetometer (VSM) is the instrument most often used for accurately measuring the magnetic properties of recording media. Figure 6.4 shows a diagram of a typical apparatus. A large electromagnet applies a uniform dc field to the sample. The resulting magnetization induced in the sample is then measured by vibrating the sample to produce a voltage in a pair of pickup coils. The coil output voltage is combined with the output from the displacement transducer to produce a magnetization signal. Variations in vibration amplitude and frequency are canceled out. The signal is then detected, usually by a lock-in amplifier, and fed along with the applied field signal to an xy plotter to generate a hysteresis loop.

Some care must be taken in the design of the instrument to ensure the uniformity of the fields and field gradients involved and to calibrate the measurements properly (Foner, 1959). Design of the pickup coils is simplified by the application of the reciprocity theorem (Mallinson, 1966), and detailed analyses of various configurations have been made (Pacyna and Ruebenbauer, 1984).

Commercially available vibrating sample magnetometers have many special features, such as computer control and automatic data acquisition, temperature control of the sample, and rotational positioning of the sample. The vibrating sample magnetometer is well suited for measuring major and minor hysteresis loops, initial magnetization curves, and remanent magnetization curves.

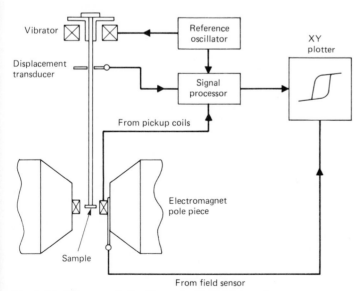

Figure 6.4 Diagram of vibrating sample magnetometer.

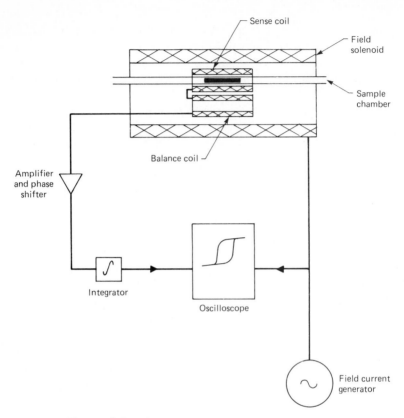

Figure 6.5 Hysteresis loop tracer.

6.1.4.2 Hysteresis loop tracers.
Hysteresis loop tracers, also known as *B-H* meters, measure steady-state hysteresis loops by applying an ac field, usually at the power-line frequency, to the sample. The induction is then measured by integrating the voltage across the sense coil. Figure 6.5 shows a schematic diagram of a representative loop tracer. With no sample present, the applied field will induce a voltage in the sense coil. This error signal must be canceled out, and usually this is done by a balancing coil which is connected in series with the sense coil. In order to obtain a correct measurement of the loop, a number of other errors must be eliminated, such as phase errors between applied current and integrator output, noise pickup, and zero drift. The reduction of these errors in the design of a 450-kA/m loop tracer has been described (Manly, 1971). Loop testers are very convenient for rapid testing of media parameters.

6.1.4.3 Torque magnetometer.
In the classical torque magnetometer, the sample is suspended from a support by a torsion wire and placed in a

uniform applied field. The torque on the sample is measured by the angular twist of the torsion wire. Figure 6.6a shows an instrument which uses electrical feedback to drive the net torque to zero in the suspension (Penoyer, 1959). This technique makes it possible to automate the measurement of torque as a function of the angle of sample orientation in the applied field. A typical torque curve measured for a sample of perpendicular recording medium is shown in Fig. 6.6b (Iwasaki and Ouchi, 1978). From interpretation of torque curves, anisotropy behavior of the sample material can be determined. Most torque magnetometers have been homemade in individual laboratories. A number of these instruments covering a wide range of torques and sensitivities have been described (Pearson, 1979).

6.1.5 Measurement techniques

6.1.5.1 Differences between particulate and film media

Sample preparation. Particulate media on flexible substrates can be measured in a number of configurations. Magnetic coating thicknesses may range from 1 to 20 μm. For the remanent magnetizations usually encountered, one thickness of material suffices for the sensitivity of most instruments. If more output signal is required, several thicknesses may be laminated together. For in-plane measurements in a loop tracer, the samples may be made long and rolled up into a tube so that skewing of the hysteresis loop by demagnetizing effects is minimized.

Particular care must be taken in preparing samples for measurement of anhysteretic remanence. In the course of this measurement, demagnetizing fields caused by the shape of the sample can mask the effects of small interaction fields to produce erroneous results. It is good practice to use a sample shape with a demagnetizing factor in the applied field direction of less than 0.0001. Effects of internal demagnetizing fields, caused by voids and particle clusters in the sample, must also be considered.

To take account of various particle loadings of the coatings, it is useful to determine the specific magnetic moment σ_s, defined as the saturation magnetization divided by the bulk material density.

Samples for the vibrating sample magnetometer must be small enough to fit within the uniform field working volume of the instrument. The shape of the sample may then require deskewing of the measured hysteresis loops to remove demagnetizing field effects. Because the samples are rarely ellipsoids with uniform demagnetizing field, care must be taken to determine an appropriate average demagnetizing factor over the entire sample. This choice is particularly important for measurements perpendicular to the plane of the medium, for which an inappropriate demagnetizing factor can result in false reentrant loops.

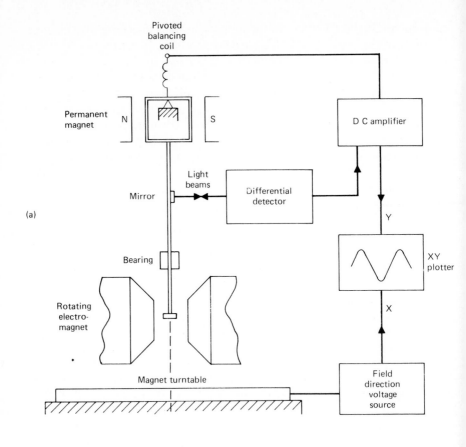

(a)

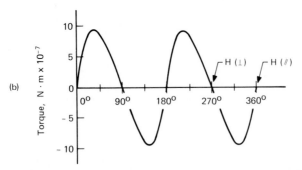

(b)

Figure 6.6 (a) Torque magnetometer *(after Penoyer, 1959)* and (b) torque curve measured for sample of perpendicular recording medium *(Iwasaki and Ouchi, 1978)*.

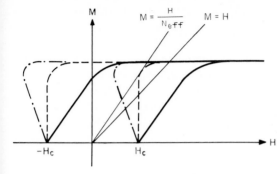

Figure 6.7 Deskewing of hysteresis loop measured for sample of film perpendicular recording media (—). Deskewing with a demagnetizing factor $N = 1$ gives reentrant loop (— • —). Deskewing with N_{eff} determined from slope of measured loop at $-H_c$ gives well-behaved loop (— — —).

Figure 6.7 shows a typical hysteresis loop of magnetization versus applied field for a sample of film perpendicular recording media measured in a vibrating sample magnetometer. The measured loop must be deskewed using the relationship

$$H = H_{appl} - NM \tag{6.6}$$

where H_{appl} is the field applied in the vibrating sample magnetometer. For a thin sample shape, N approaches a limiting value of 1. However, when $N = 1$ is used to deskew the loop, the resulting loop is a reentrant loop, as shown in Fig. 6.7. A better estimate for the effective demagnetizing factor N_{eff} can be derived from the slope of the M versus H_{appl} loop at the coercivity point, $H_{appl} = -H_c$ (Chu, 1984). In the neighborhood of this point, the magnetization of the *deskewed* loop is approximately

$$M = \chi_c(H + H_c) \tag{6.7}$$

Substituting for H from Eq. (6.6) and rearranging gives

$$M = \frac{\chi_c}{1 + \chi_c N_{eff}} (H_{appl} + H_c) \tag{6.8}$$

Differentiating with respect to H_{appl} gives the slope of the M versus H_{appl} loop

$$\left[\frac{dM}{dH_{appl}}\right]_{-H_c} = \frac{\chi_c}{1 + \chi_c N_{eff}} \tag{6.9}$$

For large χ_c, corresponding to high-coercivity squareness,

$$\left[\frac{dM}{dH_{appl}}\right]_{-H_c} \approx \frac{1}{N_{eff}} \tag{6.10}$$

Loops deskewed using N_{eff} determined from Eq. (6.10) are well behaved, as shown in Fig. 6.7. It should be noted, however, that the analysis assumes that N_{eff} is not a function of applied field.

Special measurement apparatus has been developed to study the properties of individual particles. Knowles (1978, 1980) describes sample preparation and special measurement techniques required for this work. A very dilute suspension of particles is made in a transparent viscous lacquer. Ultrasonic vibration produces wide dispersion of the particles so that interactions between individual particles or small agglomerates of particles are small. The solution is sucked into a small, flat capillary tube which is placed in a region of uniform, controllable magnetic field. By observing the particles on a television display of the image of a 10^4 magnification optical microscope, it is possible to observe single particles as distinct from small agglomerates. By aligning a particle in a field and then applying a pulsed field in the reverse direction of sufficient amplitude to rotate the particle by 180°, the remanence coercive force of the particle can be readily obtained. A remanent loop may be drawn by observing the time for the particle to rotate 90° as a function of the pulsed field amplitude.

Most film media are deposited on rigid, conducting substrates. The conducting substrate complicates the measurements with the ac field used in hysteresis loop plotters; thus the vibrating sample magnetometer is more commonly used. Although film media often have higher retentivity than particulate media, the recording layers are much thinner so that the magnetization in samples with the same area may differ by as much as a factor of 20. Samples are cut out of the medium and are typically 10-mm squares or disks. These sizes require instrument sensitivities on the order of 0.1 μmA$\cdot$m^2 (0.0001 emu) or better. Demagnetizing effects must also be taken into account for out-of-plane measurements.

Anisotropy characterization. Anisotropy is often characterized by measuring hysteresis loops in two, sometimes three, orthogonal directions. For isotropic media, such curves should be identical. The anisotropy magnitude can be inferred from the shapes and parameters of the loops. Often, a simple characterization is made in terms of a single parameter, for example, remanent magnetization M_r. This parameter leads to the orientation ratio defined earlier. Sometimes, remanence squareness is used as the single parameter.

The anisotropy constants are usually more important for describing film media performance. This is particularly so for perpendicular film media, for which torque magnetometer measurements can be used to obtain K and the anisotropy field H_k. Anisotropy field characterization of particulate media is also possible. A method for obtaining the anisotropy field from measurement of initial reversible susceptibility of ac-erased particulate media samples has been reported (Köster, 1970; Kullmann et al., 1984).

Amplitude effects. For particulate media, hysteresis loop shapes can change significantly as a function of the peak amplitude of the field, even for fields 10 times the coercivity. Figure 6.8 shows typical variations of maximum magnetization and remanent magnetization versus maximum applied field. All variables are normalized with respect to their values as $H_m \to \infty$. It is seen that $M_r(H_m)$ approaches its final value fairly rapidly, while $M(H_m)$ approaches the final saturation value more slowly. This behavior causes a wide variation in the apparent remanence squareness of the material as seen in the plot of $S(H_m)/S(\infty)$. Fortunately, the approach of $M(H_m)$ to saturation is predictable for most particulates and follows closely a $1/H$ law:

$$M(H_m) = M_s - \frac{a}{H_m} \tag{6.11}$$

where a is a constant. This behavior has several important practical consequences. The saturation magnetization can be accurately estimated by plotting $M(H_m)$ as a function of $1/H_m$ and making a linear extrapolation

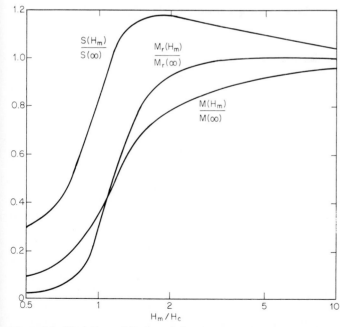

Figure 6.8 Variation of hysteresis loop parameters with maximum applied field strength.

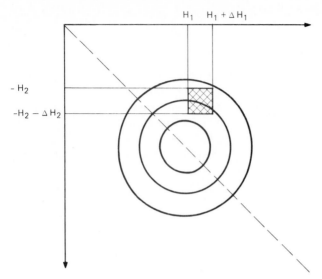

Figure 6.9 Preisach distribution contours in interaction-field plane (H_1, H_2). Particle density in shaded area can be determined from magnetization measurements.

to $1/H_m = 0$ to obtain M_s. Using this method avoids the need to produce the very high fields required to saturate the samples completely. Field values of roughly 10 times the coercivity are adequate.

To eliminate the large variations in apparent squareness as a function of the peak applied field H_m, M_s determined from the extrapolation method should be used to obtain the squareness of a particulate sample.

Film media do not appear to approach saturation in a simply predictable way. In most cases, they approach saturation faster than a $1/H$ law, and so that peak applied fields of 10 times the coercivity are adequate for characterization of saturation behavior. Because of the relatively rapid approach to saturation, squareness is not so variable with applied field as for particulate media.

6.1.5.2 Preisach diagram measurement. The Preisach diagram, treated in Chap. 2, is a useful way of depicting the distribution of interaction fields in recording media. Figure 6.9 shows a Preisach diagram in terms of interaction fields H_1 and H_2. The relative number of particles in the shaded region with area $\Delta H_1 \, \Delta H_2$ can be determined from magnetization measurements. Define the magnetization $M(H_1, -H_2)$ as that resulting from applying to the sample a field sequence of negative saturation, positive field of amplitude H_1, and negative field of amplitude H_2. The saturation magnetization $M(\infty)$ may be expressed in terms of a particle-interaction-field distribution function, $K(H_1, H_2)$, as

$$ M(\infty) = 2 \int_{-\infty}^{0} \int_{0}^{\infty} K(H_1, H_2) \, dH_1 \, dH_2 \tag{6.12} $$

The distribution function $K(H_1, H_2)$ may then be obtained from

$$K(H_1, H_2) = \frac{1}{2 \Delta H_1 \Delta H_2} \{M[(H_1 + \Delta H_1), -H_2] + M[H_1, -(H_2 + \Delta H_2)]$$

$$- M(H_1, -H_2) - M[(H_1 + \Delta H_1), -(H_2 + \Delta H_2)]\} \qquad (6.13)$$

A number of workers have used experimentally determined Preisach diagrams to characterize particulate media. Using an automatic measurement of the Preisach diagram, magnetization measurements for the field sequences shown in Fig. 6.9 are made to determine the relative number of particles in the hatched area on the diagram (Völz, 1971).

6.1.5.3 Vector magnetization measurements. In magnetic recording, the field experienced by the recording medium as it passes the head is a time-varying vector field. Consequently, scalar M versus H curves, even three curves for orthogonal components, can only coarsely predict recording performance. A true vector magnetization characterization would be superior. Several approaches to vector magnetization measurements have been made in which programmable currents were applied to two orthogonal sets of Helmholtz coils which magnetized a sample placed at their center (Lemke and McClure, 1966; Clark and Finegan, 1985). The sample experienced a rotating field history similar to what it would experience if passed under a recording head. After the completion of the vector-field program, the vector magnetization was measured by spinning the sample in the sense coils. A number of complicated rotating-vector recording fields occur in magnetic recording. The further development of instruments which can simulate a specified vector-field history and measure the resulting magnetization in a sample would find useful application.

6.1.5.4 Measurement of time-scale effects. In measuring hysteresis loops, the time dependence of switching mechanisms can influence the experimental results significantly. For example, the measured values of coercivity from a relatively long time-scale vibrating sample magnetometer may differ from the results of a 60- or 50-Hz hysteresis loop tester by as much as 10 percent. This behavior is caused by the temperature dependence of the energy barrier which must be overcome for the domains to switch.

Assuming that the magnetization reversal of an ensemble of identical, noninteracting, single-domain particles follows first-order kinetics, the rate constant r is given by

$$r = Ae^{-\Delta E/kT} \qquad (6.14)$$

where A = frequency of approach to the barrier
ΔE = height of the energy barrier
k = Boltzmann's constant
T = absolute temperature

For particles of uniaxial anisotropy, an expression for the coercivity as a function of the measurement time scale t can be derived (Kneller and Luborsky, 1963; Sharrock, 1984):

$$H_c(t) = H_k \left[1 - \left(\frac{kT}{K_u v} \ln \frac{At}{0.69} \right)^{1/2} \right] \qquad (6.15)$$

K_u is the uniaxial anisotropy constant, and v is the volume of each particle. By making measurements of H_c on two different time scales, it is possible to determine H_k and the $K_u v$ product. From these, the coercivity on any time scale may be predicted, and such a characterization is very useful for evaluating high-frequency writing performance, long-term storage stability, and print-through.

6.2   Recorded Magnetization

Measurement of recorded magnetization patterns can provide valuable information about the recording process. Most magnetization measurement techniques are indirect and measure the external fields produced by the magnetization pattern. It is not possible to infer uniquely the magnetization distribution from field measurements, no matter how accurate or complete (Mallinson, 1981). The field can be represented in terms of net dipole, quadrupole, and higher-order moments of the magnetization through a multipole expansion. The difficulty is that many magnetization distributions have the same moments. In spite of this basic limitation, field measurements on recorded media are very useful and plausible, though not unique, and magnetization distributions may be inferred using additional knowledge of media and head-field properties.

6.2.1   Bitter pattern techniques

Magnetized media are coated with a suspension of very fine magnetic particles which are attracted to regions of nonuniform field produced by the magnetization pattern in the medium. The resulting patterns of particles are then viewed under a microscope. This technique was first used to observe domain walls in ferromagnetic metals (Bitter, 1931). Several methods of preparing the colloidal magnetic suspension have been developed (Bozorth, 1951; Chikazumi, 1964). Another approach has been to use a magnetic viewer (Youngquist and Hanes, 1961) containing an aqueous dispersion of Fe_2O_3 platelets. In regions of high field normal to the medium, the platelets stand on edge and reflect less light, causing a longitudinally recorded transition to appear dark. Figure 6.10 shows a typical pattern seen using the Bitter method to observe the magnetization recorded. Bitter methods are widely used because of their convenience.

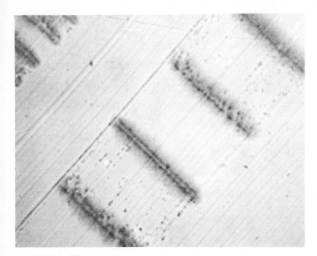

Figure 6.10 Bitter pattern of recorded magnetization transitions *(Tong et al., 1984).*

Although they give a useful qualitative picture of magnetization patterns, quantitative information is lacking. Resolution is limited by the particle size, which may be as small as 0.1 μm.

6.2.2 Fringing-field measurement

To obtain a quantitative description of the magnetization, the stray or fringing fields produced by the recorded magnetization may be measured directly. One high-resolution system uses a Hall probe (Lustig et al., 1979). The active area of the Hall probe is 2 by 2 μm, and it can be accurately positioned to a minimum separation of 1 μm from the surface of the medium. In principle, finite separation from the surface of a single-layer medium is not a problem because the Fourier transforms of the fields at any two separations are related by the exponential spacing-loss formula. Measurement accuracy imposes a practical limit on transferring field measurements from one spacing to another. High-resolution piezo-electric positioners and electronic position sensors are used to achieve 0.1-μm resolution under computer control, and the minimum detectable fields are 200 to 300 A/m (3 to 4 Oe). The system has been used to study tran-transitions recorded on longitudinal and perpendicular media (Baird et al., 1984).

6.2.3 Lorentz microscopy

In Lorentz microscopy, electrons in a beam directed at the medium sample are deflected by the Lorentz force, $\mathbf{F} = q\mathbf{V} \times \mathbf{B},$ which is produced

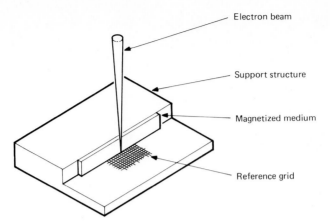

Electron beam

Support structure

Magnetized medium

Reference grid

Figure 6.11   Measurement of fringing field by deflection of electron beams *(Thornley and Hutchison, 1969)*.

by the magnetization in the medium. The deflection of the electrons is measured, and the magnetization M and fringing field H are inferred. Experiments are generally performed in commercially available scanning electron microscopes (SEM) or transmission electron microscopes (TEM).

Several techniques have been used to measure the fringing field above the surface of the medium with a scanning electron microscope. Figure 6.11 shows the electron beam directed parallel to the medium surface and normal to a magnetized track. The deflection of the beam is sensed below the recording by use of a microchannel plate as a two-dimensional detector of the beam's deflection (Elsbrock and Balk, 1984). The two components of stray field can be determined from the displacement in two dimensions. The deflection of the beam can also be determined by observing how the image of a reference grid is distorted (Thornley and Hutchison, 1969).

When the electron beam is directed so as to strike the medium sample, fringing fields or magnetization can be measured by using the SEM type 1 or type 2 methods. In the type 1 method, low-energy secondary electrons are produced when the primary electrons hit the medium surface. The detector is a collector of these secondary electrons and is sensitive to the direction from which they leave the surface (Banbury and Nixon, 1967; Joy and Jakubovics, 1968). The direction of the secondary electrons is dependent on the surface magnetic field and topography. The image produced from the collected electrons has both magnetic and topographic contrast. For a smooth, thin-film medium, the topographic contrast is small. For inhomogeneous specimens, there may be additional contrast caused by variation in the secondary emission coefficient along the surface. Several methods have been used to reduce these effects (Cort and Steeds, 1972; Griffiths et al., 1972).

In the type 2 method, higher-energy incident electrons are used, and the image is formed by detecting the number of electrons which are backscattered from inside the minimum (Fathers et al., 1973). The beam strikes the specimen at an oblique angle and penetrates inside. Depending upon the sense of the B field tangent to the surface of the medium, the electrons are forced toward or away from the surface. The electrons with shallower trajectories are more likely to be backscattered. Consequently, the detected intensity of backscattered electrons can be related to the B field inside the medium. Both magnetization and fields inside the medium contribute to this effect. Because the effect occurs inside the specimen, the image is relatively insensitive to the surface topography, in contrast to the type 1 method.

Magnetization microstructure can also be observed in thin samples (less than 1 μm) by passing high-energy electrons through them in a transmission electron microscope. Figure 6.12 shows a beam of electrons with intensity I_0 being deflected as they pass through a medium sample of thickness t. The magnetization is normal to the plane of the paper and has a reversal in the domain wall shown. This picture might represent a section cut normal to the domain wall of a sawtooth domain typically observed in transitions written on film media. The beam is focused on a plane which is a distance y below the surface of the medium. This method is referred to as the *defocused,* or *Fresnel,* mode. The relative intensity of the electrons striking the image plane is given by

$$\frac{I(x')}{I_0} = \left[1 + by \frac{dB_z(x)}{dx} \right]^{-1} \tag{6.16}$$

where $b = (|e|/m)(t/V)$, where $|e|/m$ is the electron charge-to-mass ratio and V is the electron velocity in the y direction. Note that there is

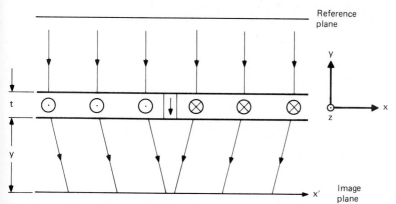

Figure 6.12 Deflection of electrons around a domain wall in a thin magnetized sample placed in a transmission electron microscope.

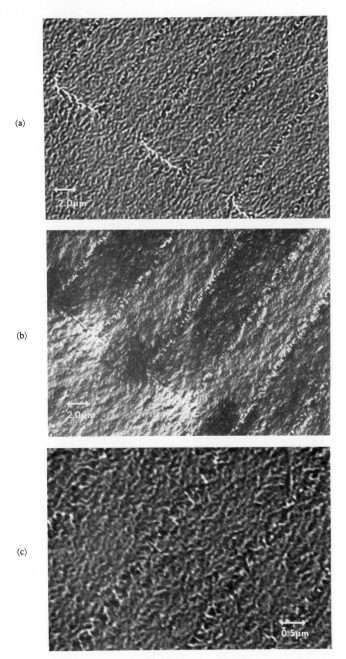

(a)

(b)

(c)

Figure 6.13 Lorentz micrographs of a portion of a recorded track on an isotropic sputtered Co-Pt alloy film medium *(Alexopoulos and Geiss, 1985).* (*a*) Fresnel mode showing contrast in regions of magnetization changes, such as domain walls. (*b*) Foucault mode showing contrast between domains of a different orientation or magnitude of magnetization. (*c*) Enlarged Fresnel micrograph showing irregular zigzag domain wall.

no variation in intensity when $y = 0$. This is why the beam must be defocused. The intensity is proportional to the derivative of the B field so that image contrast appears in regions of changing magnetization such as domain walls. Figure 6.13a and c shows typical Fresnel mode electron micrographs of a recording track region exhibiting irregular transition domain walls.

A novel method has been used to obtain both magnetization gradient and fringing-field measurements on the same magnetized sample (Chen, 1981). The sample is examined in the transmission electron microscope using the defocused mode and is then folded back on itself. At the fold, the plane of the surface fields becomes normal to the direction of the incident electron beam, and the deflection of the beam indicates the field distribution.

The transmission electron microscope may also be operated in the Foucault mode (Marton, 1948). The principle is illustrated in Fig. 6.14 in its optical version. The source illuminates the object and is focused by an objective lens onto a stop which blocks all direct rays from the source. If now some inhomogeneity is introduced in the index of refraction, some rays will miss the stop and form an image of the inhomogeneity in the image space beyond the stop. In the transmission electron microscope, inhomogeneities are produced by the magnetization in the sample, and the stopping action is achieved by a knife edge. The Foucault mode is similar to the optical schlieren method for showing shock waves in air (Wells et al., 1983). Figure 6.13b shows a Foucault image Lorentz micrograph of a recording track. Magnetization, rather than its derivative, produces contrast in this mode.

All the Lorentz microscopy techniques are labor-intensive and time-consuming in the sense that they require very careful sample preparation and adjustment of complex equipment. There is always a chance that the magnetization on the medium may change during the course of the experiments.

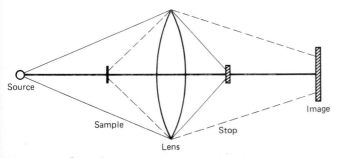

Figure 6.14 Optical magnification using the Foucault mode (*after Marton, 1948*).

This treatment of Lorentz microscopy has been based upon geometric optics. Application of geometric optics has a fundamental limit (Wohlleben, 1966). The smallest detectable flux change $\Delta\phi$ between two ray paths separated by Δx is given by $h/2e$, $\frac{1}{2}$ fluxon. To go beyond this limit, wave optics must be used (Cohen, 1967). In wave optics, the magnetization in the medium sample changes the phase of the incident electron waves, but not their amplitude. One method recently demonstrated for observing the phase change is electron holography.

6.2.4 Electron holography

The idea of holography was first proposed for electron microscopy (Gabor, 1949), but with the invention of the laser, much more work has been done with optical holography. Figure 6.15 shows the phase shift occurring when an electron wave front is incident on a magnetized medium. The difference in phase between rays 2 and 1 at a plane below the medium is given by (Aharonov and Bohm, 1959)

$$\phi_2 - \phi_1 = \frac{2\pi e}{h} \int \mathbf{A} \cdot d\mathbf{l} = \frac{2\pi e}{h} \int_s \mathbf{B} \cdot d\mathbf{s} \qquad (6.17)$$

where $\mathbf{A}$ is the vector potential.

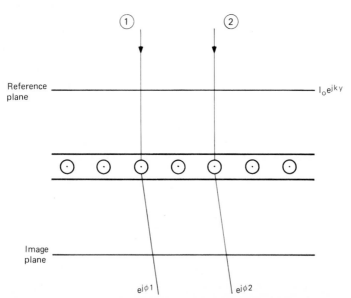

Figure 6.15 Aharonov-Bohm phase shift produced in electron wave front passing through magnetized sample. Complex representation of wave is $I_o e^{jky}$ at reference plane y. Phase shift $\phi_2 - \phi_1$ between points 2 and 1 on the image plane is proportional to flux enclosed by ray paths.

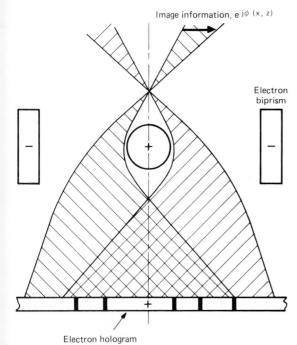

Image information, $e^{j\phi\,(x,\,z)}$

Electron biprism

Electron hologram

Figure 6.16 Hologram formation by superposing electron wave containing image phase information $\phi(x, z)$ on reference wave in an electron biprism *(after Tonomura et al., 1980).*

This phase difference is equal to 2π when

$$\int_s \mathbf{B} \cdot d\mathbf{s} = \frac{h}{e} \qquad (6.18)$$

which is 1 fluxon, approximately $4 \ \text{mT} \cdot \mu\text{m}^2$ ($40 \ \text{G} \cdot \mu\text{m}^2$). Suppose that lines of constant phase spaced apart at a constant phase interval $\Delta\phi$ were drawn on the image plane. Between any two adjacent phase contours, a tube of constant flux between the reference plane and image plane would have been enclosed by the rays which produced the contours. For thin media magnetized in plane, the **B** field may be considered to be localized within the thickness of the medium. Consequently, the equal phase contours coincide with **B**-field lines.

In order to display equal phase contours, a hologram is made by superimposing a reference wave on the wave which contains the phase information from the image plane. Figure 6.16 shows the production of the hologram by using an electron biprism to superpose the two waves (Yoshida et al., 1983). The two waves interfere to produce an intensity given by

$$I = A \cos \frac{\phi(x, z)}{2} \qquad (6.19)$$

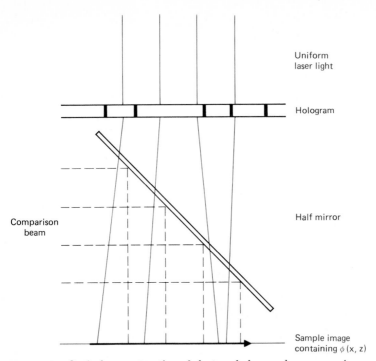

Figure 6.17 Optical reconstruction of electron hologram by superposing coherent light diffracted by hologram and uniform comparison beam to produce sample image.

where $\phi(x, z)$ is the phase shift in the image plane. This intensity variation is recorded on the hologram, which is removed from the vacuum chamber. chamber.

Reconstruction from the hologram is accomplished by uniform laser light as shown in Fig. 6.17. The light is diffracted upon passing through the hologram and acquires the phase information $\phi(x, z)$. A comparison wave superimposed upon the diffracted wave reconstructs the $\cos(\phi/2)$ variation in intensity which produced the hologram in the vacuum chamber originally. Interference fringes appear when $\phi/2 = 0, \pi, 2\pi, \ldots$, corresponding to

$$\phi = 0, 2\pi, 4\pi, \ldots$$

Two adjacent fringes enclose a flux of h/e, which is just twice the limit of detectable flux using geometric optics. In the reconstructed hologram, intensity can be measured between fringes to obtain higher resolution. By more complicated optical signal processing using multiple conjugate images, it is possible to obtain phase amplification to increase the resolution between fringes (Tonomura et al., 1980). The ability to do signal processing on the hologram outside the vacuum chamber is an important feature of electron holography.

6.3 Head Measurements

Many types of head measurements are useful in characterizing head performance. These include measurements of impedance, frequency response, noise, saturation effects, and field distributions. This section will describe several measurements developed especially for magnetic recording heads.

6.3.1 Material properties

One of the most important material properties of the soft magnetic materials used in magnetic recording heads is permeability. For conventional isotropic head materials, measurement of toroidal samples is adequate. For anisotropic films used in film ring heads or magnetoresistive heads, special permeance meters have been developed.

Calcagno and Thompson (1975) developed a permeance meter capable of measuring permeance of films with uniaxial anisotropy at frequencies up to 100 MHz. The measured permeance is the average permeability over the film cross section times its thickness. Figure 6.18 shows the permeance measuring jig into which the samples are inserted. Mutually perpendicular Helmholtz coils apply any desired preconditioning fields to the sample, for example, saturation in the easy axis. A drive strap supplies sinusoidal

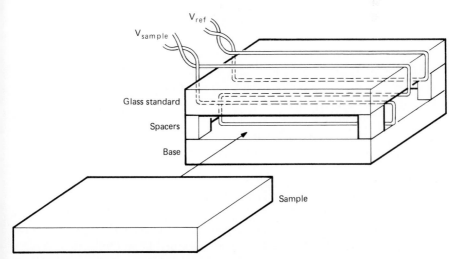

Figure 6.18 Permeance measuring jig for film samples *(after Calcagno and Thompson, 1975)*.

drive fields to the sample, inducing a voltage in the figure-eight sense coil. The coil is wound so that its induced voltage is zero with no sample inserted. An absolute permeance value is obtained by comparing the sense voltage with the voltage from the reference coil which surrounds a glass substrate of precisely known dimensions. An important feature of the technique is that the field linking the reference coil is not the applied field but the total field, which includes the demagnetizing field from the sample. The geometry has been carefully designed so that the total field in the reference coil agrees with that in the sample coil within 2 percent. This results in high accuracy and great simplification in the measurement. A number of automated procedures and data reduction techniques were designed into the apparatus.

A similar permeance meter capable of operation to 100 MHz has been reported (Kawakami et al., 1973). In addition, this meter can be used for magnetostriction measurements. The glass substrate upon which the magnetic film has been deposited is subjected to stress by means of a weighted fixture using the four-point wedge method. Magnetostriction constants as small as 0.05×10^{-6} can be measured with this meter.

6.3.2 Head fields

Many of the same techniques used to measure fringing fields from recorded magnetization patterns on media are also applicable to measuring head fields. Direct measurement of head fields is possible using miniaturized Hall probes (Lustig et al., 1979) and magnetoresistive probes (Fluitman, 1978). Lustig's apparatus is described in Sec. 2.2. Fluitman was able to use a relatively wide (3-μm) transducer to obtain useful results by shifting the position of the transducer over very small increments and processing the gathered data. He used a computer simulation which included demagnetizing effects in the magnetoresistive film to obtain the relation between the transducer output and a given applied field variation. With this approach and an air-bearing traversing table for positioning, he was able to achieve a resolution of 0.4 μm.

Another approach for measuring head fields directly is to sense the fields with an inductive pickup loop. The head gap dimensions currently used in high-density recording make it very difficult to build a pickup loop with adequate spatial resolution to resolve the head-field variations. Hoyt et al. (1984) have used a high-resolution "microloop" to measure fields from both ferrite and film heads. The microloop was fabricated using advanced processing techniques. Masks were produced by electron beam lithography. After a film of Au-Ti was deposited on a glass substrate, the final dimensions of the microloop were obtained by ion milling. The high-resolution side of the loop, which is shown in Fig. 6.19, measures 0.6 μm

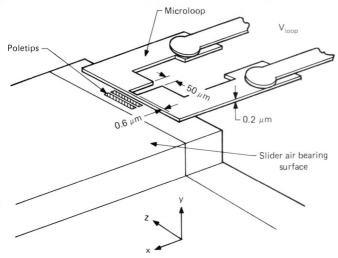

Figure 6.19 Microloop positioned for inductive sensing of fields from film head
(Hoyt et al., 1984).

wide by 50 μm long by 0.2 μm thick. The loop is accurately positioned
vertically using optical interferometry and then moved along the track to
obtain the voltage readings of the surface flux. The macroscopic portions
of the loop, including the leads, sense fields from portions of the head
structure which are far from the pole tips. Because these fields are much
more slowly varying than those emanating from the pole tips, they are
easily subtracted out. Direct measurement of the efficiency of Permalloy
film heads gave values of 75 percent at low frequencies, decreasing slowly
to 60 percent at 50 MHz. The inductive microloop technique has also been
used to study head saturation (Hoyt, 1985).

Several techniques have been developed to measure recording head
fields by Lorentz microscopy. Some of these are also applicable to measurement of fringing fields from recorded media and have already been
described. Some additional methods include the stroboscopic electron
mirror microscope (Spivak et al., 1971) and field plotting in the scanning
electron microscope (Thornley and Hutchison, 1969; Ishiba and Suzuki,
1974; Rau and Spivak, 1980; Wells and Brunner, 1983; Wells et al., 1983).

For field plotting in the scanning electron microscope, the electron
beam passes through the head field and strikes a reference grid placed
below. The beam will be deflected by the head field and cross the plane
of the reference grid at a spot displaced from the position of incidence
with no field present. This displacement of the beam will cause the image
of the reference grid to appear distorted. Figure 6.20 shows the two modes
commonly used to scan the beam in the xy plane to determine head-field

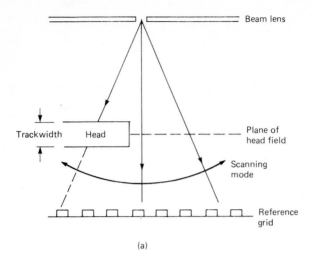

(a)

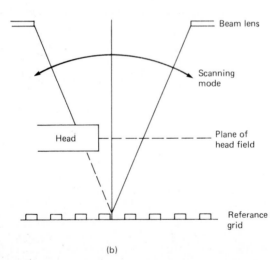

(b)

Figure 6.20 Measurement of head fields in a scanning electron microscope using (a) normal mode and (b) rocking mode *(Wells and Brunner, 1983).*

patterns. In the normal scanning mode, Fig. 6.20a, the grid distortion vector is measured at points in the xy plane and the corresponding head-field vector is calculated.

In the rocking mode, shown in Fig. 6.20b, scanning is done by rocking the beam about a fixed point in the reference grid plane, for example, the intersection of two grid conductors. In the absence of any head field, the beam will strike the conductors in the same spot during the whole scanning cycle, giving an image with constant intensity over its entire area. With field present, the image shows a distorted picture of the reference grid with the interesting and useful property that the distorted reference grid conductors now lie along contours of equal magnitude of the respective head-field components in the xy plane. Figure 6.21 shows how a rectangular reference grid would appear distorted to show contours of constant x and y head-field components. Other shapes for grids may be used. For example, a grid of concentric rings would produce a distorted image showing contours of constant vector magnitude of the head field. The field contouring property occurs in the following way. The condition that a contour will appear in the image is just that the deflection of the beam from its fixed, no-field position in the reference grid plane is sufficient for it to strike the neighboring conductor in the grid. In the rocking mode, this condition is determined only by the head-field component experienced by the beam and not by the position in the scanning cycle. Hence, as the scan is carried out, contours are mapped as the beam is deflected to strike successive conductors. Note that the mapping is inverse in that the conductor

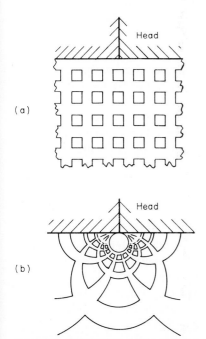

Figure 6.21   Distortion of (a) reference grid to produce (b) contours of constant x and y head-field components *(after Wells et al., 1983)*.

closest to the origin is mapped the farthest away in the xy image plane because that is where the head fields are weakest.

Camarota and Thompson (1983) made a novel measurement of recording head frequency response in a standard scanning electron microscope. They recognized that a broadband detector is not necessary because the electrons, even in a low-voltage beam of 5 kV, traverse the track of a typical high-density recording head in a very short time, on the order of a picosecond. Consequently, transit-time effects can be ignored at frequencies up to a gigahertz. A head was placed so that the electron beam passed across the head in the center of the gap and continued on to a reference grid below. The beam was then scanned in a line outwardly normal to the head. With no current applied to the head, the microscope detector current is a square wave with transitions corresponding to the edges of the reference grid conductors. When a sinusoidal current is applied to the head coil, the ac head field causes a time-varying deflection of the electron beam which shears the sharp edges of the detected square wave. The amount of shear is proportional to the peak ac field. The head coil is switched on successive line scans between a low-frequency reference current and a test current source variable in both frequency and amplitude. Response at a particular test frequency is measured by adjusting the test current amplitude to match the detector waveforms on successive line scans. The ratio of test current to reference current is the frequency response. The frequency response of film heads measured in this way was lowered by 3 dB at 100 MHz.

Each technique for head-field measurements has its particular requirements for success. Resolution problems manifest themselves in requirements for microfabrication, micromanipulation, and background cancellation for direct methods such as Hall probe or inductive sensors. Lorentz microscopy methods require careful control of head positioning and beam and detector control, alignment, and sensitivity. The experimenter has a number of attractive alternatives, but must choose carefully, consistent with requirements and resources.

6.3.3 Domain observation

The increasing interest in using film heads has led to the need for observing domain and domain wall structure in heads under both static and dynamic conditions. Bitter techniques are useful for showing static domain wall structure over reasonably large areas.

Wells and Savoy (1981) have used a scanning electron microscope to observe domains in film heads. Except for regions near the pole tips and localized regions around domain walls, the magnetic fields of a film head are largely inside the films. For an undriven head, the fringing fields near the pole tips are zero except for any residual remanent magnetization in the tips. For this reason, the type 2 magnetic contrast using backscattered electrons from the interior of the films is the most effective method for the study of domains. The nature of the film surfaces, particularly the last surface to be deposited, is such that, even after conventional techniques have been used to remove the topographic contrasts, it is still difficult to observe domain structure using type 2 magnetic contrast. Wells and Savoy overcame this problem by using a lock-in image processing technique. In this technique, a small alternating current is applied to the head coil along with a direct drive current. This results in a modulated output from the backscattered electron detector which is demodulated in a lock-in amplifier. In this way, the topographic contrast is greatly reduced. For proper signal processing, the frequency of the applied ac current must be greater than 10 times the frequency at which the picture elements are being generated in the scanning electron microscope.

Magnetooptic techniques have also been used to investigate domain behavior in film heads. Detection of the Kerr magnetooptic effect caused by the reflection of a laser beam from a magnetized head characterizes the local domain behavior. Spot sizes are on the order of 0.5 μm. Both magnetization curves and frequency responses have been measured (Narishige et al., 1984; Re and Kryder, 1984). Through the use of lock-in amplifier techniques, frequency responses up to 50 MHz have been measured. The Kerr magnetooptic effect is primarily a surface phenomenon. Consequently, the problem of interpreting surface behavior from many different locations on the head in order to predict its recording performance is a formidable one.

6.4 Large-Scale Modeling

Large-scale modeling is an attractive measurement technique for studying many recording phenomena. It serves as a useful link between highly complex computer simulations of recording using self-consistent mathematical models of media and very demanding measurements in real time and space.

6.4.1 Scaling laws

Under certain conditions, behavior of magnetic devices can be determined from scale models having the same magnetic properties as those in the modeled device (Valstyn et al., 1971). Consider Maxwell's equations:

$$\nabla \times \mathbf{E} = -\frac{\partial \mathbf{B}}{\partial t} \tag{6.20}$$

$$\nabla \times \mathbf{H} = \sigma \mathbf{E} + \frac{\partial \mathbf{D}}{\partial t} \tag{6.21}$$

The left-hand side of each equation contains first derivatives with respect to the space variables. Expanding the spatial scale by a factor of n amounts to multiplying the left-hand side by n. If the time scale is also expanded by a factor of n, the Faraday's law equation remains unchanged. In the Ampère's law equation, if the conductivity is decreased by the factor n, the equation will remain the same. Hence, the amplitudes and spatial variations of all of the vectors $\mathbf{E}$, $\mathbf{B}$, $\mathbf{H}$, and $\mathbf{D}$ will remain unchanged. If displacement current can be neglected, which is a good assumption for magnetic recording applications, then the conductivity may remain the same. That is, it is possible to use exactly the same materials in the scale model as in the simulated devices. The time scale must be expanded by the factor n^2, $\mathbf{B}$ and $\mathbf{H}$ remain the same, but $\mathbf{E}$ must be divided by the factor n.

In summary, if both space and time variables are scaled by the same factor, the scale model will faithfully simulate actual devices in both static and dynamic behavior if all electrical and magnetic material properties are identical except for conductivity, which must be reduced by the factor n in the scale model. When only static and quasi-static behavior are required, all material properties are the same, time is scaled by n^2, and the electric field is divided by n. There is no other restraint on the bulk material properties. They may be linear or nonlinear, magnetically soft or hard, and so on.

6.4.2 Large-scale modeling applications

Workers in several laboratories have built large-scale models for a number of applications. These include measurement of head fields and study of both read and write processes.

Scale models of simple head geometries have given head fields in close agreement with those calculated by numerical methods. The increasing availability of both hardware and software for head-field calculations is reducing the attractiveness of scale models for simply measuring head fields. Nonetheless, in applications where numerical studies are still difficult or time-consuming, scale models may be useful. Applications have included complex geometries (Fluitman and Marchal, 1973; Iwasaki et al., 1983; Chi and Szczech, 1984), and saturation effects (Nakagawa et al., 1972).

A scale model head can be used to study the read process (Monson et al., 1975). In this application, the recorded magnetization is simulated by a "magnetization generator" consisting of a rectangular loop of very fine wire whose plane is perpendicular to the surface of the recording medium. The outline of the loop surrounds the track width to the desired depth of recording, and its magnetic shell equivalent is an impulse of longitudinal magnetization. Exciting the loop with alternating current induces a voltage in the head which is identical, as a function of position, to that produced by a step transition in a real recording system. This approach is very useful for the study of three-dimensional effects.

Several large-scale models have been built for the study of the recording process (Tjaden and Leyten, 1965; Hersener, 1973; Iwasaki et al., 1979). The scale factors used in these models ranged from 2000 to 10,000. Tjaden and Leyten used a scale factor of 5000 in constructing a medium 50 mm thick. The medium was composed of 50-mm-wide lamina of gamma ferric oxide particles which were dispersed in plastic and longitudinally oriented. The plastic strips were then tightly stacked in a tray to form the medium. After recording with a large-scale ring head with a 20-mm gap length, the middle strip was removed from the medium to measure the recorded magnetization. This was done by punching out small disks 3 mm in diameter and measuring their remanent magnetization vectors in a specially constructed magnetometer. In this way, the vector magnetization along the track and through the thickness of the medium was determined. Experiments with both biased sine-wave recording and digital recording of isolated transitions were performed. The power line at 50 Hz provided the bias for a signal frequency of 0.1 Hz. The power dissipated in the recording head was 2.5 kW! The experiments showed significant components of perpendicular magnetization in the recorded medium. Another interesting result was the existence of a circular magnetization mode at very short wavelengths.

Hersener's model is very similar to that of Tjaden and Leyten, except for the method of determining the recorded magnetization in the medium. Hersener's medium contained slots designed so that a Hall probe could be inserted into the medium to measure internal fields. The magnetization distribution was then inferred from the field measurements. His experiments also confirmed the existence of a circular magnetization mode at short wavelengths.

Iwasaki and his coworkers built a model at 10,000 times actual scale to study perpendicular recording (Nakamura and Iwasaki, 1983). Figure 6.22 shows their apparatus. They used an array of slender cylinders of hard magnetic material to simulate the columnar structure of a Co-Cr perpendicular recording medium. Recording performance was determined from field measurements made just above the medium surface. Measurements could be made both during and after the recording process. Extensive experimentation and modification of this apparatus led to the development of Iwasaki's purely perpendicular recording system. This system, consisting of a main pole probe head driven by an auxiliary pole and a double-layer medium of Co-Cr and a soft magnetic back layer, was constructed to actual scale after large-scale model studies had shown its effectiveness.

Further improvement of scale-model studies was achieved by providing capability for both recording and reproducing (Monson et al., 1978). A

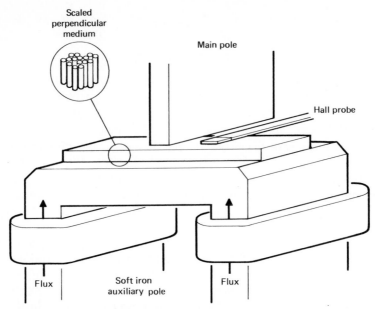

Figure 6.22 Large-scale model of perpendicular recording. Auxiliary pole windings are operated in push-push mode to write on the perpendicular medium *(after Nakamura and Iwasaki, 1983).*

5000-times scale model of a film head was used to write and read scaled-up particulate and film media. Figure 6.23 shows a block diagram of the apparatus. Writing is carried out by slowly moving the head over the stationary medium and energizing it with a programmable write-current driver. During the read operation the head again sweeps slowly over the recorded medium, but now the medium vibrates sinusoidally with an amplitude small compared with the gap length. The voltage induced in

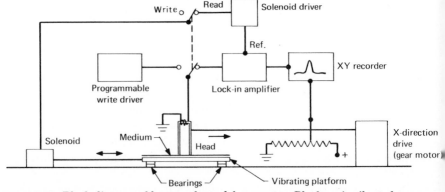

Figure 6.23 Block diagram of large-scale model apparatus. Platform is vibrated to produce readback of written magnetization transitions *(Valstyn et al., 1979).*

the head coil by this vibration is detected by a lock-in amplifier and plotted as a function of head position on an xy recorder.

An expression for the head voltage can be derived as follows. The position of the head with respect to the medium is

$$x = Vt + \Delta x \cos \omega t$$

where V is the head velocity and Δx is the peak vibration amplitude at radian frequency ω. The readback voltage from the changing flux ϕ is

$$
\begin{aligned}
E &= -\frac{d\phi}{dt} = -\frac{d\phi}{dx}\frac{dx}{dt} \\
&= (\omega \Delta x \sin \omega t - V)\frac{d\phi}{dx}
\end{aligned}
\tag{6.22}
$$

The lock-in amplifier detects the peak value of the ac component, giving

$$E_{\text{det}} = \omega \Delta x \frac{d\phi}{dx} \tag{6.23}$$

which is of the same form as that which would result from flying the head over the medium in an actual disk drive. The vibrational frequency must be chosen small enough to avoid eddy-current effects in the large-scale Ni-Fe head (Monson et al., 1975).

This scale model has been used to study transition density effects and side writing (Monson et al., 1978). On scale models it is easy to measure stray fields on both sides of a recorded medium. From such measurements it was shown (Valstyn and Monson, 1979) that flux from an isolated magnetization transition written on an isotropic particulate medium tended to concentrate on one side as predicted (Mallinson, 1973). Studies of perpendicular recording were carried out using the model with a film ring head and single-layer medium. By using the same scaled-up film medium in two configurations, it was possible to obtain a direct comparison between longitudinal and perpendicular recording modes (Monson et al., 1981).

The main limitation of scale modeling is that of simulating micromagnetic effects. In principle, it ought to be possible to make scaled-up particles and construct media with controlled switching-field distributions. Care would have to be taken to ensure that the multidomain scaled-up particles accurately simulate single-domain particle behavior. The simulation of domain behavior in film media offers additional challenges.

6.5 Recording System Performance

Measurement of recording system performance requires careful instrumentation techniques over many disciplines. The task is further complicated by the wide range of parameters used in magnetic recording systems. The head-to-medium velocities of rigid and flexible disks may range from 1.2 to 50 m/s. Data frequencies range from 100 kHz to 20 MHz, while wavelengths span 500 μm to about 1 μm. In some tape systems, frequencies as low as 1 Hz combined with a velocity of 25 mm/s produce wavelengths up to 25 mm long. High-density digital tape recorders and video recorders operate at wavelengths as short as 500 nm.

Measurement systems for evaluating recording performance should have precise speed and position control. They should be isolated from unwanted environmental disturbances. It is desirable to have separate fixtures for record and reproduce heads so that these two processes may be evaluated separately. Provision for various modes of erasure should also be made. The reproduce amplifier following the head should have a frequency response which is flat in gain and linear in phase over the required bandwidth. The amplifier should also have a noise level which is lower than the other noises in the system. Amplifier input impedance should be carefully designed for the reproduce head to optimize frequency response and noise.

6.5.1 Absolute head efficiency

Head efficiency should be measured in a low-frequency range where it is independent of frequency. Head efficiency may be defined as

$$\eta = \frac{R_g}{R_g + R_c} \qquad (6.24)$$

where R_g and R_c are the gap and core reluctances, respectively. Frequency dependence of the efficiency is caused by eddy currents and other mechanisms in the core which result in a complex, frequency-dependent core reluctance.

6.5.1.1 Input. The input to the head should be a calibrated square wave of flux recorded on a medium. Calibrated recorded media are commercially available. They may be produced on longitudinal media by recording square waves of low density at saturation so that the remanent magnetization is known approximately. To calibrate the flux precisely requires careful consideration of a number of effects (McKnight, 1970). These include fringing effects to the side of the track, head efficiency, and frequency and wavelength response factors. By using a large-gap, high-efficiency head operating at medium wavelengths of 0.25 to 2.0 mm, it is possible to achieve a flux calibration accuracy of better than 3 percent.

6.5.1.2 Output. The output voltage from the head has a fundamental component whose root-mean-square value is given by

$$e = 4 \sqrt{2} \, n f \mu_0 M_r w_{\text{eff}} \delta \eta \tag{6.25}$$

where n = number of turns
$\quad\; f$ = frequency
$\quad M_r$ = remanent magnetization
$\quad w_{\text{eff}}$ = effective track width
$\quad\; \delta$ = medium thickness
$\quad\; \eta$ = head efficiency.

From the measured voltage, the efficiency may be determined. Best accuracy is obtained by having the recorded track much wider than the head. An effective track width is then used, which takes into account flux induced in the head from its sides (Bertram, 1985):

$$w_{\text{eff}} = w + \frac{3\lambda}{4\pi} (1 - e^{-\pi \Delta w / \lambda}) \tag{6.26}$$

where Δw is the difference between the recorded track width and the head width.

As an alternative, or a check to the efficiency determined from Eq. (6.25), the isolated pulse waveform may be integrated to obtain the total flux sensed by the head:

$$\int e \, dt = n \mu_0 (2 M_r w_{\text{eff}} \delta \eta) \tag{6.27}$$

in which w_{eff} should be chosen to account for flux sensed at the sides of the head. If the observed pulse shape closely resembles a standard form such as a Lorentzian or Gaussian form, then the area under the pulse may be obtained from measurement of its peak amplitude and width.

6.5.2 Nonlinear recording

Much useful information for characterizing a recording system can be obtained from nonlinear recording measurements.

6.5.2.1 Output voltage versus write current.
Measurement of the output voltage versus write current provides information for setting system write-current levels. Square waves are recorded at the highest recording density to be used. The peak output voltage is then measured as a function of write current. Figure 6.24 shows the shape of curve which is typically found for longitudinal particulate media. The output voltage increases with increasing current and then drops, often exhibiting several sharp nulls. These can be caused by a circular magnetization mode produced at high current levels (Iwasaki and Takemura, 1975). The current for maximum peak output voltage is designated I_{opt}.

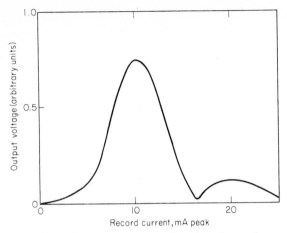

Figure 6.24 Output voltage as a function of record current for square-wave recording on a γ-Fe_2O_3 medium. Observed null can be caused by circular magnetization mode *(after Bertram and Niedermeyer, 1978)*.

6.5.2.2. Roll-off curve. To measure a roll-off curve, the write current is set to I_{opt} and square waves are written at various densities. Both the peak voltage and the rms amplitude of the fundamental component are plotted as a function of recording density, as shown in Fig. 6.25. The density at which the peak voltage falls to 50 percent of its low density limit is labeled D_{50}. At very high densities one or more read-head gap nulls are seen, and their locations may be correlated with the gap length observed optically. Also at high densities, only the fundamental component of the square wave is recorded, so the peak voltage approaches 1.414 times the fundamental rms component.

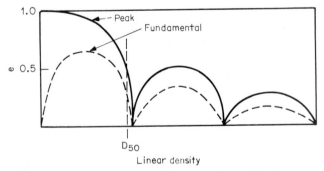

Figure 6.25 Output voltage as a function of recording density for recorded square waves. Both peak amplitude and rms amplitude of fundamental component of reproduced waveform are shown.

Roll-off curves can exhibit several interesting effects. When a film head is used to read a longitudinal medium, the negative undershoots of the isolated pulse interfere with the pulse peaks from adjacent magnetization transitions to produce a peaking of the roll-off curve. A similar, more pronounced effect is seen when a perpendicular recording medium is read with a ring head.

6.5.2.3 Isolated pulse shape. At low recording density, the isolated pulse shape from a single transition may be measured. As far as the read portion of the system is concerned, the isolated pulse shape is the impulse response of the system. By taking the Fourier transform of the isolated pulse, the system gain and phase response can be found. This information, particularly the phase portion, is very useful for designing linear post-equalization.

Under some simplifying assumptions, it is possible to obtain information about head-to-medium spacing and transition width from a measurement of the width of the isolated pulse. If the magnetization is purely longitudinal and varies only in the x direction as an arctangent function with parameter a, then it is possible to obtain an expression for the isolated pulse waveform by performing a correlation integral of the magnetization with a simple expression for the longitudinal field component of a ring head (Middleton, 1966). The isolated pulse width at 50 percent amplitude is given by

$$p_{50} = [g^2 + 4(a + d)(a + d + \delta)]^{1/2} \tag{6.28}$$

If the gap length and medium thickness are known, then $a + d$ may be found from Eq. (6.28). Note that a and d have the same effect on the pulse width as a result of the arctangent transition assumption.

6.5.2.4 Fundamental amplitude versus wave number analysis. Although the isolated pulse width expression [Eq. (6.28)] contains four important recording system parameters, it is not possible to solve for any one of them without knowing the other three. If a wave number analysis is carried out on an isolated pulse, the Fourier component of the output voltage at wave number k may be approximated by

$$e(k) = 2\,nV\mu_0 M\eta w\delta\,\frac{1 - e^{-k\delta}}{k\delta}\,e^{-k(d+a)}\,\frac{\sin\,(kg_{\text{eff}}/\,2)}{kg_{\text{eff}}/\,2} \tag{6.29}$$

where the effective gap length $g_{\text{eff}} = \lambda_{\text{null}}$, the wavelength of the first null. This expression applies for media with no soft magnetic back layer, and it is assumed that the magnetization is uniform throughout the thickness of the medium.

The form of Eq. (6.29) has the advantage that many of the system parameters are separated in product form so that they may be determined from measurement of output voltage as a function of wavelength. The measurements are carried out by recording square waves at low density and observing the output voltage on a spectrum analyzer or by varying the recording density and measuring the rms value of the fundamental component at the wavelength corresponding to each density. The methods give the same result within the assumptions of Eq. (6.29), but the latter is equivalent to multiplying (6.29) by $\sqrt{2}k/\pi$, giving higher signals at short wavelengths. At long wavelengths, or small k, the thickness loss term $(1 - e^{-k\delta})/k\delta$ may be approximated as unity. In the long-wavelength region, the other terms are essentially constant. By measuring the slope of the fundamental curve, the thickness δ may be obtained.

After the gap length and thickness are known, their wave number effects can be removed from the measured output voltage curve. Figure 6.26 shows a plot of the relative output voltage in decibels which has been corrected for medium thickness and gap loss effects. By measuring the slope of the curve, which is linear, the combined arctangent parameter and head-to-medium spacing may be determined. It is important to note the conditions for which this method is valid. The magnetization transition must be of arctangent form, and the medium must be single-layer. Some measurements have been made which indicate negative transition widths for perpendicular recording, but these have been attributed to magnetization transitions which are not arctangent in form (Yeh, 1985).

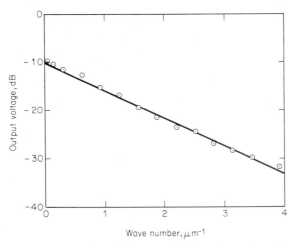

Wave number, μm^{-1}

Figure 6.26 Output voltage as a function of wave number, corrected for thickness and gap effects. Slope is proportional to $a + d$. Voltage intercept is $2nV\mu_oM\eta w\delta$ (after Bertram and Niedermeyer, 1978).

Equation (6.29) is not valid for double-layer media except in the limit of very short wavelengths. At long wavelengths, the slope of the output voltage curve can greatly exceed that shown in Fig. 6.25 (Yamamoto et al., 1984). Also it must be remembered that Eq. (6.29) is based upon a Fourier or sine-wave analysis of the output voltage. Applying Eq. (6.29) to plots of the peak of the voltage waveform instead of its sine-wave components could give misleading results.

6.5.2.5 Overwrite. Overwrite is defined and measured in several different ways depending upon the recording system application. The highest square-wave frequency used in the system is denoted as f_2, the frequency corresponding to an all-1s data pattern. At a particular write-current level, square waves are written at the frequency f_1, which is half of f_2. The amplitude of the readback voltage at f_1 is measured on a spectrum analyzer. Square waves at f_2 are then written at the same current level, and the voltages at both f_1 and f_2 are measured. *Overwrite* is most often defined as the ratio of the two f_1 voltages measured before and after overwriting with f_2. Overwrite performance is usually measured as a function of write current. This definition represents a worst case for longitudinal recording, where transitions recorded at lower frequencies are most difficult to overwrite. In perpendicular recording, the opposite seems to hold, and a more appropriate test is to record f_2 first and then overwrite with f_1 (Langland and Albert, 1981).

An overwrite measurement which emphasizes the phase modulation produced by the residual f_1 signal after overwriting with f_2 is made by measuring the peak-to-peak time jitter, or bit shift, between transitions recorded at f_2. Amplitude effects can be measured by first dc erasing the medium and then writing a low-frequency square wave to produce isolated transitions. The positive- and negative-going peak amplitudes of the readback pulses are measured separately and their ratio defined as a pulse asymmetry. When measurements are carried out as a function of write current, it is usually found that minimum pulse asymmetry and minimum bit shift do not occur at the same current level.

Care must be taken in making and interpreting overwrite measurements. Large variations in overwrite performance have been observed which depend upon gap length, write current, and transition density (Wachenschwanz and Jeffers, 1985). These appear to be related to interference effects between residual overwritten data and data which are rerecorded with the overwriting signal acting as bias. Similar effects have been observed in ac-erasure studies (McKnight, 1963).

6.5.2.6 Bit shift. *Bit shift,* also referred to as *peak shift,* is defined as the normalized shift in detected positions of readback bits with respect to the positions where they were written. The simplest measurement of bit shift is made by recording a dibit pattern, namely, two magnetization transitions spaced apart by the bit cell length. The distance between the two peaks of the readback pulses is the detected bit cell length. The bit shift is given by the difference between the detected and recorded bit cell lengths, normalized to the recorded bit cell length. Bit shift is sensitive to the recorded pattern, and the measurement is readily extended to tribit and higher order patterns of recorded transitions.

Bit shift has both linear and nonlinear components. The linear component results from superposition of isolated pulse waveforms which interfere to produce peak shift as they are spaced closer and closer together. Because it results from superposition, linear bit shift may be equalized by linear networks in the read circuits. Nonlinear bit shift arises from interbit interactions occurring in the nonlinear write process. Although nonlinear bit shift may be partially canceled by predistorting the write-current waveform, it cannot be postequalized by linear networks. Consequently, the measurement of the nonlinear component of bit shift is a very important one for evaluating high-density recording system performance.

In many applications, nonlinear writing effects extend over only a few bit cells so that nonlinear bit shift can be determined from measurements on tribit patterns (Koren, 1981). An alternative approach is to write a maximal-length pseudo-random sequence of data and observe the spectrum of the readback signal (Haynes, 1977; Wood and Donaldson, 1979). The envelope of the spectral lines follows the linear recording channel response, while nonlinearities are indicated by ripples and irregularities in the spectrum. By adjusting the predistortion of the write-current waveform to minimize the rippling in the read spectrum, nonlinear bit shift can be minimized (Coleman et al., 1984). A quantitative measure of nonlinear bit shift can be obtained by expressing the distorted readback waveform in a Volterra series expansion. The coefficients of this expansion are obtained from the measured cross-correlation function between the input pseudo-random sequence and the readback waveform (Wood and Donaldson, 1979).

6.5.3 Linear recording

Linearization of the recording process is achieved by using an ac bias whose frequency is more than three times the highest signal frequency to be recorded. Adjustment of the bias amplitude is very important for optimizing system performance with respect to frequency response, harmonic distortion, and output at short wavelengths.

To evaluate the frequency response, the output voltage is measured as a function of wave number with bias amplitude as a parameter. The signal level is kept at a low, constant level. Typical results are shown in Fig. 6.27.

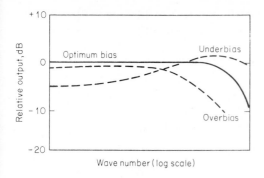

Figure 6.27 Wavelength response for ac-bias recording, showing effect of bias amplitude.

At bias levels under the optimum, short-wavelength response is improved at the expense of long wavelengths. Overbiasing gives good long-wavelength response but poor performance at short wavelengths.

The effects of bias amplitude on signal output level and distortion are measured at long wavelengths. The signal input level is held constant as bias amplitude is varied. The output level and third harmonic distortion are plotted as a function of bias level as shown in Fig. 6.28 (Mee, 1964). The long-wavelength output signal saturates and decreases as the bias field penetrates into and through the medium with increasing bias amplitude. Third harmonic distortion has a minimum value associated with the shapes of the signal output versus input curves as bias level is increased. The initial portion of these curves changes shape from concave upward to concave downward with increasing bias amplitude (Daniel and Levine, 1960). The curve is most nearly linear at the bias setting where the change occurs, giving minimum distortion (Westmijze, 1953; Fujiwara, 1979). The bias value for minimum distortion is less than that for maximum output level.

Another commonly measured curve is the peak output level at short wavelengths as a function of bias amplitude. The bias amplitude for peak output level at short wavelengths is less than that for maximum output

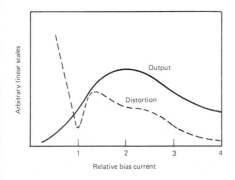

Figure 6.28 Output and distortion versus bias amplitude for long-wavelength recording *(after Mee, 1964)*.

level at long wavelengths. Consequently, the choice of bias level in a recording system must be a compromise among the various performance curves obtained from the measurements.

The linear recording counterpart to overwrite measurement in nonlinear recording is the measurement of signals remaining after ac erasure. Erasure is usually done by a separate erase head. To measure the effectiveness of erasure, signals are recorded at various wavelengths and levels and reproduced after erasure. Both frequency and wavelength effects have been observed along with interference phenomena which are caused by re-recording by the erase head (McKnight, 1963).

Gain and phase measurements on linear recorders are important for systems implementation. Gain measurements as a function of frequency or wave number are readily performed, but phase measurements using conventional methods are more difficult because of the long delays between recording and reproducing. One approach is to record long-wavelength square waves and analyze the reproduce waveforms using Fourier transform techniques to obtain the phase response. Some care must be taken in this measurement to ensure that the square-wave amplitude is small enough that the recording channel is linear, yet large enough that there is enough signal strength at short wavelengths compared with noise to give valid results.

An alternative approach which overcomes these problems is to use a very long pseudo-random binary sequence as input (Haynes, 1977). Such an input signal has a favorable spectral density for testing the channel at amplitudes which produce negligible harmonic distortion. By digitizing the output voltage and carrying out a fast Fourier transform for comparison with the Fourier transform of the input, the system transfer function, both gain and phase, is determined. Accuracy may be improved by averaging repeated measurements. Generating an accurate time base is important for sampling of the output waveform, and this can be accomplished by generating a stable trigger from a preset data pattern occurring in the input sequence.

In instrumentation recording applications, which often use several subcarriers for transmitting data, the group or envelope delay performance is more important than the overall phase function (Starr, 1965). Measurement of group delay is carried out by using sets of reference signals (IRIG standard, 1979).

6.5.4   Noise measurements

6.5.4.1   Noise spectrum measurements.   Noise arises from several sources in a magnetic recording system, and these may be isolated by careful measurement of the noise spectra under different conditions. The rms noise voltage in a very narrow bandwidth is measured, and the noise voltage density, in $V/(Hz)^{1/2}$, is plotted as a function of frequency. This measurement is sometimes referred to as a slot signal-to-noise ratio, for which the bandwidth of the frequency slot and a reference signal level are specified.

Measurements are first carried out with the medium stopped. Under this condition, the equipment, including amplifiers and other circuitry, and head produce the measured noise. This is the condition for lowest noise, as shown in Fig. 6.29. Motion of the medium induces several kinds of nonmagnetic noise. Variations in heat transfer across the head-to-medium interface can cause noise voltage spikes in magnetoresistive heads (Hempstead, 1974). In perpendicular recording, mechanically induced noise mechanisms in the permeable back layer have been observed to produce noise voltage spikes (Ouchi and Iwasaki, 1985). Magnetostrictive effects can produce noise in heads which can be measured by using a mechanically equivalent nonmagnetic medium such as one made of α-Fe_2O_3.

In addition to providing information about system noise performance, noise measurement is a useful tool for analyzing the quality of recording media. In such an application, it is helpful to subtract out on a power basis the equipment and head noise. Measurement of noise from bulk-erased media gives the spectral density of the additive noise caused by granularity of the medium. Ac-biased noise is measured after passing a record head energized with an ac-bias current over the bulk-erased medium. The noise voltage will increase over that for the bulk-erased medium. Noise from the bulk-erased medium is re-recorded by the bias field to produce additional noise.

Bulk modulation effects such as packing and coating thickness variations may be observed by measuring the noise from the medium after it has been recorded to saturation by a dc field. Usually, the applied field is made strong enough to saturate the medium through its entire thickness.

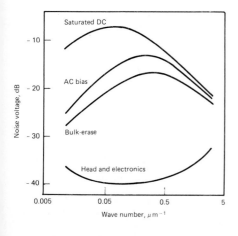

Figure 6.29 Slot noise spectrum as a function of wave number, showing noise contributions from head and equipment, bulk-erased medium, ac-biased medium, and dc-saturated medium.

Finer spatial resolution of the noise measurement may be achieved by measuring the noise using a very narrow gap head which is excited with direct current. By varying the amplitude of the current, the zone of medium saturation can be controlled to explore regions very close to the surface of the medium. The measured noise characterizes the surface asperities of the medium (Daniel, 1964; Eldridge, 1964).

Recording a signal such as a single-tone sine wave, or a square wave corresponding to an all-1s data pattern, causes some interesting changes in the observed noise spectrum. In the neighborhood of the recorded signal frequency, a sideband structure caused by noise amplitude and phase modulation is seen on a spectrum analyzer. For some particulate media, the wideband dc-noise spectrum is not greatly changed in amplitude in the presence of a recorded signal, as shown in Fig. 6.30. Changing the frequency of the recorded signal does not appreciably change the wide-band noise spectrum for particulate media, and, if anything, the noise decreases with increasing signal frequency.

For film media, on the other hand, the amplitude of the noise spectrum varies significantly with the frequency of the recorded signal. Figure 6.31 shows typical measured spectra which indicate that the noise amplitude increases with the frequency of the recorded signal. Experimental results indicate that the noise observed in film media may be localized in the magnetization transition regions (Tanaka et al., 1982; Baugh et al., 1983).

6.5.4.2 Time-domain noise measurements. The apparent nonuniform spatial distribution of noise in film media with recorded signals has motivated the development of several techniques for measuring media noise in the time domain.

One method obtains the time jitter of the transition locations of a recorded square wave (Baugh et al., 1983). The times between successive transitions are measured using a time interval meter. The root-mean-

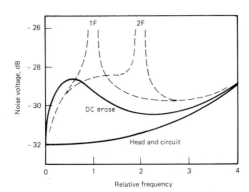

Figure 6.30 Noise spectra from particulate medium under dc-erased condition and recorded signal conditions of $1F$ and $2F$. Noise spectrum is nearly independent of recorded frequency *(after Stefanelli and Lelarge, 1984).*

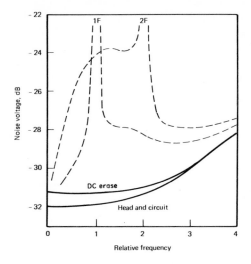

Figure 6.31 Noise spectra from film medium. Spectra vary markedly with recorded frequency *(after Stefanelli and Lelarge, 1984).*

square time jitter is then calculated. The jitter caused by only the medium noise may be obtained by subtracting on a mean-square basis the jitter from many repeated measurements on a single pair of adjacent transitions. The time jitter, Δt, of a transition detected by a zero-crossing detector may be expressed as

$$\Delta t = -e_n(t_0)\left[\frac{de_s(t_0)}{dt}\right]^{-1} \tag{6.30}$$

where e_n and e_s are the noise and signal voltages, respectively. When the noise is a stationary process and the signal is a square wave whose transition density is greater than $1/3D_{50}$, Eq. (6.30) gives for the rms jitter

$$\Delta t_{rms} = \frac{1}{2\pi f(\text{SNR})} \tag{6.31}$$

where f is the frequency of the recorded square wave, and SNR is the broadband signal-to-noise ratio expressed as a voltage ratio. Equation (6.31) holds well for particulate media, indicating that the noise is stationary. When the noise is not stationary, as observed for some film media, transition-time jitter and noise power are not simply related.

Instrumentation developed for nuclear counting experiments is very useful for time-jitter analysis (Nunnelly, 1985). The signal containing time jitter is fed into a discriminator which generates a very fast nuclear-instrumentation module logic signal at the zero crossing of the noisy transition. The logic signal controls the triggering of a time-to-amplitude converter module which produces an analog pulse whose amplitude is proportional to the jitter in the transition times. The pulses are sorted into a histogram of transition times, with resolution of 5 ps, by a multichannel analyzer. The nuclear instrumentation modules are well developed and commercially available.

A powerful approach to time-domain noise measurement is to use autocorrelation to characterize both stationary and nonstationary noise in media (Tang, 1985). The autocorrelation function is given by

$$R_n(t_1, t_2) = \frac{1}{m} \sum_{i=1}^{m} e_{ni}(t_1)e_{ni}(t_2) \qquad (6.32)$$

Noise voltages are obtained by subtracting from each measured noise-contaminated signal the average signal waveform. For stationary noise process, the autocorrelation function depends only on the difference between t_2 and t_1 and not on their values, so a measurement and plot of Eq. (6.32) will clearly distinguish between stationary and nonstationary processes. Tang observed significant differences in shape between the measured autocorrelation functions of particulate and film media.

One of the main problems in performing generalized autocorrelation in a magnetic recording system is to develop a timing technique which can be corrected for jitter. This is especially difficult because much of the jitter is caused by the noise which is to be autocorrelated. Noise-independent jitter correction may be accomplished by generating a reference timing signal which is insensitive to amplitude-modulation noise, and then delaying the beginning of waveform timing by a fixed time. If this time delay is long enough, the additive noise at that point is uncorrelated with the noise at the reference time.

References

Aharonov, Y., and D. Bohm, "Significance of Electromagnetic Potentials in the Quantum Theory," *Phys. Rev.,* **115,** 485 (1959).

Alexopoulos, P. S., and R. H. Geiss, Private Communication (1985).

Baird, A. W., R. A. Johnson, W. F. Chaurette, and W. T. Maloney, "Measurements and Self-Consistent Calculations of the Magnetic Fields near a Perpendicular Medium," *IEEE Trans. Magn.,* **MAG-20,** 479 (1984).

Banbury, J. R., and W. C. Nixon, "The Direct Observation of Domain Structure and Magnetic Fields in the Scanning Electron Microscope," *J. Sci. Instrum.,* **44,** 889 (1967).

Baugh, R. A., E. S. Murdock, and B. R. Natarajan, "Measurement of Noise in Magnetic Media," *IEEE Trans. Magn.,* **MAG-19,** 1722 (1983).

Bertram, H. N., Private Communication (1985).

Bertram, H. N., and R. Niedermeyer, "The Effect of Demagnetization Fields on Recording Spectra," *IEEE Trans. Magn.,* **MAG-14,** 743 (1978).

Bitter, F., "On Inhomogeneities in the Magnetization of Ferromagnetic Materials," *Phys. Rev.,* **38,** 1903 (1931).

Bozorth, R. M., *Ferromagnetism,* Van Nostrand, New York, 1951, p. 533.

Calcagno, P. A., and D. A. Thompson, "Semiautomatic Permeance Tester for Thick Magnetic Films," *Rev. Sci. Instrum.,* **46,** 904 (1975).

Camarota, R. C., and D. A. Thompson, "Measurement of Recording Head Frequency Response Using a Scanning Electron Microscope," *3M Conf.*, Pittsburgh, 1983.

Chen, T., "The Micromagnetic Properties of High-Coercivity Metallic Thin Films and Their Effects on the Limit of Packing Density in Digital Recording," *IEEE Trans. Magn.*, **MAG-17,** 1181 (1981).

Chi, C. S., and T. S. Szczech, "A Write-Head Design with Improved Field Profile for Perpendicular Recording," *IEEE Trans. Magn.*, **MAG-20,** 836 (1984).

Chikazumi, S., *Physics of Magnetism*, Wiley, New York, 1964.

Chu, Y., Private Communication (1984).

Clark, B. K., and J. D. Finegan, "Rotational Magnetic Measurements on Recording Media," *IEEE Trans. Magn.*, **MAG-21,** 1456 (1985).

Cohen, M. S., "Wave-Optical Aspects of Lorentz Microscopy," *J. Appl. Phys.*, **38,** 4966 (1967).

Coleman, C., D. Lindholm, D. Petersen, and R. Wood, "High Data Rate Magnetic Recording in a Single Channel," *IERE Conf. Proc.*, **59** (1984).

Cort, D. M., and J. W. Steeds, "A Liquid Helium Cooled Stage for the Scanning Electron Microscope," *Proc. Fifth Eur. Congr. Electron Micros.*, 376 (1972).

Cullity, B. D., *Introduction to Magnetic Materials*, Addison-Wesley, Reading, Mass. 1972.

Daniel, E. D., "A Preliminary Analysis of Surface-Induced Noise," *IEEE Trans. Comm. Electron.*, **83,** 250 (1964).

Daniel, E. D., and I. Levine, "Experimental and Theoretical Investigation of the Magnetic Properties of Iron Oxide Recording Tape," *J. Acoust. Soc. Am.*, **32,** 1 (1960).

Eldridge, D. F., "D-C and Modulation Noise in Magnetic Tape," *IEEE Trans. Comm. Electron.*, **83,** 585 (1964).

Elsbrock, J. B., and L. J. Balk, "Profiling of Micromagnetic Stray fields in Front of Magnetic Recording Media and Heads by Means of a SEM," *IEEE Trans. Magn.*, **MAG-20,** 866 (1984).

Fathers, D. J., J. P. Jakubovics, D. C. Joy, D. E. Newbury, and H. Yakowitz, "A New Method of Observing Magnetic Domains by Scanning Electron Microscopy," *Phys. Status Solidi (a)*, **20,** 535 (1973).

Fluitman, J. H. J., "Recording Head Field Measurement with a Magnetoresistive Transducer," *IEEE Trans. Magn.*, **MAG-14,** 433 (1978).

Fluitman, J. H. J., and J. Marchal, "Head Fields of Asymmetric Recording Heads," *IEEE Trans. Magn.*, **MAG-9,** 139 (1973).

Foner, S., "Versatile and Sensitive Vibrating-Sample Magnetometer," *Rev. Sci. Instrum.*, **30,** 548 (1959).

Fujiwara, T., "Nonlinear Distortion in Long Wavelength AC Bias Recording," *IEEE Trans. Magn.*, **MAG-15,** 894 (1979).

Gabor, D., "Microscopy by Reconstructed Wavefronts," *Proc. R. Soc. London, Ser. A*, **197,** 454 (1949).

Griffiths, B. W., P. Pollard, and J. W. Venables, "A Channel Plate Detector for the Scanning Electron Microscope," *Proc. Fifth Eur. Congr. Electron Micros.*, 176 (1972).

Haynes, M. K, "Experimental Determination of the Loss and Phase Transfer Functions of a Magnetic Recording Channel," *IEEE Trans., Magn.*, **MAG-13,** 1284 (1977).

Hempstead, R. D., "Thermally Induced Pulses in Magnetoresistive Heads," *IBM J. Res. Dev.*, **18,** 547 (1974).

Hersener, J., "Model Investigation of the Magnetization Distribution of Sinusoidal Recordings in Magnetic Tape" (German), *Wiss. Ber. AEG TELEFUNKEN*, **46,** 15 (1973).

Hoyt, R. F., "Microloop Measurements of Recording Head Fields versus Current," *J. Appl. Phys.*, **56,** 3947 (1985).

Hoyt, R. F., D. E. Heim, J. S. Best, C. T. Horng, and D. E. Horne, "Direct Measurement of Recording Head Fields Using a High-Resolution Inductive Loop," *J. Appl. Phys.*, **55,** 2241 (1984).

IRIG Document 118-79, *Test Methods for Recorder/Reproducer Systems and Magnetic Tape*, 1979, vol. III.

Ishiba, T., and H. Suzuki, "Measurements of Magnetic Field of Magnetic Recording Head by a Scanning Electron Microscope," *Jpn. J. Appl. Phys.*, **13,** 457 (1974).

Iwasaki, S., and K. Ouchi, "Co-Cr Recording Films with Perpendicular Magnetic Aniso-tropy," *IEEE Trans. Magn.*, **MAG-14**, 849 (1978).

Iwasaki, S., and K. Takemura, "An Analysis for the Circular Mode of Magnetization in Short Wavelength Recording," *IEEE Trans. Magn.*, **MAG-11**, 1173 (1975).

Iwasaki, S., Y. Nakamura, S. Yamamoto, and K. Yamakawa, "Perpendicular Recording by a Narrow Track Single Pole Head," *IEEE Trans. Magn.*, **MAG-19**, 1713 (1983).

Joy, D. C., and J. P. Jakubovics, "Direct Observation of Magnetic Domains by Scanning Electron Microscopy," *Phil. Mag.*, **17**, 61 (1968).

Kalvius, G. M., and R. S. Tebble, *Experimental Magnetism*, Wiley, Chichester, 1979, vol. 1.

Kawakami, K., S. Narishige, and M. Takagi, "A High Frequency Permeance Meter for Anisotropic Films and Its Applicaiton in the Determination of Magnetostriction Con-stants," *IEEE Trans. Magn*, **MAG-19**, 2154 (1973).

Knowles, J. E., "Measurements on Single Magnetic Particles," *IEEE Trans. Magn.*, **MAG-14**, 858 (1978).

Knowles, J. E., "Magnetic Measurements on Single Acicular Particles of *gama* Fe_2O_3," *IEEE Trans. Magn.*, **MAG-16**, 62 (1980).

Koren, N. L., "A Simplified Model of Nonlinear Bit Shift in Digital Magnetic Record-ing," *Intermag Conf. Dig.*, 2–9 (1981).

Köster, E., "Reversible Susceptibility of an Assembly of Single-Domain Particles and Their Magnetic Anisotropy," *J. Appl. Phys.*, **41**, 3332 (1970).

Köster, E., "Recommendation of a Simple and Universally Applicable Method for Mea-suring the Switching Field Distribution of Magnetic Recording Media," *IEEE Trans. Magn.*, **MAG-20**, 881 (1984).

Kullmann, U., E. Köster, and B. Meyer, "Magnetic Anisotropy of Ir-doped CrO_2," *IEEE Trans. Magn.*, **MAG-20**, 742 (1984).

Langland, B. J., and P. A. Albert, "Recording on Perpendicular Anisotropy Media with Ring Heads," *IEEE Trans. Magn.*, **MAG-17**, 2547 (1981).

Lemke, J. U., and R. J. McClure, "The Effect of Vector Field History on the Remanence of Magnetic Tape," *IEEE Trans. Magn.*, **MAG-2**, 230 (1966).

Lustig, C. D., A. W. Baird, W. F. Chaurette, H. Minden, W. T. Maloney, and A. J. Kurt-zig, "High-Resolution Magnetic Field Measurement System for Recording Heads and Disks," *Rev. Sci. Instrum.*, **50**, 321 (1979).

Mallinson, J. C., "Magnetometer Coils and Reciprocity," *J. Appl. Phys.*, **37**, 2514 (1966).

Mallinson, J. C., "One-Sided Fluxes—A Magnetic Curiosity," *IEEE Trans. Magn.*, **MAG-9**, 678 (1973).

Mallinson, J. C., "On the Properties of Two-Dimensional Dipoles and Magnetized Bod-ies," *IEEE Trans. Magn.*, **MAG-17**, 2453 (1981).

Manly, W. A., Jr., "A 5.5-kOe 60-Hz Magnetic Hysteresis Loop Tracer with Precise Dig-ital Readout," *IEEE Trans. Magn.*, **MAG-7**, 442 (1971).

Marton, L., "Electron Optical Schlieren Effect," *J. Appl. Phys.*, **19**, 687 (1948).

McKnight, J. G., "Erasure of Magnetic Tape," *J. Audio Eng. Soc.*, **11**, 223 (1963).

McKnight, J. G., "Tape Flux Measurement Theory and Verification," *J. Audio Eng. Soc.*, **18**, 250 (1970).

Mee, C. D., *The Physics of Magnetic Recording*, North-Holland, Amsterdam, 1964.

Middleton, B. K., "The Dependence of Recording Characteristics of Thin Metal Tapes on their Magnetic Properties and on the Replay Head," *IEEE Trans. Magn.*, **MAG-2**, 225 (1966).

Monson, J. E., D. J. Olson, and E. P. Valstyn, "Scale-Modeling the Read Process for a Film Head," *IEEE Trans. Magn.*, **MAG-11**, 1182 (1975).

Monson, J. E., M. A. Barker, S. T. Ritchie, and E. P. Valstyn, "Scale-Modeling the Write and Read Processes in Digital Magnetic Recording," *Intermag Conf. Dig.*, 26–5 (1978).

Monson, J. E., R. Fung, and A. S. Hoagland, "Large Scale Model Studies of Vertical Recording" *IEEE Trans. Magn.*, **MAG-17**, 2541 (1981).

Nakagawa, S., K. Kanai, and F. Kobayashi, "Investigation of Head Saturation Using a Large-Scale Ferrite Model Head," *IEEE Trans. Magn.* **MAG-8**, 538 (1972).

Nakamura, Y., and S. Iwasaki, "Perpendicular Magnetic Recording Characteristics," *Bull. Magn. Soc. Japan*, **32–2**, 7 (1983).

Narishige, S., M. Nanazono, M. Takagi, and S. Kuwatsuka, "Measurements of Magnetization Processes of Thin Film Heads Using Micro-Kerr Method," *J. Appl. Phys.*, **8**, 2 (1984).

Newman, J. J., "Magnetic Measurements for Digital Magnetic Recording," *IEEE Trans. Magn.*, **MAG-14**, 154 (1978).

Nunnelly, L., "Time-Domain Noise-Induced Jitter: Theory and Precise Measurement," Private Communication (1985).

Ouchi, K., and S. Iwasaki, "Properties of High Rate Sputtered Perpendicular Recording Media," *J. Appl. Phys.*, **57**, 4013 (1985).

Pacyna, A. W., and K. Ruebenbauer, "General Theory of a Vibrating Magnetometer with Extended Coils," *J. Phys. E: Sci. Instrum.*, **17**, 141 (1984).

Pearson, R. F., "Magnetic Anisotropy," in *Experimental Magnetism*, G. M. Kalvius and R. S. Tebble (eds.), Vol. I, Wiley, Chichester, England, 1979.

Penoyer, R. F., "Automatic Torque Balance for Magnetic Anisotropy Measurements," *Rev. Sci. Instrum.*, **30**, 711 (1959).

Rau, E. I., and G. V. Spivak, "Scanning Electron Microscopy of Two-Dimensional Magnetic Stray Fields," *Scanning*, **3**, 27 (1980).

Re, M. E., and M. H. Kryder, "Magneto-Optic Investigation of Thin-Film Recording Heads," *J Appl. Phys.*, **55**, 2245 (1984).

Sharrock, M. P., "Particle-Size Effects on the Switching Behavior of Uniaxial and Multiaxial Magnetic Recording Materials," *IEEE Trans. Magn.*, **MAG-20**, 754 (1984).

Spivak, G. V., A. E. Lukianov, E. I. Rau, and R. S. Gvosdover, "Electron Mirror Measurements of Magnetic Fields Over Audio and Video Heads and Tapes," *IEEE Trans. Magn.*, **MAG-7**, 684 (1971).

Starr, J., "Envelope Delay in a Tape Recorder System," *Proc. Int. Telem. Conf.*, **1**, 595 (1965).

Stefannelli, P., and J. L. LeLarge, "Signal-to-Noise Ratio Measurement in Digital Recording," Paper BP-21, *Intermag Conf. Dig.* (1984).

Tanaka, H., H. Goto, N. Shiota, and M. Yanagisawa, "Noise Characteristics in Plated Co-Ni-P Film For High Density Recording Medium," *J. Appl. Phys.*, **53**, 2576 (1982).

Tang, Y. S., "Noise Autocorrelation in Magnetic Recording Systems," *IEEE Trans. Magn.*, **MAG-21**, 1389 (1985).

Thornley, R. F. M., and J. D. Hutchison, "Magnetic Field Measurements in the Scanning Electron Microscope," *IEEE Trans. Magn.*, **MAG-5**, 271 (1969).

Tjaden, D. L. A., and J. Leyten, "A 5000:1 Scale Model of the Magnetic Recording Process," *Philips Tech. Rev.*, **25**, 319 (1964).

Tong, H. C., R. Ferrier, P. Chang, J. Tzeng, and K. L. Parker, "The Micromagnetics of Thin-Film Disk Recording Tracks," *IEEE Trans. Magn.*, **MAG-20**, 1831 (1984).

Tonomura, A., T. Matsuda, J. Endo, T. Arii, and K. Mihama, "Direct Observation of Fine Structure of Magnetic Domain Walls by Electron Holography," *Phys. Rev. Lett.*, **44**, 1430 (1980).

Valstyn, E. P., and J. E. Monson, "Magnetization Distribution in an Isolated Transition," *IEEE Trans. Magn.*, **MAG-15**, 1453 (1979).

Valstyn, E. P., R. I. Potter, and A. Paton, "Validity of Scale Models in Magnetics," *Comput. Dis.*, **9**, 94 (1971).

Valstyn, E. P., J. E. Monson, R. W. Akeo, and R. K. Moloney, "A Study of Transition-Density Effects with Film Heads," *Int. Conf. Video Data Recording*, 27 (1979).

Völz, H., "Device for Semi-Automatic Measurement of Preisach Values of Magnetic Tape," *Hochfrequenztech. U. Electakust.*, **79**, 101 (1971).

Wachenschwanz, D., and F. Jeffers, "Overwrite as a Function of Record Gap Length," *IEEE Trans. Magn.*, **MAG-21**, 1380 (1985).

Wells, O. C., *Scanning Electron Microscopy*, McGraw-Hill, New York, 1974.

Wells, O. C., and M. Brunner, "Schlieren Method as Applied to Magnetic Recording Heads in the Scanning Electron Microscope," *Appl. Phys. Lett.*, **42**, 114 (1983).

Wells, O. C., and R. J. Savoy, "Magnetic Domains in Thin-Film Recording Heads as Observed in the SEM by a Lock-in Technique," *IEEE Trans. Magn.*, **MAG-17**, 1253 (1981).

Wells, O. C., P. J. Coane, and C. F. Aliotta, "Measuring the Field From a Magnetic Recording Head in the Scanning Electron Microscope," *Proc. 18th Annu. Conf. MAS*, Phoenix, 1983.

Westmijze, W. K., "Studies on Magnetic Recording," *Philips Res. Rep.*, **8,** 343 (1953).

Williams, M. L., and R. L. Comstock, "An Analytic Model of the Write Process in Digital Magnetic Recording," *AIP Conf. Proc.*, **5,** 738 (1971).

Wohlleben, D., "On the Detection of Magnetic Microstructures with Charged Particles," *Phys. Lett.*, **22,** 564 (1966).

Wood, R. W., and R. W. Donaldson, "The Helical-Scan Magnetic Tape Recorder on a Digital Communication Channel," *IEEE Trans. Magn.*, **MAG-15,** 935 (1979).

Yamamoto, S., Y. Nakamura, and S. Iwasaki, " Spacing Loss in Perpendicular Magnetic Recording," *Rep. TGMR IECE Jpn.*, **84,** 27 (1984) (in Japanese).

Yeh, N. H., "A Negative Transition Width in Perpendicular Recording?" *J. Appl. Phys.*, **57,** 3940 (1985).

Yoshida, K., T. Okuwaki, N. Osakabe, H. Tanabe, Y. Horiuchi, T. Matsuda, K. Shinagawa, A. Tonomura, and H. Fujiwara, "Observation of Recorded Magnetization Patterns by Electron Holography," *IEEE Trans. Magn.*, **MAG-19,** 1600 (1983).

Youngquist, R. J., and R. H. Hanes, "Magnetic Reader," U.S. Patent 3,013,206, 1961.

Zijlstra, H., *Experimental Methods in Magnetism,* Wiley, New York, 1967, vol. 2.

Head-Medium Interface

Frank E. Talke

Center for Magnetic Recording Research
University of California, San Diego
La Jolla, California

David B. Bogy

University of California
Berkeley, California

The growth and increasing commercial importance of magnetic recording are based, to a large degree, on two implementations of the technology: first, the invention and development of the rigid-disk file, in which a magnetic read-write element is supported on a thin air film in close proximity to a rotating disk supporting a thin layer of magnetic recording media; second, the development of magnetic media using flexible substrates that allow the positioning of a magnetic head in close proximity to the magnetic medium without excessive head or medium wear as in flexible disks or tape drives.

In all magnetic recording configurations, one of the more important parameters is the spacing between the recording element and the magnetic medium. The importance of this parameter can be seen from Fig.

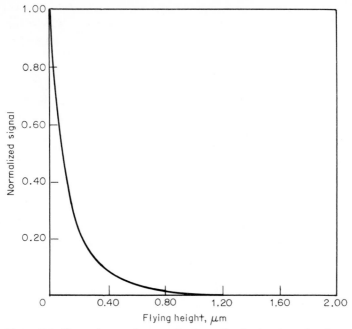

Figure 7.1 Dependence of normalized readback signal on head-to-medium spacing for a single magnetic transition.

7.1, where the output signal of a typical read-write element is plotted as a function of the head-to-medium separation (Talke and Tseng, 1973). The amplitude of the readback signal for a head-disk separation of 1 μm is less than 1 percent of the amplitude that would be obtained in the case of contact between the head and medium. The most desirable configuration from a magnetics viewpoint is the one of contact between the magnetic medium and the magnetic head. Unfortunately, contact recording is very undesirable from wear considerations; contact between sliding surfaces generally results in wear and materials interactions, and this, in turn, results in a degradation of the performance and reliability of the recording system.

In order to avoid wear, the head-disk interface in a magnetic disk file is designed so that the magnetic head is separated from the disk by a thin air film (Harker et al., 1981). The air film on which the head flies must be thick enough to prevent excessive material interactions; yet it must be thin enough to give a sufficiently large recording signal. Thus, the design of the head-disk interface is an optimization problem, which requires the understanding of not only air-bearing theory but also tribology, materials, magnetics, and mechanics.

The first commercially available disk file, made in 1957, used a head with a pressurized air supply which maintained a spacing between the magnetic element and the disk of approximately 20 μm (Harker et al., 1981). In 1962, a new concept, the hydrodynamically lubricated slider bearing, was introduced, allowing a head-disk spacing of approximately 6 μm. Since then, progressive improvements in air-bearing technology have been accomplished, with the result that air bearings in present-day disk files have a head-to-medium separation on the order of 0.3 μm, and extrapolations to spacings on the order of 0.125 μm have been made.

The theoretical analysis of the head-disk interface spacing requires use of the hydrodynamic theory of lubrication or, more specifically, of gas-bearing theory. This analysis is based on the Reynolds Lubrication equation, which is a partial differential equation relating pressure and spacing in the lubricating film to other physical design parameters. In applying this equation to the small head-disk spacings encountered in modern disk files, the assumptions made in its derivation must be reexamined. Rarefaction effects due to the increase in the Knudsen number (the ratio of mean free path of air molecules to the head-disk spacing) become important. Also, the neglect of surface roughness becomes increasingly questionable.

To understand the function and design of the head-media interface one must first study the individual components. First the in-plane stresses in rotating disks are analyzed, followed by the transverse dynamics and vibrations of rotating disks, taking into account the in-plane stresses as well as the effects of bending terms. Next the dynamics of the slider are analyzed, modeling it first as a point-mass, single-degree-of-freedom system and later as a rigid body with multiple degrees of freedom. Finally, consideration is given to the dynamics equations, including the suspension dynamics of the slider, accounting for pitch, roll, and vertical motion, and coupled with the time-dependent air-film equations for the spacing between the slider and the disk.

A study of the numerical solution of the Reynolds equation is required in order to simulate the head-disk interface. The results of this analysis can be illustrated by a summary of the slider design evolution and design curves for the three-rail and the two-rail sliders. Experimental methods are also important for understanding the head-disk interface, and a discussion of recent results obtained by use of laser-Doppler interferometry methods to study the dynamics of the interface is included. Finally, head-disk tribology is an area of great concern, but one that is still largely empirical and difficult to analyze.

Similar problem areas are addressed for flexible disks. Attention is given mainly to the high-speed, out-of-contact head-disk interface occurring in the Bernoulli disk, that is, the configuration in which an air film separates the head and disk as in the rigid-disk file. The equations governing the deflection of the disk and the numerical procedure that allows a simultaneous solution of the Reynolds equation and the disk deflection equation are discussed. Experimental results for the contour design of a flying head are then presented, and the question of substrate stability is examined. This area is important to maximum track density on flexible disks, since the anisotropic creep of the substrate over a period of 5 to 7 years must be predicted on the basis of incomplete and short-term experimental data.

Section 7.13 is devoted to the study of the elastohydrodynamic equations governing the head-media interface in linear tape drives and rotating-head devices. The modeling of the head-tape interface for both longitudinal and rotary configurations is of concern, and questions related to head contour design and tape dynamics as well as media wear and substrate stability are examined.

7.1   Hydrodynamic Theory of Lubrication

7.1.1   Boundary lubrication versus hydrodynamic lubrication

Two different processes can occur if two surfaces slide relative to each other in the presence of a thin lubricant flim. In the first process, both surfaces touch each other on a number of asperities at which the lubricant film is discontinuous; the load between the two surfaces is carried by the lubricant and asperities, and, with regard to wear and surface interaction, the properties of the sliding materials as well as those of the lubricant film are important. In the second process, the surfaces are completely separated by the fluid film; asperities do not contact each other, and the load between the two surfaces is carried by the lubricant film. In this case, the properties of the lubricant are much more important in determining the dynamics of the interface than the properties of the two surfaces. The first of the two processes is generally referred to as *boundary lubrication*. A schematic view of this situation is shown in Fig. 7.2*a*, where it is observed

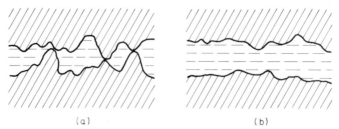

(a)   (b)

Figure 7.2   Types of lubrication. (*a*) Boundary lubrication: some materials contacts, with lubricant in intermediate space. (*b*) Fluid lubrication: no materials contacts, with lubricant completely separating surfaces.

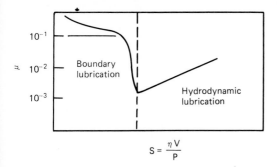

Figure 7.3 Stribeck diagram. Coefficient of friction versus generalized Sommerfeld number.

that only a relatively small number of asperities are in contact when two surfaces touch. The second situation, where the surfaces are completely separated, is called *hydrodynamic lubrication* and is shown schematically in Fig. 7.2b.

For disk file applications, hydrodynamic lubrication is preferable to boundary lubrication; thus, a sliding interface should ideally be designed to operate in the hydrodynamic lubrication mode. However, contacts between the two mating surfaces generally occur during start-stop, and, therefore, a knowledge of the behavior of the bearing in both the boundary lubrication and the hydrodynamic lubrication regimes is important.

The two different lubrication regimes are depicted in the Stribeck diagram (Moore, 1975) shown in Fig. 7.3, which is a plot of the coefficient of friction (tangential force divided by normal force) versus the generalized Sommerfeld number S (ratio of viscous to pressure forces). Hydrodynamic lubrication exists in the region to the right of the minimum friction coefficient, while boundary lubrication exists to the left. The coefficient of friction in the boundary lubrication region is substantially larger than that in the hydrodynamic lubrication region, and it increases to the value for unlubricated surfaces in the limit of small Sommerfeld numbers, where more and more material contacts occur. There is a slight increase in the coefficient of friction in the hydrodynamic lubrication region as the Sommerfeld number increases; however, this increase is small compared with that at small Sommerfeld numbers.

7.1.2 The Reynolds equation

The first theoretical treatment of fluid-film lubrication appeared in Reynolds's classic paper (1886) and a large number of papers have appeared on various aspects of fluid lubrication since that time. Lubrication theory is derived from the continuity and momentum equations, the equation of state, and the energy equation (Cameron, 1966; Constantinescu, 1969;

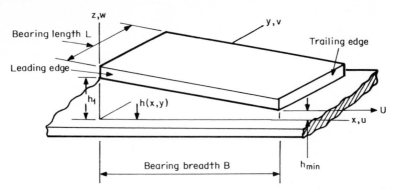

Figure 7.4 Slider-bearing coordinates and dimensions. In practice, h is of the order of micrometers, and $B \gg h$.

Gross et al., 1980). In addition, the following assumptions are made: (1) gravitational and inertial forces are negligible, (2) the fluid is Newtonian and the flow is laminar, (3) the viscosity of the fluid is constant, (4) non-slip boundary conditions are obeyed at the walls, (5) the fluid-film thickness is much smaller than other typical bearing dimensions, (6) the surfaces are smooth, and (7) surface tension effects are negligible.

A derivation of the Reynolds equation can be found in any of the references just cited. In reference to Fig. 7.4, the form to be considered is

$$\frac{\partial}{\partial x} ph^3 \frac{\partial p}{\partial x} + \frac{\partial}{\partial y} ph^3 \frac{\partial p}{\partial y} = 6U\mu \frac{\partial ph}{\partial x} + 12\mu \frac{\partial ph}{\partial t} \tag{7.1}$$

where x, y = spatial coordinates
$\quad\quad\;\; t$ = time
p and μ = pressure and viscosity, respectively, of the gas
$\quad\quad\; h$ = gas-film thickness
$\quad\quad\; U$ = velocity of moving bearing surface

It is often convenient to work with a nondimensional form of the Reynolds equation. Defining the nondimensional length scales as

$$X = \frac{x}{B} \quad\quad Y = \frac{y}{L} \quad\quad H = \frac{h}{h_{min}}$$

where B, L, and h_{min} are the breadth, length, and minimum spacing of the slider, and introducing the nondimensional pressure and time scales as

$$P = \frac{p}{p_a} \quad\quad T = \frac{t}{U/B}$$

where p_a is the ambient pressure outside the bearing, one obtains the following nondimensional form of the Reynolds equation.

$$\frac{\partial}{\partial X} PH^3 \frac{\partial P}{\partial X} + \frac{B^2}{L^2} \frac{\partial}{\partial Y} PH^3 \frac{\partial P}{\partial Y} = \Lambda \frac{\partial PH}{\partial X} + \sigma \frac{\partial PH}{\partial T} \qquad (7.2)$$

where B/L = slenderness ratio
Λ = bearing number
σ = squeeze film number, defined by

$$\Lambda = 6U\mu \frac{B}{p_a h_{min}^2} \qquad \sigma = 12\mu \frac{\omega B^2}{p_a h_{min}^2}$$

where ω is an appropriate frequency.

The above form of the Reynolds equation describes the pressure as a function of spacing in a compressible, time-dependent air bearing. In the case of steady conditions, the equation reduces to

$$\frac{\partial}{\partial X} PH^3 \frac{\partial P}{\partial X} + \frac{B^2}{L^2} \frac{\partial}{\partial Y} PH^3 \frac{\partial P}{\partial Y} = \Lambda \frac{\partial PH}{\partial X} \qquad (7.3)$$

7.1.3   Plane slider bearings

The design of fluid bearings begins in general with a calculation of the pressure distribution as a function of the bearing design parameters. To appreciate the importance of the Reynolds equation for this step, it is instructive to discuss the simple case of a plane slider bearing as shown in Fig. 7.4. The object is to calculate the pressure distribution in the fluid film as a function of the spacing h. For simplicity, it is assumed that the bearing is infinitely long ($L = \infty$), and that the flow is steady. From the solution of the pressure field it is possible to derive all other quantities needed for the design of the bearing, such as total load, pivot location, air flow, and total frictional force. If the coordinates and velocity components are as shown in Fig. 7.4, the Reynolds equation for the case of a plane slider reduces to

$$\frac{\partial}{\partial X} PH^3 \frac{\partial P}{\partial X} = \Lambda \frac{\partial PH}{\partial X} \qquad (7.4)$$

with boundary conditions

$$P = 1 \quad \text{at} \quad X = 0, 1$$

A closed-form solution of Eq. (7.4) has been obtained for arbitrary values of the bearing number. Prior to a discussion of this solution, however, the mathematically less complicated limited cases of very small and very large bearing numbers are investigated.

7.1.3.1 Small bearing-number limit. For small bearing numbers, Eq. (7.4) simplifies to the equation appropriate for incompressible fluids (Constantinescu, 1969):

$$\frac{\partial}{\partial X} H^3 \frac{\partial P}{\partial X} = \frac{\partial H}{\partial X} \tag{7.5}$$

Defining the spacing variation in the fluid film as

$$H = 1 + K - KX \tag{7.6}$$

where $\quad K = \dfrac{h_1 - h_{min}}{h_{min}} = H_1 - 1$

and integrating Eq. (7.5) twice with respect to X, using the boundary conditions of Eq. (7.4), one obtains the pressure distribution in a steady, infinitely long, incompressible bearing as

$$P = 1 + \Lambda \frac{KX(1 - X)}{(2 + K)(1 + K - KX)^2} \tag{7.7}$$

Equation (7.7) shows that the pressure increase generated in an incompressible plane inclined slider bearing is proportional to Λ, or, equivalently, to velocity U and fluid viscosity μ, and inversely proportional to the square of the minimum spacing h_{min}. Plots of the pressure distributions for several values of K are shown in Fig. 7.5.

After the pressure is obtained, the load per unit length is calculated from

$$W^* = \frac{W}{p_a LB} = \int_0^1 (P - 1) \, dX \tag{7.8}$$

resulting in

$$W^* = \frac{\Lambda}{K} \left[\frac{\ln(1 + K)}{K} - \frac{2}{2 + K} \right] \tag{7.9}$$

Furthermore, the pivot position $\bar{x}$, which should be located at the zero moment point, is given by

$$\bar{x} = \frac{1}{W} \int_0^L \int_0^B (p - p_a) x \, dy \, dx \tag{7.10}$$

resulting in

$$\overline{X} = \frac{\bar{x}}{B} = \frac{2(3 + K)(1 + K)\ln(1 + K) - K(6 + 5K)}{2K[(2 + K)\ln(1 + K) - 2K]} \tag{7.11}$$

Finally, the air flow per unit length is given by

$$\frac{q_x}{L} = Uh_{\min} \frac{1 + K}{2 + K} \tag{7.12}$$

and the frictional force by

$$F = \int_0^L \int_0^B \tau \, dy \, dx = \frac{LB\mu U}{h_{\min}} \left[\frac{4 \ln (1 + K)}{K} - \frac{6}{2 + K} \right] \tag{7.13}$$

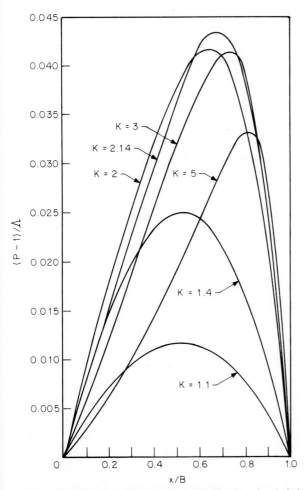

Figure 7.5 Incompressible pressure distribution for infinitely long, plane slider bearings for several values of K (*Gross et al., 1980*); $K = H_1 - 1$.

7.1.3.2 Large bearing-number limit.
As the bearing number increases to very large values, the left-hand side of Eq. (7.4) becomes very small compared with the right-hand side, and, in the limit of infinitely large Λ, this equation reduces to

$$\frac{\partial PH}{\partial X} = 0 \tag{7.14}$$

Since Eq. (7.14) is only a first-order differential equation, compared with the Reynolds equation, which is second-order, only one boundary condition can be satisfied. If this equation is integrated and the boundary condition $P = 1$ at $X = 0$ is satisfied, there results

$$P = \frac{H_1}{K + 1 - KX} \tag{7.15}$$

From Eq. (7.15) it is observed that the pressure distribution in the limiting case of large bearing numbers is independent of velocity and viscosity. Also, the load-carrying capacity of this bearing is given by

$$W^* = \frac{K + 1}{K} \ln (K + 1) - 1 \tag{7.16}$$

and it is also independent of velocity or fluid viscosity.

In Fig. 7.6 the variation of load for an infinitely long, plane slider bearing is shown as a function of the bearing number for several values of H_1. There is an incompressible region at small bearing numbers where the

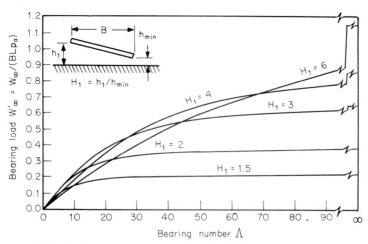

Figure 7.6 Load variation as a function of the bearing number Λ for several values of H_1; $\Lambda = 6\mu \, UB/h_{\min}^2 \, p_a$ (*Gross et al., 1980*).

load is proportional to the bearing number, and a compressible, infinitely large bearing-number region where the load is independent of the bearing number.

For a typical air bearing in a magnetic disk file, the bearing number is on the order of 100, assuming a typical breadth of 5 mm, a flying height of 1 μm, and air with a viscosity of 1.8×10^{-5} Pa · s. Thus, the bearing is in the intermediate range, where neither the incompressible nor the infinitely large bearing-number assumption is applicable. It should be pointed out that the dependence of the load-carrying capability on the bearing number is modified by the length of the bearing, because short bearings experience side flow with a decrease in pressure.

7.1.3.3 Intermediate bearing numbers. Harrison (1913) obtained a closed-form solution of the compressible Reynolds equation, (Eq. 7.4), for the case of the infinitely long, plane slider bearing. He first replaced dP/dX by $dP/dH \, dH/dX$, and then introduced the new dependent variable $\psi = PH$. Separating variables, integrating, and rearranging, one obtains in this manner a solution of the form

$$\psi^2 - \frac{\Lambda}{H_1 - 1}\psi + c_1 = c_2 H^2 f(\psi) \tag{7.17}$$

in which $f(\psi)$ is a complicated algebraic function that has different forms depending on the value of the bearing number Λ, and the constants c_1 and c_2 must be determined from the boundary conditions. A numerical solution of this equation, showing the pressure as a function of the position in the breadth direction, is plotted in Fig. 7.7 for $K = 1$ (Gross et al., 1980). Here one can observe the large bearing-number limit with $\Lambda \rightarrow \infty$, and also the small bearing-number limit with $\Lambda \rightarrow 0$. The variation of the load with bearing number is shown in Fig. 7.6.

7.1.4 The Reynolds equation in the limit of very small spacing

When the spacing between the slider and disk decreases to the submicrometer region, the Knudsen number (the ratio of mean free path to bearing separation) reaches a value of the order of 0.1 or larger. Under these conditions, the air film deviates from continuum-flow assumptions; that is, the usual no-slip boundary condition at the walls is no longer satisfied. The occurrence of slip at the boundaries implies that the velocity gradient in the air film is decreased. Thus, the apparent viscosity of the air is reduced and the load-carrying capacity of the slider is diminished.

The kinetic theory of gases represents viscosity in a gas as the tendency toward uniformity of mass velocity due to the motion of molecules from

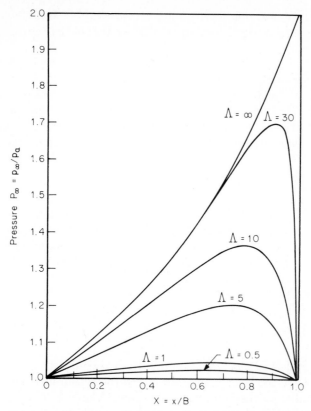

Figure 7.7   Pressure variation over the infinitely long slider bearing for $K = 1$ and several values of the bearing number *(Gross et al., 1980).*

point to point (Chapman and Cowling, 1964). As a consequence of this motion, momentum and energy are transported in a gas and the conditions at the beginning and end of each free path are equalized. At sufficiently low pressures the mean free path of molecules becomes comparable to the physical dimensions of the system, and the transport of momentum ceases to be a free-path phenomenon. A similar situation is encountered if the dimensions of a physical system decrease so that the mean free path becomes an appreciable fraction of these dimensions as in the case of a gas slider bearing at very close spacing.

If the gas departs only slightly from the uniform state, approximate expressions for viscosity can still be obtained using classical kinetic theory. Assuming flow between parallel flat plates, one of which is moving with velocity U, and considering the momentum exchange of molecules at the walls, the gas near a wall slips relative to the wall with a speed

$$k\lambda\left(\frac{2 - \theta}{\theta}\right)\frac{\partial u}{\partial z}$$

where θ is the fraction of molecules entering the wall, $2 - \theta$ is the portion of molecules that is reflected specularly, k is a numerical factor on the order of 1, λ is the mean free path, and $\partial u/\partial z$ is the velocity gradient of the gas (Chapman and Cowling, 1964). Applying these considerations to the case of a slider bearing, one obtains the Reynolds equation modified for boundary slip (Burgdorfer, 1959),

$$\frac{\partial}{\partial X}\left[PH^3\frac{\partial P}{\partial X}\left(1 + \frac{6K_\infty}{PH}\right)\right] \tag{7.18}$$
$$+ \frac{\partial}{\partial Y}\left[PH^3\frac{\partial P}{\partial Y}\left(1 + \frac{6K_\infty}{PH}\right)\right] = \Lambda\frac{\partial}{\partial X}PH + \sigma\frac{\partial P}{\partial T}$$

where $K_\infty = \lambda/h_{min}$ is the Knudsen number based on the ambient mean free path and minimum spacing.

In Fig. 7.8 the pressure distribution in a plane slider bearing is plotted with and without slip flow, assuming a minimum slider-disk spacing of 0.2 μm. A substantial reduction in the pressure occurs due to slip.

It should be pointed out that the derivation of the slip-flow boundary conditions was carried out for the case of pure shear flow between parallel plates. In the case of a slider bearing, strong pressure gradients exist and the flow is not that of two parallel plates. Thus, the slip-flow boundary conditions must be applied with caution. This point will be discussed in the context of comparison between experiment and theory in a later section.

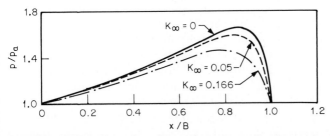

Figure 7.8 Effect of boundary slip on the pressure distribution for an infinitely long, plane slider bearing; $K_\infty = \lambda / h_{min}$; $h_1 / h_{min} = 2$ (Burgdorfer, 1959).

7.1.5 Surface roughness effects

One of the assumptions made in the derivation of the Reynolds equation was that the surfaces of the bearings are perfectly smooth. When the film thickness between the bearing surfaces becomes small enough that the surface roughness of the individual surfaces is a substantial fraction of the bearing clearance, the effect of surface roughness must be taken into account. If a special mathematical form of surface roughness such as a single sine-wave profile is postulated,

$$h = h_0 + a \sin \beta x \tag{7.19}$$

the Reynolds equation can be solved by standard methods. An alternative approach to the treatment of surface roughness effects has been suggested in which the film thickness in a rough bearing is viewed as a stochastic process, and a modified Reynolds equation is derived in the statistical mean pressure (Christensen and Tonder, 1969a, 1969b; Tonder, 1984). For the case of unidirectional furrows and ridges on one or both bearing surfaces, the following modified Reynolds equation is derived:

$$\frac{\partial}{\partial x}\left(\phi_x \overline{p}\, \frac{\partial \overline{p}}{\partial x}\, h^3\right) + \frac{\partial}{\partial y}\left(\phi_y\, \overline{p}\, \frac{\partial \overline{p}}{\partial y}\, h^3\right) = 6 U \eta\, \frac{\partial}{\partial x}\,(\phi_u \overline{p} h) \tag{7.20}$$

where ϕ_x, ϕ_y, and ϕ_u are nondimensional, film-height averages related to pressure flow in the x and y directions, and to shear flow. For the case of longitudinal striations, Tonder (1984) gives the following expressions, where overbars denote averages over an appropriate neighborhood:

$$\phi_x = \frac{\overline{H^3}}{h^3} \qquad \phi_y = \frac{1}{h^3 \overline{H^{-3}}} \qquad \phi_u = 1 \tag{7.21}$$

For the case of transverse striations the expressions are

$$\phi_x = \frac{1}{h^3 \overline{H^{-3}}} \qquad \phi_y = \frac{\overline{H^3}}{h^3} \qquad \phi_u = \frac{\overline{h^{-2}}}{h \overline{H^{-3}}} \tag{7.22}$$

An attempt at treating a random surface has been made for the case of an infinitely long, plane slider bearing and incompressible flow, and it was found that both the load-carrying capacity as well as the frictional force increases substantially owing to the roughness effect (Tzeng and Saibel, 1967). The general case of compressible finite bearings with arbitrary roughness is too difficult to solve in closed form, but numerical solutions may be feasible.

Certain basic assumptions used in the derivation of the Reynolds equation, such as unidirectional flow and no variation of pressure across the film thickness, become suspect for rough bearing surfaces in which the wavelength of the roughness becomes as small as the spacing. Fundamental developments accounting for variations in the thickness direction may be necessary in order to get an accurate solution in these cases.

7.2 Dynamics of Rotating Disks

The dynamical behavior of a spinning disk is complicated. The motions of concern are those in the plane of the disk as well as those transverse to the disk. Some important considerations in determining these motions for an annular disk are the conditions at the inner and outer boundaries, the bending stiffness, and the speed of rotation. A very stiff rotating disk may have vibrations that are essentially the same as when it is not rotating, but any disk's transverse vibrational behavior can be radically altered by spinning it sufficiently fast. This is because the restoring forces associated with transverse displacements have two sources in a rotating disk: those due to the bending stiffness and those due to the in-plane stresses that result from the inertial forces of rotation. These latter forces restore transverse displacements in a manner similar to that of a stretched membrane, and, as is shown below, the in-plane membrane stresses are proportional to the square of the rotational speed.

The analytical description of the problem is much simpler for a complete disk than for an annular one; but magnetic recording disks always have a central hole, and attention cannot be confined entirely to the simple case. Nevertheless, the physics of the problem is easier to understand for the complete disk, and hence that case is considered as a preliminary to the more complicated annular disk.

7.2.1 In-plane stresses due to disk rotation

Figure 7.9 shows a circular annular disk of uniform thickness b, outer radius r_o, and clamping radius r_i. It rotates about its central axis with angular speed ω. Polar coordinates (r, θ) locate a point in the two-dimensional plate model for the disk, where r is measured from the center and θ is the angular coordinate measured from a radial line scribed on the disk.

First, only the in-plane stresses due to the rotation are considered. The axisymmetric, in-plane stress equations of motion for steady rotation reduce to (Timoshenko and Goodier, 1974)

$$\frac{\partial \sigma_r}{\partial r} + \frac{\sigma_r - \sigma_\theta}{r} + \rho \omega^2 r = 0 \tag{7.23}$$

and the stress-displacement relations are

$$\sigma_r = \frac{E}{1 - \nu^2} \left(\frac{dU}{dr} + \nu \frac{U}{r} \right)$$

$$\sigma_\theta = \frac{E}{1 - \nu^2} \left(\frac{U}{r} + \nu \frac{dU}{dr} \right) \tag{7.24}$$

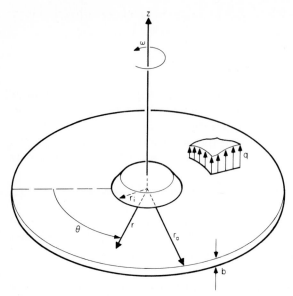

Figure 7.9 Disk geometry and coordinates.

where σ_r, σ_θ = radial and azimuthal normal stress components, respectively
U = radial displacement
E = Young's modulus
ν = Poisson's ratio of the isotropic and homogeneous elastic disk

Substitution of Eq. (7.24) into Eq. (7.23) gives

$$\frac{d^2U}{dr^2} + \frac{1}{r}\frac{dU}{dr} - \frac{1}{r^2}U + \frac{1-\nu^2}{E}\rho\omega^2 r = 0 \tag{7.25}$$

which must be solved subject to the boundary conditions of a clamped inner radius r_i and a free outer radius r_0, which are

$$U(r_i) = 0, \qquad \sigma_r(r_0) = 0 \tag{7.26}$$

The solution of this problem gives the in-plane stress components

$$\sigma_r = \frac{\rho\omega^2 r_0^2}{8}\left[-A\zeta^2 + (1+\nu)\frac{(1-\nu)\kappa^4 + A}{(1-\nu)\kappa^2 + 1 + \nu}\right.$$

$$\left. -(1-\nu)\frac{(1+\nu)\kappa^4 - A\kappa^2}{(1-\nu)\kappa^2 + 1 + \nu}\frac{1}{\zeta^2}\right.$$

$$\sigma_\theta = \frac{\rho\omega^2 r_0^2}{8}\left[-B\zeta^2 + (1+\nu)\frac{(1-\nu)\kappa^4 + A}{(1-\nu)\kappa^2 + 1 + \nu}\right. \tag{7.27}$$

$$\left. + (1+\nu)\frac{(1+\nu)\kappa^4 - A\kappa^2}{(1-\nu)\kappa^2 + 1 + \nu}\frac{1}{\zeta^2}\right]$$

in which ρ is the mass density, and κ, ζ, A, and B are defined by

$$\kappa = \frac{r_i}{r_0}, \qquad \zeta = \frac{r}{r_0}$$
$$A = 3 + \nu, \qquad B = 1 + 3\nu \tag{7.28}$$

If the disk is complete, $r_i = 0$, and Eq. (7.27) simplifies to

$$\sigma_r = \frac{1}{8} \rho \omega^2 r_0^2 A (1 - \zeta^2)$$
$$\sigma_\theta = \frac{1}{8} \rho \omega^2 r_0^2 (A - B\zeta^2) \tag{7.29}$$

These results indicate that the in-plane stresses are directly proportional to $\omega^2 r_0^2$, and therefore they increase as the squares of the rotational speed and the outer radius.

7.2.2 Transverse free vibrations of a rotating membrane disk

The in-plane stresses due to rotation act as variable membrane stress restoring forces for transverse displacements. If bending stiffness effects are neglected, the equation for the transverse motion is given by

$$\frac{1}{r} \frac{\partial}{\partial r} \left(r\sigma_r \frac{\partial W}{\partial r} \right) + \frac{1}{r^2} \sigma_\theta \frac{\partial^2 W}{\partial \theta^2} = \rho \frac{\partial^2 W}{\partial t^2} \tag{7.30}$$

where W is the transverse displacement of the disk. For simplicity, the case of the complete disk is considered so that σ_r, σ_θ are given by Eq. (7.29). Then, in order to study free vibrations it is assumed that solutions are of the form

$$W(r, \theta, t) = z(r) \sin (n\theta + \epsilon) \sin (pt) \tag{7.31}$$

which, with Eqs. (7.29) and (7.30), gives the equation for determining z,

$$\frac{A}{r} \frac{d}{dr} \left[(r_0^2 - r^2) r \frac{dz}{dr} \right] - \left(A \frac{r_0^2}{r^2} - B \right) n^2 z + \frac{p^2}{\omega^2} z = 0 \tag{7.32}$$

where p is the frequency of vibration and ϵ is a phase angle. The natural frequencies and mode shapes for free vibrations are given by (Prescott, 1961)

$$\frac{p_{n,s}^2}{\omega^2} = (n + 2s + 2)(n + 2s)A - n^2 B \tag{7.33}$$

and the series expression

$$z = C_{n,s}\zeta^n \left[1 - \frac{s(n + s + 1)}{n + 1} \zeta^2 \right.$$

$$\left. + \frac{(s - 1)s(n + s + 1)(n + s + 2)}{2(n + 1)(n + 2)} \zeta^4 + \cdots \right] \quad (7.34)$$

The form of the assumed solution in Eq. (7.31) indicates that the integer n represents the number of nodal diameters, whereas the alternating series in Eq. (7.34), which terminates to a polynomial in r of degree $2s$, indicates that s represents the number of nodal circles. Some particular examples are as follows.

1. n nodal diameters, zero nodal circles:

$$W_{n,0} = C_{n,0}\zeta^n \sin{(n\theta + \epsilon)} \sin{(p_{n,0}t)}$$

$$\frac{p_{n,0}^2}{\omega^2} = \frac{1}{4}n\,[(n + 3) - \nu(n - 1)] \quad (7.35)$$

2. Zero nodal diameters, one nodal circle:

$$W_{0,1} = C_{0,1}(1 - 2\zeta^2) \sin{(p_{0,1}t)}$$

$$\frac{p_{0,1}^2}{\omega^2} = 3 + \nu \quad (7.36)$$

The latter mode cannot be realized on a disk fixed to a central axis since it predicts a nonzero center displacement.

The vibration modes described here are in terms of coordinates rotating with the disk. But the same solution is obtained if, in Eq. (7.31), the standing waveform, relative to the disk, is replaced by the propagating waveform

$$W(r, \theta, t) = z(r) \sin{(n\theta \pm pt + \epsilon)} \quad ((7.37)$$

This means that the nodal diameters are those for which

$$n\theta \pm pt + \epsilon = m\pi \quad (7.38)$$

and they may rotate with the angular velocities given by

$$\frac{d\theta}{dt} = \pm\frac{p}{n} \quad (7.39)$$

that is, they may rotate with a speed p/n relative to the disk in the same or opposite direction. Since the angular speed of the disk is ω, it follows that the nodal diameters may rotate relative to space-fixed coordinates with angular velocities given by Ω, where

$$\Omega = \omega - \frac{p}{n} \qquad \Omega = \omega \qquad \text{or} \qquad \Omega = \omega + \frac{p}{n} \quad (7.40)$$

It is therefore possible for the n nodal diameters to be at rest if $\omega = p/n$, which means n and s are given by (Prescott, 1961)

$$n = \frac{1}{2\sqrt{3}}\,[(2 + \sqrt{3})^l - (2 - \sqrt{3})^l] \qquad l = 1, 2, \ldots$$

$$n + 2s + 1 = \frac{1}{2}[(2 + \sqrt{3})^l + (2 - \sqrt{3})^l]$$

(7.41)

The first four possibilities for stationary modes of free vibration of the rotating disk are

$$\omega = \frac{p}{n} \qquad (n, s) = (1, 0), (4, 1), (15, 5), (56, 20) \tag{7.42}$$

so that the nodal diameters n are restricted to 1, 4, 15, 56, ... , and the corresponding nodal circles are given by 0, 1, 5, 20, ...

7.2.3 Transverse free vibrations of a stationary stiff disk

The classical equation for transverse motion of an elastic plate is (Timoshenko and Woinowsky-Krieger, 1940)

$$\frac{Eb^2}{12(1 - \nu^2)}\,\nabla^4 W + \rho\,\frac{\partial^2 W}{\partial t^2} = 0 \tag{7.43}$$

where ∇ is the Laplace operator and b is the disk thickness. If normal mode solutions are assumed of the form

$$W(r, \theta, t) = z(r)\sin(n\theta + \epsilon)\sin(pt) \tag{7.44}$$

then $z(r)$ is found to satisfy Bessel's equation and solutions can be expressed in the form

$$z_n(r) = AJ_n(kr) + BY_n(kr) + CI_n(kr) + DK_n(kr) \tag{7.45}$$

where J_n and Y_n are Bessel functions of the first and second kinds, and I_n, K_n are modified Bessel functions of the first and second kinds (Abramowitz and Stegun, 1964). The parameter k is related to the frequency p by

$$k^4 = 12(1 - \nu^2)\rho\,\frac{p^2}{Eb^2} \tag{7.46}$$

For a disk with a central hole or with no central hole but with its center fixed, there are two boundary conditions at the hole, or center, and two more at the outer rim. These boundary conditions determine the frequency of vibrations, p, in Eq. (7.44), and three ratios of the four integration constants A, B, C, and D in Eq. (7.45).

One of the simplest cases of interest is the case of a complete disk that is free at its outer edge. In this case B and D are set to zero in Eq. (7.45) and the free boundary conditions are imposed at the outer edge; that is, at $r = r_0$,

$$\frac{\partial^2 W}{\partial r^2} + \frac{\nu}{r}\frac{\partial W}{\partial r} = 0$$

$$\frac{\partial}{\partial r}\left[\frac{1}{r}\frac{\partial}{\partial r}\left(r\frac{\partial W}{\partial r}\right)\right] = 0$$

(7.47)

The solution from Eqs. (7.44) and (7.45) is of the form

$$W = [AJ_n(kr) + CI_n(kr)] \sin (n\theta + \epsilon) \sin (pt)$$

(7.48)

The frequency equation determining p, with use of Eq. (7.46), is

$$\frac{k^2 J_n(k) + (1 - \nu)[k J_n'(k) - n^2 J_n(k)]}{k^2 I_n(k) - (1 - \nu)[k I_n'(k) - n^2 I_n(k)]}$$

$$= \frac{k^3 J_n'(k) + (1 - \nu)n^2[k J_n'(k) - J_n(k)]}{k^3 I_n'(k) - (1 - \nu)n^2[k I_n'(k) - I_n(k)]}$$

(7.49)

in which

$$k = kr_0$$

(7.50)

and J_n', I_n' are derivatives of the Bessel functions, expressible in terms of the Bessel functions themselves (Abramowitz and Stegun, 1964). For the case $n = 1$, Eq. (7.49) determines k and the mode is found to have one nodal circle. It is found that for $\nu = 0.25$ this natural frequency $p_{1,1}$ is given by

$$k_{1,1} = 4.518$$

(7.51)

and the corresponding mode is

$$W_{1,1} = AJ_2(k)\left[\frac{J_1(k_{1,1}\zeta)}{0.2119} - \frac{I_1(k_{1,1}\zeta)}{10.84}\right] \sin (\theta + \epsilon) \sin (p_{1,1}t)$$

(7.52)

and the nodal circle is at $\zeta = 0.781$, where ζ is the dimensionless radius r/r_0.

A mode cannot exist that has one nodal diameter and no nodal circles. The mode in which the frequency is the smallest corresponds to $n = 2$ and $s = 0$. The frequency for this mode (for $\nu = 0.25$) is determined by

$$k_{2,0} = 2.348$$

The accompanying table gives the k solutions for the first few modes ($\nu = 0.25$). It is observed that modes exist for all positive-integer values of (n, s) except for (0, 0) and (1, 0), and the natural frequencies increase for increasing values of either n or s.

		n		
s	0	1	2	3
0	· · · ·	· · · ·	2.348	3.570
1	2.982	4.518	5.940	7.291
2	6.192	7.729		

7.2.4   Dynamics of a rotating disk loaded by a transverse force

Since any rotating disk will have both bending and membrane effects present in some ratio, depending on the bending stiffness of the disk and the speed of rotation, the results above show that an accurate description of the free vibrations for the forced motion of the disk in magnetic disk files is complicated. In rigid-disk drives the bending effects probably dominate, whereas in flexible-disk applications, the membrane stresses may be more important. Some investigations have indicated that both effects must be retained in the analysis of each of these applications in order to obtain an accurate description of the motion.

In magnetic disk drives the loads that a disk experiences during operation are primarily those of the read-write head and the air flow. The latter is often associated with air turbulence and is difficult to characterize. The read-write head loads the disk over a relatively small area by the pressure forces generated in the air bearing. In rigid-disk drives this load can usually be characterized as a transverse resultant force that is fixed in space. In flexible-disk drives it is often necessary to take into account the spatial variations of the air-bearing pressure and couple the dynamics of the disk with the Reynolds equation in order to establish the "foil" bearing pressure distribution.

To describe disk dynamics when applied loads are spatially fixed, it is convenient to transfer the equations of motion of the disk, which are usually written in terms of coordinates that rotate with the disk, into coordinates that are fixed in space. If the membrane and bending effects described by Eqs. (7.29), (7.30), and (7.43) are combined and nondimensional variables are used, the following equation is obtained (Benson and Bogy, 1978):

$$\alpha \nabla^4 W + L[W] + \frac{\partial^2 W}{\partial \tau^2} + \frac{4\sqrt{2}}{\sqrt{3 + \nu}} \frac{\partial^2 W}{\partial \tau \, \partial \phi} = Q \tag{7.53}$$

in which ϕ is the space-fixed azimuthal coordinate defined by

$$\phi = \theta + \omega t \tag{7.54}$$

and the other nondimensional terms in Eq. (7.53) are

$$\alpha = \frac{2Eb^2}{3\rho\omega^2(3+\nu)(1-\nu^2)r_0^4} \qquad \text{spin stiffness}$$

$$\nabla^4 W = \left(\frac{\partial^2}{\partial\zeta^2} + \frac{1}{\zeta}\frac{\partial}{\partial\zeta} + \frac{\partial^2}{\partial\phi^2}\right)W \qquad \text{bending terms}$$

$$W = \frac{W'}{b} \qquad \text{transverse displacement}$$

$$L[W] = -\frac{1}{\zeta}\frac{\partial}{\partial\zeta}\left[+(1-\zeta^2)\frac{\partial W}{\partial\zeta}\right] + \left(\frac{3\zeta^2-1}{\zeta^2}\right)\frac{\partial^2 W}{\partial\phi^2} \qquad \text{membrane terms} \tag{7.55}$$

$$\tau = t\sqrt{\frac{1}{8}\omega^2(3+\nu)} \qquad \text{time}$$

$$\zeta = \frac{r}{r_0} \qquad \text{radial coordinate}$$

$$Q = \frac{8q}{\rho b^2\omega^2(3+\nu)} \qquad \text{pressure load}$$

The appropriate boundary conditions for a disk clamped at $\zeta = \kappa$ ($\kappa = r_i/r_0$) and free at $\zeta = 1$ are

$$W(\kappa, \phi) = 0 \qquad \frac{\partial W}{\partial\zeta}(\kappa, \phi) = 0 \qquad \text{deflection, slope}$$

$$\left[\frac{\partial^2 W}{\partial\zeta^2} + \left(\frac{\partial W}{\partial\zeta} + \frac{\partial^2 W}{\partial\phi^2}\right)\right]_{\zeta=1} = 0 \qquad \text{moment}$$

$$\left[\frac{\partial^3 W}{\partial\zeta^3} + (2-\nu)\frac{\partial^3 W}{\partial\zeta\,\partial\phi^2} + \frac{\partial^2 W}{\partial\phi^2}\right. \tag{7.56}$$

$$\left.-\frac{\partial W}{\partial\zeta} - (3-\nu)\frac{\partial^2 W}{\partial\phi^2}\right]_{\zeta=1} = 0 \qquad \text{shear}$$

$$W(\zeta, \phi) = W(\zeta, \phi + 2\pi) \qquad \text{periodicity}$$

Equations (7.53) and (7.54) show that, if $\alpha = 0$, no bending effects are present, and Eq. (7.30) is recovered. Equation (7.55) shows that the spin stiffness α is small if the intrinsic bending stiffness is small, or if $\rho\omega^2 r^4$ is large. Therefore a disk which is quite stiff at rest will tend to behave like a membrane if the rotational speed is fast enough. If α is set to zero in Eq. (7.53), a fourth-order equation is reduced to a second-order equation. Correspondingly, the number of boundary conditions in (Eq. 7.56) must be reduced from four to two. The value of α appropriate to a flexible disk with $b = 0.075$ mm, $r_o = 150$ mm, $E = 6 \times 10^9$ N/m^2, $\nu = 0.23$, $\rho = 1.47$ g/cm^3, $\omega = 190$ s^{-1}, $D = Eb^3/12(1 - \nu^2) = 0.22 \times 10^{-3}$ N·m is $\alpha = 0.00027$. The value for a rigid disk with $b = 2$ mm, $r_o = 200$ mm, $E = 20 \times 10^{10}$ N/m, $\nu = 0.3$, $\rho = 8$ g/cm^3, $\omega = 260$ s^{-1}, $D = 150$ N·m is $\alpha = 0.21$. Thus the α range of interest for magnetic recording is over two orders of magnitude.

In the remainder of this section the investigation is restricted to steady deflections of the rotating disk as viewed by a space-fixed observer. Equation (7.53) then takes the form

$$\alpha\nabla^4 W + L[W] = Q \tag{7.57}$$

In the case of flexible disks, for which $\alpha \approx 0.0003$, the implication of neglecting the $\alpha\nabla^4 W$ bending stiffness term in Eq. (7.53) is obtained by considering a spinning membrane with a space-fixed transverse load governed by

$$L[W] = Q \tag{7.58}$$

This leads to the differential operator $L[W]$ having a different character in different regions of the disk (Benson and Bogy, 1978). The operator is hyperbolic in the outer portion and elliptic in the inner portion, which means that waves can propagate in the outer region but not in the inner region. The regions are defined by

$$\frac{1}{\sqrt{3}} < \zeta < 1 \qquad \text{hyperbolic}$$

$$\zeta = \frac{1}{\sqrt{3}} \qquad \text{parabolic}$$

$$\kappa \leq \zeta < \frac{1}{\sqrt{3}} \qquad \text{elliptic}$$

The characteristics in the hyperbolic region are given by

$$\psi(\zeta, \phi) = f(\zeta) \pm \phi \tag{7.59a}$$

with $\quad f(\zeta) = \sqrt{3}\tan^{-1}\frac{1}{\sqrt{3}}\left(\frac{3\zeta^2 - 1}{1 - \zeta^2}\right)^{1/2} - \tan^{-1}\left(\frac{3\zeta^2 - 1}{1 - \zeta^2}\right)^{1/2} \tag{7.59b}$

These regions and characteristic lines are shown in Fig. 7.10. The characteristics are both orthogonal to the inner transition circle, $\zeta = 1/\sqrt{3}$, and tangent to the outer circle, $\zeta = 1$. An experimental observation on a Bernoulli flexible disk, which was loaded by a probe through the base plate, shows that the wave disturbances occur along the characteristics in

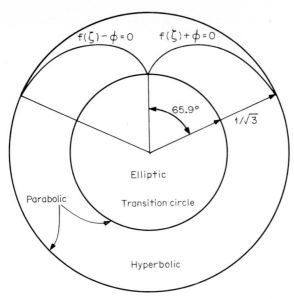

Figure 7.10 Different regions of classification of the rotating membrane equation *(Benson and Bogy, 1978).*

the outer region, but no such waves appear in the inner region (Benson and Bogy, 1978).

7.2.4.1 Solution with bending stiffness neglected.   Here the possibility is investigated of solving Eq. (7.58) for an applied transverse load Q. The approach taken is based on the eigenfunction expansion method. The general idea is to find the inverse operator L^{-1} of L, if it exists, and write the solution as

$$W = L^{-1}[Q] \tag{7.60}$$

It was shown, however (Benson, 1977), that the inverse operator L^{-1} does not exist, in general, because there is an infinite sequence of zero eigenvalues:

$$\lambda_{1,0} = 0, \quad \lambda_{4,1} = 0, \quad \lambda_{15,5} = 0, \quad \ldots \tag{7.61}$$

These values can be compared with Eq. (7.42). The zero eigenvalues for the space-fixed coordinate description correspond to eigenfunctions that have a certain number of nodal circles and diameters. The same number of nodal circles and diameters occur in the modes of free vibration in the rotating coordinate description and move relative to the disk in such a manner as to be stationary in space.

7.2.4.2 Solution with bending stiffness retained. Since Eq. (7.58) does not have a solution for arbitrary loads, Eq. (7.57) has to be solved with the bending terms retained, even though α may be as small as 10^{-4}. The solution was derived (Benson and Bogy, 1978) for a single-point load, which is represented by the function

$$Q = \frac{1}{\zeta} \delta(\zeta - \xi)\, \delta(\phi - \psi)$$

where δ is Dirac's delta function. The method employed Fourier series expansion in the azimuthal coordinate ϕ, followed by use of a finite-difference method for solving the separable radial functions. It was found that numerical difficulties arise if the value of α is less than 0.001. This means that the bending terms cannot be neglected for small α, but it is numerically difficult to solve the complete equations for values of α appropriate to flexible disks.

Figures 7.11 and 7.12 present displacement contours for $\alpha = 0.001$, $\kappa = 0.2$, and illustrate many of the solution properties that have been discussed on the basis of the membrane analysis. Even though the bending stiffness is retained, the solution shows that the deflection occurs mostly in the hyperbolic region, $\zeta > 0.577$. In Fig. 7.11 the transverse point load is located at $\zeta = 0.6$, whereas in Fig. 7.12 it is located at $\zeta = 0.4$. The

Stiffness $\alpha = 0.001$ Load radius $\zeta = 0.6$

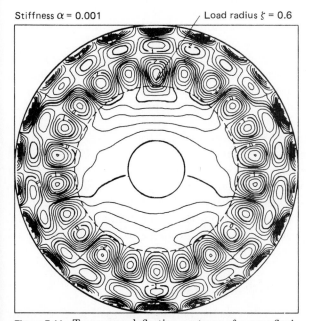

Figure 7.11 Transverse deflection contours of a very flexible spinning disk loaded by a space-fixed transverse force in the hyperbolic region *(Benson and Bogy, 1978)*.

Stiffness $\alpha = 0.001$ Load radius $\zeta = 0.4$

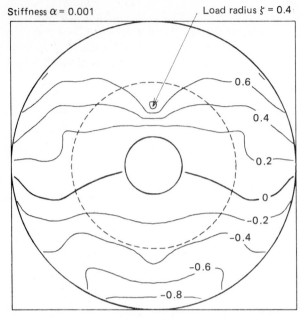

Figure 7.12 Transverse deflection contours resulting from the force located in the elliptic region *(Benson and Bogy, 1978)*.

difference in the two solutions is significant, and this result should be an important concern for disk drive designers.

A hybrid equation has been derived in which the fourth-order terms of $\alpha \nabla^4 W$ are neglected but certain second-order terms, which also contain a large separation parameter, are retained (Benson, 1983). In this manner it is possible to get tractable solutions of the problem for α as small as 0.0001.

Another approach to the related problem of free vibrations of a spinning disk with bending effects retained has been examined (Eversman and Dodson, 1969). Numerical calculations of the frequencies of free vibration were made for low modes and a wide range of disk stiffnesses. Analyses appropriate for flexible disks have been made with read-write head interactions and hydrodynamic effects from the surrounding air (Greenberg, 1978; Adams, 1980). Such dissipation mechanisms tend to make the numerical methods much more manageable.

7.3 Dynamics of the Head-Disk Interface

The need to maintain a steady spacing between the read-write element and the recording surface is the most important design criterion for a head-disk interface. A magnetic disk is not perfectly flat, and its motion deviates from true in-plane motion due to spindle runout, clamping distortions, and disk vibrations. The slider must follow the disk over these out-of-plane motions. Such amplitude variations in a typical disk file exceed the air-bearing spacing by several orders of magnitude, and are typically in the range of 10 to 30 μm, compared with a head-disk spacing of about 0.3 μm. In order to accommodate this runout with a minimum of spacing modulation, it is necessary to design the head suspension mechanism and the head-disk interface with a high degree of compliance.

7.3.1 Single mass-spring-dashpot model for slider and air bearing

In Fig. 7.13 a simple model of the equivalent dynamical system for a typical spring-loaded recording head is shown (Gross, 1984), consisting of a mass m, two linear springs of stiffness k_1 and k_2 for the air-bearing and suspension spring, and two dampers with damping constants c_1 and c_2 for the air-bearing and suspension damping, respectively. In a typical disk file, where high compliance is required, the stiffness and damping constants of the suspension are smaller by several orders of magnitude than those of the air bearing, and the former can be neglected for purposes of first-order analysis. For this model the equation of motion of the air-bearing slider is

$$m\ddot{z} + c_1\dot{z} + k_1z = c_1\dot{y} + k_1y + F(t) \tag{7.62}$$

where $F(t)$ is a time-dependent forcing function, and $y(t)$ is the time-dependent vertical motion of the disk.

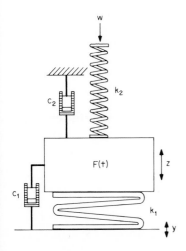

Figure 7.13 Simple model of equivalent dynamic system for head-disk interface *(Gross, 1984)*.

Although the mass-spring-dashpot model for the slider is a very simplified idealization, several important observations relevant to slider dynamics can be made from analyzing its response to the motion of the driving surface, as well as its response to arbitrary applied forces.

7.3.1.1 Harmonic excitation.

If the disk vertical motion is given by $y = A \sin \omega t$, and the damping is zero, the following solution for the response of the slider is obtained from Eq. (7.62) [with $F(t) = 0$]:

$$z(t) = \frac{A}{1 - (\omega/\omega_n)^2} \sin \omega t \tag{7.63}$$

where $\omega_n^2 = k_1/m$.

In order for the slider to follow the disk with constant spacing, it is necessary that

$$\Delta h = z - y = 0 \tag{7.64}$$

This last requirement is approximately met if

$$\frac{(\omega/\omega_n)^2}{1 - (\omega/\omega_n)^2} \ll 1 \tag{7.65}$$

which requires $\omega \ll \omega_n$; that is, in order for the slider to follow the disk, the out-of-plane motion frequency of the disk must be much lower than the air-bearing slider resonance frequency.

7.3.1.2 Periodic excitation.

A typical vertical runout profile for a rotating disk is shown in Fig. 7.14. It is observed that the runout is periodic; it repeats itself for each revolution, but its frequency may be different from the frequency of the disk rotation. The disk out-of-plane motion can be represented by the Fourier series,

$$y(t) = \frac{1}{2}a_0 + \sum_{p=1}^{\infty} (a_p \cos p\omega_0 t + b_p \sin p\omega_0 t) \tag{7.66}$$

where p is a positive integer, and a_p, b_p are the Fourier coefficients given by

$$a_p = \frac{2}{T} \int_{-T/2}^{T/2} y(t) \cos p\omega_0 t \, dt \qquad p = 0, 1, 2, \ldots$$

$$b_p = \frac{2}{T} \int_{-T/2}^{T/2} y(t) \sin p\omega_0 t \, dt \qquad p = 1, 2, \ldots$$

It is often useful to employ the equivalent complex form which is given by

$$y(t) = \text{Re}\left[\sum_{p=1}^{\infty} A_p e^{ip\omega_0 t}\right] \tag{7.67a}$$

where

$$A_p = \frac{2}{T} \int_{-T/2}^{T/2} y(t) e^{-ip\omega_0 t} \, dt \qquad p = 1, 2, \ldots \tag{7.67b}$$

and T is the period given by $T = L\pi/\omega_0$. Because Eq. (7.62) is linear, the slider response to the disk excitation given in Eq. (7.66) can be obtained by superposition,

$$z(t) = \text{Re}\left[\sum_{p=1}^{\infty} H_p A_p e^{ip\omega_0 t}\right] \tag{7.68}$$

$$H_p = \frac{1}{1 - (p\omega_0/\omega_n)^2 + i2\zeta p\omega_0/\omega_n} \qquad \text{and} \qquad \zeta = \frac{c_1}{2nw_n}$$

where H_p is the complex frequency response function. The response is periodic and has the same period as the excitation. Furthermore, if one of the higher harmonics $p\omega_0$ of the disk motion is near the air-bearing resonance frequency ω_n, this harmonic will produce a highly amplified response, especially if the damping is small. In particular, resonance would occur in the undamped case if $p\omega_0 = \omega_n$. Thus, a resonance condition of the slider may occur in spite of a very low fundamental fre-

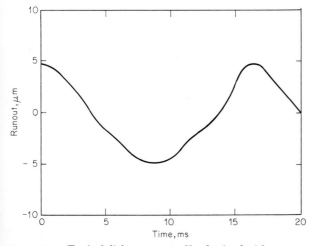

Figure 7.14 Typical disk runout profile obtained with capacitance probe.

quency, provided one of the higher harmonics of the disk out-of-plane motion coincides with the air-bearing natural frequency. To avoid such excitations, manufacturers of disk files must control the runout and vertical acceleration of the disk within close limits.

7.3.1.3   Response to impulse.   Excursions from the steady-state spacing between the slider and disk can also occur because of contacts with asperities on the disk surface. These can result from accessing of the head in the radial direction during track following or track seeking, or from impacts transmitted through the frame or the surroundings. Since these nonperiodic disturbances are generally of short time duration, it is convenient to approximate them by an impulsive force so that Eq. (7.62) takes the form

$$m\ddot{z} + c_1\dot{z} + k_1z = \hat{F}\delta(t) \tag{7.69}$$

where $\delta(t)$ is Dirac's delta function. In this case the zero initial conditions apply, and the resulting solution is

$$z(t) = \begin{cases} \dfrac{\hat{F}}{m\omega_d}e^{-\zeta\omega_n t}\sin\omega_{dt} & t > 0 \\[2mm] 0 & t < 0 \end{cases} \tag{7.70}$$

where $\qquad \omega_d = \omega_n(1-\zeta^2)^{1/2} \qquad \zeta = \dfrac{c_1}{2m\omega_n}$

Thus, the disturbance of the slider resulting from an asperity impact is a velocity discontinuity followed by an oscillation that decays in time.

7.3.1.4   Response to arbitrary nonperiodic force.   If the excitation force acting on the slider is described by an arbitrary function $F(t)$, the response of the slider determined by Eq. (7.62) (with y and $\dot{y}$ zero) is

$$z(t) = \frac{1}{m\omega_d}\int_0^t F(\tau)e^{-\zeta\omega_n(t-\tau)}\sin\omega_d(t-\tau)\,d\tau \tag{7.71}$$

which predicts a complicated convolution integral relationship between the excitation and response.

7.3.2   Rigid-body dynamical model for the slider

The suspension spring supporting the air-bearing slider permits the slider to roll and pitch in addition to its vertical translation. Thus, for a more detailed description of the slider dynamics, the Euler equations for rigid-body rotation must be used, together with the conservation of linear

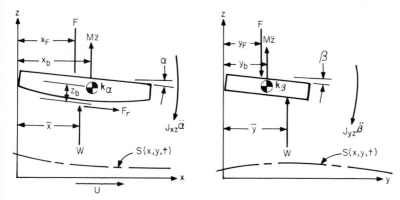

Figure 7.15 Schematic of slider showing coordinates for roll, pitch, and translation (*Tang, 1971*).

momentum in the z direction. Figure 7.15 shows a detailed view of a typical slider bearing, indicating the two additional degrees of freedom for pitch and roll motion. For this situation, assuming there is no rotation about the z axis and neglecting the nonlinear terms, the rigid-body equations of motion for the slider are

$$M\ddot{z} = W - F$$
$$J_{xz}\ddot{\alpha} = (x_b - \bar{x})W + (x_F - x_b)F - F_r z_b - k_\alpha(\alpha - \alpha_0) \qquad (7.72)$$
$$J_{yz}\ddot{\beta} = (y_b - \bar{y})W + (y_F - y_b)F - k_\beta(\beta - \beta_0)$$

where
$$W = \int_0^L \int_0^B (p - p_a)\, d\, x\, dy$$

$$\bar{x} = \int_0^L \int_0^B \frac{x(p - p_a)}{W}\, dx\, dy \qquad \bar{y} = \int_0^L \int_0^B \frac{y(p - p_a)}{W}\, dx\, dy$$

In order to have constant z for the slider, the suspension force F must balance the air-bearing pressure force W, which must be obtained from a solution of the Reynolds equation. In other words, to obtain the solution for the dynamics of a slider bearing in a disk file, simultaneous solution of the equations of motion of the slider and the Reynolds lubrication equation is required. Thus, the Reynolds equation must first be solved, starting with an initial spacing guess, to obtain the pressure distribution in the bearing. Using this pressure distribution, the motion of the slider can then be calculated by solving the dynamics equations. This, in turn, requires a new solution of the Reynolds equation, since the slider motion resulting from the solution of the dynamics equations changes the assumed initial spacing, and this, in turn, changes the pressure distribution. This iteration must be continued until convergence of the solution of the coupled system of equations is obtained. Since closed-form solutions do not exist for realistic bearing configurations, it is necessary to solve the Reynolds equation by numerical methods and couple this solution with the dynamic equations of motion of the slider. The results of such calculations will be presented in Section 7.6, after the use of numerical methods for the solution of the Reynolds equation have been discussed.

7.4   Numerical Solution of the Reynolds Equation

The numerical integration of the Reynolds equation has been the subject of numerous investigations. Finite-difference and, more recently, finite-element methods have been applied successfully to the Reynolds equation for both steady-state and time-dependent problems.

For very small spacing between the slider and the disk, the bearing number Λ is large, and steep pressure boundary layers occur at the sides and the trailing edge of the slider. It has been shown that the extent of the pressure boundary layer in the trailing-edge region is $O(1/\Lambda)$, while the extent of the side-flow boundary layers is $O(B/L\Lambda^{1/2})$, where B and L are the bearing dimensions in the x and y directions (DiPrima, 1978; White and Nigam, 1980).

Because of these steep pressure gradients, certain difficulties in the numerical solution of the Reynolds equation occur which must be dealt with to avoid instability or convergence problems. A brief outline of the numerical methods is presented here.

7.4.1   Finite-difference methods

The application of finite-difference methods for the design of slider bearings used in magnetic recording disk files was first investigated by Michael (1959). He considered a rectangular domain bounded by the edges of a bearing surface of the slider, and approximated the first and second derivatives occurring in the Reynolds equation by central differences as follows:

$$\frac{\partial}{\partial X}(X_i, Y_i) = \frac{P_{i+1,j} - P_{i-1,j}}{2\,\Delta X}$$

$$\frac{\partial}{\partial Y}(X_i, Y_i) = \frac{P_{i,j+1} - P_{i,j-1}}{2\,\Delta Y}$$

$$\frac{\partial^2}{\partial X^2}(X_i, Y_i) = \frac{P_{i+1,j} + P_{i-1,j} - 2P_{i,j}}{(\Delta X)^2} \tag{7.73}$$

$$\frac{\partial^2}{\partial Y^2}(X_i, Y_i) = \frac{P_{i,j+1} + P_{i,j-1} - 2P_{i,j}}{(\Delta Y)^2}$$

where i and j denote the grid points in the x and y directions, respectively. After substitution of these difference expressions in the Reynolds equation, Michael obtained the pressure at each grid point by the iterative solutions of quadratic equations. Steady-state solutions can generally be considered as the asymptotic limit of a time-dependent case. Therefore, much emphasis has been directed to the solution of the time-dependent Reynolds equation. A comprehensive summary of the methods is given by Castelli and Pirvics (1968).

Three methods are commonly used for solving the time-dependent Reynolds equation, namely, the explicit, the implicit, and the semi-implicit methods. In the explicit method, the time derivative $\partial P / \partial T$ is approximated by

$$\frac{\partial P}{\partial T} = \frac{P_{i,j}^{n+1} - P_{i,j}^{n}}{\Delta T} \tag{7.74}$$

where the superscripts n and $n+1$ denote the times at which the functions are evaluated. All spatial derivatives are calculated at $t = n$, so that the following difference equation is obtained for calculating P^{n+1} after substitution of Eqs. (7.73) and (7.74) into the Reynolds equation:

$$
\begin{aligned}
P_{i,j}^{n+1} = \; & P_{i,j}^{n} + \frac{\Delta T}{\sigma} \left\{ H^2 P^n \Big|_{i,j} \frac{P_{i+1,j}^n + P_{i-1,j}^n - 2P_{i,j}^n}{(\Delta X)^2} \right. \\
& + H^2 P^n \Big|_{i,j} \frac{P_{i,j+1}^n + P_{i,j-1}^n - 2P_{i,j}^n}{(\Delta Y)^2} \\
& + H^2 \Big|_{i,j} \left[\left(\frac{P_{i+1,j}^n - P_{i-1,j}^n}{2\,\Delta X} \right)^2 \left(\frac{P_{i,j+1}^n - P_{i,j-1}^n}{2\,\Delta Y} \right)^2 \right] \\
& + 3HP^n \Big|_{i,j} \left(\frac{P_{i+1,j}^n - P_{i-1,j}^n}{2\,\Delta X} \frac{\partial H}{\partial X} + \frac{P_{i,j+1}^n - P_{i,j-1}^n}{2\,\Delta Y} \frac{\partial H}{\partial Y} \right) \\
& \left. - \Lambda \left(\frac{P_{i+1,j}^n - P_{i-1,j}^n}{2\,\Delta X} + \frac{P^n}{H} \Big|_{i,j} \frac{\partial H}{\partial X} \right) - \frac{\sigma P^n}{H} \Big|_{i,j} \frac{\Delta H}{\Delta T} \right\}_{t=n}
\end{aligned}
\tag{7.75}
$$

In matrix notation this can be written as

$$L\{P^{n+1}\} = R \tag{7.76}$$

with the solution

$$P^{n+1} = L^{-1} R \tag{7.77}$$

For the explicit scheme, the matrix L is diagonal and solutions can be obtained by Gaussian elimination (DiPrima, 1978). In order for the explicit scheme to be stable, the time step ΔT must be chosen so as to satisfy the following inequality at all grid points:

$$\Delta T \leq \frac{\sigma}{2H_{i,j}^2 P_{i,j}} \frac{1}{(1/\Delta X)^2 + (1/\Delta Y)^2} \tag{7.78}$$

The above requirement for the time step causes explicit methods to consume large amounts of computer time, since very small spatial steps are needed to avoid instability of the solution because of the presence of the steep boundary layers.

In the implicit method, the time derivative in the Reynolds equation is approximated in the same way as in the explicit scheme, but, in this method, the second derivatives of the pressure are evaluated at the new time $t = n + 1$ rather than at the old time $t = n$. This leads to the following equation for the new pressure at each grid point:

$$P_{i,j}^{n+1} - \frac{\Delta T}{\sigma}\left[H^2 P|_{i,j}\, \frac{P_{i+1,j}^{n+1} + P_{i-1,j}^{n+1} - 2P_{i,j}^{n+1}}{(\Delta X)^2} \right.$$

$$\left. + H^2 P|_{i,j}\, \frac{P_{i,j+1}^{n+1} + P_{i,j-1}^{n+1} - 2P_{i,j}^{n+1}}{(\Delta Y)^2} \right] = R_{i,j}^n \quad (7.79)$$

where

$$R_{i,j}^n = P_{i,j}^n + \frac{\Delta T}{\sigma}\left\{ H^2 \right|_{i,j}\left[\left(\frac{P_{n}^i + 1,j - P_{i-1,j}^n}{2\,\Delta X} \right)^2 \right.$$

$$+ \left(\frac{P_{i,j+1}^n - P_{i,j-1}^n}{2\,\Delta Y} \right)^2 \right]$$

$$+ 3HP^n \left|_{i,j}\left(\frac{P_{i+1,j}^n - P_{i-1,j}^n}{2\,\Delta X}\,\frac{\partial H}{\partial X} + \frac{P_{i,j+1}^n - P_{i,j-1}^n}{2\,\Delta Y}\,\frac{\partial H}{\partial Y} \right) \right.$$

$$\left. - \Lambda \left(\frac{P_{i+1,j}^n - P_{i-1,j}^n}{2\,\Delta X} + \frac{P^n}{H}\right|_{i,j}\frac{\partial H}{\partial X} \right) - \frac{\sigma P^n}{H}\,\frac{\Delta H}{\Delta T} \right\}_{t=n} \quad (7.80)$$

If Eq. (7.79) is written in the matrix notation of Eq. (7.76), the operator L is not diagonal and its elements must be calculated at each new time step, since they are functions of t. The implicit scheme is unconditionally stable as long as the bearing number Λ does not assume very large values (Forsythe and Wasow, 1960). The calculation time at each time step of the implicit method is much longer than that for the explicit method, because a system of algebraic equations connecting all grid points must be solved at each time step.

In the semi-implicit method the highest derivatives are evaluated partially at the new, and partially at the old, time, and the solution procedure is similar to that in the implicit scheme. However, computation time is claimed to be comparable to that of the explicit scheme (Castelli and Pirvics, 1968).

In order to simplify the numerical calculation, a different form of the Reynolds equation is often employed in which the independent parameter is chosen to be $\psi = PH$. The Reynolds equation then takes the form

$$\sigma \frac{\partial \psi}{\partial T} + \Lambda \frac{\partial \psi}{\partial X} = H\psi \frac{\partial^2 \psi}{\partial X^2} - \psi^2 \frac{\partial^2 H}{\partial X^2} + \left(\frac{\partial \psi}{\partial X} \right)^2 H - \psi \frac{\partial \psi}{\partial X}\frac{\partial H}{\partial X}$$

$$+ H\psi \frac{\partial^2 \psi}{\partial Y^2} - \psi^2 \frac{\partial^2 H}{\partial X^2} + \left(\frac{\partial \psi}{\partial Y} \right)^2 H - \psi \frac{\partial \psi}{\partial Y}\frac{\partial H}{\partial Y} \quad (7.81)$$

The solution procedure is unaffected by the form of the Reynolds equation.

In the finite-difference schemes discussed above the truncation error is on the order of $(\Delta t)^2$. A finite-difference scheme with a truncation error of only $(\Delta t)^3$ was published by White and Nigam (1980). In this "alternating-direction factored" implicit scheme, the discretized form of the Reynolds equation is rewritten in terms of factors containing two operators $L_1 (x)$ and $L_2(y)$,

$$(1 - L_1) (1 - L_2) \Delta Z^n = \phi \tag{7.82}$$

where L_1 and L_2 are defined in terms of quantities depending only on x and y, respectively, ϕ is a function of x and y, and

$$\Delta Z^n = \Delta(PH)^n = Z^{n+1} - Z^n \tag{7.83}$$

White and Nigam rewrote this expression as

$$(1 - L_1) \Delta Z^* = \phi \qquad (1 - L_2) \Delta Z^n = \Delta Z^* \tag{7.84}$$

and they solved the two-dimensional problem expressed in Eq. (7.84) as the product of two one-dimensional problems which were solved successively. To resolve the steep pressure boundary layers at the trailing edge in the case of high bearing numbers, they used a variable-step-size grid with closer points near the trailing edge. The numerical results presented later in Section 7.6 were calculated by the factored scheme of White and Nigam (1980).

7.4.2 Finite-element methods

The finite-difference solution of the Reynolds equation discussed in the previous section is most easily obtained for rectangular bearing boundaries, where the coordinate lines and bearing boundaries conform. To solve bearing problems with irregular boundaries, approximations to the bearing contour must be made, which generally results in a loss of numerical accuracy. A numerical procedure that allows the treatment of arbitrarily shaped boundary geometries is the finite-element method, which originally was developed for the solution of structural engineering problems but recently has been applied to the solution of both the incompressible and the compressible Reynolds equation (Reddi and Chu, 1970; Huebner, 1975; Garcia-Suarez et al., 1984).

The first step in obtaining a finite-element solution of a particular problem is to divide the bearing region into a number of elements that are interconnected at nodes on the boundaries of the elements. The pressure and the resultant forces for each element are then approximated in terms of assumed interpolation functions, which are chosen so that the pressure within an element and across the boundaries of elements is continuous. The interpolation functions and the nodal values define the distribution of the pressure and the resultant forces within each element.

In the case of incompressible lubrication, the finite-element solution of the Reynolds equation is the function p that satisfies the boundary condition and minimizes the functional $\pi(p)$, where (Huebner, 1975)

$$\pi(p) = \int_A \left[\left(\frac{h^3}{24\mu} \nabla p - h\overline{U} \right) \cdot \nabla p + \frac{\partial h}{\partial t} p \right] dA \tag{7.85}$$

in which $\overline{U} = \frac{1}{2}(U_1 + U_2)i + \frac{1}{2}(V_1 + V_2)j$, and where (U_1, V_1) and (U_2, V_2) are the velocities of the bearing and stationary surfaces in the x and y directions. In the case of the compressible lubrication problem, a similar variational statement does not exist for the solution of the Reynolds equa-

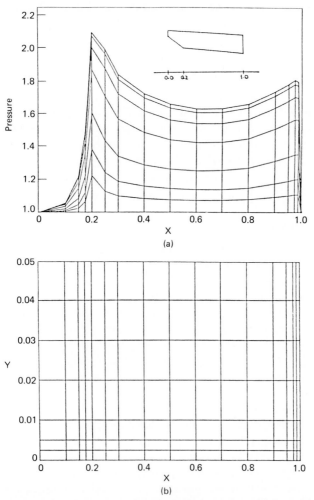

Figure 7.16 (a) Pressure distribution along the grid lines for typical three-rail slider obtained from finite-element solution; $\Lambda = 1000$; $h_o / h_{\min} = 22.2$ (*Garcia-Suarez et al. 1984*). (b) Finite-element grid.

tion, and perturbation techniques, or reduction to a series of linear equations using Newton's method in conjunction with Galerkin's method, has been applied (Reddi and Chu, 1970).

A number of incompressible and compressible lubrication problems are discussed by Huebner (1975). Recently, an upwinding technique for obtaining a finite-element solution for air-bearing problems has been applied for studying the behavior of magnetic recording sliders for intermediate and large bearing numbers (Garcia-Suarez et al., 1984). A typical result for the steady-state pressure distribution of a three-rail slider bearing is shown in Fig. 7.16. This result has been compared with the pressure distribution obtained using finite differences, and the two results are found to be in agreement.

Another example showing the usefulness of the finite-element method for irregularly shaped bearing surfaces is presented in Fig. 7.17a for the

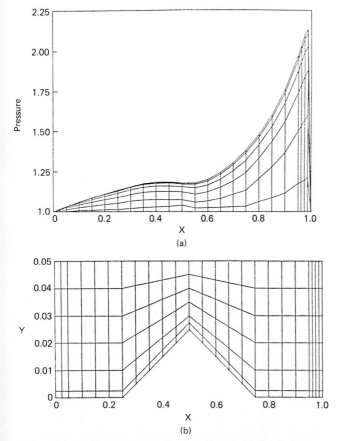

Figure 7.17 (a) Pressure distribution along grid lines of wasp waist head obtained from finite-element solution; $\Lambda = 2000$; $h_o/h_{min} = 4$ (*Garcia-Suarez et al., 1984*). (b) Finite-element grid.

TABLE 7.1 Design Data of Disk Files Relating to the Head-Disk Interface

	Year of first shipment										
	1957	1961	1962	1963	1966	1971	1973	1976	1979	1979	1981
IBM product	350	1405	1301	1311	2314	3330	3340	3350	3310	3370	3380
Recording density:											
Linear bit density, b/mm	4	8.8	20.8	41	88	161	225	257	341	485	608
Track density, t/mm	0.8	1.6	2	2	4	7.7	12	19	18	25	> 32
Key geometric parameters, μm:											
Flying height	20	16.2	6.2	3.1	2.1	1.2	0.4	0.4	0.3	0.3	< 0.3
Medium thickness	30	22.5	13.5	6.2	2.1	1.2	1.0	1.0	0.6	1.0	< 0.6
Air bearing and magnetic element:[†]											
Bearing type	HS	HS	HD	HD	HD	HD	HD	HD	HD	HD	HD
Surface contour	FL	FL	CY	CY	CY	CY	TF	TF	TF	TF	TF
Slider material	AL	AL	SS	SS	CE	CE	FE	FE	FE	CE	CE
Core material	LMM	LMM	LMM	LMM	FE	FE	FE	FE	FE	FI	FI
Slider-core bond	E	E	E	E	E	G	I	I	I	D	D

[†]Al = aluminum, CE = ceramic, CY = cylindrical, D = deposited, E = epoxy, FE = ferrite, FI = film, FL = flat, G = glass, HD = hydrodynamic, HS = hydrostatic, I = integral, LMM = laminated mu-metal, SS = stainless steel, TF = taper flat.

SOURCE: Harker et al. (1981).

"wasp waist" head design depicted in Fig. 7.17*b*. Because of the irregular boundaries, finite differences are probably more difficult to apply in this case, and finite elements are preferable.

7.5 Slider Design Evolution

The major trend in the evolution of magnetic recording sliders has been toward closer spacing between the slider and disk. This has been achieved by a progressive miniaturization of the slider and an associated reduction in the load of the air bearing (Harker et al., 1981; Gross, 1984).

The head-disk spacing and performance of several disk files are listed in Table 7.1. The spacing in the first disk file (1957) was 20 μm, achieved by using a hydrostatic (externally pressurized) air bearing. The first self-acting air bearing was introduced in 1962 and was designed to fly at a spacing of about 6 μm. The spacing has decreased in successive products over the years and is now about 0.3 μm.

Figure 7.18 shows the evolution of disk file sliders starting with the first slider and continuing with more modern designs. One can observe from the photograph the decrease in the size of the air bearing, as well as the reduction in the size of the suspension spring. The sliders used before 1973 were unloaded while not in operation, but subsequently the sliders were designed for starting and stopping in contact with the disk. The concept of start-stop in contact simplifies the loading mechanism, but intro-

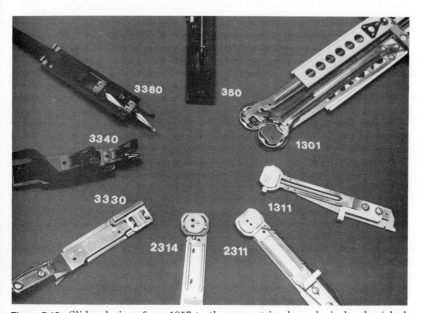

Figure 7.18 Slider designs from 1957 to the present in chronological order (clockwise) *(Gross, 1984)*.

duces problems related to friction and wear between the slider and the disk (Harker et al., 1981).

The sliders used in the early model heads had cylindrical contours. These were superseded by taper-flat, three-rail (Winchester) heads starting in 1973. The taper-flat, three-rail slider has two air-bearing rails on either side of a narrow center rail that carries the magnetic element. The outside rails develop the pressure distribution that counter-balances the suspension load, typically on the order of 0.10 N. More recent film heads have no center rail, and the magnetic element is carried on one of the pressure rails.

7.5.1 Steady-state pressure distribution of the three-rail head

The steady-state pressure distribution for a three-rail slider is shown in Fig. 7.19. The pressure increases from the loading edge toward the intersection between the taper and the flat part of the bearing; after reaching a maximum there, the pressure decreases toward the trailing edge, and, shortly before the trailing edge is reached, it increases again to attain a second maximum. The decrease of the pressure beyond the intersection of the taper and the flat portions of the head is due to the large amount of side flow that occurs in narrow slider bearings. The increase of the pressure at the trailing edge occurs because the pressure reduction from side flow is overcome by the pressure increase resulting from the decrease in the spacing toward the trailing edge of the slider.

7.5.2 Steady-state pressure distribution of the two-rail head

The two-rail head has a taper-flat design similar to the three-rail design, but it is much smaller and has no center rail. It is supported by a single load beam leaf-spring suspension that allows the head to pitch and roll. The pressure distribution of the two-rail head is similar to that of the three-rail head, featuring similar pressure peaks at the taper-flat intersection and at the trailing edge. Since the mass of the two-rail slider is substantially less than that of the three-rail slider, the resonance frequencies of the two-rail slider are substantially higher.

7.5.3 Pressure distribution of the zero-load slider

Although the natural frequencies of taper-flat slider bearings can be modified within limits, the bearing stiffness is related to the load; thus, design points may be encountered in designing very lightly loaded sliders that

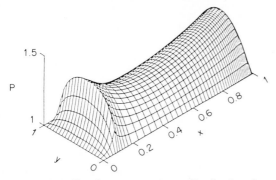

Figure 7.19 Steady-state pressure distribution for outer rail of a three-rail head *(Miu, 1985).*

produce undesirably low values of the natural frequencies. A new slider design that exhibits high stiffness even at low loads is the "zero-load" slider shown in Fig. 7.20 (White, 1983). This slider consists of two taper-flat outer bearings as in the conventional taper-flat design. Connecting these outer rails is a cross rail at the same elevation, which is abruptly recessed by a negative spacing step of about 10 μm. While the pressure distribution in the outside rail is similar to that in the conventional taper-flat slider, the pressure in the central recessed area is subambient, thereby producing a suction force that attracts the slider toward the disk. The net load of the slider is the sum of the positive load from the rails plus the negative load from the recess, and can be adjusted to very low values in spite of the fact that the stiffness of the bearing is very high. A plot of a typical pressure distribution for a zero-load slider is shown in Fig. 7.21, and the dependence of the flying height on the disk speed is compared in Fig. 7.22 for the zero-load and three-rail slider. Optimization studies of the load

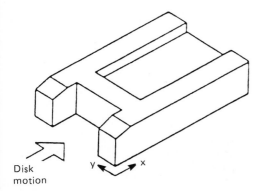

Disk
motion

Figure   7.20 Zero-load   slider design *(White, 1983).*

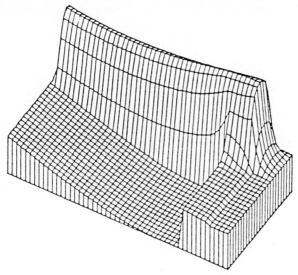

Figure 7.21 Pressure distribution of zero-load slider *(White, 1983).*

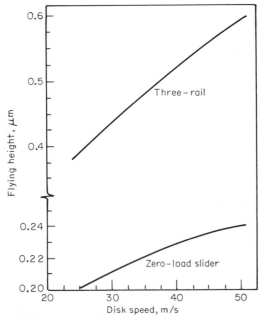

Figure 7.22 Dependence of flying height on disk speed for zero-load slider and three-rail slider *(White, 1983).*

slider indicate that the recess should have nonparallel sides for best performance (White, 1983).

7.6 Dynamic Simulation of Slider Bearings

The dynamics of slider bearings is important in the design of magnetic recording sliders, since excessive spacing modulations may cause contact between the slider and disk, thereby resulting in material interactions and wear. The first simulation of the dynamic behavior of the slider-disk interface was presented by Tang (1971). He solved simultaneously the equations of motion for the slider, using a third-order Runge-Kutta scheme, and the time-dependent Reynolds equation, using an explicit finite-difference scheme. With reference to Fig. 7.15, which shows the pitch and roll axes of a typical slider bearing, the equations of motion solved by Tang were Eq. (7.72) together with the Reynolds equation in the form

$$P_{i,j}^{n+1} = P_{i,j}^n + \frac{\Delta T}{\sigma} \left\{ H^2 P \, \nabla^2 P + H^2 \left[\left(\frac{\partial P}{\partial X} \right)^2 + \left(\frac{\partial P}{\partial Y} \right)^2 \right] \right. $$
$$\left. + 3HP \left(\frac{\partial P}{\partial X} \frac{\partial H}{\partial X} + \frac{\partial P}{\partial Y} \frac{\partial H}{\partial Y} \right) - \Lambda \frac{\partial P}{\partial X} - \Lambda \frac{P}{H} \frac{\partial H}{\partial X} - \sigma \frac{P}{H} \frac{\partial H}{\partial T} \right\}^{(n)} \quad (7.86)$$

From the numerical solution, the resonance frequency and damping properties of the air-bearing system were determined in each of its modes, and the effects of surface ripples, slider cross-curvature, and other design dimensions were also studied. In Fig. 7.23 a typical simulation result is reproduced for the case when a cosine bump on the disk moves through

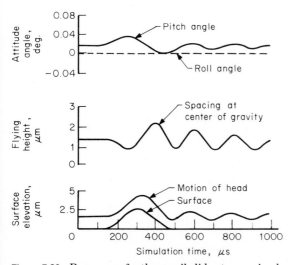

Figure 7.23 Response of a three-rail slider to a cosine bump on the disk *(Tang, 1971).*

an air bearing. Roll motion as well as translation and pitch are excited by the defect, and the roll, pitch, and translation are seen to be strongly coupled.

A different approach for simulating the head-disk interface was taken by Ono (1975), who analyzed the slider-bearing dynamic response to disk vibrations in the frequency domain using perturbation methods. Starting with the equations of motion for the slider and the time-dependent Reynolds equation, similar to those given above, it is assumed that the pressure P and the spacing H can be linearized about their respective equilibrium values P_o, H_o. Thus, introducing

$$P = P_o + \psi \qquad H = H_o + H_d$$

in Eqs. (7.2) and (7.72), and neglecting the higher-order terms in the small parameters, two sets of equations are obtained that determine the static equilibrium and the first-order deviation from it. Performing the Laplace transform on the perturbation equations and requiring that the spacing variation be written as a superposition of translation, roll, pitch, and disk motions, the dynamic response of the slider is obtained in the frequency

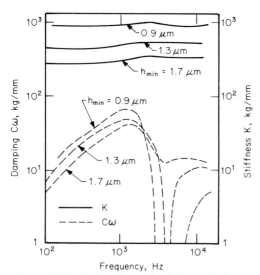

Figure 7.24 Air-bearing translation stiffness and damping as a function of frequency; $B = 10$ mm, $U = 60$ m/s, $L = 3$mm *(Ono, 1975)*.

domain. The interesting result is that both the stiffness and the damping are functions of frequency. Figure 7.24 reproduces from Ono (1975) typical results for the air-bearing translation stiffness and damping of a slider as a function of frequency, showing negative values of damping and stiffness in certain frequency ranges.

The frequency-domain approach has been applied to the case of a three-rail slider flying at submicrometer spacing (Ono et al., 1979). Because of the low flying height, rarefaction effects are taken into account in the Reynolds equation, resulting in a marked decrease in the stiffness and a substantial increase in the damping, especially in the higher-frequency region. The variation of slider stiffness and damping with load at constant frequency is reproduced in Fig. 7.25 for this slider. Damping decreases as the breadth of the bearing increases, but the stiffness of the bearing increases with load and is nearly independent of the bearing breadth.

The frequency-domain approach is limited to small slider excursions from the initial flying height on account of the linerarization used for p

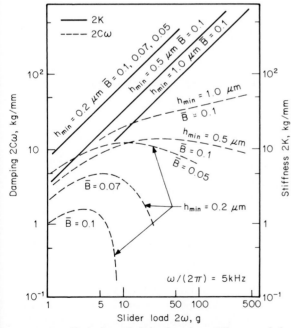

Figure 7.25 Variation of slider-bearing stiffness and damping with load at constant frequency for a three-rail slider; $h_1/h_{min} = 2$, $\overline{B} = L/B$, $U = 40$ m/s, $\omega/2\pi = 5$ kHz *(Ono et al., 1979)*.

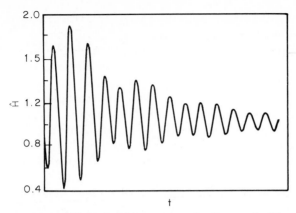

Figure 7.26   Flying-height response of a three-rail slider to vertical impulse (*White, 1984a*). H is the ratio of spacing at time t to steady-state spacing.

and h. Thus, situations in which slider-disk contacts occur cannot be described correctly. White (1984) developed a time-transient simulation model for the slider dynamics based on a factored implicit solution of the Reynolds equation, and a fourth-order Runge-Kutta scheme for the numerical integration of the slider equation of motion. With this model, flying-height response to disk flutter and impulse loading has been calculated. A typical result for the slider response to a vertical impulse is shown in Fig. 7.26 for a two-rail head. The numerical calculation predicts a substantial amount of oscillation at the resonance frequency of the bearing, which is about 17 kHz.

7.7   Measurement of Head-Disk Spacing

7.7.1   Optical interference techniques

Interferometry has become the standard tool for the measurement of steady-state air-bearing separations (Lin and Sullivan, 1972). For spacing measurements above 1 μm, monochromatic light is used, while for spacing measurements in the submicrometer range, white-light interferometry is preferable because of its better resolution.

In its simplest form the interferometry technique utilizes a monochromatic light beam that is partially reflected and partially transmitted from the first interface that bounds a thin film of thickness h. The transmitted ray is reflected at the second interface, undergoes a phase shift of 180°, and is superimposed on the beam that was reflected previously at the first interface. If the two rays are in phase after superposition, maximum reinforcement occurs. On the other hand, if the path-length difference leads to a phase difference of 180°, annihilation occurs. In the case of normal incidence, constructive interference occurs if

$$2h = (m + \tfrac{1}{2})\lambda \qquad m = 0, 1, 2, 3, \ldots$$

while annihilation occurs if

$$2h = m\lambda \qquad m = 0, 1, 2, 3, \ldots$$

where λ is the wavelength of the light.

   If the light source is white rather than monochromatic, the interference pattern consists of continuous color fringes rather than alternating dark and light fringes. By a comparison of the color of the fringes with a calibration chart (Newton's fringe chart), the absolute spacing of the air film can be determined with an accuracy of approximately 0.025 μm. A plot of the measured flying height as a function of velocity is reproduced in Fig. 7.27, together with results from a numerical solution of the Reynolds equation (Lin and Sullivan, 1972). Very good agreement between the two is observed.

   In order to apply white-light interferometry to the spacing measurement in a disk file, a transparent slider or disk must be used. This requirement poses a serious limitation, since the properties of a transparent disk or slider are different, in general, from those of the actual components. Another serious limitation of interferometry is its unsuitability for the measurement of dynamic spacing variations. A modification of white-light interferometry that allows the measurement of low-frequency spacing variations is stroboscopic interferometry, where the light source is strobed once per revolution of the disk. Thus, the head-disk spacing at a particular

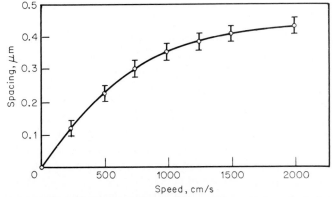

Figure 7.27 Minimum slider-to-disk spacing versus speed from white-light interference pattern *(Lin and Sullivan, 1972).*

point on the disk is measured, rather than the average of the spacing for the complete disk circumference. If the time delay between the trigger from the disk and the strobe pulse to the light source is incremented by a small amount Δt, the variation of the spacing between the previous point and the present point on the disk can be observed from the change in the fringe color between the two locations.

An application of interferometry to the measurement of dynamic spacing was suggested by Fleischer and Lin (1974) and by Nigam (1982). In the former work, an infrared He-Ne laser was used in conjunction with a quartz disk, and the intensity of the radiation of the first-order fringe was linearized to extrapolate spacings in the range below one-quarter of the wavelength of the laser light at frequencies up to 30 kHz. In the latter work, a visible laser interferometer was implemented with a photodetector to permit dynamic spacing measurements. Both techniques require a transparent slider or disk.

7.7.2 Capacitance techniques

In addition to interferometry, capacitance techniques have been applied to the measurement of the dynamic spacing h between the slider and the disk. The principle of the method is based on the fact that the capacitance C between two surfaces changes according to $C = k/h$, where k is a constant. To implement the technique in a slider-disk system, a capacitance probe must be embedded in the slider and a calibration of the capacitance variation with spacing must be obtained. Since the oxide coating in the case of a particulate disk is nonconducting, the total capacitance C depends on the capacitances of the air gap and the coating according to (Lin, 1973)

$$c = \frac{k}{\delta/\epsilon_\delta + h/\epsilon_h}$$

where ϵ_δ, ϵ_h are the dielectric constants of the coating and the air film, δ and h are the coating thickness and spacing, and k is a constant.

The first study of dynamic spacing measurements of slider bearings with capacitive probes used a shielded probe embedded in the slider and measured spacings in the range of tens of micrometers (Briggs and Herkart, 1971). In a later study, measurements of air-bearing spacings, using unshielded capacitance probes, yielded information about the roll and pitch angle in addition to flying-height data. In a more recent study, capacitance probes were used in three-rail sliders to study the dynamics of slider bearings (Ono, 1975). The method has been extended to study the head-tape spacing in a rotating-head, helical tape configuration (Feliss and Talke, 1977).

Capacitance techniques have good frequency response, generally up to about 30 kHz, and they give useful dynamic information. The disadvantage is that capacitance techniques require modification of the slider as well as auxiliary calibration based on white-light interferometry.

7.8 Measurement of Head-Disk Dynamics Using Laser-Doppler Techniques

The measurement of slider dynamics in an unmodified head-disk interface has recently been accomplished by using laser-Doppler techniques (Bogy and Talke, 1985; Bouchard et al., 1984; Miu et al., 1984). The method is based on the fact that an acoustooptical modulated light beam, which is reflected from a moving surface, is frequency-modulated by the motion of the surface. This frequency modulation is proportional to the velocity component of the moving surface parallel to the direction of the light beam, which is normally incident on the surface. In conventional interferometry, the resolution limit is related to the wavelength of light, but in the case of laser-Doppler vibrometry, the resolution limit is a function only of the resolution capability of the frequency demodulator.

The experimental setup for the laser-Doppler vibrometer is shown schematically in Fig. 7.28. A vertically polarized monochromatic light from a

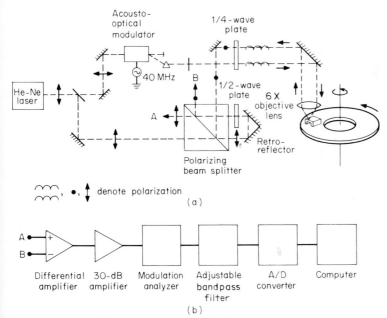

Figure 7.28 Laser-Doppler vibrometer. (a) Optical components; (b) electrical components (Miu et al., 1984).

He-Ne laser is separated into a reference beam and a signal beam. The signal beam is frequency-shifted by 40 MHz, using an acoustooptic modulator, and is circularly polarized as it passes through a quarter-wave retarding plate. The light is then focused by use of an objective lens into a 25-μm spot on the back surface of the slider. The reflected beam becomes horizontally polarized as it returns through the quarter-wave plate and is then recombined with the orthogonally polarized reference beam at the polarizing beam splitter. Thereafter, the polarization of the beam is rotated by 45°, the beam is separated into two components, and two photodetectors are used to detect the two frequency-modulated signals A and B superimposed on the carrier frequency of 40 MHz. Finally, the signals are passed through a differential amplifier and a modulation analyzer for frequency demodulation.

The transient response of a three-rail slider is shown in Fig. 7.29, after being disturbed by an artificially introduced depression in the disk (Miu et al., 1984). The slider displacement obtained by integration of the velocity signal, and the corresponding velocity spectrum, are presented. The depression causes a substantial velocity change, and the transient motion of the slider dies out after a few oscillations. This experimental result dif-

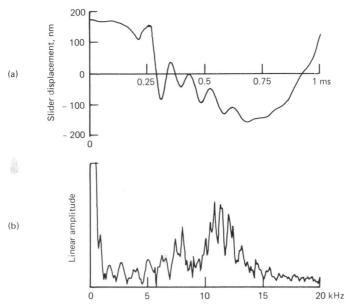

Figure 7.29 Transient motion of a three-rail slider after flying over depression. (*a*) Slider displacement obtained from integration of velocity; (*b*) velocity frequency spectrum *(Miu et al., 1984)*.

fers from numerical time-dependent slider simulation results (White, 1984) that generally predict much less damping than measured here experimentally. The spectrum of the velocity trace shows that an air-bearing resonance in the 12-kHz range is excited. Note that the measurement was performed at a point close to the trailing edge of the slider.

The transient displacements at the four corners of a three-rail slider in response to the same depression were also measured. The calculated values for the center motion, pitch, and roll are displayed in Fig. 7.30. Similar data for a two-rail head are shown in Fig. 7.31 (Bouchard et al., 1984). The corner displacements are smaller for the two-rail head, but roll and pitch are similar for both cases. The frequency spectra of the velocity sig-

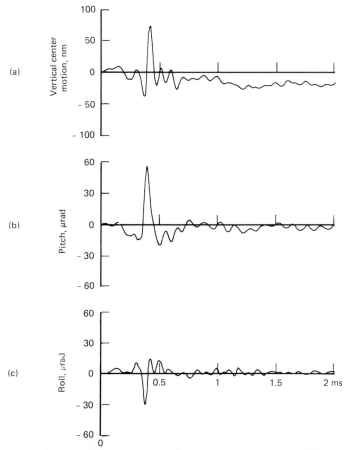

Figure 7.30 Calculated motion component of a three-rail slider. (a) Vertical center motion, (b) pitch, and (c) roll *(Miu et al., 1984)*.

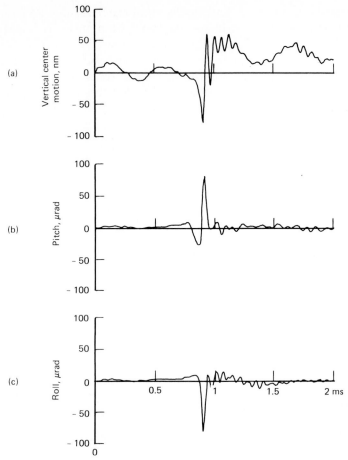

Figure 7.31 Calculated motion from measurements on two-rail slider. (*a*) Vertical center motion, (*b*) pitch, and (*c*) roll.

nal at the front and rear left corners of the two sliders are shown in Fig. 7.32. The resonance frequency of the two-rail head is much higher than that of the three-rail head, a result of the lower mass and lower flying height of the two-rail head.

The response of a two-rail slider to a step in the disk coating is shown in Fig. 7.33*a* (Miu, 1985). This step was produced by sputtering a rectangular area on the disk using a specially fabricated mask (Wood, 1984). The simulation results for the time-dependent motion of the slider caused by this step, obtained from a numerical solution of the Reynolds equation and the slider dynamics equation, is shown in Fig. 7.33*b*. The fast Fourier transforms of the measured and calculated displacements of the slider are shown in Fig. 7.34. The laser-Doppler measurement and the numerical solution are in excellent agreement.

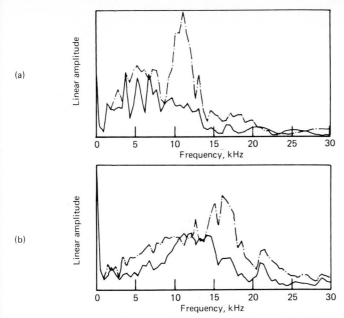

Figure 7.32 Frequency spectra of the front left corner (solid) and rear left corner (dash-dot) velocities of the (*a*) three-rail and (*b*) two-rail sliders.

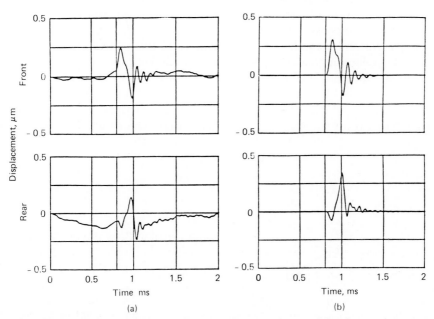

Figure 7.33 (*a*) Measured and (*b*) calculated response of two-rail head to a 0.3-μm step in the disk coating *(Miu, 1985)*.

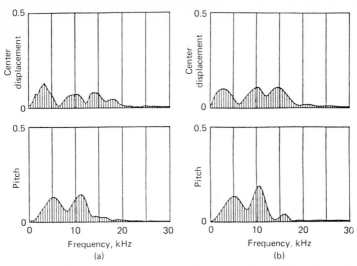

Figure 7.34 Comparison of spectra of (*a*) measured and (*b*) calculated response of two-rail slider to a 0.3-μm step in the disk coating.

In the referenced studies, the transverse motion of the slider was measured and the roll and pitch were calculated. During accessing of the head, or during intermittent contacts between the slider and the disk, in-plane transient motions of the slider also occur. These in-plane velocity components are important for the optimization of track accessing as well as the general understanding of the dynamics of the head-disk interface. A laser-Doppler anemometer has been used to detect these in-plane velocity components (Bouchard et al., 1985). The principle of the laser-Doppler anemometer is illustrated in Fig. 7.35. A He-Ne beam is split into two beams, one of which is frequency-modulated at 40 MHz. Both beams are then focused on the back surface of the head, where they interfere and

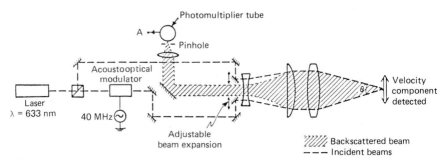

Figure 7.35 Optical components of the laser-Doppler anemometer *(Bouchard et al., 1985)*.

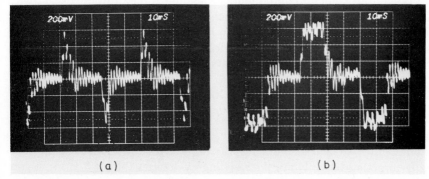

Figure 7.36 Track accessing measurement on a three-rail slider. (*a*) Radial velocity of the slider for a repetitive 2-track access, 0.0156 m/s per division; (*b*) radial velocity of the slider for a 10-track access, 0.0156 m/s per division *(Bouchard et al., 1985).*

create a sequence of moving interference fringes. The intensity of the light collected in the photomultiplier is modulated by the light scattered from the moving asperities on the back side of the slider. That is, when a particle crosses the fringe pattern with a certain velocity, it scatters light at the frequency proportional to its velocity component orthogonal to the fringe lines, and inversely proportional to the fringe spacing. Because of the 40-MHz preshift of the laser, the fringes move with a constant velocity, and the Doppler frequency appears as a frequency modulation on the 40-MHz signal after detection by the photomultiplier tube.

Figure 7.36 shows a typical velocity trace of the slider motion that occurs during track accessing. The velocity of the access reaches a constant value, and considerable vibrations exist in the system. Similar con-

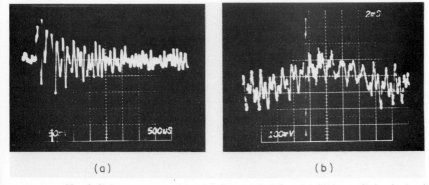

Figure 7.37 Head-disk contact motion of a three-rail slider. (*a*) Slider radial velocity (3.9 mm/s per division); (*b*) slider tangential velocity (7.8 mm/s per division).

clusions with respect to vibrations can also be made by studying the radial and tangential velocity components of a slider forced into a contact with the disk, as shown in Fig. 7.37 (Bouchard et al., 1985). For the three-rail slider, these resonances are found to be about 6 and 9 kHz in the radial direction and 0.8 to 2 kHz in the tangential direction, respectively.

7.9 Measurement of Head-Disk Contacts

Transient motion of the slider caused by disturbances from the mounting frame, or by accessing, may lead to contacts between the disk and the slider. Contacts occur also during start-up of a disk file, since the air bearing becomes stable only at relatively high velocities (about 5 m/s). In general, contacts between the slider and the disk are undesirable, since they cause friction and wear of the sliding surfaces. In recent years, several studies have been conducted to investigate the intermittent flying behavior of a slider during its transition from sliding to flying, and also its inter-

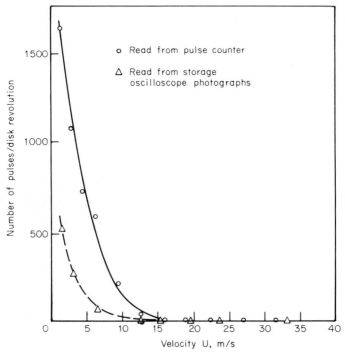

Figure 7.38 Contact pulses for three-rail slider as a function of velocity *(Tseng and Talke, 1974).*

mittent contact behavior during flying. The transition from sliding to flying of a specially designed three-pad slider was investigated as a function of velocity using the electrical resistance method (Tseng and Talke, 1974). This method consists of applying a small potential between a conducting slider and a conducting disk, and monitoring the resistance level as a function of speed or other physical parameters. Figure 7.38 shows the number of contact pulses as a function of velocity. Contacts decrease with linear speed, but exist even at speeds where the numerical solution of the Reynolds equation predicts steady-state flying heights larger than the surface roughness. Thus, it is concluded that the transition from sliding to flying is intermittent in nature, depending on surface roughness, velocity, load, and other physical parameters. Similar conclusions concerning the transition from sliding to flying have been obtained using magnetoresistive elements (Talke et al., 1975) and, more recently, using acoustic emission techniques (Kita et al., 1980).

7.10 Tribology of the Head-Disk Interface

7.10.1 Oxide-coated disks

The most commonly used magnetic medium in rigid-disk files presently is finely dispersed ferric oxide, γ-Fe_2O_3, held in a polymeric binder. In addition to the ferric oxide particles, hard aluminum oxide particles are dispersed in the coating to protect the disk from wear. Because of the hardness of the alumina, abrasive wear of ferrite sliders is the predominant mode of wear (Talke and Su, 1975; Engel et al., 1978; White, 1979).

With unlubricated disk coatings, the wear of the slider as a function of load is linear below and above a critical load, which separates a low and a high wear rate region, as shown in Fig. 7.39 (Talke and Su, 1975). Further investigations into the mechanism of wear on oxide-coated disks were undertaken using autoradiographic methods (Talke and Tseng, 1972). A substantial amount of slider material was found to be embedded in the disk coating, and autoradiographic results showed that wear products were present in the disk, even when the wear was so minute that it could not be detected optically on the disk. In general, the amount of wear debris deposited in the disk coating was found to be proportional to the amount of slider wear. Furthermore, the amount of wear was seen to depend on load, velocity, and even slider design, since the latter has a strong effect on the transition from sliding to flying. Under actual operating conditions, the wear in a disk file is even more complicated than in constant-velocity wear tests, since the disk accelerates and decelerates during start-stop. Thus, the load and velocity are functions of time and must be taken as variables.

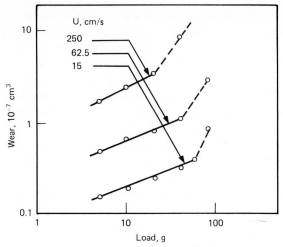

Figure 7.39 Wear on an oxide-coated disk as a function of load *(Talke and Su, 1975).*

The amounts of slider and disk wear are substantially reduced by the application to the disk of a thin lubricant film on the order of 10 nm. By empirical investigations it has been found that fluorinated oils (linear homopolymers of hexafluoropropylene epoxide) with very low vapor pressure are satisfactory for boundary lubrication of disks. These materials have the following chemical structure (DuPont, 1984):

$$F-(CF-CF_2-O)_n-CF_2\,CF_3 \qquad 10 \leq n \leq 60$$
$$|$$
$$CF_3$$

The lubricant can be applied only at the time of the manufacture of the disk and cannot be replaced at a later time; therefore extreme care has to be exercised during the application of the lubricant. Conventional techniques for lubricant application are spraying, rubbing, or dipping.

A study of the distribution of lubricant on a disk, and of lubricant retention under various experimental conditions, has been conducted using radiolabeled lubricants (Levy and Wu, 1984). A typical experimental result for the lubricant loss as a function of repeated start-stop cycles is shown in Fig. 7.40. It shows that the loss of lubricant is initially proportional to the number of start-stop cycles, but approaches a constant value after about 4000 cycles.

7.10.2   Metal-film disks

Metallic films can be produced by ion plating, sputtering, chemical plating, and evaporation. Depending on the deposition conditions and materials, metallic films may be designed for longitudinal or perpendicular recording. The most common plated materials are cobalt, cobalt-nickel, and cobalt-nickel-phosphorus (Bate, 1981; King, 1981). To protect the magnetic film, which is generally around 50 to 100 nm thick for longitudinal recording, and 1000 to 5000 nm for perpendicular recording, a hard overcoat is often required. A variety of overcoat materials have been tried. Encouraging results with hard, carbon coatings have recently been obtained by several investigators (Nyaiesh and Holland, 1984). Wear tracks on plated cobalt-phosphorous show fracture patterns indicating that the film was in tension, and that the wear probe, after breaking through the film, initiated the cracking of the coating (Keely, 1985).

To reduce the exposure to wear, lubricants have been applied to metal-film disks in a similar way as to oxide-coated disks. However, the increased smoothness of the metal disk causes the lubricant to spin off faster, in time depleting the disk of lubricant. In addition, stiction of the slider on the disk, due to the lower surface roughness, is much more pronounced and can cause a serious problem during start-up of the disk file.

The wear properties of magnetic films are strongly influenced by the adhesion of the film to the sublayer as well as by the disk substrate. Thus, in addition to protective overcoats, thin metallic films generally require a hard sublayer to isolate the magnetic film from the soft aluminum. Thus,

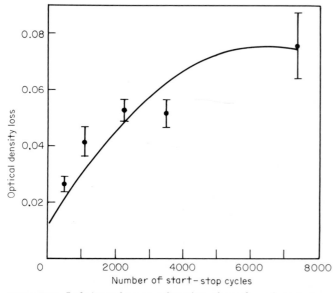

Figure 7.40 Lubricant loss as a function of number of start-stop cycles *(Levy and Wu, 1984)*.

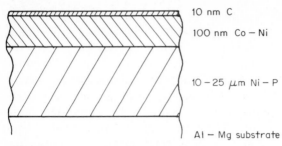

10 nm C

100 nm Co – Ni

10 – 25 μm Ni – P

Al – Mg substrate

Figure 7.41 Layers of typical thin-film disk.

the choice of the sublayer and the match of materials properties between the sublayer and the magnetic film are very important with respect to wear. The most commonly used sublayer material is nickel-phosphorous with a typical thickness of 15 to 25 μm. Since optimum magnetic properties of a metal film do not necessarily coincide with optimum wear properties, magnetic films must be carefully chosen to optimize their magnetic and tribological properties simultaneously.

A typical cross section of a metallic disk is shown in Fig. 7.41. During intermittent or frictional contacts between the slider and the disk, the highest stresses in the interface occur beneath the surface. Thus, crack nucleation, crystal plasticity, and microstructure of the subsurface layers play important roles in the wear of thin films. The type of wear that is predominant in metal-film disks is adhesive wear.

A comparison of the wear resistance of different film-disk coatings shows that, with contact between a ferrite head and a disk spinning at 2.4 m/s, a wear track developed on a carbon overcoat on cobalt-nickel after 2 h, while a rhodium overcoat showed a wear track after 90 s, and a regular oxide disk after approximately 20 s (King, 1981). Great care has to be taken with respect to defects in metal films and the possibility of their growing in time.

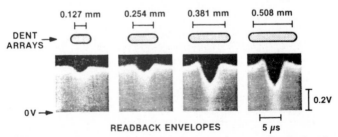

Figure 7.42 Effect of controlled disk damage on readback signal *(Chen, 1981).*

Impact wear of Ni-Co-P film disks has been investigated with a ball-and-hammer device in an attempt to model contacts between the slider and the disk during high-speed flying (Chen, 1981). Correlation was established between severity of contact, impact threshold intensity, and wear, using the modulation of the readback signal. A typical readback signal envelope disturbance, caused by a chain of overlapping defects, is shown in Fig. 7.42.

7.10.3 Slider wear

The most common magnetic slider materials are nickel-zinc and manganese-zinc ferrite. The wear pattern of these materials, in unlubricated wear, is predominantly abrasive (Talke and Su, 1975; White, 1979). The layer of material adjacent to the worn top surface appears to be amorphous as a consequence of the large plastic strains that occur during sliding (Miyoshi and Buckley, 1984); however, below this plastically deformed layer, the underlying crystalline material is undisturbed. Friction and wear of ferrites are influenced by the crystallographic orientation, by the particulate loading in the case of tape, and by environmental conditions.

7.10.4 Head-disk failure and wear testing

The mechanisms of the wear of sliders and disks are important for a basic understanding of the phenomena involved in the wear process; however, typical disk file wear does not always follow simple laboratory results. That is, wear in disk files appears often as a catastrophic event, and, up to now, very little information exists to diagnose impending failure. A variety of factors can contribute to a catastrophic event, such as airborne contaminants, debris pickup by the air-bearing surface, insufficient lubrication of the disk, and head-disk transients.

A number of tests have been developed in the industry to study the wear properties of the head-disk interface. The most commonly used test is the simple "pin-on-disk" wear test, in which a spherical probe is spring-loaded against the disk and the wear between the probe and disk is monitored. Various test devices have also been designed to study the wear of functional sliders at constant speed, in a speed range less than normal operational speeds, but at constant loads higher than normal operational loads.

A somewhat more sophisticated test is the start-stop test, where a functional slider is used and repeated start-stops are performed until the wear of the slider or the disk becomes excessive. In general, the number of successive start-stops that a slider-disk interface must undergo without degradation is on the order of 10,000. Since the time for making this large number of tests is excessive, attempts have been made to define accelerated wear tests, where either the load or the speed is substantially increased from normal operating conditions.

Although accelerated wear tests are very desirable, care has to be exercised in designing such tests and in interpreting the test results. Account must be taken of the nonlinear interaction of the parameters influencing wear, and the different types of wear that occur when the test conditions are accelerated. It is of importance that the type of wear that causes the failure in the disk file under actual functional environmental conditions is the one that is accelerated.

7.10.5   Surface roughness measurement

One of the more important tribological properties of the head-disk interface is the surface roughness of the disk coating. Although very little information is available on the relationship between surface roughness and wear, thorough characterization of surface roughness is one of the prerequisites to understanding the head-disk interface. Various techniques for measuring surface roughness are commonly used, which can be categorized as optical noncontact and stylus contact measurements. The noncontact measurements are preferable to contact profilometry, but the latter allows the specification of quantitative parameters for the surface roughness and is commonly used.

A typical trace from a profilometer is shown in the top of Fig. 7.43, obtained on a diamond-turned aluminum substrate using an Alpha Step 200 profilometer (Dao et al., 1984). It is observed that the peak-to-valley spacing is less than 25 nm. To characterize surfaces, it is useful to calculate their statistical properties from such traces. The most widely used statistical parameter is the centerline average. This parameter does not uniquely describe the surface; that is, surfaces of different texture may have the same centerline average, and additional parameters are needed to characterize a surface. Some of the parameters that are also used are the root-mean-square value, the skewness, the kurtosis, the autocorrelation function, the power spectral-density function, and the bearing-area curve (Selvam and Balakrishnon, 1977). If the data are obtained in digital form, as is the case in modern profilometers, these parameters can be defined to accommodate the discrete nature of the data (Dao et al., 1984).

Figure 7-43 shows several statistical surface parameters calculated from the profile shown there. The autocorrelation function is steep at the origin and decays to zero in a short distance with a few oscillations. The number and amplitude of the oscillations in the autocorrelation function are an indication of the periodicity of the machining process. The same information is contained in the power spectral-density curve, since it is the Fourier transform of the autocorrelation function. The steepness and

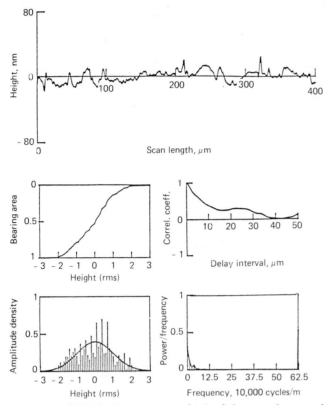

Figure 7.43 Statistical parameters obtained from surface roughness trace on diamond-turned disk *(Dao et al., 1985).*

extent of the bearing-area curve are indicative of the peak-to-valley roughness and the distribution of surface depth; the amplitude-density function is essentially the derivative of the bearing-area curve and again shows the change of the profile with depth.

Typical values for the centerline average of diamond-turned substrates, as well as burnished oxide-coated disks, is on the order of hundreds of nanometers.

7.11 Head-Disk Interface on Flexible Disks

In recent years, flexible disks have been used extensively as a storage medium for mini- and microcomputer systems. The interface between the head and the disk is similar to that in a rigid-disk file. However, a flexible disk deforms over the head, and the interface is influenced by the compliance of the medium, the head contour, and other physical design parameters. In most flexible-disk files, the angular velocity of the disk is on the order of 360 rpm. Thus, the relative velocity between the head and the disk is low, and the head does not fly. In fact, in order to increase the strength of the recorded signal, the head is usually forced to stay in contact with the disk by a pressure pad on the backside of the disk.

As discussed earlier, sliding contact between a head and a disk is undesirable from a wear viewpoint. However, since a flexible disk rotates at low speed, and since the use time of a flexible disk is less than that in rigid-disk files, the wear problem is of less concern than that in rigid-disk files.

In recent years, a flying head has been implemented on a Bernoulli disk configuration. Here, the flexible disk rotates at high speed adjacent to a rigid base plate, and the head-disk interface is designed to allow the formation of a hydrodynamic air bearing (Pelech and Shapiro, 1964; Greenberg, 1978; Adams, 1980; Talke and Tseng, 1984; White, 1984b). A schematic view of the Bernoulli disk configuration is shown in Fig. 7.44.

In order to understand the head-disk interface of a Bernoulli disk, the elasticity equations describing the deflection of a rotating flexible disk subject to a stationary load and the Reynolds equation describing the hydrodynamic pressure of the head-disk interface for a given spacing have to be solved. The hydrodynamic pressure from the Reynolds equation is a function of the head-disk spacing, which itself is a function of the deflection of the disk due to the hydrodynamic load. Thus, the solution of the head-disk interface for a Bernoulli disk requires the simultaneous solution of the Reynolds equation, Eq. (7.2), and the disk elasticity equation, Eq. (7.53).

Since the spacing between the head and the disk is the difference between the disk deflection, obtained from the equations of motion for

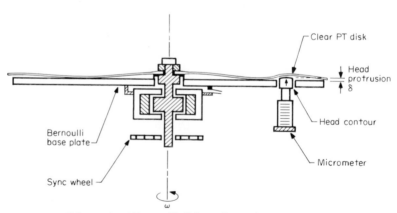

Figure 7.44 Schematic of Bernoulli disk configuration.

the disk and the fixed head contour, an iterative numerical solution can be obtained as follows (Adams, 1980). First, an initial guess is made for the hydrodynamic pressure over the head region. Then, the deflection of the disk due to this pressure is calculated, and the spacing between the disk and the head is determined. Thereafter, a new hydrodynamic pressure is calculated from the solution of the Reynolds equation. This new pressure is relaxed with the initial pressure guess, and a new disk deflection is calculated from the disk equation of motion, and so forth. This procedure is repeated until convergence of the solution is achieved, that is, until the pressures from the present and previous iteration satisfy the convergence criterion.

In Fig. 7-45 a typical numerical result is shown (Adams, 1980) for the dependence of flying height on head penetration. This result is in excellent qualitative agreement with experimental data shown in Fig. 7.46 (Talke and Tseng, 1984). A plot of the calculated pressure distribution over the head is shown in Fig. 7.47, and a map of the spacing contours is shown in Fig. 7.48. The spacing map of Fig. 7.48 agrees well with a typical white-light interference picture Fig. 7.49 (Talke and Tseng, 1984).

As with rigid disks, the head-disk spacing for flexible disks depends to a first order on the contour design of the air bearing. The most important parameter is the width of the head, as can be seen from Fig. 7.50 (Talke and Tseng, 1984). The decrease in the flying height with decreasing head widths is due to the increasing importance of side flow. Since the head-

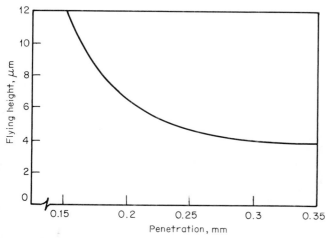

Figure 7.45 Dependence of flying height on head penetration for large-area head *(Adams, 1980).*

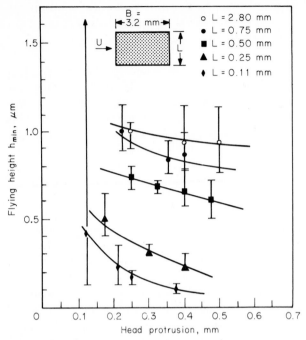

Figure 7.46 Dependence of experimentally measured flying height on head penetration; $B = 3.2$ mm, $U = 25$ m/s, $R = 19$ mm *(Talke and Tseng, 1984).*

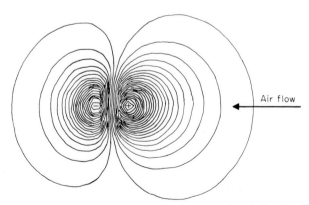

Figure 7.47 Pressure contours between head and flexible disk *(Adams, 1980).*

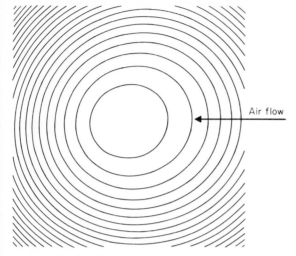

Figure 7.48 Spacing contours between flexible-disk head *(Adams, 1980)*.

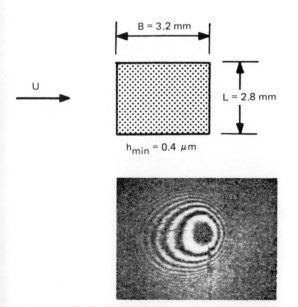

Figure 7.49 Typical white-light interference picture of flexible-disk–head interface *(Talke and Tseng, 1984)*.

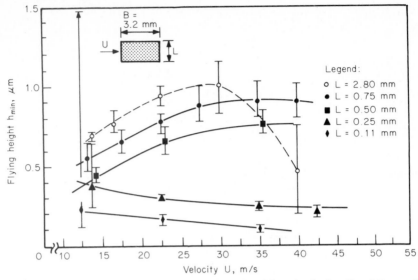

Figure 7.50 Dependence of flying height on head width and velocity; $B = 3.2$ mm, $R = 19$ mm, penetration $= 0.25$ mm *(Talke and Tseng, 1984).*

disk interface has a tendency toward instability for narrow head widths, auxiliary side air bearings, or converging-diverging contours of the wasp-waist–type head, have been used successfully to increase stability.

Recently, a new implementation of a flexible disk has been demonstrated in a "stretched-surface" configuration (Knudsen, 1985). The flexible disk is stretched at the outer rim, and an air bearing is established between the head and the disk. The main advantage of this configuration is that the dimensional stability of the flexible-disk substrate is improved dramatically, since it is determined primarily by the frame on which the disk is stretched. Thus, high track densities can be expected while a head-disk interface is maintained which is still very similar to that of a typical high-compliance flexible disk.

7.12   Dimensional Stability of Flexible Media

If flexible media are used in magnetic recording, the requirements for dimensional stability become more stringent as the track density increases. In the flexible-disk configuration, disk substrate deformation causes an initially circular track to deform so the data cannot easily be recovered except by use of expensive servo systems.

Dimensional changes occur in polymer substrates as a function of temperature, humidity, and internal structure. Furthermore, stresses due to rotation of the disk cause creep, while relaxation of the built-in stresses from the extrusion process causes shrinkage of the material. This shrinkage is generally not only time-dependent but also anisotropic, since the substrate material is stretched at different rates along different directions during its manufacture.

The most commonly used substrate for flexible disks and tapes is polyethylene terephthalate, which is extruded from the melt. The subsequent stretching orients the molecules and results in directional dependence of the physical properties of the film. The measurement of elongation and lateral contraction of polyethylene films under constant loads in temperature- and humidity-controlled environments has been made (Bogy et al., 1979) to determine the creep functions using time-temperature superposition. Typical results are shown in Fig. 7.51 for ranges of temperature (45 to 55°C) and load (1 to 1.6×10^7 N/m^2). The higher values of temperature and load are associated with the steeper curves.

Figure 7.52 shows the dimensional change of a typical flexible disk spinning in an environment of 60°C in 15 percent relative humidity for 10 days (Greenberg et al., 1978). The disk deforms in a highly anisotropic manner; that is, it shrinks at different rates along different directions. The largest shrinkage occurs along the axis of primary film stretch (MD), while the minimum shrinkage occurs along the axis of secondary stretch (TD). Typical thermal and hygroscopic deformation of polyethylene substrate

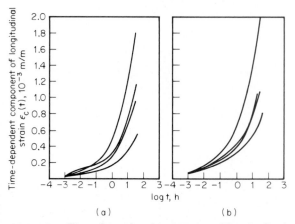

(a)   (b)

Figure 7.51 Time-dependent component of longitudinal strain under various temperature and loading conditions; relative humidity = 50 percent. (a) Uncoated tape; (b) coated tape (Bogy et al., 1979).

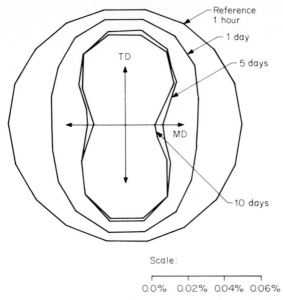

Scale:

0.0%   0.02% 0.04% 0.06%

Figure 7.52 Shrinkage (stress relaxation) in spinning flexible disk (MD = machine direction; TD = transverse direction) at 60°C, 15 percent relative humidity *(Greenberg et al., 1978).*

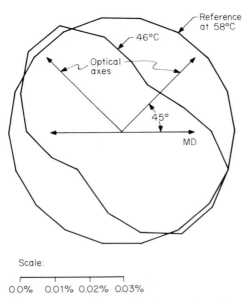

Scale:

0.0%   0.01% 0.02% 0.03%

Figure 7.53 Thermal deformation in going from 58 to 46°C at 40 percent relative humidity *(Greenberg et al., 1978).*

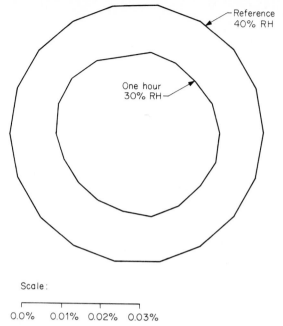

Figure 7.54 Hygroscopic deformation in going from 40 to 30 percent relative humidity *(Greenberg et al., 1978).*

is shown in Figs. 7.53 and 7.54, which show that thermal changes are strongly anisotropic, while hygroscopic changes are isotropic to a first order. Although it might appear that creep due to forces of rotation could have a major effect, it has been found experimentally (Greenberg et al., 1978) and theoretically (Bogy and Talke, 1984a) that this creep is of minor concern since it is much smaller than the effects of shrinkage and thermal expansion.

7.13 Head-Tape Interface

Magnetic tape has been used as a mass storage media from the early days of computer systems. The two major configurations using tapes are the longitudinal and rotary drives, illustrated in Figs. 7.55 and 7.56. In the longitudinal drive the magnetic tape is supported by guidance rollers and moves under tension in a straight line over a magnetic head. Because of the flexibility of the tape, an elastohydrodynamic air film is formed between the tape and the head during normal operation, and the physical condition at the head-disk interface is similar to that of the out-of-contact flexible disk discussed earlier. In the rotary tape drive application, the

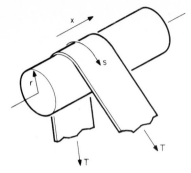

Figure 7.55 Longitudinal tape configuration *(Gross et al., 1980).*

tape is wound helically around two stationary mandrels, and a magnetic head supported on a rotor between the mandrels rotates at high speed with respect to the tape. Since the tape is flexible, an air film forms between the moving head and the tape, and elastohydrodynamic lubrication is established. Applications of rotary-head drives are found predominantly in video recording and in some large mass storage tape-cartridge systems (Harker et al., 1981).

7.13.1 Head-tape interface in a longitudinal tape drive

The equations describing the head-tape interface in a longitudinal tape drive are the elasticity equations for the deformation of a thin flexible foil and the Reynolds equation for the pressure-spacing dependence of the air film between the tape and the stationary head. For a perfectly flexible

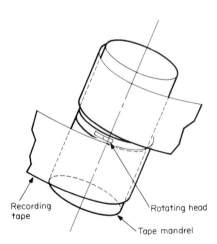

Figure 7.56 Typical rotating-head configuration *(Greenberg, 1979).*

Recording tape

Rotating head

Tape mandrel

tape and an incompressible lubricant, the equation governing the infinitely wide foil bearing is given by (Eshel and Elrod, 1965)

$$\frac{\partial}{\partial s}\left(h^3\frac{\partial^3 h}{\partial s^3}\right) = -\frac{6\mu U}{T}\frac{\partial h}{\partial s}$$

where h = spacing between head and tape
$\quad\ s$ = length dimension along the tape (Fig. 7.60)
$\quad\ \mu$ = viscosity of the air film
$\quad\ U$ = velocity of the tape
$\quad\ T$ = tape tension

This equation is nonlinear and does not allow a closed-form solution. However, solutions of the linearized form and numerical solutions can be easily obtained. Three spacing regions of different characteristics exist for large wrap angles, as shown in Fig. 7.57. These are (1) the constant gap region, where the pressure is essentially constant, (2) the entrance region, where the pressure increases from the ambient pressure to the pressure in the constant gap region, and (3) the exit zone, where the pressure decreases to ambient pressure. From the solution of the linearized equation it can further be shown that the spacing in the entrance region decreases exponentially until it attains the constant spacing value in the center region. On the other hand, the spacing in the exit region is oscillatory in nature and is attained only after going through a minimum spac-

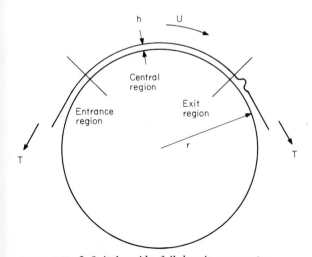

Figure 7.57 Infinitely wide foil bearing geometry.

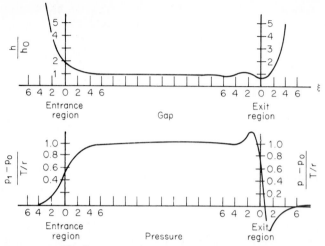

Figure 7.58 Head-tape spacing and pressure distribution for longitudinal head-tape interface; $h_o = 0.643(6\mu U/T)^{2/3}r$ *(Eshel and Elrod, 1965).*

ing region. Figure 7.58 gives numerical results for the head-tape spacing and pressure in a self-acting foil bearing (Gross et al., 1980). The three spacing regions, as well as the constant-pressure region discussed earlier, are shown. The spacing h_o in the central gap region was numerically determined to satisfy

$$h_0 = 0.643 \left(\frac{6\mu U}{T}\right)^{2/3} r$$

Extensions of the theory have been made to include the effects of compressibility (Eshel, 1968), stiffness (Eshel and Elrod, 1967), finite width (Eshel and Elrod, 1966; Licht, 1968), external air pressure (Eshel, 1974), fluid inertia (Eshel, 1970), and so forth. Analyses of the dynamic behavior of foil bearings have also been reported (Barnum and Elrod, 1971, 1972; Stahl et al., 1974). For a discussion of these and other effects the reader is referred to the sources.

7.13.2   Head-tape interface for rotating-head tape drives

Because a rotating head can be moved at a greater speed than a heavy roll of tape, much higher data rates can be achieved in a rotating-head drive than in a linear tape drive. In fact, the typical velocity in a linear tape drive is only on the order of several meters per second, while in a rotating-

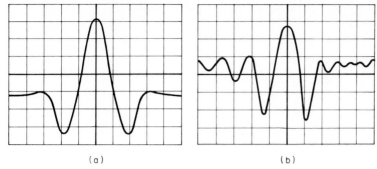

(a)   (b)

Figure 7.59 Capacitance probe traces of actual tape deflection at (a) low speed and (b) high speed *(Greenberg, 1979)*.

head system it can be as high as 40 m/s. At these speeds, sliding contact between the tape and the head would be catastrophic, and so the head must be designed to fly hydrodynamically. Similar to the head-tape interface in a linear tape drive, the spacing between the head and the tape depends on the air-bearing pressure between the head and the tape, and the equations of elasticity of the tape must be coupled with the Reynolds equation.

An additional factor to be considered in the analysis of rotating-head devices is related to the dynamics of the head-tape interface. In particular,

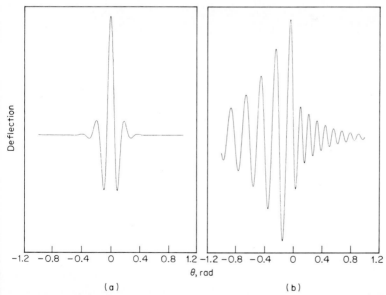

(a)   (b)

Figure 7.60 Tape deflection at (a) low speed and (b) high speed *(Greenberg, 1979)*.

(a)

(b)

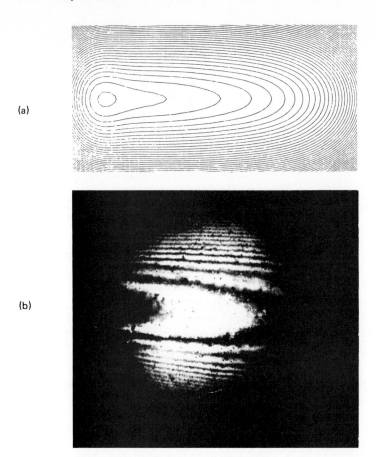

Figure 7.61 Numerical and experimental comparison for the spacing at a typical rotating-head–tape interface (*Greenberg, 1979*).

since the head enters and exits the tape at each revolution, large disturbances are created in those regions due to loss of the air bearing. These disturbances must damp out quickly, so that a uniform air bearing can be present over most of the scan of the head. In addition to the entrance and exit disturbances, waves traveling with the speed of the rotating head are present in the front and the back of the head, caused by the protrusion of the head into the tape. Below the critical speed, the wavelength of these waves is identical. As the speed increases above the critical speed, the waves change in character and shorter waves are encountered in the front of the head, while longer wavelengths are encountered in the area after the head, as shown in Fig. 7.59 (Bogy et al., 1974; Feliss and Talke, 1977; Albrecht et al., 1977; Greenberg, 1979). Since the amplitude and slope of the waves in the head region influence the boundary conditions of the head, a theoretical analysis of the head-tape interface must include the effect of these waves. The problem of a moving load on a stationary cylinder has been considered by making an inextensible bending analysis, and calculating the critical wave speeds for a load traveling on a circular cylinder (Bogy et al., 1974). An extension of this work was carried out by Greenberg (1979), who coupled the tape displacements in the area over the head with the Reynolds equation and obtained simulated results for spherical heads. A typical result from this work is shown in Fig. 7.60 for the tape deflection at low and high tape speeds, indicating the trend discussed earlier for the dependence of wave dynamics on head speed. A numerical and experimental comparison for the spacing of a typical rotating head is shown in Fig. 7.61 (Greenberg, 1979). There is good qualitative agreement between numerically calculated and experimentally determined spacing.

An alternative approach to the analysis of the rotating-head–tape interface has been carried out by Ono (1981), who determined the Green's function for the head-tape interface and solved the Green's function simultaneously with the Reynolds equation.

References

Abramowitz, M., and I. A. Stegun (eds.), *Handbook of Mathematical Functions,* National Bureau of Standards, Applied Mathematics Series, 55 (1964).

Adams, G. C., "Procedures for the Study of the Flexible-Disk to Head Interface," *IBM J. Res. Dev.,* **24**, 512, (1980).

Albrecht, D. M., E. G. Laenen, and C. Lin, "Experiments on the Dynamic Response of a Flexible Strip to Moving Loads," *IBM J. Res. Dev.,* **21**, 379 (1977).

Barnum, T. B., and H. G. Elrod, "A Theoretical Study of the Dynamic Behavior of Foil Bearings," *J. Lubr. Technol. Trans. ASME,* **93**, 133 (1971).

Barnum, T., and H. G. Elrod, "An Experimental Study of the Dynamic Behavior of Foil Bearings," *J. Lubr. Technol. Trans. ASME,* **94**, 93 (1972).

Bate, G., "Recent Developments in Magnetic Recording Materials," *J. Appl. Phys.,* **52**, 2447 (1981).

Benson, R. C., "Deflection of a Transversely Loaded Spinning Disk," Ph.D. thesis, Department of Mechanical Engineering, University of California, Berkeley, 1977.

Benson, R. C., "Observations on the Steady-State Solution of an Extremely Flexible Spinning Disk with a Transverse Load," *J. Appl. Mech.,* **50**, 525 (1983).

Benson, R. C., and D. B. Bogy, "Deflection of a Very Flexible Spinning Disk Due to a Stationary Transverse Load," *J. Appl. Mech.,* **45**, 636 (1978).

Bogy, D. B., and F. E. Talke, "Creep of Rotating Orthotropic Polymer Circular Disk," *Am. Soc. Lubr. Eng.,* **SP-16**, 115 (1984*a*).

Bogy, D. B., and F. E. Talke, "Laser Doppler Interferometry on Magnetic Recording Systems," *IEEE Trans. Magn.,* **MAG-21**, 1332 (1985).

Bogy, D. B., H. J., Greenberg, and F. E. Talke, "Steady Solution for Circumferentially Moving Loads on Cylindrical Shells," *IBM J. Res. Dev.,* **18**, 395 (1974).

Bogy, D. B., N. Bugdayci, and F. E. Talke, "Experimental Determination of Creep Functions for thin Orthotropic Polymer Films," IBM J. Res. Dev., **23**, 450 (1979).

Bouchard, G., D. K. Miu, D. B. Bogy, and F. E. Talke, "On the Dynamics of Winchester and 3370 Type Sliders Used in Magnetic Recording Disk Files," *Am. Soc. Lubr. Eng.,* **SP-16**, 85 (1984).

Bouchard, G., D. B. Bogy, and F. E. Talke, "Use of a Laser Doppler Anemometer for Studying In-Plane Motions of Sliders in Magnetic Disk Files," *Am. Soc. Lubr. Eng.,* **SP-19,** 87 (1985).

Briggs, G. R., and P. G. Herkart, "Unshielded Capacitor Probe Technique for Determining Disk File Ceramic Slider Flying Characteristics," *IEEE Trans. Magn.,* **MAG-7,** 428 (1971).

Burgdorfer, A., "The Influence of the Molecular Mean Free Path on the Performance of Hydrodynamic Gas-Lubricated Bearings," *J. Basic Eng. Trans. ASME,* **81,** 94 (1959).

Cameron, A., *Principles of Lubrication,* Wiley, New York, 1966.

Castelli, V., and J. Pirvics, "Review of Numerical Methods in Gas Bearing Film Analysis," *J. Lubr. Technol. Trans. ASME,* **90,** 777 (1968).

Chapman, S., and T. G. Cowling, *The Mathematical Theory of Non-Uniform Gases,* Cambridge University Press, Cambridge, 1964.

Chen, T. F., "Impact Wear of Thin Ni Co Film on Magnetic Recording Disk," *IEEE Trans. Magn.,* **MAG-17,** 3035 (1981).

Christensen, H., and K. Tonder, "Tribology of Rough Surfaces: Stochastic Models of Hydrodynamic Lubrication," SINTEF Rep. 10/69-18, 1969*a.*

Christensen, H., and K. Tonder, "Tribology of Rough Surfaces: Parametric Study and Comparison of Lubrication Models," SINTEF Rep. 22/69-18, 1969*b.*

Christensen, H., and K. Tonder, "The Hydrodynamic Lubrication of Rough Bearing Surfaces of Finite Width," *J. Lubr. Technol. Trans. ASME,* **93,** 324 (1971).

Constantinescu, V. N., "Gas Lubrication," American Society of Mechanical Engineers, New York, 1969.

Dao, T. T., D. B. Bogy, E. Demaray, and F. E. Talke, "Measurement and Characterization of the Surface Roughness of Rigid Disk Substrates Used in Computer Storage," TR 1984-ME-1, University of California, Berkeley, 1984.

DiPrima, R. C., "Asymptotic Methods for an Infinitely Long Squeeze Film Bearing," *J. Lubr. Technol. Trans. ASME,* **100,** 254 (1978).

E. I. du Pont de Nemours and Co., Inc., "Krytox Fluorinated Oils," Sales Literature, 1-23, 1984.

Engel, P. E., F. E. Talke, R. G. Bayer, S. Chai, J. T. Martin, C. E. Adams, and F. F. M. Lee, "Review of Wear Problems in the Computer Industry," *J. Lubr. Technol. Trans. ASME,* **100,** 189 (1978).

Eshel, A., "Compressibility Effects on the Infinitely Wide, Perfectly Flexible Foil Bearing," *J. Lubr. Technol. Trans. ASME,* **90,** 221 (1968).

Eshel, A., "On Fluid Inertial Effects in Infinitely Wide Foil Bearing," *J. Lubr. Technol. Trans. ASME,* **92,** 490 (1970).

Eshel, A., and H.G. Elrod, "Finite Width Effects on the Self-Acting Foil Bearings," Rept. 6, Lubrication Research Laboratory, Columbia Univ., New York, 1966.

Eshel, A., "Reduction of Air Films in Magnetic Recording by External Air Pressure," *J. Lubr. Technol. Trans. ASME,* **96,** 247 (1974).

Eshel, A., and H. G. Elrod, "The Theory of the Infinitely Wide, Perfectly Flexible Self-Acting Foil Bearing," *J. Bas. Eng.,* **87,** 831 (1965).

Eshel A., and H. G. Elrod, "Stiffness Effects on the Infinitely Wide Foil Bearing," *J. Lubr. Technol. Trans. ASME,* **89,** 92 (1967).

Eversman, W., and R. O. Dodson, "Free Vibrations of a Centrally Clamped Spinning Circular Disk," *AIAA J.,* **7,** 2010 (1969).

Feliss, N. A., and F. E. Talke, "Capacitance Probe Study of Rotating-Head/Tape Interface," *IBM J. Res. Dev.,* **21,** 289 (1977).

Fleisher, J. M., and C. Lin, "Infrared Laser Interferometer for Measuring Air-Bearing Separation," *IBM J. Res. Dev.,* **18,** 529 (1974).

Forsythe, G. E., and W. G. Wasow, *Finite-Difference Methods for Partial Differential Equations,* Wiley, New York, 1960.

Garcia-Suarez, C., D. B. Bogy, and F. E. Talke, "Use of an Upwind Finite Element Scheme for Air Bearing Calculations," *Am. Soc. Lubr. Eng. Publ.,* **SP-16,** 90 (1984).

Greenberg, H. J., "Flexible Disk-Read/Write Head Interface," *IEEE Trans. Magn.,* **MAG-14,** 336 (1978).

Greenberg, H. J., "Study of Head-Tape Interaction in High Speed Rotating Head Recording," *IBM J. Res. Dev.,* **23,** 197 (1979).

Greenberg, H. J., R. L. Stephens, and F. E. Talke, "Measurement of Dimensional Changes in Thin Polymer Films," *Exp. Mech.*, **18,** 115 (1978).

Gross, W. A., "Origins and Early Development of Air-Bearing Magnetic Heads for Disk-File Digital Storage Systems," *Am. Soc. Lubr. Eng. Spec. Publ.*, **SP-16,** 63 (1984).

Gross, W. A., L. Matsch, V. Castelli, A. Eshel, T. Vohr, and M. Wilamann, *Fluid Film Lubrication,* Wiley, New York, 1980a.

Harker, J. M., D. W. Brede, R. E. Pattison, G. R. Santana, and L. G. Taft, "A Quarter Century of Disk File Innovation," *IBM J. Res. Dev.*, **25,** 667 (1981).

Harrison, W. J., "The Hydrodynamical Theory of Lubrication with Special Reference to Air as a Lubricant," *Trans. Cambridge Philos. Soc.*, **22,** 39 (1913).

Huebner, K. H., "Finite Element Analysis of Fluid Film Lubrication: A Survey, in R. H. Gallagher et al. (eds)., *Finite Elements in Fluids,* Wiley, New York, 1975, vol. 2, chap. 12.

Iwasaki, S., Y. Nakamura, and K. Ouchi, "Perpendicular Magnetic Recording with a Composite Anisotropy Film," *IEEE Trans. Magn.*, **MAG-15,** 1456 (1979).

Keeley, K., "Experimental Wear Studies of Magnetic Media Disk, MS Thesis, Dept. of Mech. Engrg., Univ. California, Berkeley, 1985.

King, F. K., "Datapoint Thin Film Media," *IEEE Trans. Magn.*, **MAG-17,** 1376 (1981).

Kita, T., K. Kogure, Y. Mitsuya, and T. Nakanishi," New Method of Detecting Contact Between Floating-Head and Disk," *IEEE Trans. Magn.*, **MAG-16,** 873 (1980).

Knudsen, J. K., "Stretched Surface Recording Disk for use with a Flying Head," *IEEE Trans. Magn.*, **MAG-21,** 2588 (1985).

Kogure, K., S. Fukui, Y. Mitsuya, and R. Kaueko, "Design of Negative Pressure Slider for Magnetic Recording Disks," *Trans. ASME*, **105,** 496 (1983).

Levy, F., and A. Wu, "The Preparation and Utilization of Radiolabeled Lubricants for Determining Lubricant Distribution on Magnetic Disks," *Am. Soc. Lubr. Eng.*, **SP-16,** 49 (1984).

Licht, L., "An Experimental Study of Elastohydrodynamic Lubrication of Foil Bearings," *J. Lubr. Technol. Trans. ASME*, **90,** 199 (1968).

Lin, C., "Techniques for the Measurement of Air-Bearing Separation—A Review," *IEEE Trans. Magn.*, **MAG-9,** 673 (1973).

Lin, C., and R. F. Sullivan, "An Application of White Light Interferometry in Thin Film Measurements," *IBM J. Res. Dev.*, **16,** 269 (1972).

Michael, W. A., "A Gas Film Lubrication Study: II. Numerical Solution of the Reynolds Equation for Finite Slider Bearings," *IBM J. Res. Dev.*, **3,** 256 (1959).

Michael, W. A., "Approximate Methods for Time-Dependent Gas-Film Lubrication Problems," *J. Appl. Mech. Trans. ASME*, **30,** 509 (1963).

Miu, D. K., "Dynamics of Gas-Lubricated Slider Bearings in Magnetic Recording Disk Files: Theory and Experiment," Ph.D. dissertation, Department of Mechanical Engineering, University of California, Berkeley, 1985.

Miu, D. K., G. Bouchard, D. B. Bogy, and F. E. Talke, "Dynamic Response of a Winchester-Type Slider Measured by Laser Doppler Interferometry," *IEEE Trans. Magn.*, **MAG-20,** 927 (1984).

Miyoshi, K., and D. H. Buckley, "Properties of Ferrites Important to Their Friction and Wear Behavior," *Am. Soc. Lubr. Eng.*, **SP-16,** 27 (1984).

Moore, D. F., *Principles and Applications of Tribology,* Pergamon Press, New York, 1975).

Nigam, A., "A Visible Laser Interferometer for Air Bearing Separation Measurement to Submicron Accuracy," *J. Lubr. Technol. Trans. ASME*, **104,** 60 (1982).

Nyaiesh, A. R., and L. Holland, "The Growth of Amorphous and Graphitic Carbon Layers under Ion Bombardment in an RF Plasma," *Vacuum*, **34,** 519 (1984).

Ono, K., "Dynamic Characteristics of Air-Lubricated Slider Bearing for Noncontact Magnetic Recording," *J. Lubr. Technol. Trans. ASME*, **97,** 250 (1975).

Ono, K., "Study of Spherical Foil Bearing," *J. Soc. Mech. Eng.*, **24,** 2162 (1981).

Ono, K., K. Kogure, and Y. Mitsuya, "Dynamic Characteristics of Air-Lubricated Slider Bearings Under Submicron Spacing Conditions," *J. Soc. Mech. Eng.*, **22,** 1672 (1979).

Pelech, I., and A. Shapiro, "Flexible Disk Rotating on a Gas Film Next to a Wall," *J. Appl. Mech.*, **31,** 577 (1964).

Prescott, J., *Applied Elasticity,* Dover, New York, 1961.

Rabinowicz, E., *Friction and Wear of Materials*, Wiley, New York, 1966.

Reddi, M. M., and T. Y. Chu, "Finite Element Solution of the Steady-State Compressible Lubrication Problem," *J. Lubr. Technol. Trans. ASME*, **92**, 495 (1970).

Reynolds, O., "On the Theory of Lubrication and Its Application to Mr. Beauchamp Tower's Experiments, Including an Experimental Determination of the Viscosity of Olive Oil," *Philos. Trans. R. Soc. London*, **177**, 157 (1886).

Schmitt, J. A., and R. C. DiPrima, "Asymptotic Methods for an Infinitely Long Squeeze Film Bearing," *J. Lubr. Technol. Trans. ASME*, **100**, 254 (1978).

Selvam, M. S., and K. Balakrishnan, "The Study of Machined Surface Roughness by Random Analysis," *Wear*, **41**, 287 (1977).

Stahl, K. J., J. W. White, and K. L. Deckert, "Dynamic Response of Self-Acting Foil Bearings," *IBM J. Res. Dev.*, **5**, 513 (1974).

Talke, F. E., and J. L. Su, "The Mechanism of Wear in Magnetic Recording Disk Files," *Tribol. Int.*, **8**, 15 (1975).

Talke, F. E., and R. C. Tseng, "An Autoradiographic Investigation of Material Transfer and Wear During High Speed/ Low Load Sliding," *Wear*, **22**, 69 (1972).

Talke, F. E., and R. C. Tseng, "An Experimental Investigation of the Effect of Medium Thickness and Transducer Spacing on the Read-Back Signal in Magnetic Recording Systems," *IEEE Trans. Magn.*, **MAG-9**, 133 (1973).

Talke, F. E., and R. C. Tseng, "A Study of Elastohydrodynamic Lubrication between a Magnetic Recording Head and a Rotating Flexible Disk," *Am. Soc. Lubr. Eng.*, **SP-16**, 107 (1984).

Talke, F. E., R. C. Tseng, and G. N. Nelson, "Surface Defect Studies of Flexible Media using Magnetoresistive Sensors," *IEEE Trans. Magn.*, **MAG-11**, 1188 (1975).

Tang, T., "Dynamics of Air Lubricated Slider Bearings for Noncontact Magnetic Recording," *J. Lubr. Technol. Trans. ASME*, **93**, 272 (1971).

Timoshenko, S. P., and N. Goodier, *Theory of Elasticity*, 2d ed., McGraw-Hill, New York, 1974.

Timoshenko, S. P., and S. Woinowsky-Krieger, *Theory of Plates and Shells*, 2d ed., McGraw-Hill, New York, 1940.

Tonder, K., "Roughness Effects on Thin-Film Gas Lubrication—A State-of-the-Art Review," *Am. Soc. Lubr. Eng.*, **SP-16**, 183 (1984).

Tseng, R. C., and F. E. Talke, "Transition from Boundary Lubrication to Hydrodynamic Lubrication of Slider Bearings," *IBM J. Res. Dev.*, **18**, 534 (1974).

Tzeng, S. T., and E. Saibel, "Surface Roughness Effect on Slider Bearing Lubrication," *Am. Soc. Lubr. Eng. Trans.*, **10**, 334 (1967).

White, J. W., "Flying Characteristics of the Zero-Load Slider Bearing," *J. Lubr. Technol. Trans. ASME*, **105**, 484, (1983).

White, J. W., "Flying Characteristics of the 3370-Type Slider on a $5\frac{1}{4}$ inch Disk: II. Dynamic Analysis," *Am. Soc. Lubr. Eng. Publ.*, **SP-16**, 77 (1984a).

White, J. W., "On the Design of Low Flying Heads for Floppy Disk Magnetic Recording," *Am. Soc. Lubr. Eng. Publ.*, **SP-16**, 126 (1984b).

White, J. W., and A. Nigam, "A Factored Implicit Scheme for the Numerical Solution of the Reynolds Equation at Very Low Spacing," *J. Lubr. Technol. Trans. ASME*, **102**, 80 (1980).

White, R. D., "Abrasive Wear in Magnetic Disk Recording," *J. Appl. Phys.*, **50**,(3), 2399 (1979).

Wood, R., Ampex Corporation, Private Communication (1984).

Conversion Table

Units for Magnetic Properties

Quantity	Symbol	Gaussian & cgs emu†	Conversion factor C‡	SI§
Magnetic flux density, magnetic induction	B	gauss (G)	10^{-4}	tesla (T), Wb/m^2
Magnetic flux	Φ	maxwell (Mx), $G \cdot cm^2$	10^{-8}	weber (Wb)
Magnetic potential difference, magnetomotive force	U	gilbert (Gb)	$10/4\pi$	ampere (A)
Magnetic field strength, magnetizing force	H	oersted (Oe)	$10^3/4\pi$	A/m
Magnetization	M	emu/cm^3	10^3	A/m
Specific saturation magnetization	σ	emu/g	1	$A \cdot m^2/kg$
Magnetic moment	m	emu	10^{-3}	$A \cdot m^2$
Susceptibility	χ	dimensionless	4π	dimensionless
Permeability of vacuum	μ_o	dimensionless	$4\pi \times 10^{-7}$	$Wb/(A \cdot m)$
Permeability	μ	dimensionless	$4\pi \times 10^{-7} = \mu_o$	$Wb/(A \cdot m)$
Demagnetization factor	N	dimensionless	$1/4\pi$	dimensionless

†Gaussian units and cgs emu are the same for magnetic properties. The defining relation is $B = H + 4\pi M$.
‡Multiply a number in Gaussian units by C to convert it to SI.
§SI (Système International d'Unités) has been adopted by the National Bureau of Standards and is based on the definition $B = \mu_o(H+M)$.

Index